Smartphone-Based Detection Devices

Smartphone-Based Detection Devices

Emerging Trends in Analytical Techniques

Edited by

Chaudhery Mustansar Hussain

Department of Chemistry and Environmental Science, New Jersey Institute of Technology, Newark, N J, USA

Elsevier
Radarweg 29, PO Box 211, 1000 AE Amsterdam, Netherlands
The Boulevard, Langford Lane, Kidlington, Oxford OX5 1GB, United Kingdom
50 Hampshire Street, 5th Floor, Cambridge, MA 02139, United States

Copyright © 2021 Elsevier Inc. All rights reserved.

No part of this publication may be reproduced or transmitted in any form or by any means, electronic or mechanical, including photocopying, recording, or any information storage and retrieval system, without permission in writing from the publisher. Details on how to seek permission, further information about the Publisher's permissions policies and our arrangements with organizations such as the Copyright Clearance Center and the Copyright Licensing Agency, can be found at our website: www.elsevier.com/permissions.

This book and the individual contributions contained in it are protected under copyright by the Publisher (other than as may be noted herein).

Notices
Knowledge and best practice in this field are constantly changing. As new research and experience broaden our understanding, changes in research methods, professional practices, or medical treatment may become necessary.

Practitioners and researchers must always rely on their own experience and knowledge in evaluating and using any information, methods, compounds, or experiments described herein. In using such information or methods they should be mindful of their own safety and the safety of others, including parties for whom they have a professional responsibility.

To the fullest extent of the law, neither the Publisher nor the authors, contributors, or editors, assume any liability for any injury and/or damage to persons or property as a matter of products liability, negligence or otherwise, or from any use or operation of any methods, products, instructions, or ideas contained in the material herein.

British Library Cataloguing-in-Publication Data
A catalogue record for this book is available from the British Library

Library of Congress Cataloging-in-Publication Data
A catalog record for this book is available from the Library of Congress

ISBN: 978-0-12-823696-3

For Information on all Elsevier publications visit our website at
https://www.elsevier.com/books-and-journals

Publisher: Susan Dennis
Acquisition Editor: Kathryn Eryilmaz
Editorial Project Manager: Sara Pianavilla
Production Project Manager: Joy Christel Neumarin Honest Thangiah
Cover Designer: Matthew Limbert

Typeset by Aptara, New Delhi, India

Dedication

I would like to dedicate this book to My beloved GOD
"Mere Pyare Allah"

Contents

Contributors	xiii
About the Editor	xvii
Preface	xix

1 Smartphone: A new perspective in analysis   1
Chaudhery Mustansar Hussain, İbrahim Dolak, Fatemeh Ghorbani-Bidkorbeh, Rüstem Keçili

1.1	Introduction	1
1.2	Applications of smartphone-based sensor systems	2
1.3	Conclusions	13
	References	14

2 Smartphone-based optical and electrochemical sensing   19
Rüstem Keçili, Fatemeh Ghorbani-Bidkorbeh, İbrahim Dolak, Gurbet Canpolat, Chaudhery Mustansar Hussain

2.1	Introduction	19
2.2	Optical sensing based on smartphone technology	20
2.3	Electrochemical sensing based on smartphone technology	27
2.4	Conclusions	30
	References	31

3 Optical methods using smartphone platforms for mycotoxin detection   37
Diana Bueno, Anais Gómez, RB Dominguez, JM Gutiérrez, Jean Louis Marty

3.1	Introduction	37
3.2	Mycotoxins	39
3.3	Colorimetric detection	41
3.4	Smartphone as a portable detection	44
3.5	Conclusions	50
	References	53

4 Fluorescence measurements, imaging and counting by a smartphone   57
Tianran Lin, Danxuan Lin, Li Hou

4.1	Introduction	57
4.2	The applications of smartphone for the construction of biosensors	58
4.3	Conclusion and prospects	68
	References	69

5	**Spectrometric measurements** *Sibasish Dutta*	**73**
	5.1 Introduction	73
	5.2 Transmission grating configured smartphone spectrometers	73
	5.3 Reflection diffraction grating configured smartphone spectrometers	76
	5.4 Smartphone-based Raman spectroscopy	80
	5.5 Conclusion	82
	References	82
6	**Smartphone as barcode reader** *Arpana Agrawal, Chaudhery Mustansar Hussain*	**85**
	6.1 Introduction	85
	6.2 Smartphone based applications in analytical chemistry	87
	6.3 Smartphones as barcode reader	89
	6.4 Conclusions	98
	Acknowledgement	99
	References	99
7	**Current applications of colourimetric microfluidic devices (smart phone based) for soil nutrient determination** *Ying Cheng, Reuben Mah Han Yang, Fernando Maya Alejandro, Feng Li, Sepideh Keshan Balavandy, Liang Wang, Michael Breadmore, Richard Doyle, Ravi Naidu*	**103**
	7.1 Introduction	103
	7.2 Colourimetric methods applied to soil chemistry detection	104
	7.3 Digital image capture and image processing for smartphone analysis	112
	7.4 Microfluidic devices for colourimetric nutrient analysis	115
	7.5 Conclusion	123
	Acknowledgement	123
	References	123
8	**Smartphones as Chemometric applications** *Taniya Arora, Rohini Chauhan, Vishal Sharma, Raj Kumar*	**129**
	8.1 Introduction	129
	8.2 Application of smartphones in chemical sciences	130
	8.3 Chemometric analysis using smartphones	146
	8.4 Conclusion and future trends	150
	Acknowledgments	152
	References	152
9	**Reconstruction of human movement from large-scale mobile phone data** *Haoran Zhang, Yuhao Yao*	**159**
	9.1 Introduction	159
	9.2 Data filtering	161

	9.3	Mode detection and trip segmentation	163
	9.4	Map-matching and interpolation	167
	9.5	Validation and analysis	171
		References	182
10	**Chemical analysis**		**185**
	Eman El-Kimary, Marwa Ragab		
	10.1	Introduction	185
	10.2	Colorimetric-based techniques	185
	10.3	Fluorescence-based techniques	191
	10.4	Foam measurement technique	193
	10.5	Electrochemical-based techniques	194
	10.6	Conclusion	195
		References	196
11	**Applications of smartphones in analysis: Challenges and solutions**		**199**
	Jemmyson Romário de Jesus, Marco Flôres Ferrão, Adilson Ben da Costa, Gilson Augusto Helfer, Marco Aurélio Zezzi Arruda		
	11.1	Introduction	199
	11.2	Development of a mobile colorimetric analysis tool: challenges and solutions	200
	11.3	Applications of smartphones in analysis	210
	11.4	Conclusions	240
		List of Abbreviation	240
		References	242
12	**Applications of smartphones in food analysis**		**249**
	Adriana S. Franca, Leandro S. Oliveira		
	12.1	Introduction	249
	12.2	Food quality and authenticity	250
	12.3	Food safety	258
	12.4	Concluding remarks	264
		Acknowledgements	264
		References	265
13	**Smartphone-based detection devices for the agri-food industry**		**269**
	Aprajeeta Jha, J.A Moses, C. Anandharamakrishnan		
	13.1	Introduction	269
	13.2	Biosensors and their amalgamation with smartphones	271
	13.3	Application of smartphone- based services in agri-food processing	292
	13.4	Conclusions	302
		References	303

14	**Point-of-need detection with smartphone**	**311**
	Nuno M. Reis, Isabel Alves, Filipa Pereira, Sophie Jegouic, Alexander D. Edwards	
	14.1 Introduction	311
	14.2 Modern needs in communicable diseases and bacteria detection	312
	14.3 Modern needs in non-communicable diseases	317
	14.4 Point-of-need integration of conventional analytical techniques	319
	14.5 Point-of-need trends in immunoassays miniaturization	331
	14.6 Modern point-of-need fluidic capabilities	334
	14.7 Positioning of smartphone technology in point-of-need testing	343
	14.8 Camera requirements for smartphone diagnostics and digital imaging of microfluidic bioassays	346
	14.9 Example of point-of-need smartphone tests developed by our research team	351
	14.10 Conclusions	354
	References	354
15	**Point-of-care diagnostics with smartphone**	**363**
	Haleh Ayatollahi	
	15.1 Introduction	363
	15.2 Point-of-care testing (POCT)	364
	15.3 Detection methods	365
	15.4 Applications of smartphone-based POCD/POCT	365
	15.5 Benefits of smartphone-based POCD/POCT	366
	15.6 Development considerations	367
	15.7 mHealth and smartphone-based POCD/POCT	368
	15.8 Challenges of smartphone-based POCD/POCT	368
	15.9 Future development of smartphone-based POCD/POCT	370
	15.10 Conclusion	371
	References	371
16	**Smartphone-based sensors in health and wellness monitoring – perspectives and assessment of the emerging future**	**375**
	Himadri Sikhar Pramanik, Arpan Pal, Manish Kirtania, Tapas Chakravarty, Avik Ghose	
	16.1 Introduction to smartphone-based sensing of health and wellness	375
	16.2 Key observables – sensing body health and wellness signals and beyond	378
	16.3 Sensing capability and mechanism of smartphones, allied wearables and implantables	380
	16.4 Classification of smartphone and wearable applications for healthcare	383
	16.5 Highlights of key case-studies in smartphone-based health & wellness sensing	387

	16.6	Focus on emergent technologies	390
	16.7	Conclusion	391
	References		392
17	**Smartphone-based detection of explosives**	**399**
	Arpana Agrawal, Chaudhery Mustansar Hussain	
	17.1	Introduction	399
	17.2	Types of explosives and explosive detection	401
	17.3	Spectroscopic techniques to detect explosives	403
	17.4	Smartphone-based detection of explosives	405
	17.5	Conclusions	411
	Acknowledgement		411
	References		411
18	**Future of smartphone-based analysis**	**417**
	Rüstem Keçili, Fatemeh Ghorbani-Bidkorbeh, Ayhan Altıntaş, Chaudhery Mustansar Hussain	
	18.1	Introduction	417
	18.2	Conclusions and future perspectives	425
	References		426

Index   **431**

Contributors

Arpana Agrawal Department of Physics, Shri Neelkantheshwar Government Post-Graduate College, Khandwa, India

Fernando Maya Alejandro The School of Nature Science, University of Tasmania, Sandy Bay, TAS, Australia; CRC for High Performance Soils, University Drive, Callaghan, NSW, Australia

Ayhan Altıntaş Anadolu University, Yunus Emre Vocational School of Health Services, Eskişehir, Turkey; Anadolu University, Faculty of Pharmacy, Department of Pharmacognosy, Eskişehir, Turkey

Isabel Alves Loughborough University, United Kingdom

C. Anandharamakrishnan Computational Modeling and Nanoscale Processing Unit, Indian Institute of Food Processing Technology (IIFPT), Thanjavur, Tamil Nadu, India

Taniya Arora Institute of Forensic Science & Criminology, Panjab University, Chandigarh

Marco Aurélio Zezzi Arruda Universidade de Campinas, Instituto de Química, Departamento de Química Analítica, Grupo de Espectrometria, Preparo de amostras e Mecanização – GEPAM, Campinas, São Paulo, Brazil

Haleh Ayatollahi Health Management and Economics Research Center, Health Management Research Institute, Iran University of Medical Sciences, Tehran, Iran; Department of Health Information Management, School of Health Management and Information Sciences, Iran University of Medical Sciences, Tehran, Iran

Sepideh Keshan Balavandy The School of Nature Science, University of Tasmania, Sandy Bay, TAS, Australia; CRC for High Performance Soils, University Drive, Callaghan, NSW, Australia

Michael Breadmore The School of Nature Science, University of Tasmania, Sandy Bay, TAS, Australia; CRC for High Performance Soils, University Drive, Callaghan, NSW, Australia

Diana Bueno Bioprocess Department, UPIBI-IPN, Mexico City, Mexico

Gurbet Canpolat Siirt University, Department of Chemistry, Siirt, Turkey

Tapas Chakravarty TCS Research and Innovation, Tata Consultancy Services, Bengaluru, India

Rohini Chauhan Institute of Forensic Science & Criminology, Panjab University, Chandigarh

Ying Cheng Global Centre for Environmental Remediation, College of Engineering, Science and Environment, University of Newcastle, Callaghan, NSW, Australia; CRC for High Performance Soils, University Drive, Callaghan, NSW, Australia

Adilson Ben da Costa Universidade de Santa Cruz do Sul, Departamento de Ciências da Vida, Santa Cruz do Sul, RS, Brazil

İbrahim Dolak Dicle University, Vocational School of Technical Sciences, Diyarbakır, Turkey

RB Dominguez CONACyT- CIMAV S.C., Chihuahua, CHIH, Mexico

Richard Doyle CRC for High Performance Soils, University Drive, Callaghan, NSW, Australia; Tasmanian Institute of Agriculture, Private Bag 98, Hobart, TAS, Australia

Sibasish Dutta Department of Physics, Pandit Deendayal Upadhyaya Adarsha Mahavidyalaya (PDUAM), Eraligool, Karimganj, Assam, India

Alexander D. Edwards Capilary Film Technology Ltd, United Kingdom; Reading University, United Kingdom

Eman El-Kimary Faculty of Pharmacy, Department of Pharmaceutical Analytical Chemistry, University of Alexandria, El-Messalah, Alexandria, Egypt

Marco Flôres Ferrão Universidade Federal do Rio Grande do Sul, Departamento de Química Inorgânica, Laboratório de Instrumentação Analítica e Químiometria – LAQIA, Porto Alegre, RS, Brazil

Adriana S. Franca Universidade Federal de Minas Gerais, Av. Pres. Antônio Carlos, Belo Horizonte, MG, Brazil

Fatemeh Ghorbani-Bidkorbeh Shahid Beheshti University of Medical Sciences, School of Pharmacy, Department of Pharmaceutics, Tehran, Iran

Avik Ghose Tata Consultancy Services Research & Innovation, Kolkata, West Bengal, India

JM Gutiérrez Bioelectronics Section, Department of Electrical Engineering, CINVESTAV-IPN, Mexico City, Mexico

Anais Gómez Bioelectronics Section, Department of Electrical Engineering, CINVESTAV-IPN, Mexico City, Mexico

Gilson Augusto Helfer Universidade de Santa Cruz do Sul, Engenharias, Arquitetura e Computação, Santa Cruz do Sul, RS, Brazil

Li Hou School of Chemistry and Pharmaceutical Science, State Key Laboratory for the Chemistry and Molecular Engineering of Medicinal Resources, Guangxi Normal University, Guilin, P. R. China

Chaudhery Mustansar Hussain Department of Chemistry and Environmental Science, New Jersey Institute of Technology, Newark, NJ, USA

Sophie Jegouic Reading University, United Kingdom

Jemmyson Romário de Jesus Universidade Federal de Viçosa, Deparatmento de Química, Laboratório de Pesquisa em bionanomateriais, Viçosa, Minas Gerais, Brazil

Aprajeeta Jha Department of Agricultural and Food Engineering, Indian Institute of Technology Kharagpur, West Bengal, India

Rüstem Keçili Anadolu University, Yunus Emre Vocational School of Health Services, Department of Medical Services and Techniques, Eskişehir, Turkey

Manish Kirtania Marketing Transformation – Research, Tata Consultancy Services, Salt Lake City, Kolkata, India

Contributors

Raj Kumar State Forensic Science Laboratory, Madhuban, Karnal, Haryana

Feng Li The School of Nature Science, University of Tasmania, Sandy Bay, TAS, Australia; CRC for High Performance Soils, University Drive, Callaghan, NSW, Australia

Tianran Lin School of Chemistry and Pharmaceutical Science, State Key Laboratory for the Chemistry and Molecular Engineering of Medicinal Resources, Guangxi Normal University, Guilin, P. R. China

Danxuan Lin School of Chemistry and Pharmaceutical Science, State Key Laboratory for the Chemistry and Molecular Engineering of Medicinal Resources, Guangxi Normal University, Guilin, P. R. China

Jean Louis Marty IMAGES group, Universite de Perpignan Via Domitia, Perpignan, France

J.A. Moses Computational Modeling and Nanoscale Processing Unit, Indian Institute of Food Processing Technology (IIFPT), Thanjavur, Tamil Nadu, India

Ravi Naidu Global Centre for Environmental Remediation, College of Engineering, Science and Environment, University of Newcastle, Callaghan, NSW, Australia; CRC for High Performance Soils, University Drive, Callaghan, NSW, Australia

Leandro S. Oliveira Universidade Federal de Minas Gerais, *Av.* Pres.Antônio Carlos, Belo Horizonte, MG, Brazil

Arpan Pal Tata Consultancy Services Research & Innovation, Kolkata, West Bengal, India

Filipa Pereira Capilary Film Technology Ltd, United Kingdom

Himadri Sikhar Pramanik Marketing Transformation - Research, Tata Consultancy Services, Salt Lake City, Kolkata, India

Marwa A.A. Ragab Faculty of Pharmacy, Department of Pharmaceutical Analytical Chemistry, University of Alexandria, El-Messalah, Alexandria, Egypt

Nuno M. Reis University of Bath, United Kingdom; Capilary Film Technology Ltd, United Kingdom

Vishal Sharma Institute of Forensic Science & Criminology, Panjab University, Chandigarh

Liang Wang Global Centre for Environmental Remediation, College of Engineering, Science and Environment, University of Newcastle, Callaghan, NSW, Australia; CRC for High Performance Soils, University Drive, Callaghan, NSW, Australia

Reuben Mah Han Yang The School of Nature Science, University of Tasmania, Sandy Bay, TAS, Australia; CRC for High Performance Soils, University Drive, Callaghan, NSW, Australia

Yuhao Yao Center for Spatial Information Science, The University of Tokyo, Tokyo, Japan

Haoran Zhang LocationMind Inc, Japan

About the Editor

Chaudhery M. Hussain, PhD is an Adjunct Professor and Director of Labs in the Department of Chemistry & Environmental Sciences at the New Jersey Institute of Technology (NJIT), Newark, New Jersey, USA. His research is focused on the applications of nanotechnology and advanced technologies and materials, analytical chemistry, environmental management, and various industries. Dr. Hussain is the author of numerous papers in peer-reviewed journals as well as prolific author and editor of several (more than 60 books) scientific monographs and handbooks in his research areas published with Elsevier, Royal Society of Chemistry, John Wiley & sons, CRC, Springer, etc.

Preface

Advancement in analytical methods and techniques is not only imperative to analytical chemists but also to environmentalists, biotechnologists, pharmacists, forensic scientists, toxicologists, etc., who require to accurately analyze of their on/off site samples. This process of analysis includes sampling, extraction, identification, quantification, and data handling. Samples can be gaseous, liquid, or solid in nature and can range in complexity from a simple blend of two components to a multicomponent mixture containing widely differing species. Therefore, use of advanced and modern instruments, materials and methodologies to accurately and precisely measure and characterize the solid, liquid, and gaseous content always need improvement. An important aim of modern analysis/detection devices is to develop accurate, sensitive, high-throughput, and reliable methodologies for analysis with high quality assurance. Conventional analytical methods are becoming insufficient in terms of accuracy, selectivity, sensitivity, reproducibility, and speed.

Smartphones have been used for ample applications in diverse fields and for detection of different samples to analyze the quality of a product. Smartphones can be used as an investigation tools in multidisciplinary areas to analyze and detect. One major thing that differentiates a smartphone from other devices is its movability due to its size and its cost compared with commercials devices. Sensors in smartphones can efficiently behave as portable instruments for a range of applications. As a result, smartphone-based detection devices have opened up the new perspectives for analysis and generated a large number of detection methods with improved analytical performance. It is expected that these developed analytical methods may overcome drawbacks of the conventional methods, such as expensive instrumentation, complex sample pretreatment steps, and time-consuming procedures. This book gathers all these modern developments in analytical methods with respect to smartphones.

Smartphone usage has created new means of opportunities for detection, analysis, diagnosis and monitoring with new apps. New apps are used to run, detect, analyze, and observe results of the tests. It is important to pin point the issues with the sampling of actual samples to compare the results obtained from traditional devices with that of smartphones. A smartphone can be attached to a device to get results in a simplified format. As a result, smartphones are more reliable and consistent for detection and analysis in comparison with the devices used in routines. The different methodologies of detection with smartphones include optical methods, colorimetric method, fluorescence measurements, spectrometric measurements, and barcode reading. Overall, the aim of this book is to deliver the recent advancements in a various analytical method for detection, analysis, diagnostic techniques by using smartphone.

To apprehend a comprehensive overview of various analytical methods for detection, analysis, diagnostic techniques by using smartphone in analytical chemistry and to provide reader a sound and well-expressed portrait, the book is divided in to five parts containing different chapters. First part, smartphone as an interface for instruments has a chapter on optical sensing and electrochemical sensing. Second part, smartphones as microscopes or test result readers consists of chapters like optical methods using smartphone platforms for mycotoxin detection, fluorescence measurements, imaging and counting by a smartphone, spectrometric measurements and barcode reading. Third part that is about chemometric applications of smartphones has chapters on current applications of colorimetric microfluidic devices (smart phone based) for soil nutrient determination. Fourth part of applications of smartphones in analysis has chapters on reconstruction of human movement from large-scale mobile phone data, chemical analysis, applications of smartphones in chemical analysis: challenges and solutions and applications of smartphones in food analysis. Then, fifth part of detection with smartphones contains chapters on smartphone-based detection devices for agrifood processing industry, point-of-need detection with smartphone, point-of-care diagnostics with smartphone, and smartphone-based sensors in health and wellness monitoring, perspectives, and assessment of the emerging future. Sixth part, commercialization and standardization challenges, has chapters like smartphone-based detection of explosives and conclusion: future of analysis with smartphone. We believe the present book provides invaluable insights of the smartphone in analytical chemistry practiced so far by various world-class researchers.

Overall, this book is designed to be a reference guidebook for experts, researchers, and scientists who are searching for new and modern development in analytical chemistry. The editors and contributors are prominent researchers, scientists, and professionals from academia and industry. On behalf of Elsevier, we are very pleased with all contributors for their exceptional and passionate hard work in making of this book. Remarkable acknowledgement to Kelsey Connors (acquisition editor), Sara Pianavilla (editorial project manager), and Joy Christel Neumarin Honest Thangiah (production manager) at Elsevier for their dedicated support and help during this project. In the end, whole appreciation to Elsevier for publishing the book.

Chaudhery M. Hussain, PhD

(Editor)

Smartphone: A new perspective in analysis

Chaudhery Mustansar Hussain[a], İbrahim Dolak[b], Fatemeh Ghorbani-Bidkorbeh[c], Rüstem Keçili[d]
[a]New Jersey Institute of Technology, Department of Chemistry and Environmental Science, Newark, NJ, United States
[b]Dicle University, Vocational School of Technical Sciences, Diyarbakır, Turkey
[c]Shahid Beheshti University of Medical Sciences, School of Pharmacy, Department of Pharmaceutics, Tehran, Iran
[d]Anadolu University, Yunus Emre Vocational School of Health Services, Department of Medical Services and Techniques, Eskişehir, Turkey

1.1 Introduction

The history of wireless phones dates back to the beginning of the 20th century. The studies were focused on the wireless phone technology for automobiles in the 1940s and commercially available devices were in the market [1]. The early design and fabrication approaches of the wireless phones were much bulkier and needed more energy as well as being analog, making these devices very far away from the today's modern smartphones which almost everyone carries in their pockets. However, in those days, it was predicted that the technology mobile phones will rapidly develop, widespread and used by billions of the people all over the world. In 1926, a German painter Karl Arnold designed and drew an interesting visionary cartoon in a magazine to demonstrate this prediction on wireless phones [1,2] as shown in Fig. 1.1. This visionary cartoon clearly shows that the idea and the excitement on wireless phone technology were out there for almost a century.

Innovative and efficient miniaturized devices having superior features such as facile operation, portability and fast response time have the great potential to ensure the execution of complicated and time-consumed analytical processes in the field without the need for use of expensive analytical instruments or the researcher who has high levels of expertise [3–8]. Nanotechnology and nanomaterial's [9–18] exhibit great superiorities and have also crucial roles to design and fabrication of these portable systems (i.e. lab-on-a-chip) [19–21].

The application of a smartphone-based sensor can be particularly attractive [22–26]. The smartphone technology has found prominence and wide range of successful applications in different fields [27–40].

In this chapter, we provide an overview of the latest progresses on the design, development and applications of smartphone-based sensor platforms. Due to their so many advantages including portability, operability, connectivity and integration into the sensor systems, smartphones become powerful tools in the design and fabrication

Fig. 1.1 A visionary cartoon published in 1926 by Karl Arnold to demonstrate the future use of mobile phones (Republished with permission from [1]).

of novel sensor systems. Combination of smartphones with different types of selective and sensitive sensor platforms enables develop user-friendly, portable and sensitive devices. This chapter starts with the history and importance of the smartphones in our life's and description of their potential applications in different fields of science. Then, various interesting and of successful applications of smartphone-based sensor systems (i.e. smartphone-based colorimetric sensors, smartphone-based luminescence & fluorescence sensors and smartphone-based spectrometers) reported in the literature are presented.

1.2 Applications of smartphone-based sensor systems

1.2.1 Smartphone-based colorimetric sensors

A smartphone-based colorimetric sensor can perform as a versatile analytic instrument due to its great features such as high resolution of camera, fast computing power and telecommunication ability. Therefore, many researchers put so much effort for the utilization of the colorimetric assays. In an interesting study reported by Dong et al., an ultra sensitive smartphone-based sensor platform was designed and developed for the efficient recognition of pyrophosphate in environmental and food samples [41]. For this purpose, a colorimetric sensor system based on smartphone through anti-aggregation of silver nanoparticles modified with polyvinylpyrrolidone caused by Pb^{2+} was designed and fabricated. Fig. 1.2 shows the schematic depiction of the proposed mechanism for the detection of the target pyrophosphate by employing the developed colorimetric sensor based on smartphone

Pb^{2+} has ability to effectively induce the aggregation of AgNPs/ polyvinylpyrrolidone because of its cross-link impact that lead to color change of AgNPs/ polyvinylpyrrolidone dispersion. However, once Pb^{2+} ions are added into the AgNPs/ polyvinylpyrrolidone dispersion containing pyrophosphate, the AgNPs/polyvinylpyrrolidone can keep well dispersed without color change. It is because that pyrophosphate is more prone to chelate with Pb^{2+} via coordination bond. Therefore, it leads to a formation of a stable complex. This sensing mechanism was verified by UV–vis, transmission electron microscope (TEM) and dynamic light scattering (DLS) techniques. The achieved results exhibited that the developed smartphone-based sensor leads to a change of color from blue to yellow for the sensitive detection of pyrophosphate in canned meat and water samples. The obtained data displayed that the developed smartphone-based colorimetric sensor exhibits great sensitivity towards the target compound pyrophosphate in environmental and food samples. The limit of detection was achieved as 1.0 µM and 0.2 µM analyzed by naked-eyes and smartphone-based sensor, respectively. Confirmatory experiments indicated that the developed smartphone-based colorimetric sensor displayed great feasibility and real time sensing efficiency towards the target compound pyrophosphate in real samples including canned meat and tap water samples. In another interesting research, Oncescu et al. reported some experimental results which tested a smartphone-based self diagnostic analytical tool for the sensitive monitoring of cholesterol in blood samples [42]. In their research, an enzymatic test strip was applied for the successful

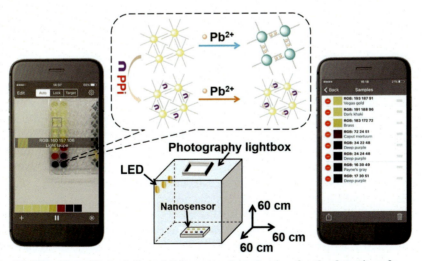

Fig. 1.2 The schematic depiction of the proposed mechanism for the detection of the target pyrophosphate by employing the developed colorimetric sensor based on smartphone (Republished with permission from [41]).

detection of the target cholesterol in blood samples in the concentration level from a low level of the cholesterol (100 $mg\ dL^1$) to high level of cholesterol (400 $mg\ dl^{-1}$). The researchers employed a smartphone's flash as their light source for the monitoring of the colorimetric reaction of the test strip and made an accessary case to prevent the colorimetric imaging from an exterior light. In this research, the Hue Saturation Lightness color coordinate was applied for the sensitive detection of color rather than Red Green Blue used as smartphone colorimetric imaging. In the conducted experiment accuracy, it was determined a regular time for the effective monitoring of the strip considering the reaction time of the enzymatic reaction. The achieved results confirmed that the developed smartphone-based technology can be successfully employed for the determination of the quantity of the total cholesterol levels in blood samples.

Chaisiwamongkhol et al. reported the design and development of a smartphone-based sensor for the colorimetric and sensitive detection of sibutramine, an appetite suppressant used for many years but it was banned by The United States Food and Drug Administration (FDA) and European Medicines Agency (EMA) due to its cardiovascular risks, in suspected food supplement products using Au nanoparticles [43]. The schematic demonstration of the developed smartphone sensor-based colorimetric technique for the sensitive recognition of the sibutramine in adulterated food supplement products is shown in Fig. 1.3. In this study, the aggregation of Au nanoparticles stabilized with citrate in the existence of the target compound sibutramine caused to a color change from wine red to blue that was successfully verified by using transmission electron microscopy (TEM) and UV–Vis spectrophotometry techniques. The change in the color of the solution was visible by the naked eye and could be effectively monitored using a smartphone. The achieved results indicated that the ratio of green and red colors in a photo using a smartphone is linearly related to the concentration of the target sibutramine in the range between 5 µM and 15 µM with the $R^2 = 0.979$. The quantification limit and detection limit values were achieved as 3.47 µM and 1.15 µM, respectively.

In another crucial work carried out by Coşkun and co-workers [44], a smartphone-based personalized food allergen testing system which employed the colorimetric assays of test samples was designed and developed. Using the android software exists on the Samsung Galaxy S II model smartphone, a 3D-printed attachment was designed for operating at the field conditions and LED (Light Emitting Diodes, a peak wavelength: 650 nm, the bandwidth: 15 nm) was utilized to illuminate both a test tube and a control tube. Arbitrary rectangular type of the frame (i.e., 300 × 300 pixels) of acquired images of the tube were processed to determine the value of relative allergen concentration which is in the sample. For the depiction of the ability of this form, a standard food allergy test kit was applied. For the system calibration, different amount of food samples should be digitally quantified before using the device. Through the real food test (i.e., peanut included in the food), the quantity of the allergen in the food samples was measured and an allergy test kit exhibited a similar result of that of smartphone. These food allergen testing results can be uploaded to servers for sharing the information and it may be very useful for allergic individuals globally.

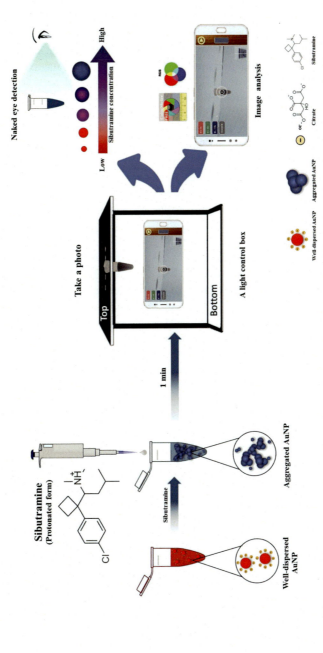

Fig. 1.3 The schematic demonstration of the developed smartphone sensor-based colorimetric technique for the sensitive recognition of the sibutramine in adulterated food supplement products (Republished with permission from [43]).

Shahvar and co-workers designed and developed a developed a cheap and simple smartphone-based colorimetric sensor for the efficient determination of water content in ethanol [45] (Fig. 1.4). In their study, the developed approach was based on the changing of the color-of $CoCl_2$ in ethanol followed by the sensitive recognition and data evaluation by employing the developed smartphone-based colorimetric sensor. In this study, smartphone efficiently used as the detector, illumination source and data processor in the developed sensor system. The impacts of crucial experimental parameters on the sensor response were carefully tested. The obtained results confirmed that a linear response was achieved in the range from 0.05 to 2.00 percent v/v. The developed smartphone-based colorimetric sensor was efficiently employed for the determination of water content in ethanol with recovery higher than 88 percent. The obtained quantification and detection limit values were 0.05 percent v/v and 0.02 percent, respectively.

In another interesting work performed by Li et al. [46], a fast, low-cost and efficient smartphone-based sensor for the bacteria including S. salivarius and S. sanguinis in saliva samples. smartphone-based sensor towards these bacteria was fabricated by using the red-emitting gold nanoclusters (AuNCs) and blue-emitting silicon carbide quantum dots (SiC Qds) to obtain a series of test strips. Fig. 1.5 shows the schematic representation of the sensing process for the target bacteria in saliva samples by using the fabricated smartphone-based sensor. For the preparation of the smartphone-based dual-emission fluorescent sensor, the syntheses of AuNCs having red fluorescence and SiC Qds having blue fluorescence were carried out due to their high quantum yields and excellent photostability. SiC Qds conjugated with the antibacterial peptide of S.salivarius (AMP1) showed an emissive peak at 440 nm by applying a single excitation at 365 nm. On the other hand, AuNCs conjugated with the antibacterial peptide of S.sanguinis (AMP2) exhibited an emissive peak at 660 nm. The prepared smartphone-based dual-emission fluorescent sensor was effectively applied for the sensitive and selective recognition of the target bacteria S. salivarius and S. sanguinis in saliva samples.

In a crucial study reported by Shen and colleagues [47], a smartphone-based sensor system was designed and developed for the determination of pH by quantifying colorimetric paper test strips. In their study, the researchers applied a CIE 1931 color space for the efficient quantification of multiple color elements since RGB color space could not show accurate data in different light conditions. Since the CIE color system just represented 2-D color space regardless of brightness, it was especially effective to the change of ambient light. By using the CIE 1931 color space, they could quantify pH measurement more accurately and compensate the ambient light effect. For compensating the error caused by different ambient light conditions, the researchers proposed a new color mapping approach. In their study, they light intensity of different situations was measured and made a color reference chart which exhibits a linear relationship between 2 light intensities. For proving this approach, a detection algorithm for pH measurement at 5000 K and 3500 K of light condition was used and the achieved results indicated that the developed approach can be very useful for compensating the error by different light situation.

Smartphone: A new perspective in analysis

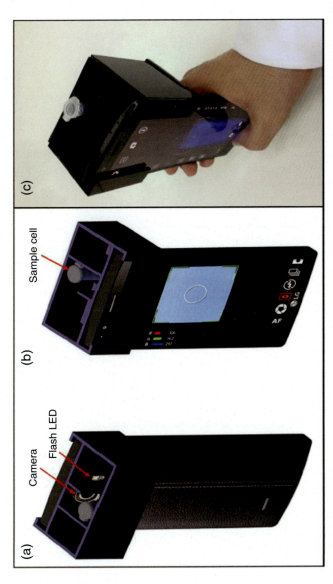

Fig. 1.4 The schematic representation (a, b) and photo of the developed smartphone-based sensor towards water in ethanol (Republished with permission from [45]).

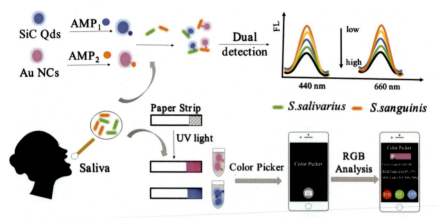

Fig. 1.5 The schematic representation of the sensing process for the target bacteria in saliva samples by using the fabricated smartphone-based sensor (Republished with permission from [46]).

1.2.2 Smartphone-based luminescence and fluorescence sensors

Among smartphone based-sensor technologies, smartphone-based luminescence and fluorescence sensors were successfully designed and fabricated. Some interesting examples reported in the literature are demonstrated in the following:

In an interesting work conducted by Roda and colleagues [48], the design and development of an effective smartphone-based sensor was performed for the monitoring of the enzymatic reaction of biochemi-luminescence. To sequester from the exterior light, they made a low cost 3D printed accessory for hosting the experiment. They used two assays – bio-luminescence and chemi-luminescence for validating their device. In the bioluminescence assay, they measured the total bile acids using 3 α -hydroxyl steroid dehydrogenase coimmobilized with the bacterial luciferase system. On the other hand, they detected the amount of total cholesterol using the specific assay. Reasonable results were shown in the experiments using the serum samples. The back-illuminated complementary metal-oxide semiconductors integrated into the smartphone was shown for the efficient detection of the biochemiluminescence in the medium-abundant concentrations (e.g., at micromolar level).

Same research groups also reported the fabrication of a portable luminescence sensor based on the smartphone platform to detect chemiluminescence of enzymatic reaction [49]. Using the 3D printing technology, the disposable case was prepared and it could be easily transformed for any kind of smartphone. Since chemiluminescence detection should have an ability of highly sensitive measurement for its low light intensity, a battery-powered blue LED (λmax = 466 *nm*) was installed. The enzymatic reaction, especially in chemiluminescence detection, was affected by environmental variables such as temperature. To demonstrate the feasibility of the developed sensor, the lactate determination ratio was tested with horseradish peroxidase. The detection limit of lactate in oral fluid was 0.5 mmol L^{-1} and in sweat samples was achieved as 0.1 mmol L^{-1}, respectively.

In another interesting work [50], Hao et al. reported the design and development of a smartphone-based sensor towards cephalexin which is an antibiotic used for the treatment of certain infections caused by bacteria such as pneumonia and other respiratory tract infections. For this purpose, a dual-emission ratiometric fluorescence probe was combined with a smartphone for specifically and visibly detection of the target compound cephalexin. Fig. 1.6 shows the schematic depiction of the detection process of the target cephalexin using the prepared smartphone-based sensor platform. In this study, the synthesis of blue-emitting fluorescent carbon dots (CDs) was carried out and covered with a thin layer of silica spacer. Then, grafting of red-emitting fluorescent CdTe QDs (r-QDs) on the surface of the silica nanospheres as an analytical probe was successfully performed. In the next step, the cephalexin antibody was covalently grafted to the sensor to enhance and the sensitivity and selectivity. The ratio of fluorescence intensity of r-QDs and CDs was quenched with the increasing of cephalexin concentration. The achieved results indicated that the developed smartphone-based sensor exhibited a good response towards cephalexin in the range between 1 and 500 μM. The limit of detection was obtained as 0.7 μM.

Cevenini and colleagues reported the design and development of a smartphone-based bioluminescence biosensor and demonstrated its analytical performance [51]. The 3D printed cartridges were used to assay the sentinel cells (Hek293T). Basic optical components including smartphone adapters were fabricated and optical interface was set for analysis of the enzymatic reaction. The FV-5 Lite app was applied to catch

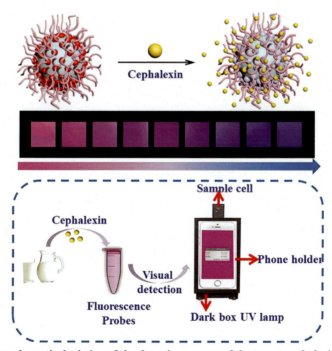

Fig. 1.6 The schematic depiction of the detection process of the target cephalexin using the prepared smartphone-based sensor platform (Republished with permission from [50]).

the images and ImageJ software was the main software to analyze the images. Finally, they developed the Android-based software app named 'Tox-App' which provides the result of toxicity. After the 150 × 150 pixels of test image was analyzed in the RGB color scheme, the 'Cell viability' was indicated as "Safe", "Harmful", and "Highly toxic".

In another important research reported by Liu and colleagues [52], the sensitive and onsite detection of toxic Hg^{2+} ions was successfully conducted by using the developed smartphone- based optical fiber fluorescence sensor (Fig. 1.7). In their research, the smartphone-based sensor was composed of a semiconductor laser for fluorescence signal excitation, a fiber probe modified with CdSe/ZnS quantum dots (QDs) towards the target Hg^{2+} ions, a laser and smartphone. The obtained data exhibited that the developed smartphone- based optical fiber fluorescence sensor showed excellent sensitivity and recognition behavior for Hg^{2+} ions with a quite wide range of detection between 1 nM and 1000 nM. The detection limit was obtained as 1 nM.

Xu et al. reported the development of a smartphone-based fluorescence sensor for the sensitive detection of tetracycline [53]. In their research, a laponite-based dual-channel fluorescent nanoprosensor was designed and fabricated by applying a facile assembly technique for the sensitive detection of tetracycline. The achieved results confirmed that the developed smartphone-based fluorescence sensor exhibited a great recognition and selectivity behavior towards the target compound tetracycline. The limit of detection was obtained as 9.5 nM.

Zangheri and co-workers studied a smartphone-based biosensor using the chemiluminescence lateral flow immunoassay (LFIA) technique [54]. The developed smartphone-based the sensor was successfully used for the detection of cortisol, which was usually used as a biomarker of depression or stress, in saliva samples by employing the chemiluminescent substrate lumino / enhancer / hydrogen peroxide. In their study,

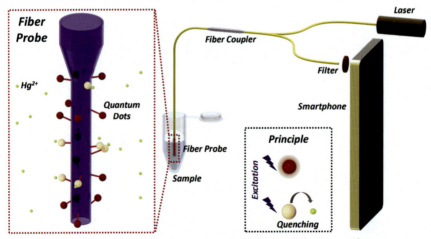

Fig. 1.7 The schematic demonstration of the detection process of the Hg^{2+} ions by applying the developed smartphone- based optical fiber fluorescence sensor (Republished with permission from [52]).

the researchers prepared a simple accessory for housing the LFIA strip using a 3D printing technique. This could fix an optical space from the camera to device and prevent the interface between the ambient light. The range for quantitative analysis was '0.3–60 ng /ml' and the experimental saliva samples were collected from volunteers considering the cortisol concentration difference. The images acquired from both the smartphone camera and Berthold Nigh Owl (LB-981) luminograph were analyzed to evaluate the feasibility of new device. From the comparison with the commercial ELISA kit, the accuracy of the technology was demonstrated and can be used in various diagnostic field.

1.2.3 Smartphone-based spectrometers

There are a number of reported studies on the design and transformation of a smartphone to a spectrometer to measure the spectrum from a light emitter. By designing an optical chamber, they modified the camera of the smartphone to detect the spectrum of incoming light. The ELISA assay which compares the ability of device was used instead of a liquid based assay [55]. In an interesting research carried out by Long et al. [55], the design and fabrication of a smartphone-based spectrometer that can be effectively employed for the determination of the concentration variation of the Enzymatic Linked Immunosorbent Assays (ELISA) was reported In this research, they made a cradle which consisted of a diffraction grating and a collimating lens in front of the smartphone camera. To illuminate the sample, a broadband light source was utilized and the light source was collected by an optical fiber. A lens with the cylindrical form was utilized to focus on the light and the diffraction grating which has 1200 groove/mm dispersed the inlet light. The wavelength range extended from 400 to 700 nm. Before starting the experimental studies, two laser pointers (red laser pointer: $\lambda = 656.26$ nm, green laser pointer: $\lambda = 532.10$ nm) were used for calibration of the device. There were two example biomarkers to assay in ELISA protocols. First of all, they studied of IL-6, a protein used to analyze the type of cancer in serum. Second, they detected a peanut allergen for food safety application. By the result of conventional ELISA microplate reader, the researchers demonstrated the feasibility of the spectral smartphone ELISA platform.

Kwon and colleagues developed a smartphone-based spectrometer that can successfully used for the analysis of the metabolomics [56]. In their work, a diffraction grating film was used for the spectrum of the light and the connector case was fabricated with acrylic plates. In addition, a fluorescent light in front of the smartphone was installed to discern the material. Utilized visible light range was from 400 nm to 700 nm and the total cost to make all of platform was under US $20. Using the Learn Light application, the value of intensity in RGB color space were extracted and measured data was analyzed with PLS-DA algorithm. The data processing was finally analyzed using the statistical method in MATLAB (Mathworks, Natick, MA, United States). The test results showed a good correlation compared with the standard UV–vis spectrometer (Hitachi, U-3010, Tokyo, Japan) and this integrated metabolomics approach with the smartphone could be applied to various medical areas.

In another study, Gallegos and co-workers reported the fabrication of a smartphone-based spectrophotometer for the label-free photonic crystal sensor [57]. This maintained a fixed alignment of optical components which made an accurate measurement. The light which was passed through the pinhole (dimeter = 100 μm) became subsequently collimated and polarized for the collimating lens (focal length = 75 mm). All of the components were prepared by using plastic materials and this accessory designed to easily attachable with smartphone. A software application was developed to change images into a transmission spectrum and the computed accuracy of the device was 0.009 nm. For the calibration of the wavelength, two colors of laser pointer (λ = 533.91 nm and λ = 656.26 nm, respectively) and calibrated spectrometer were used. For demonstrating the device and its detection system, they analyzed an immobilized protein monolayer and the result showed that the device could be a good handheld biosensor.

In another interesting work performed by Kong and co-workers [58], a novel smartphone-based compact disc (CD) spectrometer was designed and fabricated for the sensitive and efficient colorimetric analysis with the superiority of simplicity and low-cost Fig. 1.8. The optimization of the various critical parameters including the

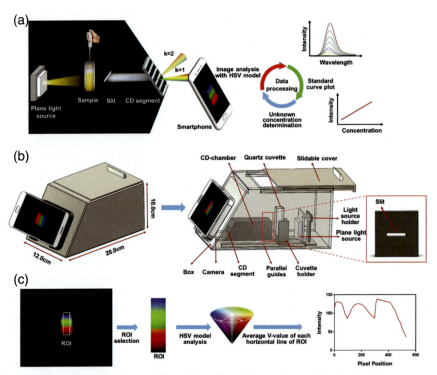

Fig. 1.8 The schematic representation of the (a) working process of the developed smartphone-based CD spectrometer, (b) the structure of the smartphone-based CD spectrometer. and (c) the image processing of the smartphone-based system (Republished with permission from [58]).

distance between the light source and slit, the structure of smartphone-based CD spectrometer and the parameters of camera in the smartphone was carefully performed to achieve the best analytical performance.

The analytical performance of the smartphone-based CD spectrometer was first validated for colorimetric detection of bovine serum albumin (BSA). The achieved results from these experiments showed that the developed smartphone-based CD spectrometer can be successfully used for the sensitive detection of BSA with a low limit of detection (0.0073 mg/mL) that is superior compared to that of the microliter plate reader. Moreover, the developed smartphone-based system by combining with 3,3',5,5'-tetramethylbenzidinemanganese dioxide (TMB-MnO_2) nanosheets reaction displayed excellent sensitivity towards ascorbic acid and the detection limit was achieved as 0.4946 mM.

In a work reported by Yu et al. [59], a smartphone based spectroscopy for the fluorescence based assay was demonstrated. For this purpose, the researchers placed a diffraction grating (1200 lines/mm, GT13-12; Thorlabs Inc., NJ, USA) in front of the smartphone camera and a spectrum was distributed to a CMOS sensor. The cradle consisted of some optical components such as a cylindrical lens (NT 48–354; Edmund Optics), a pinhole (NT56-291; Edmund Optics), and optical fiber (P1000-2-UV–VIS; Ocean Optics, FL, USA). To provide the photon intensity of pixel, The RGB color space was changed to HSV color map. An incandescent lamp (150 W) was utilized as a broadband illumination spectrum since it emitted smooth and continuous spectrum. The calibration was used by two color of laser pointers (red and green) and fiber-optic spectrophotometer A sample for specific nucleic acid sequence was applied for the demonstration of the feasibility of the developed smartphone- device and the performance was compared with the conventional laboratory fluorimeter.

1.3 Conclusions

Smartphone technology is currently an undeniable part of everyone's daily life and the successful applications of smartphones are rapidly increasing. Smartphones was mainly reported as miniuaturized systems for optical detection such as colorimetric, fluorescence, chemiluminescence detection etc. Superiorities were taken of smartphone data processor capacity, their camera capabilities and their light source. Smartphone-based image analysis is a powerful tool for visualization and quantitative analysis.

The fast increasing number of studies in which smartphones were successfully used for the different applications (i.e. biological, environmental and food samples) indicated that the smartphone technology with excellent superiorities is a promising approach for the development of novel sensing and analysis systems such as smartphone-based colorimetric sensors, smartphone-based luminescence & fluorescence sensors and smartphone-based spectrometers.

References

[1] A. Özcan, Mobile phones democratize and cultivate next-generation imaging, diagnostics and measurement tools, Lab Chip 14 (17) (2014) 3187–3194.

[2] https://historyoftelephony.wordpress.com/2017/02/14/the-mobile-phone/ (accessed on 01 February 2021).

[3] K. Khachornsakkul, W. Dungchai, Development of an ultrasound-enhanced smartphone colorimetric biosensor for ultrasensitive hydrogen peroxide detection and its applications, RSC Adv. 10 (2020) 24463–24471.

[4] S. Büyüktiryaki, Y. Sümbelli, R. Keçili, C.M. Hussain, Lab-on-chip platforms for environmental analysis, in: Paul Worsfold, Colin Poole, Alan Townshend, Manuel Miró (Eds.), Encyclopedia of Analytical Science 3rd Edition, Academic Press, 2019, pp. 267–273.

[5] R. Keçili, F. Ghorbani-Bidkorbeh, İ. Dolak, C.M. Hussain, Chapter 1 - Era of nano-lab-on-a-chip (LOC) technology, in: Chaudhery Mustansar Hussain (Eds.), Handbook on Miniaturization in Analytical Chemistry, Elsevier, 2020, pp. 1–17.

[6] Ö. Biçen Ünlüer, F. Ghorbani-Bidkorbeh, R. Keçili, C.M. Hussain, Chapter 12 - Future of the modern age of analytical chemistry: nanominiaturization, in: Chaudhery Mustansar Hussain (Eds.), Handbook on Miniaturization in Analytical Chemistry, Elsevier, 2020, pp. 277–296.

[7] J. Sengupta, C.M. Hussain, Graphene and its derivatives for analytical lab on chip platforms, TrAC Trend. Anal. Chem. 114 (2019) 326–337.

[8] R. Keçili, S. Büyüktiryaki, C.M. Hussain, Chapter 8 - Micro total analysis systems with nanomaterials, in: Chaudhery Mustansar Hussain (Eds.), Handbook of Nanomaterials in Analytical Chemistry, Elsevier, 2020, pp. 185–198.

[9] C.M. Hussain, Magnetic nanomaterials for environmental analysis, in: C.M. Hussain, B. Kharisov (Eds.), Advanced Environmental Analysis-Application of Nanomaterials, The Royal Society of Chemistry, 2017.

[10] R. Keçili, C.M. Hussain, Recent progress of imprinted nanomaterials in analytical chemistry, Int. J. Anal. Chem. (2018) 8503853.

[11] R. Keçili, S. Büyüktiryaki, C.M. Hussain, Advancement in bioanalytical science through nanotechnology: past, present and future, TrAC Trend. Anal. Chem. 110 (2019) 259–276.

[12] S. Büyüktiryaki, R. Keçili, C.M. Hussain, Functionalized nanomaterials in dispersive solid phase extraction: advances & prospects, TrAC Trend. Anal. Chem. 127 (2020) 115893.

[13] D. Sharma, C.M. Hussain, Smart nanomaterials in pharmaceutical analysis, Arab. J. Chem. 13 (2020) 3319.

[14] C.M. Hussain, Nanomaterials in Chromatography: Current Trends in Chromatographic Research Technology and Techniques, Elsevier, 2018.

[15] C.M. Hussain, R. Keçili, Modern Environmental Analysis Techniques for Pollutants, 1st Edition, Elsevier, 2019.

[16] D. Rawtani, P.K. Rao, C.M. Hussain, Recent advances in analytical, bioanalytical and miscellaneous applications of green nanomaterial, TrAC Trends Anal. Chem. 133 (2020) 116109.

[17] R. Keçili, G. Arli, C.M. Hussain, Chapter 14 - Future of analytical chemistry with graphene, in: Chaudhery Mustansar Hussain (Eds.), Comprehensive Analytical Chemistry, 91, Elsevier, 2020, pp. 355–389.

[18] S. Büyüktiryaki, R. Keçili, C.M. Hussain, Chapter 2- Modern age of analytical chemistry: nanomaterials, in: Chaudhery Mustansar Hussain (Eds.), Handbook of Nanomaterials in Analytical Chemistry, Elsevier, 2020, pp. 29–40.
[19] C. Kiang Chua, A. Ambrosi, M. Pumera, Graphene based nanomaterials as electrochemical detectors in Lab-on-a-chip devices, Electrochem. Commun. 13 (5) (2011) 517–519.
[20] A. Kobuszewska, D. Kolodziejek, M. Wojasinski, E. Jastrzebska, T. Ciach, Z. Brzozka, Lab-on-a-chip system integrated with nanofiber mats used as a potential tool to study cardiovascular diseases (CVDs), Sens. Actuators B 330 (2021) 129291.
[21] C.I.L Justino, T.A.P Rocha-Santos, A.C. Duarte, Chapter 14 - Nanomaterials in Lab-on-Chip Chromatography, in: Chaudhery Mustansar Hussain (Eds.), Nanomaterials in Chromatography, Elsevier, 2018, pp. 387–400.
[22] I. Hussain, K. Ahamad, P. Nath, Water turbidity sensing using a smartphone, RSC Adv. 6 (2016) 22374–22382.
[23] J.L.D. Nelis, A.S. Tsagkaris, M.J. Dillon, J. Hajslova, C.T. Elliott, Smartphone-based optical assays in the food safety field, TrAC, Trends Anal. Chem. 129 (2020) 115934.
[24] Y.Y. W.Chen, T. Chen, S.T. W.Shen, H.K. Lee, Application of smartphone-based spectroscopy to biosample analysis: a review, Biosens. Bioelectron. 172 (2021) 112788.
[25] M. Rezazadeh, S. Seidi, M. Lid, S. Pedersen-Bjergaard, Y. Yamini, The modern role of smartphones in analytical chemistry, TrAC, Trends Anal. Chem. 118 (2019) 548–555.
[26] S.E. Seo, F. Tabei, S.J. Park, B. Askarian, K.H. Kim, G. Moallem, J.W. Chong, Oh Seok Kwon, Smartphone with optical, physical, and electrochemical nanobiosensors, J. Ind. Eng. Chem. 77 (2019) 1–11.
[27] K.C. Majhi, P. Karfa, R. Madhuri, Chapter 6 - Smartphone-based nanodevices for in-field diagnosis, in: Suvardhan Kanchi, Deepali Sharma (Eds.), Nanomaterials in Diagnostic Tools and Devices, Elsevier, 2020, pp. 159–187.
[28] A. Hasanzadeh, Chapter 12 - Smartphone-based microfluidic devices, in: Michael R. Hamblin, Mahdi Karimi (Eds.), Biomedical Applications of Microfluidic Devices, Academic Press, 2021, pp. 275–288.
[29] Z. Li, S. Zhang, Q. Wei, Chapter 5 - Smartphone-based flow cytometry, in: Jeong-Yeol Yoon (Eds.), Smartphone Based Medical Diagnostics, Academic Press, 2020, pp. 67–88.
[30] S.K. Vashist, J.H.T Luong, C 16 - S-B Immunoassays, E(s: S K Vashist, J.H.T Luong, Handbook of Immunoassay Technologies, Academic Press, 2018, pp. 433–453.
[31] A. Pyayt, Chapter 6 - Smartphones for rapid kits, in: Jeong-Yeol Yoon (Eds.), Smartphone Based Medical Diagnostics, Academic Press, 2020, pp. 89–102.
[32] D.S.Y. Ong, M. Poljak, Smartphones as mobile microbiological laboratories, Clin. Microbiol. Infect. 26 (4) (2020) 421–424.
[33] M.-.T. Liu, J. Zhao, S.-.P. Li, Application of smartphone in detection of thin-layer chromatography: case of salvia miltiorrhiza, J. Chromatogr. A 1637 (2021) 461826.
[34] T.J. Moehling, D.H. Lee, M.E. Henderson, M.K. McDonald, P.H. Tsang, S. Kaakeh, E.S. Kim, S.T. Wereley, T.L. Kinzer-Ursem, K.N. Clayton, J.C. Linnes, A smartphone-based particle diffusometry platform for sub-attomolar detection of Vibrio cholerae in environmental water, Biosens. Bioelectron. 167 (2020) 112497.
[35] Z. Wu, J. Lu, Q. Fu, L. Sheng, B. Liu, C. Wang, C. Li, T. Li, A smartphone-based enzyme-linked immunochromatographic sensor for rapid quantitative detection of carcinoembryonic antigen, Sens. Actuators B 329 (2021) 129163.
[36] W. Luo, J. Deng, J. He, Z. Han, C. Huang, Y. Li, Q. Fu, H. Chen, A smartphone-based multi-wavelength photometer for on-site detection of the liquid colorimetric assays for clinical biochemical analyses, Sens. Actuators B 329 (2021) 129266.

[37] R. Bandi, M. Alle, C.-.W. Park, S.-.Y. Han, G.-.J. Kwon, N.-.H. Kim, J.C. Kim, S.H. Lee, Cellulose nanofibrils/carbon dots composite nanopapers for the smartphone-based colorimetric detection of hydrogen peroxide and glucose, Sens. Actuators B 330 (2021) 129330.

[38] X. Kou, L. Tong, Y. Shen, W. Zhu, L. Yin, S. Huang, F. Zhu, G. Chen, G. Ouyang, Smartphone-assisted robust enzymes@MOFs-based paper biosensor for point-of-care detection, Biosens. Bioelectron. 156 (2020) 112095.

[39] P. Teengam, W. Siangproh, S. Tontisirin, A. Jiraseree-amornkun, N. Chuaypen, P. Tangkijvanich, C.S. Henry, N. Ngamrojanavanich, Orawon Chailapakul, NFC-enabling smartphone-based portable amperometric immunosensor for hepatitis B virus detection, Sens. Actuators B 326 (2021) 128825.

[40] Z. Liu, Q. Hua, J. Wang, Z. Liang, J. Li, J. Wu, X. Shen, H. Lei, X. Li, A smartphone-based dual detection mode device integrated with two lateral flow immunoassays for multiplex mycotoxins in cereals, Biosens. Bioelectron. 158 (2020) 112178.

[41] C. Dong, X. Ma, N. Qiu, Y. Zhang, A. Wu, An ultra-sensitive colorimetric sensor based on smartphone for pyrophosphate determination, Sens. Actuators B 329 (2021) 129066.

[42] V. Oncescu, M. Mancuso, D. Erickson, Cholesterol testing on a smartphone, Lab on a Chip, 14 (2014) 759–763.

[43] K. Chaisiwamongkhol, S. Labaidae, S. Pon-in, S. Pinsrithong, T. Bunchuay, A. Phonchai, Smartphone-based colorimetric detection using gold nanoparticles of sibutramine in suspected food supplement products, Microchem. J. 158 (2020) 105273.

[44] A.F. Coskun, J. Wong, D. Khodadadi, R. Nagi, A. Tey, A. Ozcan, Cell-Phone Based Food Allergen Testing, 7, Optical Society of America, 2013 CTu2M.

[45] A. Shahvar, D. Shamsaei, M. Saraji, A portable smartphone-based colorimetric sensor for rapid determination of water content in ethanol, Measurement 150 (2020) 107068.

[46] X. Li, J. Li, J. Ling, C. Wang, Y. Ding, Y. Chang, N. Li, Y. Wang, J. Cai, A smartphone-based bacteria sensor for rapid and portable identification of forensic saliva sample, Sens. Actuators B 320 (2020) 128303.

[47] L. Shen, J.A. Hagen, I. Papautsky, Point-of-care colorimetric detection with a smartphone, Lab Chip 12 (2012) 4240–4243.

[48] A. Roda, E. Michelini, L. Cevenini, D. Calabria, M.M. Calabretta, P. Simoni, Integrating biochemiluminescence detection on smartphones: mobile chemistry platform for point-of-need analysis, Anal. Chem. 86 (2014) 7299–7304.

[49] A. Roda, M. Guardigli, D. Calabria, M.M. Calabretta, L. Cevenini, E. Michelini, A 3D-printed device for a smartphone-based chemiluminescence biosensor for lactate in oral fluid and sweat, Analyst 139 (2014) 6494–6501.

[50] A.-.Y. Hao, X.-.Q. Wang, Y.-.Z. Mei, J.-.F. Nie, Y.-.Q. Yang, C.-.C. Dai, A smartphone-combined ratiometric fluorescence probe for specifically and visibly detecting cephalexin, Spectrochim. Acta Part A 249 (2021) 119310.

[51] L. Cevenini, M.M. Calabretta, G. Tarantino, E. Michelini, A. Roda, Smartphone-interfaced 3D printed toxicity biosensor integrating bioluminescent "sentinel cells, Sens. Actuators B 225 (2016) 249–257.

[52] T. Liu, W. Wang, D. Jian, J. Li, H. Ding, D. Yi, F. Liu, S. Wang, Quantitative remote and on-site Hg^{2+} detection using the handheld smartphone based optical fiber fluorescence sensor (SOFFS), Sens. Actuators B 301 (2019) 127168.

[53] J. Xu, S. Guo, L. Jia, T. Zhu, X. Chen, T. Zhao, A smartphone-integrated method for visual detection of tetracycline, Chem. Eng. J. (2020) 127741.

[54] M. Zangheri, L. Cevenini, L. Anfossi, C. Baggiani, P. Simoni, F. Di Nardo, A. Roda, A simple and compact smartphone accessory for quantitative chemiluminescence-based

lateral flow immunoassay for salivary cortisol detection, Biosens. Bioelectron. 64 (2015) 63–68.
[55] K.D. Long, H. Yu, B.T. Cunningham, Smartphone instrument for portable enzyme-linked immunosorbent assays, Biomed. Opt. Express 5 (2014) 3792–3806.
[56] H. Kwon, J. Park, Y. An, J. Sim, S. Park, A smartphone metabolomics platform and its application to the assessment of cisplatin-induced kidney toxicity, Anal. Chim. Acta 845 (2014) 15–22.
[57] D. Gallegos, K.D. Long, H. Yu, P.P. Clark, Y. Lin, S. George, P. Nath, B.T. Cunningham, Label-free biodetection using a smartphone, Lab Chip 13 (2013) 2124–2132.
[58] L. Kong, Y. Gan, T. Liang, L. Zhong, Y. Pan, D. Kirsanov, A. Legin, H. Wan, P. Wang, A novel smartphone-based CD-spectrometer for high sensitive and cost-effective colorimetric detection of ascorbic acid, Anal. Chim. Acta 1093 (2020) 150–159.
[59] H. Yu, Y. Tan, B.T. Cunningham, Analytical chemistry, Smartphone Fluoresc. Spectrosc. 86 (2014) 8805–8813.

Smartphone-based optical and electrochemical sensing

Rüstem Keçili[a], Fatemeh Ghorbani-Bidkorbeh[b], İbrahim Dolak[c], Gurbet Canpolat[d], Chaudhery Mustansar Hussain[e]
[a]Anadolu University, Yunus Emre Vocational School of Health Services, Department of Medical Services and Techniques, Eskişehir, Turkey
[b]Shahid Beheshti University of Medical Sciences, School of Pharmacy, Department of Pharmaceutics, Tehran, Iran
[c]Dicle University, Vocational School of Technical Sciences, Diyarbakır, Turkey
[d]Siirt University, Department of Chemistry, Siirt, Turkey
[e]New Jersey Institute of Technology, Department of Chemistry and Environmental Science, Newark, N J, United States

2.1 Introduction

Smartphones can be successfully employed for sensing applications because of their portability, accessibility and multi-functionality [1]. In particular, these devices offer a user-friendly interface, rapid data processing and wireless communication facilities. A useful and effective component of modern smartphones is a high-resolution camera that can be efficiently applied for imaging applications. However, the camera quality varies on different brands and models of these smartphones that can hinder the consistency of the achieved data. Smartphones can also be employed as an efective and portable power source [2] that can be employed for the manipulation of liquids in microfluidic chip platforms through electrokinetics. A smartphone phone has 2 possible electrical connection points which are audio jack and USB port that can be applied for transmission of the obtained data. During the wireless data transmission process, smartphones offer a number of routes, such as Wi-Fi, Bluetooth or cellular networks (3G/4G), that provides fast transmission of the achieved results in real-time [3]. These superior capabilities can also ensure for the achieved results to be stored in the cloud platforms.

The powerful miniaturized devices (i.e. smartphone-based sensor platforms) exhibit a number of advantages including facile operation, portability and rapid analysis time have the great potential to ensure the execution of complicated and time-consumed analytical processes in the field without the need for use of expensive analytical instruments or the researcher who has high levels of expertise [4–8]. In addition, nanomaterials [9–18] display unique properties and play very important role in the design and development of these miniaturized systems such as lab-on-a-chip [19–21]. Rather than simply using the advanced capabilities for voice communications and web browsing, smartphones technology has found prominence and wide range of successful applications in different fields [22–35].

In this chapter, the recent progresses on the design, preparation and crucial applications of smartphone-based optical sensors including colorimetric, fluorescence, chemiluminescence-based sensor etc. and electrochemical sensors were highlighted and overviewed. It starts with the sifnigifance and superior features of the smartphones in the introduction part. Then, smartphone-based optical sensors and electrochemical sensors and their various successful application reported in the literature are presented.

2.2 Optical sensing based on smartphone technology

Smartphone-based colorimetric approaches investigate the changes in the absorbance or reflected intensity of analyte-reagent complexes. These changes are mainly because of the structural shifts or plasmon resonance phenomena that causes a shift in the sample's optical features typically over a broad range of wavelengths [36,37]. smartphone technology is excellent approach for efficient monitoring of these broad color shifts because it does not require stringent controls or filters, that can be needed in fluorescence detection to screen for fine changes at a peak wavelength. Therefore, smartphone-based colorimetric sensor systems only require lighting and image processing for the unfiltered and effective detection. More advanced smartphone-based colorimetric sensor platforms may also have extra lenses and separate lighting sources that are more specifically attuned to the absorbance spectrum. of the analyte-reagent complex.

In a study performed by Su and colleagues [38], a colorimetric sensor platform based on smartphone was designed and developed for the efficient and sensitive detection of various marine toxins such as saxitoxin and okadaic acid. For this purpose, the developed system called "bionic electronic eye (Bionic e-Eye)" was designed and fabricated for the in-field rapid measurement and real-time on-line analysis for the target toxins. Bionic e-Eye installed the homemade software—iPlate and completed the integration of image acquisition and further processing of the achieved data. With the integration of ELISA system, two color model which are HSV (hue, saturation, value) and RGB (Red Green Blue) in Bionic e-Eye were applied for the evaluation of the sensing performance of the developed smartphone-based Bionic e-Eye towards the target toxins saxitoxin and okadaic acid. The schematic representation of the workflow of the developed smartphone-based iPlate including calibration, sample measurement and data sharing steps is given in Fig. 2.1. The achieved results showed that the developed smartphone-based Bionic e-Eye exhibited a linear response towards saxitoxin and okadaic acid in the concentration ranges from 0.02 to 0.32 μg L^{-1} and 0.2 to 5 μg L^{-1}, respectively.

Chaisiwamongkhol et al. reported the design and development of a smartphone-based sensor for the colorimetric and sensitive detection of sibutramine, an appetite suppressant used for many years but it was banned by The United States Food and Drug Administration (FDA) and European Medicines Agency (EMA) due to its cardiovascular risks, in suspected food supplement products using Au nanoparticles [39,40].

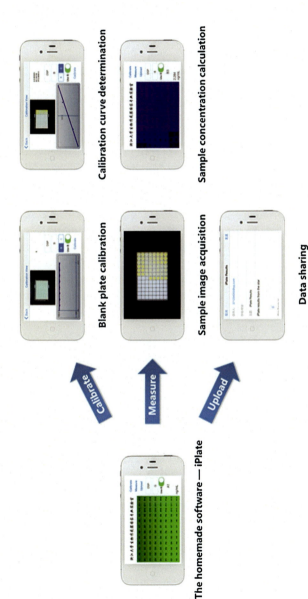

Fig. 2.1 The schematic representation of the workflow of the developed smartphone-based iPlate including calibration, sample measurement and data sharing steps (Republished with permission from [38]).

The schematic demonstration of the developed smartphone sensor-based colorimetric technique for the sensitive recognition of the sibutramine in adulterated food supplement products is shown in Fig. 2.3. In this study, the aggregation of Au nanoparticles stabilized with citrate in the existence of the target compound sibutramine caused to a color change from wine red to blue that was successfully verified by using transmission electron microscopy (TEM) and UV–Vis spectrophotometry techniques. The change in the color of the solution was visible by the naked eye and could be effectively monitored using a smartphone. The achieved results indicated that the ratio of green and red colors in a photo using a smartphone is linearly related to the concentration of the target sibutramine in the range between 5 µM and 15 µM with the $R^2 = 0.979$. The quantification limit and detection limit values were achieved as 3.47 µM and 1.15 µM, respectively,

In another work reported by Gan co-workers [41], a smartphone-based colorimetric sensor towards Cd^{2+} ions in tap water samples was designed and prepared. In their work, Au nanoparticles were functionalized with nucleic acid aptamer for the sensitive detection of target Cd^{2+} ions. The interaction of Cd^{2+} ions with the aptamers causes a reduce of free aptamers that lead to a weakness of the stability of the Au nanoparticles. This results a change of the color in the solution. This colorimetric change could be successfully monitored and analyzed by the developed smartphone-based colorimetric sensor within a short time (10 min) that implements the efficient quantification of the target Cd^{2+} ions Fig. 2.2. The obtained results displayed that the developed smartphone-based colorimetric sensor showed a great sensitivity and selectivity for Cd^{2+} ions in a linear concentration range between 2 and 20 µg L^{-1}. The detection limit was achieved as 1.12 µg L^{-1}.

Among smartphone based-optical sensor systems, smartphone-based luminescence and fluorescence sensors were designed, developed and successfully used for the sensitive detection of various compounds in complex matrices such as environmental, food and biological samples [41–43]. Some interesting examples reported in the literature are briefly described in the following:

Fahimi-Kashani and Hormozi-Nezhad reported the design and fabrication of a smartphone-based fluorescence nanosensor for the sensitive detection of methyl parathion, an organophosphate pesticide, in various food and environmental samples (i.e. wheat, rice and corn flours and tap water samples) [44]. In their research, self-assembly of cetyltrimethylammonium bromide (CTAB) on the surface of CdTe quantum dots (QDs) was carried out. This lead to a quenching of CTAB-QDs upon addition of the target pesticide methyl parathion while the fluorescence intensity of CdTe QDS remains constant. The obtained results confirmed that the developed smartphone-based optical nanosensor exhibited great sensing behaviour towards methyl parathion with a broad linear concentration range varied between 0.001 µg.mL^{-1} and 10 µg.mL^{-1}. The achieved detection limit was 1.2 ng mL^{-1}.

In another crucial research performed by Li and co-workers [45], a novel smartphone-based fluorescence sensor towards Cd^{2+} ions was developed. For this purpose, a dual-emission ratiometric fluorescence probe was prepared based on the Cu nanoclusters coated with SiO_2 and CdTe quantum dots as the signal reference and signal response, respectively. The obtained outcomes indicated that the developed

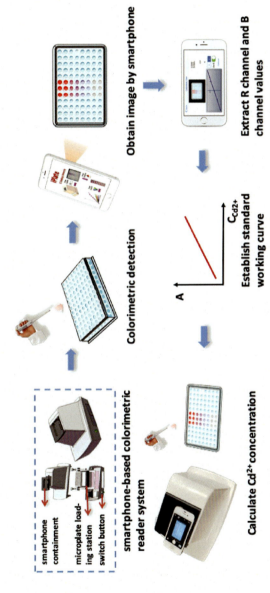

Fig. 2.2 The schematic depiction of the sensing process of prepared smartphone-based colotimetric sensor towards Cd^{2+} ions (Republished with permission from [40]).

smartphone-based fluorescence sensor can be successfully employed for the sensitive detection of the target Cd^{2+} ions in oyster samples. The achived sensor response was linear in the range between 0.010 mg L^{-1} and 2.0 mg L^{-1}. The detection limit was found as 1.1 μg L^{-1} (2.75 μg kg^{-1}).

In an interesting work conducted by Roda and colleagues [46], the design and development of an effective smartphone-based sensor was performed for the monitoring of the enzymatic reaction of biochemi-luminescence. To sequester from the exterior light, they made a low cost 3D printed accessory for hosting the experiment. They used two assays – bio-luminescence and chemi-luminescence for validating their device. In the bioluminescence assay, they measured the total bile acids using 3 α -hydroxyl steroid dehydrogenase coimmobilized with the bacterial luciferase system. On the other hand, they detected the amount of total cholesterol using the specific assay. Reasonable results were shown in the experiments using the serum samples. The back-illuminated complementary metal-oxide semiconductors integrated into the smartphone was shown for the efficient detection of the biochemiluminescence in the medium-abundant concentrations (e.g., at micromolar level).

Same research groups also reported the fabrication of a portable luminescence sensor based on the smartphone platform to detect chemiluminescence of enzymatic reaction [47]. Using the 3D printing technology, the disposable case was prepared and it could be easily transformed for any kind of smartphone. Since chemiluminescence detection should have an ability of highly sensitive measurement for its low light intensity, a battery-powered blue LED (λmax = 466 *nm*) was installed. The enzymatic reaction, especially in chemiluminescence detection, was affected by environmental variables such as temperature. To demonstrate the feasibility of the developed sensor, the lactate determination ratio was tested with horseradish peroxidase. The detection limit of lactate in oral fluid was 0.5 mmol L^{-1} and in sweat samples was achieved as 0.1 mmol L^{-1}, respectively.

Xu et al. reported the development of a smartphone-based fluorescence sensor for the sensitive detection of tetracycline [48]. In their research, a laponite-based dual-channel fluorescent nanoprosensor was designed and fabricated by appying a facile assembly technique for the sensitive detection of tetracycline. The achieved results confirmed that the developed smartphone-based fluorescence sensor exhibited a great recognition and selectivity behaviour towards the target compound tetracycline. The limit of detection was obtained as 9.5 nM.

In a study carried out by McCracken and co-workers [49], sensitive detection of bisphenol A (BPA), an emerging environmental contaminant, in water samples was successfully carried out by using the developed smartphone-based fluorescence sensor. For this purpose, 8-hydroxypyrene-1,3,6-trisulfonic acid (HPTS) was chosen as an efficient fluorescent probe. The obtained results exhibited that the developed smartphone-based fluorescence sensor displayed a great recognition behavior towards the target compound BPA. It was also obtained that there is a significant binding interactions between the target compound BPA and fluorescent probe HPTS (K_{SV} = 2040 M^{-1}). The detection limit was achieved as 4.4 mM.

In another interesting work [50], a smartphone-based fluorescence sensor for the sensitive detection of albumin in urine samples was demonstrated by Coşkun and colleagues. the developed fluorescence sensor system employed a fluorescence detector affixed to the smartphone's camera and a transparent tube which contains the urine sample. The fluorescent signal is processed through an App on the phone and generates the albumin concentration based on a calibration curve. In this work, a lower detection limit of 10 μg mL^{-1} was obtained with a linear range between 0 and 200 μg mL^{-1}. Although this device is fast and fairly simple, the optical components are bulky and complicated which limits its usefulness for point of care (POC) testing.

In a study carried out by Stemple et al. [51], a smartphone-based POC device for protein quantification based on a microbead immunoagglutination assay combined with an optic-scatter system used to detect malaria biomarker was demonstrated. Briefly, the presence of the target antigen within the channel causes the light to scatter, which is captured by the smartphone's camera. The concentration of analyte in the sample was successfully determined based on the intensity of the scattered light. A detection limit of 1 pg mL^{-1} was obtained by using diluting blood with a detection range between 1 pg mL^{-1} and 10 ng mL^{-1}. While these results are comparable to the sensitivity of previously demonstrated POC protein assays, addititonal effort is needed to make this system compatible with raw biofluids, such as whole blood.

Lillehoj et al. developed a smartphone-based POC platform through electrochemical detection for the fast protein quantification in human plasma samples [52]. In the presence of the target antigen, an electrochemical current was generated that is proportional to the protein concentration in the plasma samples. The achieved results indicated that the lowest detection limit could be down to 16 ng mL^{-1} with a linear concentration range between 0 and 1.024 ng mL^{-1}. Due to its portability, simplicity and capability to perform analytical measurements in raw clinical samples, this device offers great potential for POC testing. While promising, this assay requires several sample loading steps and uses an enzymatic substrate which offers limited room temperature stability. For the design and development of a room temperature-stable approach, Laksanasopin et al. prepared a mobile phone system to perform ELISA for the sensitve detection of HIV protein [53]. Dried reagents were stored within a microfluidic cartridge and become rehydrated by the diluted blood sample, which is transported by negative pressure generated from a mechanical vacuum spring. The presence of the target analyte causes a change in the optical density. Therefore, assessments of analytes over a dilution range of 1:128 was demonstrated in human blood samples. The obtaned results showed that the dveloped sensor system could remain stable at room temperature for up to one month and does not require any external equipment. However, this system only provides a qualitative result, and the optical signals can be affected by blood samples. Thus, there is a need for a shelf stable smartphone-based sensor platform that can effectively produce quantitative and reliable measurements of protein biomarkers in blood samples.

Various applications of smartphone-based optical sensors are shown in Table 2.1.

Table 2.1 Various applications of smartphone-based optical sensors.

Analyte	Type of detection	Sample	Range of detection	Detection time	Reference
Chlorine	Colorimetric	Water	0.3 to 1.0 mg/L	1.9 s	[54]
Fluoride	Colorimetric	Groundwater	0.1 to 2.0 mg/L	less than 1 min	[55]
Cholesterol	Colorimetric	Blood	140 to 400 mg/dL	1 min	[56]
Formaldehyde	Colorimetric	Air	1 to 600 µM	less than 4 min	[57]
Glucose	Colorimetric	Blood	3 to 1000 mg/dL	10 min	[58]
Furfural	Colorimetric	Beer	39 to 500 µg/L	60 min	[59]
Collegenase, trypsin	Fluorescence	Buffer solution	3.75 to 40 µg/mL; 3.72 to 1200 ng/mL	less than 90 min	[60]
HSV-2 virus	Fluorescence	Dulbecco's Modified Eagle's medium	21 to 2100 PFU/mL	less than 60 min	[61]
Ochratoxin A	Fluorescence	Beer	2 µg/L	5 min	[62]
Thiram	Fluorescence	Apple juice	0.1 µM to 1 mM	less than 1 min	[63]
Nanoparticles, viruses	Fluorescence microscopy	Deionized water	>100 nm	0.5 s	[64]
H_2O_2	Bio/chemiluminescence	10 mM Bis(2,4,6-trichlorophenyl) Oxalate	250 nM to 100 µM	5–30 s	[65]
HIV1-p17 IgG	Bio/chemiluminescence	Blood plasma	10 pM to 5 nM	30 min	[66]
Bovine serum albumin (BSA)	Spectroscopy	Deionized water	>0.1 mg/mL	10 min	[67]
Paraoxon	Spectroscopy	Phosphate buffered saline	5 nM to 25 µM	30 min	[68]
Immunoglobulin G (IgG)	Spectroscopy	Phosphate buffered saline	>4.25 nM	40 min	[69]
miRNA	Spectroscopy/fluorescence	Phosphate buffered saline	10 pM to 1 µM	1 s	[70]

2.3 Electrochemical sensing based on smartphone technology

The main challenge for the design and development of electrochemical sensor systems is the is the improvement of the sensitive recognition performance towards the target compound/s and construction of the compact miniaturized electrode [71–74]. To overcome this challenge, the combination of the smartphone with the electrochemical sensing platform as the effetrive sensing approach is a powerful technology. The integration of the smartphone as a reader into the sensor platform, the ability of sensing signals from the target compound/s can be more sensitive, specific and the achieved signal can be successfully transferred into the analysis device without aneed of any extra equipment. Some interesting examples of smartphone-based eletrochemical sensors are demonstrated in the following:

Low et al. reported the design and fabrication of a smartphone-based electrochemical sensor platform for the sensitive detection of microRNA-21, a cancer biomarker which is quite stable in body fluids enabling sensitive noninvasive detection, in saliva samples [75]. For this purpose, the researchers prepared a reduced graphene oxide/Au naoparticles-modified disposable screen-printed electrode and integrated into a circuit board for detection and a Bluetooth-enabled smartphone. The hybridization process between ssDNA probe and microRNA-21 lead to a reduce in the current of the obtained peak with the rising in the cıncentration of the target microRNA-21. The achieved results exhibited that te fabricated a smartphone-based electrochemical sensor platform displayed excellent selectivity and recognition perforance towards microRNA-21 in saliva samples. The obtaind detection range was between 1×10^{-4} and 1×10^{-12} M

In an interesting work reported by Kwon and colleagues, a smartphone-based electrochemical sensor system was designed and developed for the sensitive detection of dopamine based on electrochemiluminescence [76]. For this purpose, screen-printed electrodes were prepared and integrated into a smartphone. The schematic demonstration of the developed smartphone-based electrochemşcal sensor system towards dopamine is shown in Fig. 2.3. The detection of dopamine was based on the its electrochemiluminescence-based quenching feature. electrochemiluminescence systems was prepared by using tris(2,2′-bipyridine) ruthenium(II) ($Ru(bpy)_3^{2+}$) and coreactant tri-n-propylamine (TPrA). The investigaton of the quenching mechanism of $Ru(bpy)_3^{2+}$/TPrA by dopamine was conducted by estimation of the constants of the Stern-Volmer equations. The achieved experimental results under optimum sensing conditions exhibited that the developed smartphone-based electrochemical sensor system can be successfully used for the sensitive detection of the target compound dopamine. The developed smartphone-based electrochemical sensor displayed a great recognition behaviour for dopamine in a linear concenration range between 1.0 and 50 µM. The detection limit was obtained as 500 nM.

Wang and colleagues reported the preparation of a smart-phone-based electrochemical sensor platform for white blood cell-counting system on micro porous paper using patterned gold microelectrodes [77]. In their research, white blood cells separated from whole blood were successfully trapped by the paper employing

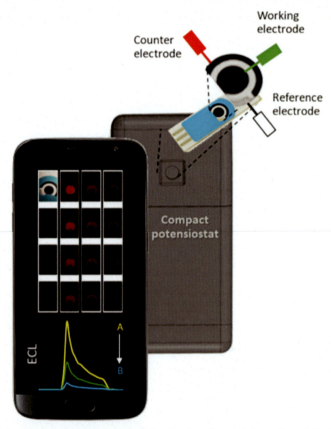

Fig. 2.3 The schematic demonstration of the developed smartphone-based electrochemical sensor system towards dopamine (Republished with permission from [76]).

microelectrodes (Fig. 2.4). Thus, white blood cells trapped on the paper caused the ion diffusion blockage on the microelectrodes and the concentration of the target cells was effivciently determined by the current of the obtained peak on the surfe of microelectrodes through differential pulse voltammeter measurements. The quantitative data were successfully achieved and collected using a smart phone within a very short time (1 min).

In a crucial work reported by Chandra [78], a promising personalized smartphone-based electrochemical sensor system for the sensitive and efficient detection of COVID-19 was proposed. According to the authors it can possible to design and fabricate a powrful biosensor towards the target virus COVID-19 by combination of a complete disposable sensing module with a smartphone-based application platform for personalized diagnosis (Fig. 2.5). Sensing surface can be efficiently optimized according to the marker molecules, then, it can be improvised for point of care (POC) diagnosis in real clinical samples. This type of miniaturized system can provide a rapid and affordable sensing process not only to sentively detect but also for efficiently monitoring of the COVID-19 in large scale. A smartphone-based "cloud"

Smartphone-based optical and electrochemical sensing

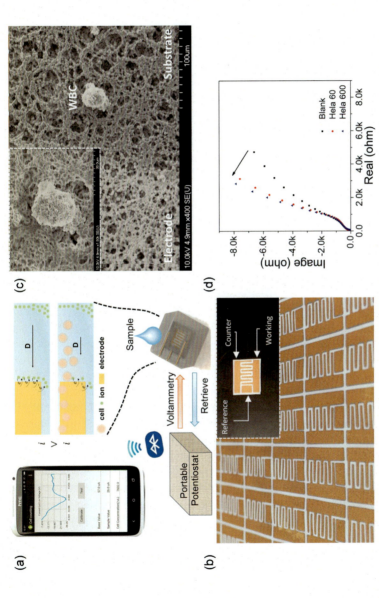

Fig. 2.4 The schematic depiction of the smartphone-based sensor for cell counting. (a) the developed system diagram and the principle of the electrochemical process for white blood cells counting (b) Au electrodes on polyvinylidene fluoride membrane paper (c) Scanning electron microscopy (SEM) image of the trapped WBC white blood cells on membrane paper (d) Nyquist plot of electrochemical impedance spectroscopy demonstrating the diffusion impedance upward bending by adding more Hela cells on the membrane electrode (Republished with permission from [77]).

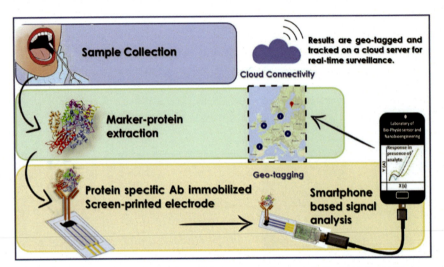

Fig. 2.5 The schematic representation of the suggested smartphone-based sensitive detection of COVID-19 and cloud-based real-time surveillance (Republished with permission from [78]).

directory can also enable real-time surveillance through geo-tagging which constitutes the process of defining, creating and provisioning a set of geolocation information to a smartphone.

An effective smartphone-based electrochemical sensor towards nitroaromatic explosives including 2,4,6-trinitrotoluene (TNT), 2,4-dinitrotoluene (DNT) and 4-nitrotoluene was designed and developed by Li et al. [79]. For this purpose, electrochemiluminescence system enhanced with silica nanopores was prepated on smartphone A bifunctional polypeptide, Lys-Trp-His-Trp-Gln-Arg-Pro-Leu-Met-Pro-Val-Ser-Ile-Lys, was prepared and then immobilized on the surface of screen printed electrodes (SPE) modified with silica nanopores. Tris(2,2′-bipyridyl) ruthenium(II) ($Ru(bpy)_3^{2+}$) was chosen as the electrochemiluminescence label for the sensitive detection of the target. With the positive charged of this label lead to an increase in luminescence intensity was achieved by the well-conductive electron channels and the selective ion channels present in the negative-charged silica nanopores. The obtained results displayed that the developed smartphone-based electrochemiluminescence sensor showed excellent sensing response for the target explosive compounds in a linear range between 10^{-7} mg mL^{-1} and 10^{-3} mg mL^{-1}. The detechin limit was achieved as 2.3×10^{-9} mg mL^{-1}.

2.4 Conclusions

The smartphone technology is undoubtly a revolutionary invention. The rapid progresses in the development of electronics technology also lead to signiicant reduce of the size but increase of the functions of smartphones that enabes them to be highly portable and allows the combination of various sensors in them. These superiorities features caught the interest of scientists who work in analytical and sensing areas. In this chapter, we have overviewed and highlighted the latest advancements on the design, development and successful applications of smartphone-based optical (colorimetric, fluorescence, chemiluminescence etc.) and electrochemical sensor platforms

The rapidy growing number of research in which smartphone technology was successfully employed for the development of novel sensing platforms confirmed that this great technology having excellent advantages is a promising and powerful approach for the design and fabrication of novel sensing and analysis systems including optical and electrochemical sensots.

References

[1] X. Liu, T.-.Y. Lin, P.B. Lillehoj, Smartphones for cell and biomolecular detection, Ann. Biomed. Eng. 42 (11) (2014) 2205–2217.

[2] D. Quesada-González, A. Merkoçi, Mobile phone–based biosensing: an emerging 'diagnostic and communication' technology, Biosens. Bioelectron. 92 (2017) 549–562.

[3] J. Jiang, et al., Smartphone based portable bacteria pre-concentrating microfluidic sensor and impedance sensing system, Sens. Actuators B Chem. 193 (2014) 653–659.

[4] S. Büyüktiryaki, Y. Sümbelli, R. Keçili, C.M. Hussain, Lab-on-chip platforms for environmental analysis, in: Paul Worsfold, Colin Poole, Alan Townshend, Manuel Miró (Eds.), Encyclopedia of Analytical Science3rd Edition, Academic Press, 2019, pp. 267–273.

[5] R. Keçili, F. Ghorbani-Bidkorbeh, İ. Dolak, C.M. Hussain, Chapter 1 - Era of nano-lab-on-a-chip (LOC) technology, in: Chaudhery Mustansar Hussain (Eds.), Handbook on Miniaturization in Analytical Chemistry, Elsevier, 2020, pp. 1–17.

[6] Ö. Biçen Ünlüer, F. Ghorbani-Bidkorbeh, R. Keçili, C.M. Hussain, Chapter 12 - Future of the modern age of analytical chemistry: nanominiaturization, in: Chaudhery Mustansar Hussain (Eds.), Handbook on Miniaturization in Analytical Chemistry, Elsevier, 2020, pp. 277–296.

[7] J. Sengupta, C.M. Hussain, Graphene and its derivatives for analytical lab on chip platforms, TrAC Trend Anal. Chem. 114 (2019) 326–337.

[8] R. Keçili, S. Büyüktiryaki, C.M. Hussain, Chapter 8 - Micro total analysis systems with nanomaterials, in: Chaudhery Mustansar Hussain (Eds.), Handbook of Nanomaterials in Analytical Chemistry, Elsevier, 2020, pp. 185–198.

[9] C.M. Hussain, Magnetic nanomaterials for environmental analysis, in: C.M. Hussain, B. Kharisov (Eds.), Advanced Environmental Analysis-Application of Nanomaterials, The Royal Society of Chemistry, 2017.

[10] R. Keçili, C.M. Hussain, Recent progress of imprinted nanomaterials in analytical chemistry, Int J Anal Chem (2018) 8503853.

[11] R. Keçili, S. Büyüktiryaki, C.M. Hussain, Advancement in bioanalytical science through nanotechnology: past, present and future, TrAC Trend Anal Chem 110 (2019) 259–276.

[12] S. Büyüktiryaki, R. Keçili, C.M. Hussain, Functionalized nanomaterials in dispersive solid phase extraction: advances & prospects, TrAC Trend Anal Chem 127 (2020) 115893.

[13] D. Sharma, C.M. Hussain, Smart nanomaterials in pharmaceutical analysis, Arab. J. Chem. 13 (2020) 3319.
[14] C.M. Hussain, Nanomaterials in Chromatography: Current Trends in Chromatographic Research Technology and Techniques, Elsevier, 2018.
[15] C.M. Hussain, R. Keçili, Modern Environmental Analysis Techniques for Pollutants, 1st Edition, Elsevier, 2019.
[16] D. Rawtani, P.K. Rao, C.M. Hussain, Recent advances in analytical, bioanalytical and miscellaneous applications of green nanomaterial, TrAC, Trends Anal. Chem. 133 (2020) 116109.
[17] R. Keçili, G. Arli, C.M. Hussain, Chapter 14 - Future of analytical chemistry with graphene, in: Chaudhery Mustansar Hussain (Eds.), Chapter 14 - Future of analytical chemistry with graphene, Compr. Anal. Chem. 91 (2020) 355–389.
[18] S. Büyüktiryaki, R. Keçili, C.M. Hussain, Chapter 2- Modern age of analytical chemistry: nanomaterials, in: Chaudhery Mustansar Hussain (Eds.), Handbook of Nanomaterials in Analytical Chemistry, Elsevier, 2020, pp. 29–40.
[19] C. Kiang Chua, A. Ambrosi, M. Pumera, Graphene based nanomaterials as electrochemical detectors in Lab-on-a-chip devices, Electrochem. Commun. 13 (5) (2011) 517–519.
[20] A. Kobuszewska, D. Kolodziejek, M. Wojasinski, E. Jastrzebska, T. Ciach, Z. Brzozka, Lab-on-a-chip system integrated with nanofiber mats used as a potential tool to study cardiovascular diseases (CVDs), Sens. Actuators B 330 (2021) 129291.
[21] CelineI.L Justino, Teresa A.P Rocha-Santos, ArmandoC. Duarte, Chapter 14 - Nanomaterials in Lab-on-Chip Chromatography, in: Chaudhery Mustansar Hussain (Eds.), Chapter 14 - Nanomaterials in Lab-on-Chip Chromatography, Nanomater. in Chromatogr. (2018) 387–400.
[22] KartickChandra Majhi, Paramita Karfa, Rashmi Madhuri, Chapter 6 - Smartphone-based nanodevices for in-field diagnosis, (Eds.), Suvardhan Kanchi, Deepali Sharma, Nanomaterials in Diagnostic Tools and Devices, Elsevier, 2020, Pages 159-187.
[23] Akbar Hasanzadeh, Chapter 12 - Smartphone-based microfluidic devices, in: Michael R. Hamblin, Mahdi Karimi (Eds.), Biomedical Applications of Microfluidic Devices, Academic Press, 2021, pp. 275–288.
[24] Zheng Li, Shengwei Zhang, Qingshan Wei, Chapter 5 - Smartphone-based flow cytometry, in: Jeong-Yeol Yoon (Eds.), Smartphone Based Medical Diagnostics, Academic Press, 2020, pp. 67–88.
[25] SandeepK. Vashist, JohnH.T Luong, Chapter 16 - Smartphone-based immunoassays, in: Sandeep K. Vashist, John H.T. Luong (Eds.), Handbook of Immunoassay Technologies, Academic Press, 2018, pp. 433–453.
[26] Anna Pyayt, Chapter 6 - Smartphones for rapid kits, in: Jeong-Yeol Yoon (Eds.), Smartphone Based Medical Diagnostics, Academic Press, 2020, pp. 89–102.
[27] D.S.Y. Ong, M. Poljak, Smartphones as mobile microbiological laboratories, Clin. Microbiol. Infect. 26 (4) (2020) 421–424.
[28] Mei-.Ting Liu, Jing Zhao, Shao-.Ping Li, Application of smartphone in detection of thin-layer chromatography: case of salvia miltiorrhiza, J. Chromatogr. A 1637 (2021) 461826.
[29] Taylor J. Moehling, Dong Hoon Lee, Meghan E. Henderson, Mariah K. McDonald, Preston H. Tsang, Seba Kaakeh, Eugene S. Kim, Steven T. Wereley, Tamara L. Kinzer-Ursem, Katherine N. Clayton, Jacqueline C. Linnes, A smartphone-based particle diffusometry platform for sub-attomolar detection of Vibrio cholerae in environmental water, Biosens. Bioelectron. 167 (2020) 112497.

[30] Ze Wu, Jinhui Lu, Qiangqiang Fu, Lianghe Sheng, Bochao Liu, Cong Wang, Chengyao Li, Tingting Li, A smartphone-based enzyme-linked immunochromatographic sensor for rapid quantitative detection of carcinoembryonic antigen, Sens. Actuators B 329 (2021) 129163.

[31] Wenfeng Luo, Jie Deng, Jinhua He, Zeping Han, Chen Huang, Yuguang Li, Qiangqiang Fu, Hanwei Chen, A smartphone-based multi-wavelength photometer for on-site detection of the liquid colorimetric assays for clinical biochemical analyses, Sens. Actuators B 329 (2021) 129266.

[32] Rajkumar Bandi, Madhusudhan Alle, Chan-.Woo Park, Song-.Yi Han, Gu-.Joong Kwon, Nam-.Hun Kim, Jin-Chul Kim, Seung-Hwan Lee., Cellulose nanofibrils/carbon dots composite nanopapers for the smartphone-based colorimetric detection of hydrogen peroxide and glucose, Sens. Actuators B 330 (2021) 129330.

[33] Xiaoxue Kou, Linjing Tong, Yujian Shen, Wangshu Zhu, Li Yin, Siming Huang, Fang Zhu, Guosheng Chen, Gangfeng Ouyang, Smartphone-assisted robust enzymes@MOFs-based paper biosensor for point-of-care detection, Biosens. Bioelectron. 156 (2020) 112095.

[34] Prinjaporn Teengam, Weena Siangproh, Sitt Tontisirin, A.morn Jiraseree-amornkun, Natthaya Chuaypen, Pisit Tangkijvanich, Charles S. Henry, Nattaya Ngamrojanavanich, Orawon Chailapakul, NFC-enabling smartphone-based portable amperometric immunosensor for hepatitis B virus detection, Sens. Actuators B 326 (2021) 128825.

[35] Zhiwei Liu, Qicheng Hua, Jin Wang, Zaoqing Liang, Jiahao Li, Jinxiao Wu, Xing Shen, Hongtao Lei, Xiangmei Li, A smartphone-based dual detection mode device integrated with two lateral flow immunoassays for multiplex mycotoxins in cereals, Biosens. Bioelectron. 158 (2020) 112178.

[36] C.V. Sapan, R.L. Lundblad, N.C. Price, Colorimetric protein assay techniques, Biotechnol. Appl. Biochem. 29 (1999) 99–108.

[37] D. Vilela, G.M. C., A. Escarpa, Sensing colorimetric approaches based on gold and silver nanoparticles aggregation: chemical creativity behind the assay. A review, Anal. Chim. Acta 751 (2012) 24–43.

[38] K. Su, X. Qui, J. Fang, Q. Zou, P. Wang, An improved efficient biochemical detection method to marine toxins with a smartphone-based partable system - Bionic e-Eye, Sensor Actuat B-Chem 238 (2017) 1165–1172.

[39] Korbua Chaisiwamongkhol, Shakiroh Labaidae, Sunisa Pon-in, Sakchaibordee Pinsrithong, Thanthapatra Bunchuay, Apichai Phonchai, Smartphone-based colorimetric detection using gold nanoparticles of sibutramine in suspected food supplement products, Microchem. J. 158 (2020) 105273.

[39] Y. Gan, T. Liang, Q. Hu, L. Zhong, X. Wang, P.Wang H.Wan, In-situ detection of cadmium with aptamer functionalized gold nanoparticles based on smartphone-based colorimetric system, Talanta 208 (2020) 120231.

[41] O. Keller, M. Benoit, A. Müller, S. Schmeling, Smartphone and tablet-based sensing of environmental radioactivity: mobile low-cost measurements for monitoring, citizen science, and educational purposes, Sens. (Basel) 19 (19) (2019) 4264.

[42] Gi Rateni, P. Dario, F. Cavallo, Smartphone-based food diagnostic technologies: a review, Sens. 17 (2017) 1453.

[43] S. Kanchi, M.I. Sabela, P.S. Mdluli, K.Bisetty Inamuddin, Smartphone based bioanalytical and diagnosis applications: a review, Biosens. Bioelectron. 102 (2018) 136–149.

[44] N. Fahimi-Kashani, M.R. Hormozi-Nezhad, A smart-phone based ratiometric nanoprobe for label-free detection of methyl parathion, Sens. & Actuators: B. Chem. 322 (2020) 128580.

[45] Wenting Li, Xinai Zhang, Xuetao Hu, Yongqiang Shi, Zhihua Li, Xiaowei Huang, Wen Zhang, Di Zhang, Xiaobo Zou, Jiyong Shi., A smartphone-integrated ratiometric fluorescence sensor for visual detection of cadmium ions, J. Hazard. Mater. 408 (2021) 124872.

[46] A. Roda, E. Michelini, L. Cevenini, D. Calabria, M.M. Calabretta, P. Simoni, Integrating biochemiluminescence detection on smartphones: mobile chemistry platform for point-of-need analysis, Anal. Chem. 86 (2014) 7299–7304.

[47] A. Roda, M. Guardigli, D. Calabria, M.M. Calabretta, L. Cevenini, E. Michelini, A 3D-printed device for a smartphone-based chemiluminescence biosensor for lactate in oral fluid and sweat, Analyst 139 (2014) 6494–6501.

[48] Jun Xu, Shengli Guo, Lei Jia, Tinghui Zhu, Xiangzhen Chen, Tongqian Zhao, A smartphone-integrated method for visual detection of tetracycline, Chem. Eng. J. (2020) 127741.

[49] K.E. McCracken, T. Tat, V. Paz, J.-.Y. Yoon, Smartphone-based fluorescence detection of bisphenol A from water samples, RSC Adv. 7 (2017) 9237–9243.

[50] A.F. Coskun, R. Nagi, K. Sadeghi, S. Phillips, A. Ozcan, Albumin testing in urine using a smart-phone, Lab. Chip 13 (21) (2013) 4231–4238.

[51] C.C. Stemple, S.V. Angus, T.S. Park, J.-.Y. Yoon, Smartphone-based optofluidic labon-a-chip for detecting pathogens from blood, J. Lab. Autom. 19 (1) (2014) 35–41.

[52] P.B. Lillehoj, M.-.C. Huang, N. Truong, C.-.M. Ho, Rapid electrochemical detection on a mobile phone, Lab Chip 13 (15) (2013) 2950–2955.

[53] T. Laksanasopin, et al., A smartphone dongle for diagnosis of infectious diseases at the point of care, Sci. Transl. Med. 7 (273) (2015) 273re1.

[54] S. Sumriddetchkajorn, K. Chaitavon, Y. Intaravanne, Mobile device-based selfreferencing colorimeter for monitoring chlorine concentration in water, Sens. Actuat BChem 182 (2013) 592–597.

[55] S. Levin, S. Krishnan, S. Rajkumar, N. Halery, P. Balkunde, Monitoring of fluoride in water samples using a smartphone, Sci. Total Environ. 551-552 (2016) 101–107.

[56] V. Oncescu, D. O'Dell, D. Erickson, Smartphone based health accessory for colorimetric detection of biomarkers in sweat and saliva, Lab Chip 13 (2013) 3232–3238.

[57] X. Yang, Y. Wang, W. Liu, Y. Zhang, F. Zheng, S. Wang, D. Zhang, J. Wang, A portable system for on-site quantification of formaldehyde in air based on G-quadruplex halves coupled with a smartphone reader, Biosens. Bioelectron. 75 (2016) 48–54.

[58] J.P. Devadhasan, H. Oh, C.S. Choi, S. Kim, Whole blood glucose analysis based on smartphone camera module, J. Biomed. Opt. 20 (11) (2015) 117001.

[59] A. Rico-Yuste, V. González-Vallejo, E. Benito-Peña, T D L C Engel, G. Orellana, M.C. Moreno-Bondi, Furfural determination with disposable polymer films and smartphone-based colorimetry for beer freshness assessment, Anal. Chem. 88 (2016) 3959–3966.

[60] P. Wargocki, W. Deng, A.G. Anwer, E.M. Goldys, Medically relevant assays with a simple smartphone and tablet based fluorescence detection system, Sensors 5 (2015) 11653–11664.

[61] S.-.C Liao, J. Peng, M.G. Mauk, S. Awasthi, J. Song, H. Friedman, H.H.L.C. Bau., Smart cup: a minimally-instrumented, smartphone-based point-of-care molecular diagnostic device, Sens Actuat B-Chem 229 (2016) 232–238.

[62] D. Bueno, R. Muñoz, J.L. Marty, Fluorescence analyzer based on smartphone camera and wireless for detection of Ochratoxin A, Sens Actuat B-Chem 232 (2016) 462–468.

[63] Q. Mei, H. Jing, Y. Li, W. Yisibashaer, J. Chen, B.N. Li, Y. Zhang., Smartphone based visual and quantitative assays on upconversional paper sensor, Biosens. Bioelectron. 75 (2016) 427–432.

[64] Q. Wei, H. Qi, W. Luo, D. Tseng, S.J. Ki, Z. Wan, Göröcs, Fluorescent imaging of single nanoparticles and viruses on a smart phone, ACS Nano 7 (10) (2013) 9147–9155.

[65] E. Lebiga, R.E. Fernandez, A. Beskok, Confined chemiluminescence detection of nanomolar levels of H_2O_2 in a paper–plastic disposable microfluidic device using a smartphone, Anal. 140 (2015) 5006–5011.

[66] R. Arts, I. den Hartog, S.E. Zijlema, V. Thijssen, S.H.E. van der Beelen, M. Merkx, Detection of antibodies in blood plasma using bioluminescent sensor proteins and a smartphone, Anal. Chem. 88 (2016) 4525–4532.

[67] C. Zhang, G. Cheng, P. Edwards, M.-.D Zhou, S. Zheng, Z. Liu, G-Fresnel smartphone spectrometer, Lab Chip 16 (2016) 246–250.

[68] L.-J Wang, Y.-.C Chang, X. Ge, A.T. Osmanson, D. Du, Y. Lin, L. Li, Smartphone optosensing platform using a DVD grating to detect neurotoxins, ACS Sens 1 (2016) 366–373.

[69] D. Gallegos, K.D. Long, H. Yu, P.P. Clark, Y. Lin, S. George, P. Nath, B.T. Cunningham, Label-free biodetection using a smartphone, Lab Chip 13 (2013) 2124–2132.

[70] H. Yu, Y. Tan, B.T. Cunningham, Smartphone fluorescence spectroscopy, Anal. Chem. 86 (2014) 8805–8813.

[71] K.G. Im, D.N. Nguyen, S. Kim, H.J. Kong, Y.K. Kim, C.S. Park, O.S. Kwon, H.S. Yoon, Graphene-embedded hydrogel nanofibers for detection and removal of aqueous-phase dyes, ACS Appl. Mater. Interfaces 9 (12) (2017) 10768–10776.

[72] I. Khalil, S. Rahmati, N.M. Julkapli, W.A. Yehye, JGraphene metal nanocomposites — Recent progress in electrochemical biosensing applications, Ind. Eng. Chem. 59 (2018) 425–439.

[73] W.J. Na, J.S. Lee, J.M. Jun, W.Y. Kim, Y.K. Kim, J.S. Jang, Highly sensitive copper nanowire conductive electrode for nonenzymatic glucose detection, J. Ind. Eng. Chem. 69 (2019) 358–363.

[74] L. Li, Y. Liu, L. Ai, J. Jiang, J. Ind, Synthesis of the crystalline porous copper oxide architectures derived from metal-organic framework for electrocatalytic oxidation and sensitive detection of glucose, Eng. Chem, 70 (2019) 330–337.

[75] S.S. Low, Yixin Pan, Daizong Ji, Yaru Li, Yanli Lu, Yan He, Qingmei Chen, Qingjun Liu, Smartphone-based portable electrochemical biosensing system for detection of circulating microRNA-21 in saliva as a proof-of-concept, Sens. Actuators B 308 (2020) 127718 2020.

[76] H.J. Kwon, ElmerCcopa Rivera, MabioR.C Neto, Daniel Marsh, JonathanJ. Swerdlow, RodneyL. Summerscales, Padma P.Tadi Uppala, Development of smartphone-based ECL sensor for dopamine detection: practical approaches, Results in Chem. 2 (2020) 100029.

[77] X. Wang, G. Lin, G. Cui, X. Zhou, G.L. Liu, White blood cell counting on smartphone paper electrochemical sensor, Biosen. Bioelectron. 90 (2017) 549–557.

[78] P. Chandra, Miniaturized label-free smartphone assisted electrochemical sensing approach for personalized COVID-19 diagnosis, Sens. Int. 1 (2020) 100019.

[79] S. Li, D. Zhang, J. Liu, C. Cheng, L. Zhu, C. Li, Y. Lu, S.S. Low, B. Su, Q. Liu, Electrochemiluminescence on smartphone with silica nanopores membrane modified electrodes for nitroaromatic explosives detection, Biosens. Bioelectron. 129 (2019) 284–291.

Optical methods using smartphone platforms for mycotoxin detection

Diana Bueno[a], Anais Gómez[b], RB Dominguez[c], JM Gutiérrez[b], Jean Louis Marty[d]
[a]Bioprocess Department, UPIBI-IPN, Mexico City, Mexico
[b]Bioelectronics Section, Department of Electrical Engineering, CINVESTAV-IPN, Mexico City, Mexico
[c]CONACyT- CIMAV S.C., Chihuahua, CHIH, Mexico
[d]IMAGES group, Universite de Perpignan Via Domitia, Perpignan, France

3.1 Introduction

Mycotoxins are toxic compounds produced naturally by some types of molds. Mycotoxin-producing molds grow on many foodstuffs, such as maize, sorghum, soybeans, groundnuts, cereals, and other food and feed crops under natural conditions. They are present on a wide range of substrates under a wide range of environmental conditions (especially in warm and humid environments). Its spread can take place before or after harvest, during storage, or in the food itself. Most mycotoxins are chemically stable and persist after food processing [1–3].

Plenty of mycotoxins have been identified, but the major concerns for human health and livestock focus on ochratoxin A, aflatoxins, fumonisins, patulin, zearalenone and nivalenol/deoxynivalenol. Most mycotoxins are presented in the food chain because of mold infection of crops, either before or after harvest. Exposure to mycotoxins can occur directly from eating infected food or indirectly from animals fed with contaminated food [4].

Mycotoxins can be the cause of various diseases and poisonings due to the different chemical structures that differentiate them from each other. Acute mycotoxin poisoning occurs when the ingested product presents a high dose of these substances, so this condition is frequent in less developed countries, where the resources for its control are limited. On the other hand, the effects or chronic diseases are worrisome for the population's health in the long term and are perhaps more important because they occur in much lower doses. Mycotoxins' most common effects are carcinogenic, genotoxic, or can affect the kidney, liver, or immune system.

Both national and international organizations are permanently evaluating the risk that mycotoxins pose to humans. For some mycotoxins, these studies have led to the establishment of maximum legal limits. Many countries today have maximum legal limits on aflatoxins that are the most toxic and frequent; although these limits are not uniform, the limits are mandatory in the regulation in food or feed. Thus, regardless of the world's region, mycotoxin analysis is necessary during any production process that includes raw materials that could be contaminated.

Also, there must be an adequate material handling program that avoids contamination of both the inputs and the final product that will reach the consumer. Within this program, it is relevant to have reliable chemical analysis methods. Therefore, mycotoxins' analysis becomes very important since any action taken during food production will depend on the laboratory results. The determination of mycotoxins is not simple since they are heterogeneously distributed in the raw material. Briefly, the analysis consists of three stages: (1) sampling in the batch, (2) sample homogenization and pre-treatment, and (3) quantifying mycotoxins.

The principal analytical methods for the quantification of mycotoxins include chromatographic techniques like liquid chromatography (LC), thin-layer chromatography (TLC) and gas chromatography (GC). However, its use requires sophisticated instrumentation and highly qualified personnel for its operation, and time-consuming analyzes. In the last two decades, several immunochemical methods have been developed for mycotoxin detection taking advantage of the highly specific and sensitive reaction between an antigen and an antibody. The methods present advantages such as the speed of analysis, simplified operation, and considerable reduction in instrumentation requirements.

Unfortunately, although their benefits are many, they have a strong susceptibility to respond to interfering elements that cause false positives that require confirmations using a reference method [5]. Commercial immunochemical methods can be classified by those that use an immunoaffinity column and enzyme-linked immunosorbent assay (ELISA) methods. Although the immunoaffinity column method was initially developed for fluorometry quantification of compounds from biological samples, columns have been used to purify and concentrate mycotoxins for subsequent detection by high performance liquid chromatography (HPLC), GC, and TLC. In counterpart, ELISA methods are commonly for rapid monitoring.

However, the developing of accurate, simple, and cost-effective techniques for mycotoxin analysis is of major concern. Optical biosensors coupled with knowledge areas such as optics, electronics, fluidics, and biochemistry, have become a powerful screening and analysis tool for the rapid estimation of mycotoxin contamination in raw materials. Optical biosensors are based on the measurement of an optical phenomenon (e.g., light absorption, fluorescence, surface plasmon resonance (SPR), or other optical properties) upon the biorecognition event to detect and identify the target molecules.

Therefore, optical detection is desirable as it can be applied to multiplexed applications. Considering that optics instrumentation is easily miniaturized at a relatively low cost, it gets systems with low power consumption and high stability, as cheap and sophisticated light sources and detection devices have become available recently [6]. In this sense, the development of optical systems applied to analytical chemistry using smartphones has increased considerably in recent years [7]. Smartphones are powerful storage, data processing, and interconnectivity tools which can be used to build low-cost and real-time analytic systems.

Their infrastructure offers robust hardware and software platforms for chemical and biochemical measurements that often demand sophisticated laboratory instruments.

Several authors have investigated detection methods with smartphones using them directly and/or coupling small instruments designed to operate with various principles whose interaction with the device's camera allows analyte detection and quantification tasks [8].

In this trend, commonly smartphones, built-in cameras are used as spectrometers that employ the spectral information of collimated light that is scattering after interaction with samples. Alternatively, smartphones as colorimeters can be used to quantify the color concentration in solid samples such as paper-based and liquid samples such as colored solutions. The computational cost required for image analysis is directly associated with the quality of the captured image. Thus, the digital image processing proposal range from the use of specialized software (available in the cloud or a desktop) to analyze content and after communicating the result to the user via Wi-Fi or Bluetooth, or the complete integration of a programmed application through some algorithms associated within the hardware/software platform of the smartphone itself.

There is a growing need for low-cost technologies, portable and rapid devices with the potential to detect, monitor, or quantify analytes, compounds, solutions, or analytical development, focusing this chapter on the detection of mycotoxins in food, without the need for expensive equipment or trained personal.

A powerful tool for environmental and food sectors, medical diagnosis, bioassay or bioanalysis, or an alternative to spectrophotometric measurements is smartphones' use due to their several advantages and their greater use in everyday life. According to Statista, the current number of mobile phone users in 2020 is 4.78 billion, indicating that 61.67 percent of people worldwide are cell phone owners. Global smartphone users increased by 40 percent between 2016 and 2020. By 2025, 1.74 billion new users will come online.

Smartphones are considered an ideal platform for colorimetric or spectrophotometric measurements due to their low cost, portability, and image quality. The colorimetric methods have grown considerably in allowing direct observation of any variation.

3.2 Mycotoxins

Mycotoxins are naturally occurring toxins produced as secondary metabolites of diverse fungi, with Fusarium spp., Aspergillus spp., and Penicillium spp., as the most common. Mycotoxins grow on various foodstuffs under different conditions, producing diverse toxic effects against organisms. There are between 300 and 400 types of mycotoxins. Some factors like age, nutrition, and exposure length can affect the susceptibility of animals and humans. According to information of the Food and Agriculture Organization (FAO), around 25 percent of crops are contaminated by mycotoxins causing a high economic loss and human health damage [9]. The chemical structure of the most prevalent mycotoxins is represented in Fig. 3.1.

Aflatoxins (AF) are one of the most potent toxic substances produced as secondary metabolites of Aspergillus species and are liable for around 25 percent of animal

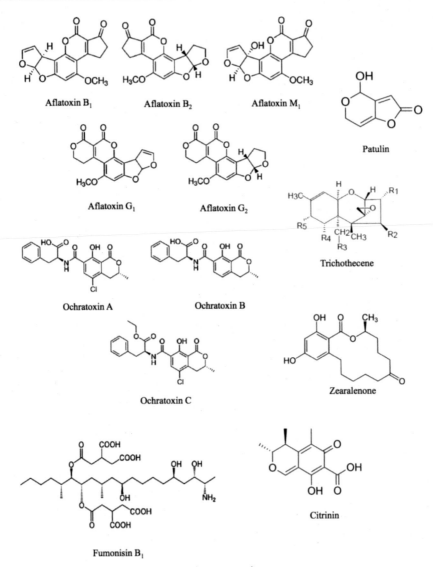

Fig. 3.1 Chemical structure for the main mycotoxins.

mortality. AF B1, AFB2, AFG1, and AFG2 are the most significant aflatoxins in foods and feeds. AFB1 and AFB2 are presented mainly in contaminated milk-based products.

Ochratoxins are the most dangerous mycotoxins found in food and beverages; Ochratoxin A (OTA) is the most common mycotoxin and can be found in cereals, dried fruits, coffee, cocoa, spices, wheat, barley, rice, oats, rye, beans, soy, peas, or

Table 3.1 The classification of mycotoxins and physiological effects.

Mycotoxins	Effects on mammals	Group (IARC)
Aflatoxins	carcinogenic, acute hepatitis, impaired immune system	1, 2B
Ochratoxins	carcinogenic, hepatotoxic, nephrotoxic, teratogenic	2B
Patulin	lung and brain hemorrhaging	3
Fumonisins	carcinogenic, hepatotoxic, causative agent in leukoencephalomalacia in horses	2B
Zearalenone	estrogenic activity	3
Trichothecenes	immunodepressants, gastrointestinal hemorrhaging	3
Citrinin	nephrotoxic	3

peanuts. Patulin (PAT) is a mycotoxin produced by fungi commonly found in vegetables and fruits, mainly in apples.

Fumonisins (FB) are water-soluble polar compounds. Found in grains, such as wheat, corn, barley, oat, rice, or sorghum, and in derived products (flour, grits, whole grain, extract, bran). Zearalenone (ZEN), Fumonisin B1 and Trichothecenes (deoxynivalenol (DON) and T-2 toxin [T-2]) are among the most toxic Fusarium mycotoxins. Zearalenone is a mycotoxin that produces estrogenic effects. Trichothecenes, unlike other mycotoxins, can act directly through the skin.

Citrinin (CIT) was one of the earliest antibiotics, but it was never used because of its mammalian toxicity. Citrinin can be co-occurring with ochratoxin A in grains and coconut products. CIT and PAT can infect apples with rotten spots. The most common ways of exposure to mycotoxins are consumption of contaminated foods, direct skin contact and inhalation [6].

Mycotoxins are potent toxins and cause serious damage to animals and humans. Table 3.1 shows the classification of the main mycotoxins, their physiological effects on mammals, and the classification of the International Agency for Research on Cancer (IARC).

3.3 Colorimetric detection

Colorimetric detection is the simplest approach to detection, quantification with a smartphone, based on digital images. This method is based on assessing the change in absorbance or reflectance of an analyte. The main tests are summarized in Table 3.2.

Chemical imaging exploits the spectroscopic techniques that provide information about the composition, structure, dynamics, and pixel concept, which is used as the measurement unit for the light emitted by an analyzed sample. The differences in the light suggest the presence of a given structure or substance, and these differences can be identified or quantified through the color models.

The model color Red-Green-Blue (RGB) is widely used for colorimetric quantification due to its simplicity. The color is decomposed and quantified in a specific range

Table 3.2 Description of the main tests [10].

Principle	Description
Brightfield test	The sample is illuminated, and the light is transmitted through the sample.
Colorimetric test	Measure concentrations of a compound in a solution with the aid of a color reagent.
Fluorescence test	It is the absorbance at a wavelength to be emitted at a different wavelength.
Electrochemical test	An electric current generates a chemical reaction.

of wavelengths related to the presence of reddish, greenish, and blueish regions of the visible light spectrum [11]. RGB model can be associated with color changes. Images described by RGB is considered a 3D arrays composed of three 2D planes, which belong to the red, green, and blue channel. Fig. 3.2 shows a 3D image as a blend of its three-color components, and it can be decomposed by the RGB model in three 2D planes separate. In the figure, the RGB color space is represented as a cube with axes R, G, and B; the black color is positioned in the cube's origin (0,0,0), meaning the absence of colors. However, the white color is placed in the opposite corner (1,1,1), corresponding to the maximum value of all three colors [12].

A sensor is a device that transforms information about a chemical or physical property of the system into a signal that can be registered by a device. The interactions of the sample with the medium and the sensor leading to a change in optical properties that can be registered. A biosensor is an analytical tool that measures biological reactions and generates signals proportional to the analyte concentration. Three main components conform to these:

1. a recognition element (receptor)
2. a signal transducer to get a measurable signal
3. a reader that register the output

Colorimetric sensors are based on measuring the intensity of color after a chemical reaction, and the concentration of an analyte could be determined using the absorption of the colored compound, that can be detected by the naked eye within visible (400–800mm) range or can be identified and quantified by instrumentation [13].

Fluorescence is the molecules' property to absorb light at a wavelength and emit light at a different wavelength; after the samples are excited, fluorescence can be distinguished as a signal. Some works described the fluorescence in mycotoxins as the absorbed radiation in the ultraviolet region, and emitted in the visible region and can be detected by a smartphone, employing the pixel information's quantification image processing. In the aflatoxins, letter B indicates blue fluorescence (blue), and the letter G indicates yellow-green fluorescence (green) to ultraviolet light. OTA under acidic conditions generates blue or green fluorescence (465 nm). The main mycotoxins present fluorescence.

Optical methods using smartphone platforms for mycotoxin detection 43

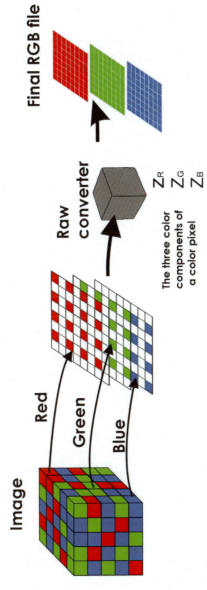

Fig. 3.2 Schematic of the implementation of the RGB color model.

3.4 Smartphone as a portable detection

The IBM Simon was the first smartphone in history, manufactured in 1992 and distributed in the US between August 1994 and February 1995. However, the technological leap in these devices occurred when Apple CEO Steve Jobs introduced the iPhone in June 2007. The first iPhone was a rectangular device, with a touch screen, without a physical keyboard or mouse, with an Internet connection and a camera. The key to its success, according to experts, was the combination of hardware and software in an effortless device to use.

Since then, both users and competitors have guessed the potential of the new smartphone. Today, most mobile phones integrate enough technology like a computer. They include not only a microprocessor with communication capabilities but also processing speed and storage. Besides, they have high-resolution digital cameras, autofocus and digital zoom, and an independent power source. Everything is managed by specialized software in which users can include applications to carry out many tasks with different purposes.

According to the World Advertising Research Center, 72.6 percent of all internet users will be accessing websites through a smartphone [14,15]. In Table 3.3, the countries are shown according to the highest number of cell phone users. Android or iOS is the operating system for smartphones.

Smartphone's level of electronic sophistication has allowed them to be viable alternatives for developing new instrumental systems. Based on the fact that a very high percentage of its integrated infrastructure can be used to reproduce most of the steps of an analytical measurement process, allowing real-time processing of data as well as transmission of the collected results (with clear advantages for *in-situ* studies due to its portability). Smartphones have been successful in applications that require diagnostic, detection, monitoring, and quantification tasks. In this way, it is possible to find scientific contributions reporting their use in environmental studies, chemical and biochemical analysis, food industry, medical and health sciences. Some of these contributions are represented in Table 3.4.

Table 3.3 Cell Phone Usage Worldwide, by Country.

Country	Smartphone users (Millions of persons)	Smartphone penetration (percent)
China	851	59.9
India	346	25.3
United States	260	79.1
Brazil	96.9	45.6
Russian Federation	95.4	66.3
Indonesia	83.9	31.1
Japan	72.6	57.2
Germany	65.9	79.9
Mexico	65.6	49.5
United Kingdom	55.5	82.9
France	50.7	77.5

Table 3.4 Smartphones as spectrometer or colorimeter for diverse applications.

Reference	Analyte	Smartphone as Spectrometer or Colorimeter	Smartphone as Detector or Instrument	LOD (limit of detection)	Detection medium
[11]	Tuberculosis	Colorimeter	Detector	–	Microplate/imagen
	Nitrite concentration and pH	Colorimeter	Instrument	0.04 units of pH; 0.52 mg L^{-1} nitrite	Image
	pH	Colorimeter	Detector	–	Image
	Lead ions	Colorimeter	Detector	20.0 ng mL^{-1}	Image
	Phenazopyridine	Colorimeter	Detector	1.0 µg L^{-1}	Image
	ELISA with microfluidic device	Colorimeter	Detector	0.89 ng ml^{-1}	Image
	Blood ketone	Colorimeter	Instrument	0 to 4 mmol/L	Test strip
[16]	Dye adsorption for field deployable environmental and wastewater management	Spectrometer	Instrument	0.1 to 10 ppm	Cuvette
[17]	Chlorophyll, hemoglobin	Spectrometer	Instrument	–	Image
[7]	Vitamin D	Colorimeter	Detector	15 nM	Test strip
	Salivary glucose	Colorimeter	Detector	22.2 mg/dL	Test strip
	Blood hematocrit	Spectrometer	Instrument	0.1 percent of hematocrit	Image
	Thyroid stimulating hormone	Spectrometer	Instrument	0.31 mIU/L	Strip
	Cholesterol in blood	Spectrometer	Instrument	140 mg dl^{-1} to 400 mg dl^{-1}	Strip
	Heavy metals	Colorimetric	Detector	Cu(II) =0.29 ppm Ni(II) =0.33 ppm Cd(II) =0.19 ppm Cr(VI) =0.35 ppm	Strip
	Salmonella	Spectrometer	Instrument	10^2 CFU mL^{-1}	Strip
	Amines	Colorimeter	Detector	≤1ppm	Image
	Mercury contamination in water	Colorimetric	Instrument	3.5 ppb	Image
	Bacteria in field water	Colorimeter	Instrument	10 bacterial cells per milliliter	Well
	Trinitrotoluene in soil	Colorimeter	Detector	50 mgL^{-1}	Image

3.4.1 Recommend hardware requirements

The evolution of smartphones' hardware is related to the miniaturization of the electronic components that make it up and improve the production/manufacturing process. Although the trend in the design of smartphones seems to focus on double, triple, and even quadruple the core of the central process unit (CPU) that integrates it, most user's applications available perform properly using two processing cores.

However, an important point to consider in the design of optical instrumentation based on these devices, far from the processing capacity, is the integration of an independent graphics processing unit (GPU) since it is responsible for efficiently displaying the graphics of the user interface, 2D and 3D effects, video playback in HD Ready (720p) or full HD (1080p) and the reproduction of advanced 2D and 3D graphics in animations.

In this way, while the GPU processes the graphics tasks, the CPU can dedicate to performing other types of calculations associated with data management/storage and the control/operation of the rest of the sensors included in the device (fingerprint reader, proximity sensor, capacitive sensors, accelerometer, gyroscope, GPS).

Another relevant element is related to the high-resolution camera. Thus, it is pertinent to consider the quality of the captured images, focal lens length, the performance under different lighting intensities (aperture), photoreceptor sensor size and type, ISO sensitivity, image stabilizer, and finally but significant the location of the phone camera.

Features such as digital zoom or increased resolution (in megapixels) are not indicative of camera quality. The number of pixels that current phone cameras use to produce an image is sufficient for most applications. Nevertheless, not all devices have powerful enough processors to perform image processing.

In this sense, it is mandatory to remember that the camera has been designed to capture images, and the inclusion of one or more of the mentioned characteristics can help quantify the intensity of the color more real.

The most relevant characteristics are the quantification of color intensity, the capacity of real-time processing of data, and the transmission of the results for storage. Due to the Beer-Lambert law, smartphone cameras can differentiate small differences in color, which can relate the concentration of analyte in solution to the intensity of its color by absorbance or transmittance [12].

The previous features enable smartphone to be used as an extremely useful analytical tool for diagnosis. The digital colorimetric determination can be performed computing the RGB components of the digital image. Fig. 3.3 shows a smartphone-based optical device for spectroscopic techniques, with a light source (external or integrated into the smartphone), a wavelength device, and the signal or image obtained directly or employing Wi-Fi (Wireless Fidelity) or Bluetooth.

3.4.2 Smartphone applications in food analysis

World Health Organization (WHO) have related food and waterborne diseases to the ingestion of harmful pathogens or toxic chemicals [5]. Some causative agents in

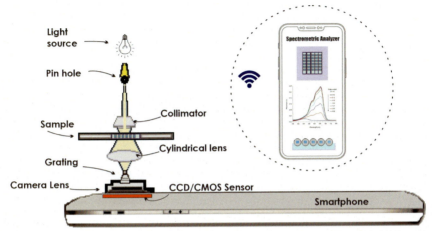

Fig. 3.3 Smartphones to be an optical device tool for analysis and diagnosis in chemistry applications.

food poisoning include proteinaceous infectious particles (prions), organic pollutants, heavy metals, bacterial agents, or natural toxins as mycotoxins.

Smartphones as detectors or instruments based on spectrometer and colorimeter have had high acceptance in point-of-care analyzes, chemical and biological sensing, electrochemical sensing, immunoassays, also in sectors such as medicine, food, environmental and agriculture, due to their relatively low cost, portability, easy to use interface and the compatibility with systems for acquisition and processing of data. Another advantage of smartphones is the versatility of samples that can be analyzed and sensors that can be employed. Also, smartphones provide a tool for imaging analysis or processing [18].

Several advances have been reported that use the smartphone to identify contaminants that cause food poisoning have been reported. Most of these methods use the smartphone camera for colorimetric detection, including quantification of the target analyte by transmittance, absorbance, or reflectance. Some examples of the smartphone-based monitoring of food safety are presented in Table 3.5 [19].

Some applications of the use of a smartphone to detect mycotoxins in food are mentioned here below.

Machado et al., proposed a lab-on-a-chip systems based on capillarity for fluidic manipulation, and employing a smartphone camera to analyze OTA, AFB1, and DON simultaneously. The portable and simple multiplexed assay achieved good sensitivities; less than 40 ng/mL for OTA, less than 0.1–0.2 ng/mL for AFB1 and 10 ng/mL for DON [21].

Smartphone as a detector in lateral flow immunoassays (LFIA) for multiplex mycotoxins was presented by Liu et al. For that purpose, to detect AFB1, ZEN, DON, T-2 toxin, and fumonisin B1 was used gold nanoparticles (GNPs) and time-resolved fluorescence microspheres with LFIA, achieving quantify low limits of detection.

Table 3.5 Use of smartphone for monitoring food safety.

Device	Principle	Sample	Target	LOD or linear range
Smartphone	ELISA	Cookies	Allergen	~1 ppm
		Milk	Salmonella enteritidis	94.2–101.2 CFU/mL
		Cheese		
		Water		
	Biochemical principle using enzymes	Spice mixtures and bouillon	(L)-glutamate	0.028 mmol/L
	Chromatography	Milk	Tetracycline	0.5 mg/mL
		Milk	Alkaline phosphatase	0.1 U/L
			Tetracycline Quinolones	0.2 – 2.3 ng/mL 0.15e3.6 ng/mL
		Water	Cr (VI)	2.5 ppm
			Total chlorine	1.5 ppm
			Caffeine	100 ppm
			E. coli K12	10 CFU/mL
	Chemical reaction	River water	Mercury ion	80 nM
		Fruits	Calcium carbide	–
		Minced meat	Isobutylamine	~1 ppm
			Isopentylamine	
			Triethylamine	
		Water	Bisphenol	4.4 µM
			Formaldehyde	0.2–2.5 ppm
			Fluoride	0– mg/L
			pH	1–12
		Fat-free milk	E. coli O157:H7	10 CFU/mL

Absorbance	Vitamin C	Ascorbic acid	5 mg/mL
	Mineral water	Oxygen, pH	7.8–8.8 mg/L O_2; 7.1–7.5 pH
	Rainwater		3.7–4.7 mg/L O_2; 5.8–6.2 pH
	Pure water		6.6–7.6 mg/L O_2; 6.6–7.0 pH
	Lake water		3.1–4.1 mg/L O_2; 7.2–7.6 pH
Lateral Flow aptamer assay	Tap water	Mercury ions	5 ppb
		Ochratoxin A	3 ng/mL
		Salmonella ATCC 50,761	85 CFU/mL
PiBA assay aggregation	–	E. coli O157:H7	YES/NO
		Salmonella enterica	YES/NO
Mie scattering	Ground beef	E. coli K12	10 CFU/mL
Simultaneous reaction and electrokinetic stacking	Water	Copper ion	30 mM

The recoveries of both LFIAs ranged from 84 to 110 percent, and the results were validated by LC-tandem mass spectrometry. [22].

A multiplex LFIA-GNPs was developed based on the identification of a single test line color. Antibodies were linked to red and blue GNPs to be evaluated for two different analytes, in which the mixture of two antigens formed the test line. Therefore, the test line assumed different colors depending on the analyte or quantity of analyte presented in the sample. The contaminant was identified based on the label's color; the strip images were acquired by a smartphone camera for obtaining RGB intensities, allowing a semi-quantitative estimation of AFB1 and FMs in wheat and derivate products [23].

A device based on a smartphone integrated with an immunochromatography platform was developed to detect and quantify ZEN in cereals and feed. The system quantified ZEN with LOD of 0.08 and 0.18 μg/kg in cereals and feed, respectively [24]. Various applications in the chemistry of digital colorimetric determination using a smartphone have widespread the use of smartphones to the smartphone camera and the software with the design and incorporation of apps.

Li et al. developed an integrated, smartphone-app-chip (SPAC) system for on-site quantitation of aflatoxins in spiked and moldy corn samples. The detection was based on an indirect competitive immunoassay and directly processed using a custom-developed Android app. The results obtained were comparable with conventional enzyme-linked immunosorbent assay kits. [25].

Table 3.6 summarized the last five years where the smartphone was employed to detect mycotoxins in food, the table details the sample detected, as well as the type of smartphone used and its application in detection.

3.5 Conclusions

We can conclude that the use of smartphones had extended beyond a spectrophotometer or colorimeter, as mentioned in the detailed publications in this paper, or the design of an app. Some publications already detail the use of smartphones to develop a complete device or as rapid kits analysis such as sperm testing system, quick sandwich ELISA for human C-reactive protein, to measure the amount of free hemoglobin in plasma analyze the color of dipsticks used for urinalysis.

Barragan et al. obtained a promising device by coupling a small potentiostat to an electronic micropipette [36], Hutchison et al. have developed a bright field smartphone microscope in conjunction with a microfluidic incubation device to detect Bacillus anthracis [37] or Ji and co-workers reported a smartphone-based cyclic voltammetry system for the electrochemical detection has great potential in public health, water, and food quality monitoring [38].

Smartphones are an excellent platform for running several applications due to the facility and low cost to integrate elements into a sensing system. It provides numerous applications in health, food, microfluidic systems, electrochemical, spectrophotometric, fluorescence systems or to develop an app or built portable device.

Table 3.6 Smartphone as a detector on mycotoxins in food.

Role of smartphone	Analyte	Detection	LOD	Working range	Sample	Smartphone	Ref.
Fully integrated analytical system	AFB1	LFIA	5 μg·kg^{-1}	5–1000 μg·kg^{-1}	maize	Samsung Galaxy S2	[26]
	AFB1	Colorimetric PDMS-μFC	3 μg·kg^{-1}	0.5–250	corn	Huawei Honor 3C	[25]
	ZEN	Colorimetric ELISA kit	2.12 ng·mL^{-1}	3.13–50 ng·mL^{-1}	corn	Huawei Honor 6	[27]
	AFB1 OTA ZEN	Fluorescence	1 ng	1–200 ng	corn	NA	[28]
	ZEN	LFIA	0.08 μg·kg^{-1} 0.18 μg·kg^{-1}	0–3.5 μg·kg^{-1}	Cereals feed	Huawei Maimag 6	[24]
	OTA AFB1	Fluorescence	0.1 ng·mL^{-1}	0.1–500 ng·mL^{-1} 0.1–200 ng·mL^{-1}	corn, flour	NA	[29]
	OTA	LFIA Fluorescence	10 μg·mL^{-1}	NA	NA	NA	[30]
	AFB1	MIP Fluorescence	20 ng·mL^{-1}	20–100 ng·mL^{-1}	maize	Meizu U 10	[31]
	AFB1 ZEN DON T-2 FB1	LFIA	0.27 μg·kg^{-1} 0.24 μg·kg^{-1} 0.9 μg·kg^{-1} 0.32 μg·kg^{-1} 0.59 μg·kg^{-1}	0–0.5 μg·kg^{-1} 0–5 μg·kg^{-1} 0–20 μg·kg^{-1} 0–1 μg·kg^{-1} 0–10 μg·kg^{-1}	maize wheat bran	Maintag 6 Huawei	[22]

(*continued*)

Table 3.6 (Cont'd)

Role of smartphone	Analyte	Detection	LOD	Working range	Sample	Smartphone	Ref.
Image Reader	OTA	Fluorescence	2 µg·L^{-1}	2–20 µg·L^{-1}	wine, beer	Iphone 4S	[32]
	OTA	Colorimetric P-µFC	40 ng·mL^{-1}	10–100 ng·mL^{-1}	Corn	Samsung Galaxy grand prime	[21]
	AFB1		0.1–0.2 ng·mL^{-1}	0.1–1 ng·mL^{-1}			
	DON		10 ng·mL^{-1}	10–100 ng·mL^{-1}			
	AFB1	ECL	0.01 ng·mL^{-1}	5–150 ng·mL^{-1}	Milk	NA	[33]
	AFM1	LFIA Colorimetric	0.016 ng·mL^{-1}	0–1 ng·mL^{-1}	Milk	NA	[34]
	AFB1	LFIA colorimetric	0.5 ng·mL^{-1}, 20 ng·mL^{-1}	0–10 ng·mL^{-1}, 0–500 ng·mL^{-1}	wheat, pasta, pastry	Huawei P9 Lite	[23]
	Type B fumonisin						
	AFB1	LFIA	0.3 ng·g^{-1}	2–16 ng·mL^{-1}	peanut bread corn rice	NA	[35]

LFA: Lateral flow assay
MIP: Molecularly Imprinted Polymers
ECL Electrochemiluminescence
µFC: Microfluidic chip
P-µFC: parallel microfluidic chip.

Significant advances have been made incorporating smartphones for detecting mycotoxins in food, mainly with colorimetric detection. Also, the smartphone's use allows timely detection of the mycotoxin in food or feed, reduction in the number of samples required, and it is not necessarily trained personal. Also, smartphone use allows the detection limits indicated by the regulatory organisms for each mycotoxin or food to be obtained in a compact system with a daily use device.

References

[1] Kharayat, B.S. and Y. Singh, Chapter 13 - Mycotoxins in foods: mycotoxicoses, detection, and management, in Microbial Contamination and Food Degradation, in: A.M. Holban and A.M. Grumezescu (Eds.), Academic Press. 2018 p. 395-421.
[2] M. Weidenbörner, Mycotoxins in Foodstuffs, Springer-Verlag, New York. XIV, 2008, p. 504.
[3] (FAO/WHO), Codex International Food Standards, WHO, 2018.
[4] International Programme on Chemical, S. and O. World Health, Selected Mycotoxins: Ochratoxins, Trichothecenes, Ergot /Published Under the Joint Sponsorship of the United Nations Environment Programme, the International Labour Organisation, and the World Health Organization, World Health Organization, Geneva, 1990.
[5] J. Singh, A. Mehta, Rapid and sensitive detection of mycotoxins by advanced and emerging analytical methods: a review, Food Sci. Nutr. 8 (5) (2020) 2183–2204.
[6] D. Bueno, et al., Determination of Mycotoxins in Food: a Review of Bioanalytical to Analytical Methods, Appl. Spectrosc. Rev. 50 (9) (2015) 728–774.
[7] D. Bueno, J.L. Marty, R. Muñoz, Smartphone as a portable detector, analytical device or instrument interface. Smartphones from an Applied Research Perspective, 2017.
[8] A. García, et al., Mobile phone platform as portable chemical analyzer, Sens. Actuators B 156 (1) (2011) 350–359.
[9] D. Liu, et al., Recent progress on electrochemical biosensing of aflatoxins: a review, TrAC, Trends Anal. Chem. (2020) 115966.
[10] J. Liu, et al., Point-of-care testing based on smartphone: the current state-of-the-art (2017–2018), Biosens. Bioelectron. 132 (2019) 17–37.
[11] M. Rezazadeh, et al., The modern role of smartphones in analytical chemistry, TrAC - Trends in Analytical Chemistry 118 (2019) 548–555.
[12] D. Bueno, et al., Colorimetric analysis of ochratoxin a in beverage samples, Sensors 16 (11) (2016) 1888.
[13] I.I. Ebralidze, et al., Chapter 1 - Colorimetric sensors and sensor arrays, in: O.V. Zenkina (Ed.), Nanomaterials Design for Sensing Applications, Elsevier, 2019, pp. 1–39.
[14] Special reports. Digital 2019, Global Internet Use Accelerates, 2020. [cited 2020 27/07/2020]; Available from https://wearesocial.com/blog/2019/01/digital-2019-global-internet-use-accelerates.
[15] A. Turner, BankMyCell, 2020. [cited 2020 27/07/2020]; Available from https://www.bankmycell.com/blog/how-many-phones-are-in-the-world.
[16] G.K. Özdemir, et al., Smartphone-based detection of dyes in water for environmental sustainability, Anal. Methods 9 (4) (2017) 579–585.

[17] J. Chen, et al., Experimental demonstration of remote and compact imaging spectrometer based on mobile devices, Sensors (Switzerland) 18 (7) (2018).

[18] Volkan, K., H. Nesrin, and S. Mehmet Ertugrul, From Sophisticated Analysis to Colorimetric Determination: Smartphone Spectrometers and Colorimetry, in Sophisticated Analysis to Colorimetric Determination: Smartphone Spectrometers and Colorimetry, Color Detection, in: L.-W. Zeng and S.-L. Cao, (Eds.), 2020, IntechOpen.

[19] D.D. Uyeh, et al., Chapter 11 - Food safety applications, in: J.-Y. Yoon (Ed.), Smartphone Based Medical Diagnostics, Academic Press, 2020, pp. 209–232.

[20] J.M.D. Machado, et al., Multiplexed capillary microfluidic immunoassay with smartphone data acquisition for parallel mycotoxin detection, Biosens. Bioelectron. 99 (2018) 40–46.

[21] Z. Liu, et al., A smartphone-based dual detection mode device integrated with two lateral flow immunoassays for multiplex mycotoxins in cereals, Biosens. Bioelectron. (2020) 158.

[22] F. Di Nardo, et al., Colour-encoded lateral flow immunoassay for the simultaneous detection of aflatoxin B1 and type-B fumonisins in a single Test line, Talanta 192 (2019) 288–294.

[23] X. Li, et al., A smartphone-based quantitative detection device integrated with latex microsphere immunochromatography for on-site detection of zearalenone in cereals and feed, Sens. Actuators B 290 (2019) 170–179.

[24] X. Li, et al., Integrated smartphone-app-chip system for on-site parts-per-billion-level colorimetric quantitation of aflatoxins, Anal. Chem. 89 (17) (2017) 8908–8916.

[25] S. Lee, G. Kim, J. Moon, Performance improvement of the one-dot lateral flow immunoassay for aflatoxin b1 by using a smartphone-based reading system, Sensors (Switzerland) 13 (4) (2013) 5109–5116.

[26] Y. Chen, et al., A smartphone colorimetric reader integrated with an ambient light sensor and a 3D printed attachment for on-site detection of zearalenone, Anal. Bioanal. Chem. 409 (28) (2017) 6567–6574.

[27] M. Yang, et al., A smartphone-based quantitative detection platform of mycotoxins based on multiple-color upconversion nanoparticles, Nanoscale 10 (33) (2018) 15865–15874.

[28] W. Ji, et al., Shape coding microhydrogel for a real-time mycotoxin detection system based on smartphones, ACS Appl. Mater. Interfaces 11 (8) (2019) 8584–8590.

[29] Y. Gong, et al., A portable and universal upconversion nanoparticle-based lateral flow assay platform for point-of-care testing, Talanta 201 (2019) 126–133.

[30] T. Sergeyeva, et al., Development of a smartphone-based biomimetic sensor for aflatoxin B1 detection using molecularly imprinted polymer membranes, Talanta 201 (2019) 204–210.

[31] D. Bueno, R. Muñoz, J.L. Marty, Fluorescence analyzer based on smartphone camera and wireless for detection of Ochratoxin A, Sens. Actuators B 232 (2016) 462–468.

[32] S.M. Khoshfetrat, H. Bagheri, M.A. Mehrgardi, Visual electrochemiluminescence biosensing of aflatoxin M1 based on luminol-functionalized, silver nanoparticle-decorated graphene oxide, Biosens. Bioelectron. 100 (2018) 382–388.

[33] M. Han, et al., An octuplex lateral flow immunoassay for rapid detection of antibiotic residues, aflatoxin M1 and melamine in milk, Sens. Actuators B 292 (2019) 94–104.

[34] T. Sojinrin, et al., Developing gold nanoparticles-conjugated aflatoxin B1 antifungal strips, Int. J. Mol. Sci. 20 (24) (2019).

[35] J.T.C. Barragan, L.T. Kubota, Minipotentiostat controlled by smartphone on a micropipette: a versatile, portable, agile and accurate tool for electroanalysis, Electrochim. Acta (2020) 341.

[36] J.R. Hutchison, et al., Reagent-free and portable detection of Bacillus anthracis spores using a microfluidic incubator and smartphone microscope, Analyst 140 (18) (2015) 6269–6276.

[37] D. Ji, et al., Smartphone-based cyclic voltammetry system with graphene modified screen printed electrodes for glucose detection, Biosens. Bioelectron. 98 (2017) 449–456.

Fluorescence measurements, imaging and counting by a smartphone

Tianran Lin, Danxuan Lin, Li Hou
School of Chemistry and Pharmaceutical Science, State Key Laboratory for the Chemistry and Molecular Engineering of Medicinal Resources, Guangxi Normal University, Guilin, P. R. China

4.1 Introduction

Smartphone has played a key role in our everyday lives. A smartphone is a mobile device that is widely used all over the world. In the past decade, due to the rapid improvement in both hardware and software for smartphones, it is a favorable choice to allow a smartphone to take the place of complex instruments and to be a part of the sensor or as the sensor itself. Researchers have attempted to create conventional devices which have compact dimensions and can be suitable for smartphones. Such devices were developed to link controlled laboratory experiments to public health, which would make it easy for everyone to enjoy health care and rapid diagnosis services, no matter people in developed cities or rural areas of the world [1].

Smartphones have useful features like internal memory for data storage, user-friendly interface, high-quality cameras, smooth operating systems and wireless data communication which attract more and more researchers to integrate smartphones with biosensing. Besides, in comparison to previous cellphones, smartphones have stronger computing ability and better operating system. With these advantages, sensors based on smartphone were designed to be easier and more useful. With the application of smartphone, sensing systems can replace the original readout devices such as screen display, data analyzer, input button and even detectors, and it decreased dimension, reduced the cost of the systems and simplified electronic design, which makes it a possible to allow on-site test. Smartphones have been applied for analytical chemistry, and play a more and more important role in this area in the future.[2] smartphone-based sensing systems act as a compact and low-cost sensing platform for rapid and accurate detection and analysis in disease diagnosis, environmental monitoring and so on.

Simultaneously, smartphone-based fluorescence detection platforms are increasingly attractive for the development of portable sensors, because the sensitivity of the fluorescent sensing system is usually 2–3 orders of magnitude higher than that of the UV–vis sensing system. Moreover, the popularity and improvement of smartphone technology provide appropriate fluorescence response sensitivity and convenience of the operation. It has been widely used in health inspection, environmental and food analysis, drug analysis, biochemical and clinical detection. Smartphones are commonly used in conjunction with fluorescence analysis to detect light emitted by target fluorophore complexes during radiation excitation and decay.

Smartphones detect the light which is emitted from the target or fluorescence probe under the excitation of radiation by combining with fluorescence-based assays. Small molecule fluorophores, quantum dots (QDs), and fluorescent nanoparticles are the common fluorescent sensing materials. Small molecule fluorophores have several features such as generally low molecular weight and water-soluble and have reactive groups which are capable of combining with capturing agents like antibodies, peptides, and nucleic acids. Because of their low molecular weights, they tend to attach biomolecules rather than the fluorophore itself when conjugated to the larger biomolecules like antibodies. QDs are semiconductor nanoparticles that can emit fluorescence when excited. QDs are short of water solubility and biocompatibility, therefore, they generally are encapsulated with functional polymers to obtain these lacking properties and retain their optical features. Most fluorescent nanoparticles are fluorescence materials with a big scale than QDs. These materials were used for fluorescent biosensing in many ways through fluorescence resonance energy transfer (FRET) or electron-transfer mechanism or just as a probe for optical imaging. Optical imaging can offer microscopic and macroscopic information which is real-time and high-resolution for a rapid and accurate diagnosis [3].

4.2 The applications of smartphone for the construction of biosensors

With the rapid development of biosensors, smartphone-based sensing systems combine the advantages of biosensors with various functions of smartphones, which offer an inexpensive, portable, and rapid way for detection. In these systems, smartphones can detect fluorescence response as a spectrometer, capture fluorescent images as cameras, observe and count cells as microscopes, and communicate with doctors as an effective tool [1, 4].

4.2.1 Smartphone-based fluorescence spectrometer

As the traditional quantitative method of fluorescence sensor, fluorophotometer has the advantages of accuracy and stability. The disadvantage is that the fluorophotometer is large and lacks portability which is not benefited for the point-of-care test. Smartphone has the advantage of portability and can be used for portable detection by combining with a designed external component.

Yuehe Lin group has developed a aptamer-based lateral flow biosensor (APTA-LFB) based on a fluorescent APTA that is integrated with a fluorogroup-densifier nanometer pair and a smartphone spectrum reader to accomplish the three-target detection of poisoning chlorpyrifos, diazinon and malathion. In this work, they used aptamer instead of the antibodies in LFB as an alternative recognition element, providing better specificity and stability. A new type of fluorescence quenching agent nanopair (quantum-dot nanocrystals and gold nanocrystals) is implemented as "signal on" rather than "signal off". The combination of improved signals and zero background may greatly improve sensitivity. This innovative cascade strategy can reduce the detection limit of pesticide residue levels in food. The modified method can detect a variety of pesticides in food samples. This sensitive handheld system promises to be a powerful tool for practical field applications of a variety of pesticide quantification programs (Fig. 4.1) [5].

Yu et al. showed a smartphone fluorescence spectrometer that could be used for fluorescence-based assays. In their work, a transmission diffraction grating was placed before the camera of the smartphone, and then the system can emit the fluorescence light which can be processed into a colored band recorded by the smartphone. The group has demonstrated that this system can detect microRNA sequences with fluorescence probes. In the matter of fact, with a diffraction grating, the system builds a relationship between the spectrum changes and the images, finally used a smartphone for capture and analysis [4,6,7].

Smith group reported a system by using a smartphone as a spectrometer [8]. In this system, a slit-shaped hole was formed over the grating, two parallel strips of the band were used and separated by around 1 mm. A tube with dark foil was installed above the camera, and adjusted its tilt angle to about 45° to illuminate the sensor through the grating's first-order diffraction output. Besides, the other slit was situated at the non-spectrometer end of the tube, ensuring that only the quasi-collimated light can pass through the grating, allowing the system to effectively capture the diffraction image of the slit to form a spectral simulation of the pinhole camera. The authors proposed a method to demonstrate concept functionality through collecting spectra of light transmitted from human tissues and fluorescence of rhodamine 6G solution. The method demonstrated good performance when compared with spectra which were obtained simultaneously from commercially available spectrometers [8].

Ding et al. have designed a smartphone spectrometer and adopted a specially designed calibration approach which could improve spectral accuracy. The effect of nonlinear sensitivity come from CMOS camera was eliminated by calibrating wavelength and correcting intensity. Through comparing the fluorescence as well as the absorption spectra of Rhodamine B with standard commercial spectrometers, the group tested the calibrated smartphone spectrometer. In the meanwhile, the device was successfully validated to be feasible in mHealth application through quantitative detection of creatinine [9].

Bi group showed a handheld micro-fluorescence spectrometer, which was integrated with an ultraviolet light-emitting diode. In this device, as a dispersive element, transmission grating was utilized to disperse light. And a CCD array was used to detect the spectrum. The instrument can be coupled with a smartphone via a USB cable to process data and display a spectrum. The wavelength range of the spectrometer was 380 nm to 750 nm, its spectral resolution was 3 nm. Bi group used the spectrometer to measure the fluorescence spectra of edible oil, chlorophyll, and paper [10].

The advantages of a smartphone-based fluorescence spectrometer are its small size, low cost, fast detection speed, and lightweight. However, in the design process

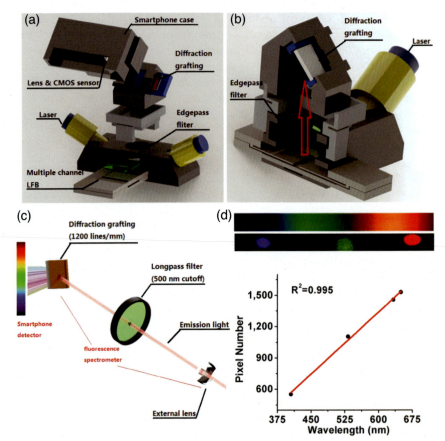

Fig. 4.1 Mobile phone fluorescence spectrum reader. (a) Anatomy of the mobile device. (b) Schematic side view showing inner structure. (c) Smartphone-based fluorescence detection principle. Excitation laser (405 nm) illuminates test zone containing the probe target duplexes; a portion of the emission is gathered via a collecting lens oriented perpendicular to the laser and sent through a long pass filter, where it is further collimated before incidence upon a diffraction grating placed directly in front of the camera- (d) Images of broadband light, red, green, and blue laser pointers over smartphone screen. Calibration of pixel versus wavelength by three laser pointers. Reproduced with the permission from the Elsevier. (e) From https://doiorg/10.1039/cOlc00358a N. Cheng, Y. Song, Q. Fu, D. Du, Y. Luo, Y. Wang, W. Xu, Y. Lin, Aptasensor based on fluorophore- quencher nano-pair and smartphone spectrum reader for on-site quantification of multi-pesticides, Biosens. Bioelectron. 117 (2018) 75–83, https://doi.org/10.1016/j.bios.2018.06.002.

of smartphone-based fluorescence spectrometers, it is still necessary to consider the design of traditional optical components, such as light sources, spectroscopic devices, optical path design, and detectors. This brings certain difficulties to general users, because users may need to purchase corresponding external equipment, and even mobile phones need to be replaced to achieve better spectrometer detection results.

4.2.2 Smartphone-based imaging and quantitative analysis

Smartphone-based imaging and quantitative analysis applications are mainly divided into two types: one is to observe the morphology of target through the imaging analysis of the smartphone; the other is to analyze the image through the smartphone and process the color channel of the imaged picture to achieve quantitative analysis. There is no doubt that both applications reflect the advantages of portable, low-cost, and fast detection of mobile phones. Applications that use smartphone imaging analysis to observe the morphology of target generally need to design some external components to improve resolution, such as magnifying glass, light source, etc. The application of quantitative analysis using the color channel of the photo is generally to take a picture directly through the mobile phone, and then import the photo into the computer or analyze it through the designed smartphone software. By analyzing the value of various color channels such as RGB channels (red, green, and blue), HSL (hue, saturation and lightness), CMYK (cyan, magenta, yellow and black), etc., the linear relationship of the values of channels and the concentration of target was constructed and used for quantitative analysis [4].

Lee and Yang developed a compact lensless microscope with a smartphone that can be applied to mobile healthcare as well as environmental monitoring (Fig. 4.2). As for the light source of this microscopy, they firstly used ambient illumination instead of a dedicated light source. The samples' direct shadow can be captured at various angles of illumination to make the image reconstruction with a super-resolution. Both blood smear and freshwater microorganisms could be successfully imaged with a good resolution of 500 nm by this system. The results showed that microscopy devices on smartphones had a good application prospect [11].

Wei et al. used an optomechanical attachment to integrate with a smartphone-based fluorescence microscope. The attachment included laser diode, focus adjustment stage, interference filter, and external lens. Objects like nanoparticles, viruses, and DNA can be imaged by the fluorescence microscope with resolution in the nanometer scale. In particular, this smartphone-based microscope can also measure the length of individual DNA, and the measurement accuracy is less than 1 kg base-pairs [12]. Yetisen group also developed a smartphone application and utilized smartphone in urine tests with good reproducibility and accuracy [13].

Petryayeva and Algar synchronously quantified fluorescent quenching signals by monitoring the red (R), green (G), and blue (B) channels of QDs with a smartphone. Multichannel sensing methods were developed for detecting proteolytic activities, the fluorescent intensity of QDs could be directly monitored with the help of the phone camera. After being labeled with dark quenchers, peptides were conjugated with R-, G-, and B-emitting QDs generating FRET pairs that were sensitive to proteolytic activities. Therefore, with the help of peptides which can be cleaved by proteases to form changes in ODs photoluminescence and RGB channels of the smartphone images, the group demonstrated an assay for activities of three enzymes (trypsin, chymotrypsin, and enterokinase) with RGB imaging which made it possible for smartphone devices to act as independent fluorescence detectors [14].

Bueno et al. used a smartphone as a fluorescence device for the quantitative detection of concentrations of ochratoxin A. The sample is excited by ultraviolet (UV) light

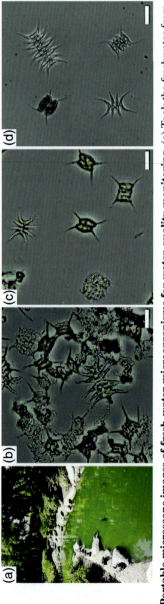

Fig. 4.2 Portable microscope images of fresh-water microorganisms for water quality monitoring. (a) Took the freshwater from a koi pond. 20 μL of the sample was dispensed on the image sensor and the particles were let settle down for few minutes before image acquisition. (b) and (c) Reconstructed images of green algae found in the pond. (d) Conventional microscope images of the same sample taken with a 20x (0.4 Numerical Aperture) objective lens. The green algae found in the sample are different species of *Scenedesmus*, a genus of *Chlorophyceae*. All scale bars indicate 20 μm. Reproduced with the permission from Royal Society of Chemistry. (e) From S.A. Lee, C. Yang, A smartphone-based chip-scale microscope using ambient illumination, Lab. Chip. 14 (16) (2014) 3056–3063. https://doi.org/10.1039/c4lc00523f.

and emitted fluorescence which can pass to smartphone camera via a lens. At last, it is wirelessly sent to a personal computer for analysis. The computer process the fluorescence image captured by the smartphone camera and showed in Red, Green and Blue (RGB) components [15].

Lihachev et al. proposed a method that used a smartphone to realize autofluorescence imaging in vivo skin. In this study, the smartphone RGB camera periodically captured the filtered autofluorescence images of the same tissue area, and then detect the fluorescence intensity decreasing at every image pixel, so that the planar distribution of those values can be imaged. This method was clinically validated with 13 basal cell carcinoma as well as 1 atypical nevus [16].

Knowlton et al. showed a multiplex tool that could sort the cells based on density through magnetic focusing and fluorescence imaging, which provided highly specific clinical analysis. The device was low-cost, portable and had compact sizes. According to density, the cells can be separated by the device, which can identify cell type as well as cell activity. Through the smartphone camera, the cells also can be imaged in either darkfield, brightfield or fluorescent imaging modes [17].

Tian et al. presented an intelligent method that was coupled with smartphone image processing technology for point-of-care testing. In this platform, multiple targets could be detected simultaneously and sensitively. Through flow lithography, Hydrogel microparticles of various coding modes like shapes and numbers were prepared for the detection of different miRNAs. Different shapes of hydrogels presented different fluorescence intensities after sandwich immunoassays. A smartphone was used to capture the images, and smartphone software was used for image recognition processing. The result can be obtained within 10 s, and the whole reaction process took less than 2 h [18].

Yang et al. proposed a convenient and economical method for the detection of water-based on the ratiometric fluorescence sensing strategy which consists of fluorescent Cu nanoclusters (CuNCs) as well as carbon dots (CDs). CuNCs have solvent-dependent effects and respond differently to water in different solvents. The ratiometric fluorescence probe showed a distinct color variation from orange to blue when water content increase in organic solvents. Through fluorescence intensity ratios (orange/blue), the water contents can be detected. Finally, a smartphone was used to capture fluorescence images, read changing colors and analyze data [19].

Luchiari et al. showed a small fluorimetric device based on a smartphone for the detection of caffeine. In this device, the images were captured by the OpenCamera app and processed by ImageJ software for the acquisition of RGB channels values. The feasibility of this method is demonstrated for the detection of caffeine in an energy drink as well as medicines [20].

Hou et al. designed a visual method for the detection of alkaline phosphatase (ALP) based on a cascade reaction between ALP and NH_2Cu-MOFs nanozyme. pyrophosphate (PPi) can inhibit the catalytic activity of NH_2Cu-MOFs and a ratiometric sensing platform through the controllable catalytic activity of NH_2Cu-MOFs was realized. RGB analysis of the fluorescent sample images was adopted for ALP quantitative analysis. Besides, a hydrogel test kit and mobile app for ALP detection were designed as conceptual products for point-of-care. This method was applied to ALP detection in serum samples with satisfying results and showed a promising application in other biosensing areas like ALP-mediated clinical diagnosis [21]. Yachana Upadhyay et al. also used lysozyme-stabilized fluorescent gold nanoclustersand pyridoxal 5′-phosphate as a monophosphate ester substrate to develop a highly selective and cost-effective bioassay for the detection of alkaline phosphatase activity [22].

Zhang et al. proposed a ratiometric fluorescence sensing platform with a smartphone imaging for the detection of fluoride ion (F^-) to realize high sensitivity and accuracy (Fig. 4.3). The images of fluorescence from samples were captured by the smartphone camera, and their Red, Green and Blue (RGB) channel values were read out by the Color Picker APP which was installed in smartphone. The ratio of Red to Green (R/G) showed linearity against F^- concentration with the range of 0–70.0 μM. And the limit of detection (LOD) was 2.0 μM [23]. Xiaoliang Zeng et al. also provided a visual Detection of F^- by using mixed lanthanide metal– organic frameworks with a smartphone [24].

Chu et al. developed a smartphone sensing platform that was combined with a paper strip and coupled with a UV lamp as well as a dark cavity through 3D-printing technology. The platform showed a quantitative analysis that was sensitive, instrument-free, rapid, and visual in real-time/on-site conditions. Red-emitting CdTe quantum dots, as an internal reference, were embedded in the silica nanoparticles (SiO_2 NPs). Besides, as a signal report unit, blue-emitting carbon dots were covalently connected to the exterior of SiO_2 NPs. Then, gold nanoparticles (Au NPs) quenched the blue fluorescence which recovered with pesticide finally. A color recognizer application which was installed in the smartphone determined the red (R), green (G) as well as blue (B) channel values of the images generated. The R/B values could be applied to the quantitative analysis of pesticide, and the sensitive LOD was 59 nM. [25]

Yang et al. showed a platform for the quantitative detection of mycotoxins. Based on a smartphone, the platform combined fluorescence image processing with multicolor upconversion nanoparticle barcode technology. Different mycotoxins can be simultaneously detected through the encoded signals – multi-colored upconversion nanoparticle encoded microspheres (UCNMs). After using UCNMs for indirect competitive immunoassays, images were captured by the camera of a smartphone. Finally, a self-written Android application was installed on a smartphone and used for analyzing images and producing a result within 1 min [26]. Several other works about the detection of food safety related hazards like organophosphate pesticides, [27] tetracycline, [28] plastic pollutant polystyrene, [29] estradiol, [30] and heavy metal ions [31] have been reported and constructed based on color channel changes for quantitative analysis.

4.2.3 Smartphone-based microscope for particle counting

The Central Processing Unit, as well as Graphics Processing Unit, which installed in a smartphone improve the computing power and make the captured images analyzed, corrected and improved for detection via such computational microscopy without using external processing on PC. Combined with attachments like a light-emitting

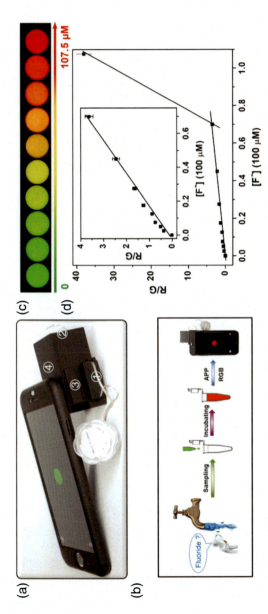

Fig. 4.3 The smartphone imaging-based sensing platform for F⁻. (a) Picture of the smartphone imaging-based sensing platform. (b) Steps for detecting F⁻ by using of smartphone-based sensing platform. (c) The luminescent images of testing system upon additions of different amounts of F⁻, which were directly taken by the smartphone digital camera. (d) The color change and the ratio of R/G values versus F⁻ concentration in the range of 0-107.50 μM. The inset figure is linear relationship of mean luminescent intensities with F⁻ concentrations from 0 to 70 μM. Reproduced with the permission from the Elsevier. (e) From J. Zhang, J. Qian, Q. Mei, L. Yang, L. He, S. Liu, C. Zhang, K. Zhang, Imaging-based fluorescent sensing platform for quantitative monitoring and visualizing of fluoride ions with dual-emission quantum dots hybrid, Biosens. Bioelectron. 128 (2019) 61–67. https://doi.org/10.1016/j.bios.2018.12.044.

diode, the smartphone-based microscope can identify the morphological characteristics of samples like size, shape as well as color, which can be used to detecting tumor cells, microorganisms, blood cells, nanoparticles, DNA as well as virus [1,19,32].

Zhu et al. showed a smartphone microscope with the help of a microfluidic device to count the cells. Firstly, they set the flow rate of the additional pump to 1 μL min^{-1} and injected cells labeled by fluorescence into the entrance of the microfluidic device. Secondly, to excite the fluorescence labeling, they aligned two blue LEDs on both sides of the microfluidic channel. Finally, they developed an imaging processing algorithm. A smartphone microscope was able to record the flow videos at a resolution of 2 μm and a rate of 7 frames s^{-1}. Through white blood cell count in human blood, this smartphone-based counting system was successfully proved to be feasible. The results of the test were consistent with the measurements by a commercial hematology analyzer [33].

Shrivastava et al. demonstrated a smartphone-based sensor for Staphylococcus aureus (S. aureus). The sensor was culture-free, rapid and used aptamer-functionalized fluorescent magnetic nanoparticles to detect S. aureus from minimally processed liquid samples (Fig. 4.4). The labeled S. aureus cells could be captured in a detection cassette. Under the excitation from a light-emitting diode, the fluorescence was imaged by a smartphone camera. The results showed the ability for quantitative detection and the minimum detectable concentration was 10 CFU mL^{-1}, S. aureus cells could be captured from a peanut milk sample in less than 10 min. In this system, the fluorescently labeled bacteria could be excited by the smartphone microscope configured with a white LED, a 10 × objective, a cassette, and a dichroic mirror. The field of view of the smartphone microscope is 1.76 mm^2, which can distinguish 1 μm microsphere. But the images still needed to use ImageJ in an external PC for analysis. When the range of concentration is 50–2000 CFU mL^{-1}, S. aureus cells were imaged and counted, which validated the effectiveness of this smartphone system [34].

Zhu et al. developed smartphone-based fluorescence microscopy by using butt-coupled LEDs to excite the fluorescent probes and through an external lens that was placed above the lens of the smartphone camera, they captured the fluorescent images from target samples. Both white blood cells labeled and Giardia Lamblia cysts in water could be successfully imaged by this prototype, it was useful for quality monitoring. When the field of view was 81 mm^2, 10 μm resolution could be achieved by utilizing digital image processing that was based on compressive sampling. As for the detection of nanoparticle, viruses, or molecular biological samples like DNA, this system would achieve a limit of detection with the nanoscale [33].

Wei et al. developed a smartphone-based optomechanical attachment. They adjusted the large incidence angle of a 450 nm laser diode to around 75 °C to illuminate the sample. Then, the scattering light was rejected by a long-pass interference filter. Next, to adjust focus, they used an external low numerical aperture lens to generate a 2 × magnification with a stage that can be translated. Such design efficiently makes the background rejection rejected, in this way, the weak fluorescent signal can be successfully isolated from nanoscale samples. Through scanning electron microscopy, this imaging system by smartphone was validated to image fluorescent particles in 100 nm scale and human cytomegalic viruses [12].

Fluorescence measurements, imaging and counting by a smartphone

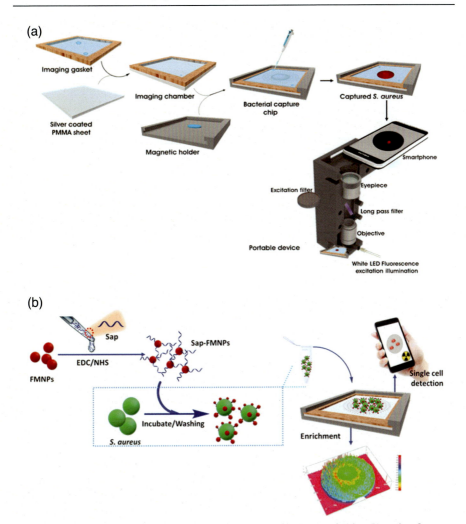

Fig. 4.4 Schematic representation of bacterial capture and quantitative detection from minimally processed samples. (a) A highly reflective, silver-deposited polycarbonate sheet was attached to a gasket to construct a sealed chamber with inlets at two corners for sample loading and washing. A magnet-mounted holder was 3D-printed, and the sealed chamber was inserted to create a bacterial detection cassette. Once loaded with liquid sample, the bacterial detection cassette was allowed to capture the target pathogen, which was followed by washing and imaging with a smartphone fluorescence microscope specifically designed for fluorescence imaging with a light emitting diode. (b) An S. aureus specific aptamer (Sap) was covalently attached to fluorescent magnetic nanoparticles (FMNPs). The Sap-conjugated FMNPs (Sap-FMNPs) were used to tag target bacteria, and those fluorescently tagged bacteria were then spiked in a minimally processed, complex liquid sample. The spiked samples were loaded into the bacterial detection cassette for capture and subsequent quantification using smartphone imaging. Reproduced with the permission from the Elsevier. (c) From S. Shrivastava, W.I. Lee, N.E. Lee, Culture-free, highly sensitive, quantitative detection of bacteria from minimally processed samples using fluorescence imaging by smartphone, Biosens. Bioelectron. 109 (2018) 90–97. https://doi.org/10.1016/j.bios.2018.03.006.

In 2017, Kühnemund M. et al. presented a microscopic biosensor based on a smartphone that could target next-generation DNA sequencing as well as point mutation analysis. This biosensor can not only diagnose POC molecules but also has the advantage of low cost [35]. Also, a thin metal-film based surface-enhanced fluorescence was proposed by the Wei Q. et al. to enhance the sensitivity of fluorescent detection. The laser diode was excited and the laser beam was filtered by a linear polarizing mirror. They optimized the experimental conditions like film thickness and excitation angle and finally achieved good experimental results: their fluorescence intensity was about 10 times higher than that of bare glass substrates. In this way, both fluorescent particles with a diameter of 50 nm and single quantum-dots can be imaged [36].

Slusarewicz et al. used a cellular smartphone to imaging fluorescent eggs and process the image to count the eggs. The result generated by the smartphone system showed obvious linearity against manual McMaster counts ($R(2) = 0.98$). However, the coefficient of variation decreased obviously ($P = 0.0177$). This showed that the method was simple and field-portable, and can detect and count the parasite eggs in mammalian faeces without a laboratory microscope [37].

Ulep et al. developed a smartphone-based fluorescence microscope and software which can automatically process the image and count the particle, which allows particles to count on paper, and the limit of detection was 1 cell/µL. In this study, ROR1+ cancer cells of a complex sample matrix can be captured, imaged on-chip, and the cell concentration can be quantified in a point-of-care way [38].

Koydemir et al. showed a platform to detect and quantify Giardia lamblia cysts, which had low cost and was portable. The platform was composed of a smartphone and an optomechanical attachment. A hand-held fluorescence microscope design was aligned with the smartphone's camera unit, which allows the platform to image customized disposable water sample cassettes. The fluorescence images from the filter surface can be captured and transmitted to the servers through the smartphone, and then can be quickly processed by a machine learning algorithm. Because the algorithm has trained on the statistical characteristics of Giardia cysts, the cysts can be automatically detected and counted. The results are shown in less than 2 min via a smart application on the smartphone [39].

Minagawa et al. presented a smartphone-based mobile imaging platform that could be used for digital bioassays. the group completed a digital enzyme analysis of bovine alkaline phosphatase via this platform. They also demonstrated the digital influenza virus counting, which was applied for the virus's neuraminidase activity. Through the mobile imaging platform, obvious fluorescence spots from a single virus particle could be observed. The linear relationship between the number of fluorescent spots and the virus titer is good and indicates that this method had high sensitivity and quantification [40].

4.3 Conclusion and prospects

In these systems, smartphones can observe and count cells as microscopes, capture fluorescent images as cameras, analyze experimental data as an analytical platform, and connect detection devices and online doctors as an effective tool. The studies show that smartphone-based sensing systems can detect and analyze various analytes rapidly and accurately in disease treatment, disease diagnosis, food quality monitoring and environmental quality monitoring. In addition, compared with traditional biomedical detection devices, smartphone-based sensing systems are low-cost, convenient and easy-to-use, providing sustainable biomedical technologies in resource-limited regions of the world. However, there are still some challenges for researchers to overcome in the design and application of smartphone fluorescence.

Firstly, the performance of the smartphone camera plays an important role in the sensing system. To obtain higher-quality images getting higher-quality detection results, it is quite significant to improve the focusing time, pixel density as well as image stability of the cameras of smartphones. Secondly, The detection results are easily affected by the external light, which is a limitation for the system's application in outdoor testing. So alternative tags like quantum dots can be utilized to enhance photostability, or 3D printing technology can be used to design a more opaque detection accessory for avoiding light exposure comes from the environment. thirdly, it is necessary to study image processing algorithms as well as data processing algorithms based on smartphone, which can improve the quality of captured images and accuracy for data analysis in biomedical detection. Finally, the combination of sensing systems based on smartphone with Big Data technology as well as fifth-generation wireless communication technology has a favorable prospect. The experimental data can be analyzed through Big Data technology, while real-time remote detection and diagnosis can be performed via fifth-generation wireless communication technology, which does help to get better diagnostic results.

In a word, because of the popularity of smartphone in people's daily life, the continuous progress of smartphone-based sensing technology as well as a full exploration of smartphone fluorescence, we have reasons to believe that there will be a revolution both in the healthcare systems and centralized diagnostic.

References

[1] E. Aydindogan, E. Guler Celik, S. Timur, Paper-based analytical methods for smartphone sensing with functional nanoparticles: bridges from smart surfaces to global health, Anal. Chem. 90 (21) (2018) 12325–12333. https://doi.org/10.1021/acs.analchem.8b03120.

[2] M. Rezazadeh, S. Seidi, M. Lid, S. Pedersen-Bjergaard, Y. Yamini, The modern role of smartphones in analytical chemistry, Trends Analyt. Chem. 118 (2019) 548–555. https://doi.org/10.1016/j.trac.2019.06.019.

[3] H. Zhu, S.O. Isikman, O. Mudanyali, A. Greenbaum, A. Ozcan, Optical imaging techniques for point-of-care diagnostics, Lab. Chip. 13 (1) (2013) 51–67. https://doi.org/10.1039/c2lc40864c.

[4] D. Zhang, Q. Liu, Biosensors and bioelectronics on smartphone for portable biochemical detection, Biosens. Bioelectron. 75 (2016) 273–284. https://doi.org/10.1016/j.bios.2015.08.037.

[5] N. Cheng, Y. Song, Q. Fu, D. Du, Y. Luo, Y. Wang, W. Xu, Y. Lin, Aptasensor based on fluorophore-quencher nano-pair and smartphone spectrum reader for on-site

quantification of multi-pesticides, Biosens. Bioelectron. 117 (2018) 75–83. https://doi.org/10.1016/j.bios.2018.06.002.

[6] A.J.S. McGonigle, T.C. Wilkes, T.D. Pering, J.R. Willmott, J.M. Cook, F.M. Mims, A.V. Parisi, Smartphone spectrometers, Sensors 18 (1) (2018) 223. /1-223/15. https://doi.org/10.3390/s18010223.

[7] H. Yu, Y. Tan, B.T. Cunningham, Smartphone fluorescence spectroscopy, Anal. Chem. 86 (17) (2014) 8805–8813. https://doi.org/10.1021/ac502080t.

[8] Z.J. Smith, K. Chu, A.R. Espenson, M. Rahimzadeh, A. Gryshuk, M. Molinaro, D.M. Dwyre, S. Lane, D. Matthews, S. Wachsmann-Hogiu, Cell-phone-based platform for biomedical device development and education applications, PLoS One 6 (3) (2011) e17150. https://doi.org/10.1371/journal.pone.0017150.

[9] H. Ding, C. Chen, S. Qi, C. Han, C. Yue, Smartphone-based spectrometer with high spectral accuracy for mHealth application, Sens. Actuators A Phys 274 (2018) 94–100. https://doi.org/10.1016/j.sna.2018.03.008.

[10] Z. Bi, Y. Zhang, S. Zhang, L. Wang, E. Gu, Z. Tian, A handheld miniature ultraviolet LED fluorescence detection spectrometer, J. Appl. Spectrosc. 86 (3) (2019) 538–541. https://doi.org/10.1007/s10812-019-00855-9.

[11] S.A. Lee, C. Yang, A smartphone-based chip-scale microscope using ambient illumination, Lab Chip 14 (16) (2014) 3056–3063. https://doi.org/10.1039/c4lc00523f.

[12] Q. Wei, H. Qi, W. Luo, D. Tseng, S.J. Ki, Z. Wan, Z. Gorocs, L.A. Bentolila, T.T. Wu, R. Sun, A. Ozcan, Fluorescent imaging of single nanoparticles and viruses on a smart phone, ACS Nano 7 (10) (2013) 9147–9155. https://doi.org/10.1021/nn4037706.

[13] A.K. Yetisen, J.L. Martinez-Hurtado, A. Garcia-Melendrez, F. da Cruz Vasconcellos, C.R. Lowe, A smartphone algorithm with inter-phone repeatability for the analysis of colorimetric tests, Sens. Actuators B Chem. 196 (2014) 156–160. https://doi.org/10.1016/j.snb.2014.01.077.

[14] E. Petryayeva, W.R. Algar, Multiplexed homogeneous assays of proteolytic activity using a smartphone and quantum dots, Anal. Chem. 86 (6) (2014) 3195–3202. https://doi.org/10.1021/ac500131r.

[15] D. Bueno, R. Muñoz, J.L. Marty, Fluorescence analyzer based on smartphone camera and wireless for detection of Ochratoxin A, Sens. Actuators B Chem. 232 (2016) 462–468. https://doi.org/10.1016/j.snb.2016.03.140.

[16] A. Lihachev, A. Derjabo, I. Ferulova, M. Lange, I. Lihacova, J. Spigulis, Autofluorescence imaging of basal cell carcinoma by smartphone RGB camera, J. Biomed. Opt. 20 (12) (2015) 120502. https://doi.org/10.1117/1.JBO.20.12.120502.

[17] S. Knowlton, A. Joshi, P. Syrrist, A.F. Coskun, S. Tasoglu, 3D-printed smartphone-based point of care tool for fluorescence- and magnetophoresis-based cytometry, Lab. Chip. 17 (16) (2017) 2839–2851. https://doi.org/10.1039/c7lc00706j.

[18] Y. Tian, L. Zhang, H. Wang, W. Ji, Z. Zhang, Y. Zhang, Z. Yang, Z. Cao, S. Zhang, J. Chang, Intelligent detection platform for simultaneous detection of multiple MiRNAs based on smartphone, ACS Sens. 4 (7) (2019) 1873–1880. https://doi.org/10.1021/acssensors.9b00752.

[19] J. Yang, Z. Li, Q. Jia, Design of dual-emission fluorescence sensor based on Cu nanoclusters with solvent-dependent effects: visual detection of water via a smartphone, Sens. Actuators B Chem. 297 (2019). https://doi.org/10.1016/j.snb.2019.126807.

[20] N dC Luchiari, G.A. da Silva, C.A. Marasco Júnior, P C F d Lima Gomes, Development of miniaturized fluorimetric device for caffeine determination using a smartphone, RSC Adv. 9 (60) (2019) 35033–35038. https://doi.org/10.1039/c9ra06220c.

[21] L. Hou, Y. Qin, J. Li, S. Qin, Y. Huang, T. Lin, L. Guo, F. Ye, S. Zhao, A ratiometric multicolor fluorescence biosensor for visual detection of alkaline phosphatase activity via a smartphone, Biosens. Bioelectron. 143 (2019) 111605. https://doi.org/10.1016/j.bios.2019.111605.

[22] Y. Upadhyay, R. Kumar, S.K. Sahoo, Developing a cost-effective bioassay to detect alkaline phosphatase activity and generating white light emission from a single nano-assembly by conjugating vitamin B6 cofactors with lysozyme-stabilized fluorescent gold nanoclusters, ACS Sustain. Chem. Eng. 8 (10) (2020) 4107–4113. https://doi.org/10.1021/acssuschemeng.9b06563.

[23] J. Zhang, J. Qian, Q. Mei, L. Yang, L. He, S. Liu, C. Zhang, K. Zhang, Imaging-based fluorescent sensing platform for quantitative monitoring and visualizing of fluoride ions with dual-emission quantum dots hybrid, Biosens. Bioelectron. 128 (2019) 61–67. https://doi.org/10.1016/j.bios.2018.12.044.

[24] X. Zeng, J. Hu, M. Zhang, F. Wang, L. Wu, X. Hou, Visual detection of fluoride anions using mixed lanthanide metal-organic frameworks with a smartphone, Anal. Chem. 92 (2) (2020) 2097–2102. https://doi.org/10.1021/acs.analchem.9b04598.

[25] S. Chu, H. Wang, X. Ling, S. Yu, L. Yang, C. Jiang, A portable smartphone platform using a ratiometric fluorescent paper strip for visual quantitative sensing, ACS Appl. Mater. Interfaces 12 (11) (2020) 12962–12971, https://doi.org/10.1021/acsami.9b20458.

[26] M. Yang, Y. Zhang, M. Cui, Y. Tian, S. Zhang, K. Peng, H. Xu, Z. Liao, H. Wang, J. Chang, A smartphone-based quantitative detection platform of mycotoxins based on multiple-color upconversion nanoparticles, Nanoscale 10 (33) (2018) 15865–15874. https://doi.org/10.1039/c8nr04138e.

[27] R. Jin, D. Kong, X. Yan, X. Zhao, H. Li, F. Liu, P. Sun, Y. Lin, G. Lu, Integrating target-responsive hydrogels with smartphone for on-site ppb-level quantitation of organophosphate pesticides, ACS Appl. Mater. Interfaces 11 (31) (2019) 27605–27614. https://doi.org/10.1021/acsami.9b09849.

[28] T. Wang, Q. Mei, Z. Tao, H. Wu, M. Zhao, S. Wang, Y. Liu, A smartphone-integrated ratiometric fluorescence sensing platform for visual and quantitative point-of-care testing of tetracycline, Biosens. Bioelectron. 148 (2020) 111791. https://doi.org/10.1016/j.bios.2019.111791.

[29] X. Lian, T. Miao, X. Xu, C. Zhang, B. Yan, Eu3+ functionalized *Sc*-MOFs: turn-on fluorescent switch for ppb-level biomarker of plastic pollutant polystyrene in serum and urine and on-site detection by smartphone, Biosens. Bioelectron. 97 (2017) 299–304. https://doi.org/10.1016/j.bios.2017.06.018.

[30] W.-I. Lee, S. Shrivastava, L.-T. Duy, B.Y. Kim, Y.-M. Son, N.-E. Lee, A smartphone imaging-based label-free and dual-wavelength fluorescent biosensor with high sensitivity and accuracy, Biosens. Bioelectron. 94 (2017) 643–650. https://doi.org/10.1016/j.bios.2017.03.061.

[31] M. Xiao, Z. Liu, N. Xu, L. Jiang, M. Yang, C. Yi, A smartphone-based sensing system for on-site quantitation of multiple heavy metal ions using fluorescent carbon nanodots-based microarrays, ACS Sens. 5 (3) (2020) 870–878. https://doi.org/10.1021/acssensors.0c00219.

[32] X. Huang, D. Xu, J. Chen, J. Liu, Y. Li, J. Song, X. Ma, J. Guo, Smartphone-based analytical biosensors, Analyst 143 (22) (2018) 5339–5351. https://doi.org/10.1039/c8an01269e.

[33] H. Zhu, S. Mavandadi, A.F. Coskun, O. Yaglidere, A. Ozcan, Optofluidic fluorescent imaging cytometry on a cell phone, Anal. Chem. 83 (17) (2011) 6641–6647. https://doi.org/10.1021/ac201587a.

[34] S. Shrivastava, W.I. Lee, N.E. Lee, Culture-free, highly sensitive, quantitative detection of bacteria from minimally processed samples using fluorescence imaging by smartphone, Biosens. Bioelectron. 109 (2018) 90–97. https://doi.org/10.1016/j.bios.2018.03.006.

[35] M. Kuhnemund, Q. Wei, E. Darai, Y. Wang, I. Hernandez-Neuta, Z. Yang, D. Tseng, A. Ahlford, L. Mathot, T. Sjoblom, A. Ozcan, M. Nilsson, Targeted DNA sequencing and in situ mutation analysis using mobile phone microscopy, Nat. Commun. 8 (2017) 13913. https://doi.org/10.1038/ncomms13913.

[36] Q. Wei, G. Acuna, S. Kim, C. Vietz, D. Tseng, J. Chae, D. Shir, W. Luo, P. Tinnefeld, A. Ozcan, Plasmonics enhanced smartphone fluorescence microscopy, Sci. Rep. 7 (1) (2017) 2124. https://doi.org/10.1038/s41598-017-02395-8.

[37] P. Slusarewicz, S. Pagano, C. Mills, G. Popa, K.M. Chow, M. Mendenhall, D.W. Rodgers, M.K. Nielsen, Automated parasite faecal egg counting using fluorescence labelling, smartphone image capture and computational image analysis, Int. J. Parasitol. 46 (8) (2016) 485–493. https://doi.org/10.1016/j.ijpara.2016.02.004.

[38] T.H. Ulep, R. Zenhausern, A. Gonzales, D.S. Knoff, P.A. Lengerke Diaz, J.E. Castro, J.Y. Yoon, Smartphone based on-chip fluorescence imaging and capillary flow velocity measurement for detecting ROR1+ cancer cells from buffy coat blood samples on dual-layer paper microfluidic chip, Biosens. Bioelectron. 153 (2020). 112042 https://doi.org/10.1016/j.bios.2020.112042.

[39] H.C. Koydemir, Z. Gorocs, D. Tseng, B. Cortazar, S. Feng, R.Y. Chan, J. Burbano, E. McLeod, A. Ozcan, Rapid imaging, detection and quantification of Giardia lamblia cysts using mobile-phone based fluorescent microscopy and machine learning, Lab. Chip. 15 (5) (2015) 1284–1293. https://doi.org/10.1039/c4lc01358a.

[40] Y. Minagawa, H. Ueno, K.V. Tabata, H. Noji, Mobile imaging platform for digital influenza virus counting, Lab Chip 19 (16) (2019) 2678–2687. https://doi.org/10.1039/c9lc00370c.

Spectrometric measurements

Sibasish Dutta
Department of Physics, Pandit Deendayal Upadhyaya Adarsha Mahavidyalaya (PDUAM), Eraligool, Karimganj, Assam, India

5.1 Introduction

An estimation of over 7.8 billion global population out of which nearly 50 percent being smartphone users, shows that smartphone is one of the most popular handheld gadget currently running in the market [1]. With different embedded sensors, in-built software features and high-end processors, smartphones are portable supercomputers in the hands of common people. Taking into consideration the current popularity of mobile phones, researchers across the globe have demonstrated the possibility of turning smartphones into microscopes, spectrophotometer and different types of luminescence, SPR and colorimetric devices [2–4].

Spectrophotometer used for optical spectroscopy finds a wide range of usage in different fields especially in need of quick and non-invasive sensing investigations [5]. A spectrophotometer is an assembly of optical elements namely, light source, a dispersive element, sample holder, detector, signal processing system and read out unit. Commercial spectrometers are bulky, expensive, non-portable and require trained personnel to handle, hence remain constrained in laboratory settings. Furthermore, such spectrometers cannot function standalone and often require external PC for data analysis which further limits their usage for in-field sensing. In order to address such issues, optical elements integration with smartphone can practically promote the concept of pocket spectrometer for real-time in-field sensing applications. This article provides a comprehensive review of spectrometers developed on a smartphone platform. The camera module of the smartphone has been widely used to capture the grating-induced dispersed spectrum of the incoming radiation. As of now, smartphone-based spectrophotometers have been realized primarily using transmission grating and reflection grating.

5.2 Transmission grating configured smartphone spectrometers

The first ever spectrometer on a cellphone platform was reported by Smith et al. [6]. A PVC tube lined with darkened foil, housing a transmission grating (1000 lines/mm) and a slit of 1 mm width on the grating was attached to the camera of an iPhone 2 to capture the dispersed spectrum. With a 60 W light source, transmission spectrum of finger and rohodamine 6G were successfully recorded with wavelength resolution of 5 nm. Further, Gallegos et al. [7] reported a smartphone spectrometer of pixel resolution 0.333 nm/pixel by integrating a cradle consisting of optical components. Herein, a broadband light source entering the optical system passed through a series of optical elements namely collimating lens, cylindrical lens and dispersed by a transmission grating (1200 lines/mm) before being finally collected by a iPhone 5s rear camera. With a custom designed app, the designed system could successfully quantify protein monolayer coated over a photonic crystal incorporated into the optical path of the system. Similar optics designs were utilized for measuring absorption band of colored dyes [8], pH monitoring [9,10], size monitoring of gold nanoparticles and biomolecular quantification [11]. Using acrylic plates, Kwon et al. [12] fabricated a connector case housing a diffraction grating and a cuvette for sample holding. The connector case attached to the rear camera of a phone was tested for measurement of toxicity in kidney induced by cisplatin within the visible domain of 400–700 nm. Using compact disc (CD) as dispersing element, Kong et al. [13] developed a compact spectrometer for colorimetric detection of ascorbic acid. Similarly, Wang et al. [14] reported a DVD transmission grating based smartphone spectrometer for detection of neurotoxins within visible domain of 400–700 nm of the electromagnetic spectrum. Using similar technique [7], the same group reported a smartphone based ELISA platform for detection of two different biomarkers (IL-6 and Ara h 1) and shown to be in consistent with conventional ELISA platform [15]. Further, incorporating transmission filters, they have developed a smartphone-based fluorimeter which showed similar performance like that of a conventional fluorimeter for measuring the fluorescence spectra of T-miRNAs of different concentration [16]. Furthermore, the research group has developed a smartphone biosensing system (TRI-analyzer) capable of measuring spectrum due to colorimetric absorption, fluorescence emission and resonant reflection using flash of the smartphone as light source [17]. The designed system was investigated for spectrometric detection of ascorbic acid [18] and analysis of urinanalysis test strips by recording the scattered light [19]. The functionality of smartphone spectrometer has been extended to multichannel with the use of micro prism array that increases the field of view sufficient enough to scan a 96-well microplate. With a developed smartphone app, the performance of the multichannel spectrometer was evaluated for protein quantification [20] with pixel resolution of 0.2521 nm per pixel. An effort towards miniaturization of smartphone spectrometer was made using G-Fresnel, a diffractive element performing the function of collimation as well as dispersion of light into the CMOS sensor of smartphone camera. The designed G-Fresnel spectrometer of spectral resolution of 1.6 nm could measure the transmission spectra for Bradford assay-based BSA protein quantification with nanometer resolution similar to that of a commercial spectrometer [21]. The same research group further reported a standalone G-Fresnel spectrograph module with detection hardware that can be connected to any smartphone variant via micro-USB port. The designed spectrometer could record spectrum from 400 to 1000 nm with spectral resolution < 1 nm. The utility of this device was demonstrated by measuring the haemoglobin concentration by collecting light in the form of diffused reflected spectrum from tissue phantoms [22]. Due to change in camera position of different smartphone variants and consequently inconvenience in adjusting the diffraction grating and other optical elements, attachable optomechanical module with smartphone camera may not be the best solution. To mitigate these limitations, Cai et al. [23] reported a Wi-fi enabled handheld spectrometer module housing necessary optical elements for wireless integration with a smartphone giving a spectral resolution of 17 nm and spectral range of 400 nm–675 nm. The proposed device system was used for wireless monitoring of chlorophyll in banana, Myoglobin in pork and hemoglobin in blood by capturing the reflectance spectrum. Table 5.1 summarizes the advancement of smartphone spectrometers employing transmission grating as dispersive element.

Table 5.1 Transmission grating coupled smartphone spectrometers.

Spectrometer device	Developer	Detection characteristics	References
Cellphone spectrometer	University of California Davis, California, United States of America	Transmission spectrum of human tissue with spectral resolution of 5 nm	[6]
PC biosensor	University of Illinois at Urbana-Champaign, USA	Photonic crystal based detection of porcine immunoglobulin G with min. detectable shift of 0.009 nm	[7]
Smartphone based pH sensing	Tezpur University, Assam, India	pH sensing with resolution of 0.12 pH unit	[10]
LSPR sensor	Tezpur University, Assam, India	BSA protein and Trypsin enzyme detection with LoD value 0.28 μM and 1.10 μM respectively.	[11]
Smartphone metabolomics platform	Seoul National University, South Korea	Cisplatin detection in kidney with LoD in $mgdL^{-1}$	[12]
Smartphone-based CD spectrometer	Zhejiang University, Hangzhou, China	Ascorbic acid detection in the dynamic range 0.6250 μM to 40 μM with LoD of 0.4946 μM	[13]
Smartphone Optosensing Platform	The Washington State University, United states	Paraoxon (neurotoxin) detection in the dynamic range 5 nM to 25 μM with LOD of 0.29 nM	[14]
Smartphone ELISA	University of Illinois at Urbana-Champaign, USA	i) Human IL-6 detection in the dynamic range of 2 $pgmL^{-1}$ to 125 $pgmL^{-1}$ with LoD of 2 $pgmL^{-1}$ ii) Ara h 1 detection in the dynamic range of 2.5 ppm to 20 ppm with LoD of 0.107 ppm.	[15]

(continued)

Table 5.1 (Cont'd)

Spectrometer device	Developer	Detection characteristics	References
TRI analyzer	University of Illinois at Urbana-Champaign, USA	i) Ascorbic acid detection in the dynamic range 20 to 80 µgmL^{-1} with LoD of 5 µgmL^{-1}. ii) pH Sensing in the range 6–8.4 iii) Glucose sensing in the dynamic range 0 to 400 mgdL^{-1} with LoD of 60 mgdL^{-1}	[17–19]
Multichannel spectrometer	Washington State University, United States	i) BSA detection in the range 2–25 µgmL^{-1} with LoD of 4.4 µgmL^{-1}. ii) IL-6 (cancer biomarker) detection with LoD of 10.6 pgmL^{-1}	[20]
G-Fresnel spectrometer	The Pennsylvania State University, USA	i) BSA detection in the range 0.1 mgmL^{-1} to 1 mgmL^{-1} ii) Haemoglobin detection in the range 5.39 to 36.16 µM	[21,22]
Pencil spectrometer	Beijing University of Chemical Technology, Beijing, China	i) Chlorophyll monitoring via white light reflectance ii) Myoglobin detection white light reflectance iii) Hemoglobin monitoring in blood white light reflectance	[23]

5.3 Reflection diffraction grating configured smartphone spectrometers

Reflective diffraction grating based spectrometers have been reported in similar fashion for performing wide range of analytical investigations. Unlike transmission grating that are coated with an antireflection coating, reflection grating are coated with a reflective coating for use within domain of UV–VIS to NIR. Jian et al. [24] reported an automated spectrometer for detection of avian influenza virus (AIV) H7N9 and porcine circovirus type 2 (PCV2) antibodies using sunlight as light source. Light falling on the CMOS image sensor of smartphone after getting dispersed by a reflective diffraction grating produced pixel resolution of 0.276 nm/pixel. Further, Ding et al. [25] reported a smartphone based spectrometer where light is collected by an optical fiber and after passing through a series of optical elements, got recorded by the smartphone camera after getting dispersed by a reflection grating. With pixel resolution of 0.305 nm/pixel, the accuracy of the obtained spectrum was enhanced by wavelength calibration as well as intensity correction. The performance of the designed spectrometer was tested by measuring the absorption and fluorescence spectrum of Rhodamine B along with Creatinine detection. In similar line of work, Hossain et al. [26] reported a dual functional automated smartphone spectrometer using flash of the smartphone and UV LED for absorption and fluorescence measurement respectively of samples. Using a nano-imprinted reflection grating, the system could measure spectrometric data spanning over a range of 300 nm with pixel resolution of 0.42 nm/pixel. Integrating an endoscopic fiber bundle to the existing design, the scope of ambient illumination was removed that further improved the recorded spectrum by the smartphone camera [27]. Reflection grating based smartphone spectrometer was further realized using Compact Disk (CD) as dispersing unit for detection of glucose and cardiac human troponin I using peptide functionalized gold nanoparticles [28]. In an effort to increase the spectral efficiency and reduction in focusing aberration along with optical complexity, Lo et al. [29] demonstrated the possible use of concave blazed grating in conjugation with smartphone's camera for spectrophotometric application. With a custom designed software interface, the designed system of spectral resolution 1.075 nm was evaluated for surface plasmon resonance (SPR) based refractive index sensing application. Due to inherent limitations in terms of color overlaps in Bayers filters, default focusing and white balance adjustment and absence of rigourous calibration of the smartphone's image sensor, smartphone cameras are not specifically designed for spectral measurements. To address these issues, Das et al. [30] developed a standalone spectrometer platform (containing Hamamatsu: C12666MA model chip) employing blazed concave reflection grating that can be wirelessly connected to smartphone and interfaced using a software application for data collection and analysis. The designed device of spectral resolution 15 nm could record fluorescence emission spectrum of chlorophyll upon excited with UV light and yielded similar results in the range 340 nm to 780 nm when compared with compact commercial spectrometers. Similar spectrometer module (containing Hamamatsu: C12880MA model chip) with bluetooth connectivity to a smartphone giving spectral sensitivity from 450 nm to 750 nm and spectral resolution of 15 nm was reported by Laganovska et al. [31] for detection of analytes. Fig. 5.1 shows the transmission and reflection grating configured smartphone spectrometers for different types of sensing applications.

Due to limitations posed by Bayer's filter arrays on CMOS image sensor of smartphone, the applications of smartphone have been limited within visible domain only. However, this limitation was addressed with the use of a Rasberry Pi camera module whose Bayer filter and IR filter were removed in order to increase its optical throughput from UV to NIR domain as reported by Wilkes et al. [32]. Herein, the spectrometer sensor comprising of reflective diffraction grating, filters and focusing optics providing a spectral resolution of 1.0 nm is connected to a smartphone for measuring SO_2 concentration. Table 5.2 summarizes the advancement of smartphone spectrometers employing reflective grating as dispersive element.

78   Smartphone-Based Detection Devices

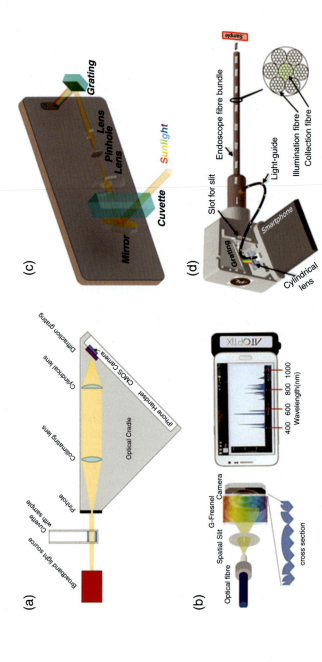

Fig. 5.1 Schematic representation of (a) spectrometer for ELISA. Reprinted with permission from [15] © The Optical Society, (b) G-Fresnel diffuse reflectance spectrometer reproduced from [22], (c) Sunlight spectrometer. Reproduced with permission from [24] and (d) fiber optics spectrometer. Reprinted with permission from [27] © The Optical Society, realized using smartphone camera.

Table 5.2 Reflecting grating coupled smartphone spectrometers.

Spectrometer device	Developer	Device characteristics & detection	References
Sunlight smartphone spectrometer	Nanjing Agricultural University, Nanjing, China	i) AIV H7N9 detection with LoD of 3×10^5 in the dilution ratio range 10^3 to 10^5 ii) PCV2 detection with LoD of 12,800 in the dilution ratio range 400 to 25,600	[24]
mHeath application spectrometer	Jiaotong University, Xi'an, China	i) Fluorescence spectrum detection of RhB ii) Creatinine detection in the range 50 μmol/L to 250 μmol/L with LoD of 50 μmol/L	[25]
Absorption and fluorescence spectrometer	The University of Sydney, NSW, Australia	pH sensing in the range 6.09, 7.20 and 8.25 and Zn^{2+} ion detection in the range ~0 to 50 μM through fluorescence spectrum.	[26]
Fiber optic spectrometer	The University of Sydney, NSW, Australia	Rh B and pigment content detection using fluorescence and reflectance spectra.	[27]
Spectrometer for Colorimetric biosensing	Laboratoire de Réactivité de Surface, France	i) Glucose detection in the range 0.1 to 50 mM with LoD of 0.2 mM ii) Human troponin I detection in the range 500 ngmL^{-1} to 2 μg mL^{-1} with LoD 50 ngmL^{-1}	[28]
Concave blazed grating spectrometer	Academia Sinica, Taipei, Taiwan	RI sensing with wavelength sensitivity of 461.16 nm/RIU and FoM of 0.08 (1/RIU)	[29]
Wireless spectrometer with spectrometer chip	Massachusetts Institute of Technology, USA	Chlorophyll monitoring by UV fluorescence	[30]
	University of Latvia, Riga, Latvia	i) Vit. B_{12} detection in the range 3 ppm to 24 ppm with LoD of 2.7 ppm. ii) Phosphate detection in the range 0.2 ppm to 2 ppm with LoD of 0.2 ppm. iii) HRP detection in the range 22 μM to 264 μM with LoD of 1.9 μM.	[31]
UV-Spectroscopy	The University of Sheffield, Sheffield, UK	SO2 detection with std. dev. in the range 20 to 60 ppm.m	[32]

5.4 Smartphone-based Raman spectroscopy

Raman spectroscopy is a popular method providing characteristic fingerprint for identification of molecules. The technique, owing to its high sensitivity, specificity, quick and being non-invasive gained large scale popularity for sample characterization in the field of pharmaceuticals, cosmetics, mineralogy and life science industry. However, the Raman signal being weak, requires sophisticated optical set-ups and detector, which leads to increase in overall size as well as cost of the entire instrument. These factors bring limitations on the use of Raman spectroscopy for on-site sensing investigations and in resource limited conditions. Hence the need of low-cost and compact Raman spectrometers is the need of analytical and bio-analytical sensing domain. Towards this end, Ayas et al. [33] demonstrated the possible use of smartphone camera for capturing weak Raman signals with resolution of 20 cm^{-1}. With the phone's camera they have succeeded in their effort to detect Raman signals from plasmonic substrate for ethanol detection and displayed sensitivity slight lower than that of a standard CCD device in terms of relative photon count per *sec*. Towards development of miniaturized Raman systems, Barnett et al. [34] reported a spatial heterodyne Raman spectrometric (SHRS) module of spectral resolution 9–11 cm^{-1} that can be coupled with a smartphone camera. The miniaturized SHRS unit in combination with cellphone camera was successfully implemented to obtain Raman spectrum of potassium perchlorate, ammonium nitrate and sodium sulphate. Raman spectrometer module that can be interfaced on a smartphone platform has been successfully commercialized and made available by Cloudminds XI™ Inc. The smartphone integrable Raman unit provided by XI™ was utilized by Zeng et al. [35] for analyzing common analyte samples on custom designed paper-based SERS substrates made of silver nanoparticles embedded in a nylon filter membrane. The group further modified this readymade Raman module by facilitating attachment option to a smartphone via smartport interface and controlled by software application installed in the phone itself. The integrated Raman spectrometer was assessed by sensing common Raman active dye samples in low concentrations [36]. In similar line of work, Mu et al. [37] reported a cellphone based Raman system for detection of different types of pesticides using SERS substrate. Further, Mu et al. [38] and Liang et al. [39] reported the integration of similar Raman module with smartphone where the phone was used for displaying the Raman spectral information. Further, in conjugation with customized software applications, the phone acted as wireless data transmission medium to a cloud server. Employing matching algorithm, the Raman spectrum was compared to the standard database spectrum in the server to identify the target sample and display the information in the phone itself. Table 5.3 summarizes the advancement of smartphone-based Raman spectrometers.

Table 5.3 Smartphone-based Raman Spectrometers.

Spectrometer device	Developer	Detection characteristics	References
Raman spectroscopy with plasmonic substrate	Bilkent University, Ankara, Turkey	Ethanol detection with 10 nM methylene blue solution and 1 µM cresyl violet solution	[33]
Spatial heterodyne Raman spectroscopy	University of South Carolina, Columbia, USA	Suphur, potassium perchlorate, ammonium nitrate and sodium sulphate detection with S/N values of 18, 37, 12 and 16 respectively	[34]
Raman spectroscopy with paper substrate	University of Electronic Science and Technology of China, Chengdu, China	Rhodamine 6G detection with LoD of 1pmol and Crystal Violet & Malachite green detection with LoD of 10 pmol	[35]
Raman spectroscopy with SERS chip	University of Electronic Science and Technology of China, Chengdu, China	Rhodamine 6G and Crystal violet detection with LoD of 10^{-4}M and 10^{-5}M respectively.	[36]
		Pesticides detection with LoD < 10ppm	[37]
Cloud architechture based Raman system	Beijing Information Science and Technology University, Beijing, China	Identification of ethanol, acetone and their mixtures	[38]
Deep learning integrated cloud architecture based Raman system	Beijing Information Science and Technology University, Beijing, China	Identification of mixture of alcohols and mixture of white powders	[39]

5.5 Conclusion

Spectrophotometry on a cell phone platform has made tremendous advancement in developing absorption, reflection, fluorescence and Raman type spectrometers. Smartphone-based spectrophotometers have been realized by attaching simple optical elements such as aperture, focusing lens and grating with camera module of the phone. Using two different types of grating configurations; transmission and reflection, it has been possible to capture modulated light spectrum either in transmission or reflection mode from the sample under investigation. Although current generation smartphones are built with good megapixel cameras coming with advanced optics, but these cameras are made for photography and not specifically for laboratory grade applications. Before proceeding for developing laboratory grade spectrometers using smartphones, we need to understand the inherent shortcomings of image sensor of the phone. The CMOS image sensor of smartphone's camera has several limitations such as overlapping in Bayer filters, fixed focusing, non-linearity in image response and non rigorous image sensor calibration along with other factors such as ISO level, shutter speed, white balance adjustment. These limitations can be circumvented by developing external spectrometer module that can be interfaced and controlled by a smartphone connected to the system via USB port or other wireless modules. Affordable spectrometer module that can be connected to a smartphone camera for measuring absorption, emission and transmission spectra has been made available by GoyaLab Inc. Two different varieties of GoyaLab products are available namely; GoSpectro whose sensitivity lays within 400 nm to 750 nm with spectral resolution of 10 nm and IndiGo whose sensitivity lays within 370 nm to 810 nm and spectral resolution of 5 nm. Also hand-held Raman spectrometer module with smartphone as interfacing device with additional cloud server facilities are currently made commercially available by Cloudminds X1 Inc. Such spectrophotometers built on a smartphone platform are light weight, portable, handy and brings opportunities in different field of sensing investigations.

References

[1] https://www.statista.com/statistics/330695/number-of-smartphone-users-worldwide/.
[2] S. Dutta, Point of care sensing and biosensing using ambient light sensor of smartphone: critical review, TrAC, Trends Anal. Chem. 110 (2019) 393–400.
[3] J. Liu, Z. Geng, Z. Fan, J. Liu, H. Chen, Point-of-care testing based on smartphone: the current state-of-the-art (2017–2018), Biosens. Bioelectron. 132 (2019) 17–37.
[4] D. Quesada-González, A. Merkoçi, Mobile phone-based biosensing: an emerging "diagnostic and communication" technology, Biosens. Bioelectron. 92 (2017) 549–562.
[5] J.C. Lindon, G.E. Tranter, D. Koppenaal, Encyclopedia of Spectroscopy and Spectrometry, Academic Press, 2016.
[6] Z.J. Smith, K. Chu, A.R. Espenson, M. Rahimzadeh, A. Gryshuk, M. Molinaro, S. Wachsmann-Hogiu, Cell-phone-based platform for biomedical device development and education applications, PLoS One 6 (3) (2011) e17150.

[7] D. Gallegos, K.D. Long, H. Yu, P.P. Clark, Y. Lin, S. George, B.T. Cunningham, Label-free biodetection using a smartphone, Lab Chip 13 (11) (2013) 2124–2132.

[8] S. Dutta, A. Choudhury, P. Nath, Evanescent wave coupled spectroscopic sensing using smartphone, IEEE Photon. Technol. Lett. 26 (6) (2014) 568–570.

[9] S. Dutta, D. Sarma, P. Nath, Ground and river water quality monitoring using a smartphone-based pH sensor, AIP Adv 5 (5) (2015) 057151.

[10] S. Dutta, D. Sarma, A. Patel, P. Nath, Dye-assisted pH sensing using a smartphone, IEEE Photon. Technol. Lett. 27 (22) (2015) 2363–2366.

[11] S. Dutta, K. Saikia, P. Nath, Smartphone based LSPR sensing platform for bio-conjugation detection and quantification, RSC Adv. 6 (26) (2016) 21871–21880.

[12] H. Kwon, J. Park, Y. An, J. Sim, S. Park, A smartphone metabolomics platform and its application to the assessment of cisplatin-induced kidney toxicity, Anal. Chim. Acta 845 (2014) 15–22.

[13] L. Kong, Y. Gan, T. Liang, L. Zhong, Y. Pan, D. Kirsanov, P. Wang, A novel smartphone-based CD-spectrometer for high sensitive and cost-effective colorimetric detection of ascorbic acid, Anal. Chim. Acta 1093 (2020) 150–159.

[14] L.J. Wang, Y.C. Chang, X. Ge, A.T. Osmanson, D. Du, Y. Lin, L. Li, Smartphone optosensing platform using a DVD grating to detect neurotoxins, ACS Sensors 1 (4) (2016) 366–373.

[15] K.D. Long, H. Yu, B.T. Cunningham, Smartphone instrument for portable enzyme-linked immunosorbent assays, Biomed. Opt. Express 5 (11) (2014) 3792–3806.

[16] H. Yu, Y. Tan, B.T. Cunningham, Smartphone fluorescence spectroscopy, Anal. Chem. 86 (17) (2014) 8805–8813.

[17] K.D. Long, E.V. Woodburn, H.M. Le, U.K. Shah, S.S. Lumetta, B.T. Cunningham, Multimode smartphone biosensing: the transmission, reflection, and intensity spectral (TRI)-analyzer, Lab Chip 17 (19) (2017) 3246–3257.

[18] M.Á. Aguirre, K.D. Long, A. Canals, B.T. Cunningham, Point-of-use detection of ascorbic acid using a spectrometric smartphone-based system, Food Chem. 272 (2019) 141–147.

[19] E.V. Woodburn, K.D. Long, B.T. Cunningham, Analysis of paper-based colorimetric assays with a smartphone spectrometer, IEEE Sens. J. 19 (2) (2018) 508–514.

[20] L.J. Wang, Y.C. Chang, R. Sun, L. Li, A multichannel smartphone optical biosensor for high-throughput point-of-care diagnostics, Biosens. Bioelectron. 87 (2017) 686–692.

[21] C. Zhang, G. Cheng, P. Edwards, M.D. Zhou, S. Zheng, Z. Liu, G-Fresnel smartphone spectrometer, Lab Chip 16 (2) (2016) 246–250.

[22] P. Edwards, C. Zhang, B. Zhang, X. Hong, V.K. Nagarajan, B. Yu, Z. Liu, Smartphone based optical spectrometer for diffusive reflectance spectroscopic measurement of hemoglobin, Sci. Rep. 7 (1) (2017) 1–7.

[23] F. Cai, D. Wang, M. Zhu, S. He, Pencil-like imaging spectrometer for bio-samples sensing, Biomed. Opt. Express 8 (12) (2017) 5427–5436.

[24] D. Jian, B. Wang, H. Huang, X. Meng, C. Liu, L. Xue, S. Wang, Sunlight based handheld smartphone spectrometer, Biosens. Bioelectron. 143 (2019) 111632.

[25] H. Ding, C. Chen, S. Qi, C. Han, C. Yue, Smartphone-based spectrometer with high spectral accuracy for mHealth application, Sens. Actuators A 274 (2018) 94–100.

[26] M.A. Hossain, J. Canning, S. Ast, K. Cook, P.J. Rutledge, A. Jamalipour, Combined "dual" absorption and fluorescence smartphone spectrometers, Opt. Lett. 40 (8) (2015) 1737–1740.

[27] M.A. Hossain, J. Canning, K. Cook, A. Jamalipour, Optical fiber smartphone spectrometer, Opt. Lett. 41 (10) (2016) 2237–2240.

[28] Y. Wang, X. Liu, P. Chen, N.T. Tran, J. Zhang, W.S. Chia, B. Liedberg, Smartphone spectrometer for colorimetric biosensing, Analyst 141 (11) (2016) 3233–3238.

[29] S.C. Lo, E.H. Lin, K.L. Lee, T.T. Liang, J.C. Liu, P.K. Wei, W.S. Tsai, A concave blazed-grating-based smartphone spectrometer for multichannel sensing, IEEE Sens. J. 19 (23) (2019) 11134–11141.

[30] A.J. Das, A. Wahi, I. Kothari, R. Raskar, Ultra-portable, wireless smartphone spectrometer for rapid, non-destructive testing of fruit ripeness, Sci. Rep. 6 (2016) 32504.

[31] K. Laganovska, A. Zolotarjovs, M. Vázquez, K. Mc Donnell, J. Liepins, H. Ben-Yoav, K. Smits, Portable low-cost open-source wireless spectrophotometer for fast and reliable measurements, HardwareX (2020) e00108.

[32] T.C. Wilkes, A.J. McGonigle, J.R. Willmott, T.D. Pering, J.M Cook, Low-cost 3D printed 1 nm resolution smartphone sensor-based spectrometer: instrument design and application in ultraviolet spectroscopy, Opt. Lett. 42 (21) (2017) 4323–4326.

[33] S. Ayas, A. Cupallari, O.O. Ekiz, Y. Kaya, A. Dana, Counting molecules with a mobile phone camera using plasmonic enhancement, ACS Photonics 1 (1) (2014) 17–26.

[34] P.D. Barnett, S.M. Angel, Miniature spatial heterodyne Raman spectrometer with a cell phone camera detector, Appl. Spectrosc. 71 (5) (2017) 988–995.

[35] F. Zeng, W. Duan, B. Zhu, T. Mu, L. Zhu, J. Guo, X. Ma, Paper-Based versatile surface-enhanced Raman spectroscopy chip with smartphone-based Raman analyzer for point-of-care application, Anal. Chem. 91 (1) (2018) 1064–1070.

[36] F. Zeng, T. Mou, C. Zhang, X. Huang, B. Wang, X. Ma, J. Guo, Paper-based SERS analysis with smartphones as Raman spectral analyzers, Analyst 144 (1) (2019) 137–142.

[37] T. Mu, S. Wang, T. Li, B. Wang, X. Ma, B. Huang, J. Guo, Detection of pesticide residues using Nano-SERS chip and a smartphone-based Raman sensor, IEEE J. Sel. Top. Quantum Electron. 25 (2) (2018) 1–6.

[38] T. Mu, S. Li, H. Feng, C. Zhang, B. Wang, X. Ma, L. Zhu, High-sensitive smartphone-based raman system based on cloud network architecture, IEEE J. Sel. Top. Quantum Electron. 25 (1) (2018) 1–6.

[39] J. Liang, T. Mu, Recognition of big data mixed Raman spectra based on deep learning with smartphone as Raman analyzer, Electrophoresis (2019).

Smartphone as barcode reader

Arpana Agrawal[a], Chaudhery Mustansar Hussain[b]
[a]Department of Physics, Shri Neelkantheshwar Government Post-Graduate College, Khandwa, India
[b]Department of Chemistry and Environmental Science, New Jersey Institute of Technology, Newark, USA

6.1 Introduction

Thanks to the technological advancements over the last decades in the field of competent smartphones, smartphone-based barcode readers are now capable of scanning barcodes and can therefore be considered as a game changer for commercial world [1]. Now-a-days, people are well aware of the capabilities of the smartphone and instead of merely using it for capturing photographs, people are now a day using their smartphones for barcode reading purposes to collect the valuable informations of the products from their database. Prior to the advancement in the field of smartphone-based barcode technologies, the more common types of barcode reader technologies that are available includes: pen-type barcode reader, laser scanners and light emitting diode (LED) scanners (also known as a charge-coupled device) which are used to capture the barcode patterns. Pen type barcode reader is much more economic and the simplest one for barcode reading purposes where the tip of the pen serving as photodiode when moved horizontally over the barcode lines, light will return a waveform. In a very similar fashion, in a laser scanner, a laser light is shined on the barcode and mirrors are use to collect the reflected light intensity of barcode pattern. In contrast to these two barcode readers, LED scanner detects the light intensity of ambient lights from the barcodes using several tiny light sensors and then generates digital illustration of the barcodes. In contrast to all the above mentioned barcode readers, camera-based readers or smartphone based barcode readers are much more fascinating.

Smartphone based barcode reading technology has simplified our day-to-day life and is involved in almost every aspect of our life including the industries [2], health care: detection and clinical diagnostic systems [3-5], education [6], food and beverage producers [7,8], microscopy/imaging [9], etc. Barcode is simply a well defined pattern of either lines or squares on a contrast background and used to store the information of the products. The barcode when scanned using the smartphone's camera, the encoded information will be then decoded to provide the stored information and analyzed using several smartphone apps. It is worthy to mention here that there is also an increasing interest of the app developers in developing smartphone apps to scan the barcode reader which can facilitate the users in getting several detailed informations about the product by scanning the barcodes. In logistic industries, barcode reading solutions allow consumers to read the barcode in order to compare the price of items and also enable them to provide the product review. Barcode reading technology is also extensively used at the airports to help keeping track of the baggage and security at the airports; in automobile sector, it helps to read one dimensional (1D) and two dimensional (2D) codes on replacement parts, equipments etc. and assists to follow production status; in food and beverage producer industries, barcode reading technology helps to serve the special dietary needs [10], monitoring the aging and quality of food products [7], food allergen management system [11,12], minimizing waste by providing a reminder about the expiry date of the food products [13,14], allows to get aware of the nutritional contents of the food products [15], etc. In health care sector, this barcode technology is highly important for elderly care, for reminding about the expiry dates of the medicines [16], for component traceability, for diagnostic of special hormones such as pregnancy hormones [17], to diagnose various viruses [18,19], residue pesticides [20], blood typing evaluation [21] etc. This technology also gives rise to an advanced attendance management system which is much superior to that of the existing traditional attendance system [22], visual secret sharing [23], trip field [24], and teaching [25]. Fig. 6.1 summarizes few of the smartphone based applications in our day-to-day life. Looking at all these enormous advantages of smartphone based barcode reading, smartphones can be considered as boon for consumer technology in the age of modern information. The various features of smartphones such as the in-built camera, high-resolution display, a variety of remote sensors, various advanced audio-visual features, GPS systems, Bluetooth, internet connectivity and their integration along with several developing software or allied applications has made this smartphone to be well thought-out of as the mini portable personal computers or mobile computing device and is an important part of everyday item for majority of global population.

However, there are several obstacles that may affect the successful scanning of a barcode using a smartphone barcode reading application including the physical properties of the product containing barcode to be scan such as the weight or shape, size/type/contrast and reflection of the barcode to be scan, light conditions while scanning

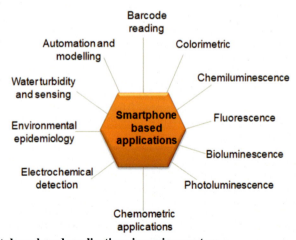

Fig. 6.1 Smartphone based applications in various sectors.

a barcode because of being an optical imaging process, diversities of the hardware/software including camera of the smartphone and the scanning person. Scanning a barcode image using the in-built camera of the smartphone requires the user to properly align the barcode within the specifically chosen area on the display of the smartphone which is sometime an ill at ease process because of the other unfavorable conditions such as low light or unfocussed image etc. Consequently, more than one attempt is often required to capture the proper image of the barcode appropriate for decoding.

Apart from all the above mentioned complication for the successful capturing of the barcode, smartphone-based barcode reading is a pinnacle of end user technology of the modern information age. Accordingly, the present chapter summarizes the various applications of smartphone in the field of analytical chemistry with a particular focus on barcode reading. However, it is impossible to comprehensively discuss all the aspect of the utility of smartphone as barcode reader; an attempt has been made to introduce the field (smartphone as barcode reader). Various smartphone apps developed to serve this purpose has also been discussed in details along with recent results of its use in our day-to-day life.

6.2 Smartphone based applications in analytical chemistry

Smartphone has a wide range of applications including optical detection [26,27], electrochemical detection [28], barcode reader [29], label free biosensors/biodetection [30-32], chemometric, luminescence, chemiluminescence, bioluminescence, photoluminescence applications, or fluorescence microscopy [33,34]. Optical detection relies on the change in color, pixel count/pixelation, change in the reflected/scattered light intensities, refractive index variations or smartphone based microscopy such as luminescence, chemiluminescence, bioluminescence or photoluminescence [35]. This section provides a flavor of various utilities of smartphone in the field of analytical chemistry. For optical detections, the built-in camera of smartphone serves as an input device to capture the digital image which is then processed by various smartphone apps for detection. Under colorimetric detection, the change in color is observed while the light intensities are measured in fluorescence detection. Colorimetric quantification in general is done via Red-Green-Blue (RGB) analysis where the color is decomposed in its component colors e.g. red, green and blue. Another way for colorimetric quantification is the CIE 1931 color coding system where the color is represented by parameters x and y and is used for smartphone based urine testing and commercially available pH strips. Hong et al. [36], have employed smartphone app for simultaneous detection of red blood cells, ascorbic acid, leucocytes, glucose, nitrite, ketone and proteins via the colorimetric evaluation of urine strips. Simultaneous detection of pH and nitrides in water has also been done by Lopez-Ruiz et al. [37]., by using in housing fabricated paper based microfluidic device. Smartphone based clinical diagnostic is a promising and developing field. Imaging capabilities of smartphone can also be coupled with fluorescence microscopy and are widely used in the field of biochemistry. Orth et al. [38], have reported a smartphone based microscope which employs the internal flash as the illumination source and is capable of dual mode microscopy e.g. bright-field and dark-field imaging modes and hence facilitates the visualization of several cells by simply inserting the objective lens of smartphone's camera into the microscope chip. Fig. 6.2(A) shows the image of Lilium ovary under bright-field imaging mode using iphone camera with an exposure time of 1/4808s. Smartphone based infrared G-Fresnel spectrometer for diffusive reflectance spectroscopic measurements of hemoglobin has been reported by Edward et al. [39]. Fig. 6.2(B) schematically shows the experimental arrangement of diffuse reflectance spectroscopy using G-Fresnel spectrometer comprising of an optical fiber to deliver light onto the slit which was then passed through the G-Fresnel device having a grating pattern with 600 line per mm in one side and a Fresnel lens on the other side to collimate the light coming from the slit and disperse the different wavelengths across the mobile camera. Yeo et al. [40], have illustrated the smartphone based fluorescent diagnostic system for the detection of highly pathogenic H5N1 viruses by means of efficient reflective light collection module as shown in Fig. 6.2(C). They have employed a coumarin-derived dendrimer based fluorescent lateral immunoassay to detect three different avian influenza (H5N3, H7N1 and H9N2) subtypes. When the fluorescent lateral immunoassay strip is inserted into the measurement module followed by the excitation using LED, the fluorescence emission is then filtered by the emission filter and collected and detected by the non-imaging reflector module and smartphone's camera. The utility of smartphone as a microscope has also been demonstrated by Jung et al. [41]. They have reported the smartphone based portable multi-contrast microscopy based on color-coded light emitting diode microscopy which is capable of creating the bright-/dark-field imaging as well as differential phase contrast image of biological samples. Fig. 6.2(D) schematically depicts the arrangement of smartphone module system. Wei et al. [9], have also captured the optical image of nanoscale objects based on field portable fluorescence microscopy platform installed on smartphone for imaging single nanoparticle and viruses employing optomechanical interfacing with the existing Smartphone's camera.

Pixelation allows converting any change in the color or illumination into digits and is reported by Lee et al. [42]., to read the grayscale intensity from lateral flow immunoassay. In some cases, variations in the refractive index or the intensities of reflected or scattered light is examined. For electrochemical applications, smartphone can be interfaced with external electrochemical cell while for barcode reading, the smartphone-based built-in camera is use to scan the 1D/2D barcodes patterns and are then processed by various barcode reading apps. Furthermore, smartphones can also be used for chemometric applications to differentiate multiple raw data of multiple analytes [43,44]. Hernandez-Neuta et al. [4]., have provides a comprehensive overview of all the parallel advancements where smartphone is adapted as a protable and versatile clinical diagnostic tool within the field of biosensing, mathematical algorithms, molecular analysis, microfabrications, microfluidic three dimensional printing etc. Mosa et al. [45]., have also systematic reviewed healthcare applications for smartphones. Alawsi and Al-Bawi [46]., have also presented a detailed review

Smartphone as barcode reader 89

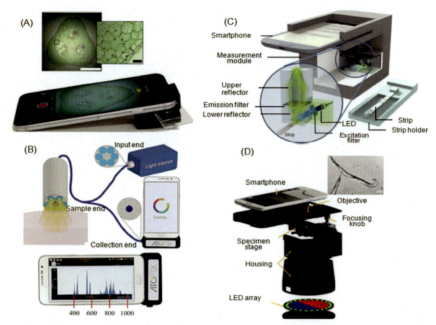

Fig. 6.2 (A) Bright-field image of Lilium ovary using smartphone by inserting sample slide and attaching the clip to an iphone camera [38]. (B) Experimental setup for diffuse reflectance spectroscopy using the G-Fresnel smartphone spectrometer [39]; (C) Schematic description of a smartphone-based fluorescence detector with a reflective light concentrator module [40]; (D) Portable cLEDscope along with smartphone module system arrangement for multi-contrast imaging [41].

on potential of smartphone-based applications. Among all the applications of smartphone, smartphone as a barcode reader is quite fascinating as it has become a part of our day-to-day life.

6.3 Smartphones as barcode reader

A barcode is a graphical image which can store data in special patterns either of unequally spaced lines having variable thickness or vertical and horizontal squares. In other words, it can be defined as a common identifier which can store a barcode data/identifier. The encoded data can be decoded to get the useful stored information using a barcode scanner machine or a smartphone having high resolution camera and specific barcode reader application. Symbolically, a barcode can be classified in two categories- (i) one dimensional barcode which is a series of parallel lines (bars) separated by some distance, with varying widths and required to be scan along a single direction to obtain accurate data, and (ii) two dimensional barcode containing black modules arranged in a square pattern on a white background consisting of encoding region and the function patterns and are widely used after the invention of

Fig. 6.3 (A) One dimensional (linear) barcode and (B) two dimensional (Quick response) barcode patterns.

efficient smartphones. Figs. 6.3(A) and (B) show the linear and two dimension barcode patterns, respectively. In contrast to 1D linear barcode systems, the 2D barcodes are economical tools in food industries or business marketing, which facilitate them to take the pre-sale and post-sale follow-up of their products. This 2D barcode are also termed as QR code and is an advanced version of its predecessor (1D barcodes) because of the greater storage capacity and fast readability. For geometrical corrections and encouraging identification, function patterns are used and the error correction levels, mask patterns and versions are enclosed in formats. The error correction level enables QR code readers to acceptably decode data, even of dirty or damaged symbols. So far, there are 4 levels of error correction (L, M, Q, and H) and several versions of QR codes from (version 1 to version 40) have been launched depending upon the size of the modules. Version 1 is the smallest QR code having size of 21×21 modules and for each successive versions, the size of the QR code is increased by 4 modules. The version 40 of QR code is the largest one with 177×177 modules. The readability of QR codes depends on several factors including size of the QR code, scanning distance, number of modules, that is, more the stored information denser will be the QR codes which makes reading more difficult. However, the storage capacity of a QR code depends upon the version, error correction levels and the data type that is to be encoded. In general, the data can be encoded in three types of data modes namely, numeric, alphanumeric and byte and each of these can be encoded to certain characters. For example, for only one numerical data, the created QR code can be encoded upto 7089 characters while the alphanumeric and byte data type can be encoded to 4296 and 2953 characters, respectively. In contrast to this QR codes, iQR code was developed with initial composition of 9×9 cells which is more advanced version of QR codes and can save much more information quantities as compared to the existing QR codes. For a same sized QR and iQR codes, iQR code can store upto 80 percent more information quantity and for same amount of information, the size of iQR code can decrease upto 30 percent and hence create a smaller code as compared to QR codes.

It is noteworthy to mention here that the utility of smartphones as barcode reader also relies on the barcode reading apps which were installed on the smartphones and allows the camera of the smartphone to look into the barcode (either in 1D or 2D) of several items/products, decipher the concealed data into constructive informations and implement a lookup from the allied database. Consequently, there is a always an encouragement among the app developers to develop much more efficient apps that can be potentially used for barcode reading using the smartphone camera and hence, can contribute in building a trouble-free life of the user. The built-in camera of a smartphone serves as an input device for the fast scanning of the barcode and hence overrules the possibility of typing errors and this when coupled with the smartphone barcode reading app will make available the information regarding the items on the screen of the smartphone. There are several smartphone apps that can be utilized to serve the purpose of barcode reading for the comparison of prices of products among various retailers, to assist visually impaired individuals, for monitoring the aging or quality of food products, for reminding the elderly people about their medicines, alarming of the expiry dates of medicines as well as several food products, assisting in maintaining the special dietary need, etc.

For visually impaired individuals, the smartphone based barcode reading app is developed by the Smith-Kettlewell Eye Research Institute and named as BLaDE which is an abbreviation of Barcode Localization and Decoding Engine and works on Andriod smartphones [47]. This smartphone app provides real-time audio feedback which allows the visually impaired person to locate and read the product barcodes. There are a number of smartphone apps that have make it straightforward to compare the prices, including ShopSavvy which allows the users to get the product prices from various retailers/stores, LiveCompare and Ubira that allows the price comparison of grocery items and products, scanning the barcode using BEEP smartphone app allows to keep track of expiration dates by register the expiration dates. SCAN4CHEM smartphone app was developed by EU LITE project allowing the consumers to effortlessly appeal the product information included in the product database. Myhalal or MyMobiHalal 2.0 are the smatphone based barcode reading apps that accomplish the special dietary needs of Muslim consumers by verifying and confirming the Halal status of the food products from Halal authority database. Other available smartphone apps to assist the consumers needs include- Product Empire, a gaming smartphone app where players were rewarded and can public their achievements; GiftScroll smartphone app that assists in organizing the "wish list" of gifts for several events; TUMEDICINA smartphone app that takes care of the elderly patients by assisting them in taking their proper medication by providing verbal instructions; FridgePal recommends recepies to users based on grocery list recorded in application database; Pervasive Fridge helps to avoid food wastage by sending a prompt of the expiry dates of the food products; ShopSocial, a product information system that enable users to explore the product information and pertinent videos from the users social graph; My2cents allows buyer to share the product reviews; Foodswitch provides nutritional contents of food, etc. Blood culture evaluation can also be done using smartphone based barcode reader using a smartphone app named as "Blood Culture" which is based on the identification of patient's barcode using smartphone camera and reminds the physician about the proper method and time stamp to sample the patient's blood for blood culture [3]. It should be noted that for the blood culture evaluations, the patient's blood is required to be scan for two or more times at certain time spans. For this, the user is require login the Blood Culture app using hospital staff ID and scan the blood sample barcode at each step, manually entering the sample site and volume which is then simultaneously transmitted to the legacy laboratory information system. Figs. 6.4(A) and (B) shows the barcode reading menu, barcode reading phase of the "Blood Culture" smartphone application. Fig. 6.4(C) depicts the reading results and manual enter of the additional data (sample site and volume) and Fig. 6.4(D) show the simultaneous transmission to the legacy laboratory information system.

There is a wide range of utility of smartphones including the field of education, healthcare, retail shops, sensors to monitor the quality of food items, etc. In the field of education, smartphones are reported to be used either to keep the attendance records, for the identification of various flora and fauna species by scanning their QR codes, for teaching chemistry using student-created videos and photo blogs or for the authentication of degree certificate awarded to the students by the institutes. Lee at al. [24], have reported the usage of smartphone to scan the QR codes and develop a teaching process for biology field study. For such purposes, the instructor generates the QR codes for flora or fauna species at biology field site and students use to decode this QR codes using QR code decoding app on their smartphones and hence explore and identify the species using QR code sheets and finally share the data via social network systems. On the other hand, Cho and Bae [22]., have also utilized smartphone based barcode reading technique in educational field. They have developed an automatic attendance check system to overcome the limitations of traditional attendance management system such as marking in the attendance book. QR codes can also be employed in improved visual secret sharing scheme [23]. Benedict and Pence [25], have utilized the 2D barcodes accessed by using camera of smartphone for teaching chemistry using student-created videos and photo blogs.

In the field of health care, barcode reading technology is of great importance and a lot of work has been reported on smartphone based barcode reading. Guan et al. [21], have demonstrated the fabrication and working of paper based diagnostic device containing barcode like pattern for blood typing test via the integration of smartphone based barcode technology. Their device is based on the principle of haemagglutination reaction between the red blood cells and the antibodies. Initially, hydrophilic bar channels were made by printing technology which were then treated with anti-A, -B and –D antibodies followed by the introduction of the blood samples in the sample sites for blood type testing. Phosphate-buffered physiological salt solution (PBS) acting as eluting buffer is then added for 1 min to identify the blood type depending upon the eluting lengths in the bar channels. A simple and low cost chip based wireless multiplex diagnostic device has also been reported by Ming et al. [48]., via the integration of quantum dot barcode technology with the smartphone and isothermal amplification which is capable of diagnosis of HIV or hepatitis B infected patients by detecting down to 1000 viral genetic copies per mL.

For rapid detection, barcodes are generally engineered using inorganic nanoparticles. However, for large scale readout systems, such barcode designs are not generally preferred for clinical purposes. Accordingly, barcode assay are considered to the most potential way for such purposes. Ding et al. [49], have illustrated an electrochemical bio barcode assay based on DNA modified gold nanoparticles which serves as a

Fig. 6.4 Smartphone app for "Blood Culture" evaluation. Barcode reading menu (A); Barcode reading Phase (B); Reading results and manually entry of sample site and volume (C) and saving data for pseudonymized ID and patient [3].

non-enzymatic method for the quantitative detection of human α-fetoprotein. This detection method is established by sensitive sandwich immunoassay relying on barcode method and can directly detect α-fetoprotein with a low detection limit ~ 9.6 pg/mL. Wong et al. [17]., have demonstrated the adaptation of smartphones for diagnostic purposes where smartphones are utilized for direct reading of BonaFide barcode containing patterned immunoassay strips. These strips can be validated for the real world problems such as detection of pregnancy hormones or human chorionic gonadotropin (hCG). hCG hormone is mainly produced during pregnancy and its level in the blood and the urine goes on varying drastically as pregnancy progresses. For the detection of hCG, Wong et al. have prepared two assay, (i) Biotin-Streptavidin binding assay serving as a reference system assay to examine the barcode assay principle and (ii) hCG assay. Both the assays were prepared on polycarbonate plates using a polydimethylsiloxane (PDMS) chip having embedded microchannels to form the binding strips (test and control lines) where the reagents can be casted to create the barcode pattern. This patterns were designed in the most popular linear barcoding format e.g. Code 39 using two characters of Code 39 as barcodes namely "+" and "-" characters. Figs. 6.5(A) and (B) shows the barcodes with "– " and "+ " symbols as generated using Code 39 and enlarged image representing the design of the assay strips, as patterned on a polycarbonate plate using a microfluidic PDMS chip, respectively. The red bracket shows the two identical characters that are apart from the four elements at the middle of each character as shown by the dashed green box and the blue brackets show the start and stop characters designated by symbol "*". The "– " character bar is further divided into four separate binding strips where the first and third are the test lines while the second and fourth are the control lines for the test. The control line serves as a validity test and the state of no signal indicates the invalid test with null result. If the test is valid, then a signal will appear in the control line irrespective of whether the planned antigen binds. In contrast to this control lines, test lines serves to qualitatively quantify the positive and negative results provided the test is valid. When a specific antibody is immobilized in the test line to detain the analyte, a visible signal appear in the test line because of the presence of analyte, provided the test is valid and enable the scanner to read the barcode. Accordingly, for a valid positive test, a visible signal will appear in both the test and control lines, and the reader will scan the test as "-" while in case of valid negative test, a "+" character will appear. This test and control regions are then evaluated colorimetrically (based on RGB analysis) to examine the change in intensity of the color (grayscale change) owing to the binding reactions and is compared with the intensity of the background color. Optical density ratio (ODR) can also be calculated from the averaged grayscale intensities using the expression [17];

$$ODR = \frac{I_b - I_s}{I_b} \quad (6.1)$$

Here I_b and I_s are the luminosity of the background and the assay strip (binding site). Figs. 6.5(C) and (D) shows the dependence of ODR on the concentration of nanogold streptavidin conjugate following silver enhancement for biotin-streptavidin

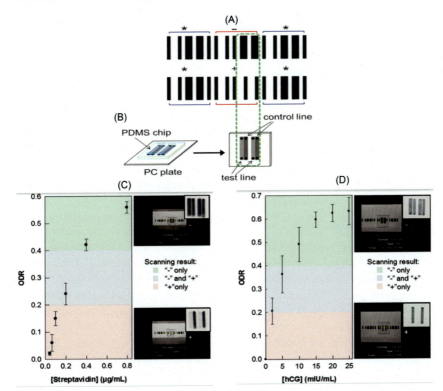

Fig. 6.5 (A) Barcodes with "– " and "+ " symbols as generated using Code 39. The red bracket shows the two identical characters apart from the four elements at the middle of each character as shown by the dashed green box and the blue brackets show the start and stop characters designated by symbol "*"; (B) Enlarged image illustrating the design of assay strips, as patterned on a polycarbonate plate using a microfluidic PDMS chip. (C) and (D) Dependence of ODR on concentration of Nanogold streptavidin conjugate following silver enhancement for biotin-streptavidin assays prepared as "–" and "+" and hCG with the barcode assay, respectively. Plot with different highlighting colors indicate their feasibility to be scanned by the barcode app [17].

assays prepared as "–" and "+" and hCG assay, respectively. Fig. 6.5(C) clearly shows an increasing trend of ODR reaching saturation at ODR of 0.50–0.60 and corresponds to high concentrations (0.8 µg/mL) of the conjugate. ODR also increases with increasing the concentrations of hCG as shown in Fig. 6.5(D) and reaches to saturation limit at 15 mIU/mL and shows that the barcode app can also read a low level (5 mIU) of hCG assay for a positive test e.g. "–" symbol.

It is a common practice to check the aging of the food products by simply looking for the "best use by" date on the packaging of the food product while buying from the market. It is worth mentioning here that a smartphone camera can be utilized to quantitatively estimate the aging and quality of the food by reading the color information from the food's barcode that can be used as a colorimetric sensor. Such a food's barcode based colorimetric sensor has been demonstrated by Chen et al. [7]. They have integrated the qualities of optical dyes for sensing and the camera of smartphone for detection to fabricate a paper-based colorimetric geometric barcode sensor to monitor the aging and quality of the food product (mainly the chicken meat). This paper based barcode sensor is fabricated by casting three different optical dyes (Nile red, Zinc-Tetraphenylporphyrin (Zn-TPP) and Methyl red) onto various geometrical patterns resembling a QR code and is sensitive to the pH and volatile gases from the food product. In order to sense the emanating gases from the chicken meat, the dye based sensor is attached to the food surface and the color of the dye changes as a consequence of volatile emanating gases from the food product. Any change in the color can be then examined by the sensing barcode using the camera of a smartphone and built-in barcode reading apps.

For the fabrication of dye-based sensor, microbeads containing various optical dyes (Nile red, Zn-TPP and Methyl red) were prepared. A similar approach of preparation of Nile red and methyl red was followed by dissolving 25 mg of dye in 5 ml of deionized water with continued magnetic stirring for 20 min followed by the addition of 256 mg of anion exchange resin microbead which again undergoes stirred at room temperature for about 2 hr to allow the optical dye to occupy the resin microbeads. On the other hand, the Zn-TPP dye microbead was prepared by dissolving 25.6 mg Zn-TPP in 3 ml Toluene via continued magnetic stirring for 2 hr. This mixture was them poured into a 256 mg cation exchange resin microbead with stirring for 2 hr at room temperature and was left overnight to allow the residues of toluene to evaporate completely. After the preparation of this three optical dye microbeads, a polymer mold is prepared containing different shaped grooves (triangular, square and circular grooves) using laser cutting method. Each of these grooves is then assigned to particular sensing elements using shape based encoding and hence is the geometrical alternate for QR codes. Prior to casting the optical dyes into these geometrical shapes, the polymer mold is poured by the De-gassed PDMS followed by curing at 80 °C for 5 hr. These cured shapes were then peeled-off the mold and finally the dye-contained microbeads were loaded onto the different shaped patterns. The geometrical patterns containing dye-casted microbeads are then transferred from the PDMS mold using the peeled off method onto a double-sided tape serving as a sensing substrate and then finally attached to filter paper which is acting as a protective cover for the fabricated device. Fig. 6.6 schematically shows the whole process of device fabrication including; the preparation of microbeads of optical dyes (Fig. 6.6(A)), formation of triangular, circular and square shaped groove on polymer mold using laser cutting technology (Fig. 6.6(B)), curing and peeling-off the mold to get the geometrical patterns (Fig. 6.6(C)), pouring of optical dyes in the geometrical patterns (Fig. 6.6(D)), pasting of double-sided tape onto the dye-stamped geometrical patterns (Fig. 6.6(E)), transferring of dye-casted microbeads onto double sided tape (Fig. 6.6(F)), attachment of filter paper (Fig. 6.6(G)) and the final device (Fig. 6.6(H)).

Chen at el [7], have employed smartphone diagnostic sensing system where the camera of Apple smartphone to capture the colorimetric images of geometric barcode sensors under ambient illumination and then imported into MATLAB for data processing. The channel intensities of the three optical sensing dyes were then quantify

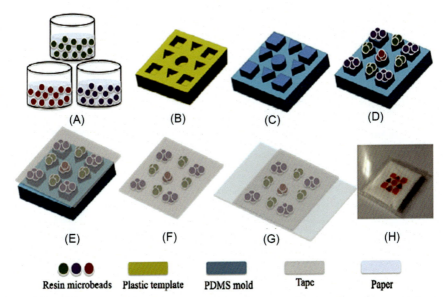

Fig. 6.6 Schematic illustration of the fabrication of paper-based colorimetric geometric barcode sensor. (A) resin microbeads of optical dye, each with different color; (B) Polymer mold (plastic template) grooved with different shapes by laser cutter; (C) De-gassed PDMS pouring on polymer mold; (D) casting of dye-contained microbeads onto the different shaped patterns; (E) attachment of a double-sided tape acting as a sensor substrate onto the PDMS mold; (F) the tape with beads are peeled off from the PDMS mold; (G) the tape with beads attached to a piece of filter paper; (H) fabricated device. (Reprinted with permission from ref. [7]).

to monitor the aging and quality of packed chicken using RGB analysis by averaging over the square, triangular and circular patterned sensing areas of that dye at different times after carful calibration and subtraction of background variations from the scanned images of sensing regions. Extracted R, G, B values of the explicit optical dyes on barcode sensors will be then compared with the pre-obtained R, G, B characterization curves of all dyes to determine the freshness of the chicken whether it is suitable for consumption or should be discarded. Fig. 6.7 schematically illustrates the paper-based colorimetric geometric barcode device as a food quality sensor.

Guo et al. [20]., have demonstrated a colorimetric barcode assay by employing smartphone based barcode reading technology for the detection of pesticide residues particularly methyl parathion residue which is a toxic organophosphorus pesticide and is commonly used in agriculture and fish hatcheries. Zhang et al. [18]., have also presented the design and fabrication of a microfluidic barcoded microchip with several channels having different widths for biomolecular assay. They have also reported the applicability of this barcoded microchip for the detection of several pathogen specific oligonucleotides and for multiplexed human immunodeficiency virus (HIV) immunoassay testing which were easily readable using smartphones. Two identical channels of each of the channel are loaded with same sample solution and hence enable duplicate experiment. Dark and light codes were generated by negative and positive controls

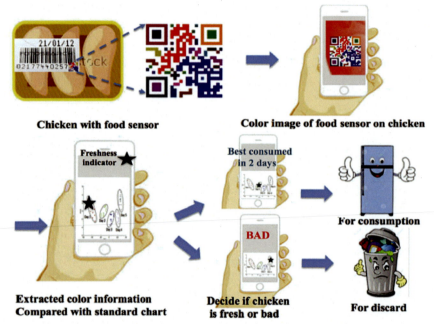

Fig. 6.7 Schematic illustration of paper-based colorimetric geometric barcode device application as a food quality sensor. The paper-based colorimetric geometric barcode sensor is attached to the surface of the meat or onto the inside lining of the package from where the quality and aging of the product can be quantify using smart phone by capturing a photo of the sensor using the extracted image upon comparison with the standard chart. (Reprinted with permission from ref. [7]).

of the chip, respectively which are then captured using smartphone camera and then decoded to obtain the useful information. Arens et al. [12], have demonstrated the concept of mobile electronic patient diaries having barcode based food identification which help to avoid the food allergies of the patients. For systematic implementation, MENSSANA (Mobile Expert and networking System for Systematical Analysis of Nutrition-based Allergies) project was developed based of modern mobile computing and database technology [12]. Patients with food allergies are required to scan the food package barcode using smartphone based barcode reader and make entries in the mobile electronic patient diary and then the diary is synchronizes with MENSSANA electronic patient record for allergies. This personal allergy assistant then stores the patient's data regarding several known food products and the patient's forbearance to them.

6.4 Conclusions

In conclusion, the present chapter provides an overview of various applications of smartphone in the field of analytical chemistry with particular focus on smartphones as barcode reader. 1D (linear barcode) and 2D (QR code) barcodes are comprehensively discussed along with the several smartphone based apps to decode the hidden data encoded in the barcode. The chapter also summarizes few recent results in the field of smartphone based barcode readers that were employed for education, healthcare, food and beverages, etc. Overall, the smartphone based barcode reader technology is excellent and software develop should be encouraged to develop much more efficient barcode reader apps for smartphone to further enhance the utility of smartphone based barcoders.

Acknowledgement

I heartily thank Mr Jatin Gandhi for his constant support and encouragement.

References

[1] www.waspbarcode.com/buzz/how-to-use-a-smart-phone.
[2] https://www.cognex.com/en-in/blogs/industrial-barcode-reader/why-smartphone-based-barcode-scanners-are-a-business-game-changer.
[3] Y.R. Park, Y. Lee, G. Lee, J.H. Lee, S.-.Y. Shin, Smartphone applications with sensors used in a tertiary hospital—current status and future challenges, Sensors 15 (5) (2015) 9854–9869.
[4] I. Hernandez-Neuta, F. Neumann, J. Brightmeyer, T. BaTis, N. Madaboosi, Q. Wei, et al., Smartphone-based clinical diagnostics: towards democratization of evidence-based health care, J. Intern. Med. 285 (1) (2019) 19–39.
[5] W. Zhao, S. Tian, L. Huang, J. Guo, K. Liu, L. Dong, Smartphone-based Biomedical sensory system, Analyst 145 (8) (2020) 2873–2891.
[6] A. Singhal, R.S. Pavithr, Degree certificate authentication using QR Code and smartphone, J. Comput. Appl. 120 (16) (2015) 38–43.
[7] Y. Chen, G. Fu, Y. Zilberman, W. Ruan, S.K. Ameri, Y.S. Zhang, et al., Low cost smart phone diagnostics for food using paper-based colorimetric sensor arrays, Food Control 82 (2017) 227–232.
[8] J.L.D. Nelis, A.S. Tsagkaris, M.J. Dillon, J. Hajslova, C.T. Elliott, Smartphone-based optical assays in the food safety field, Trends Anal. Chem. 129 (2020) 115934.
[9] Q. Wei, H. Qi, W. Luo, D. Tseng, S.J. Ki, Z. Wan, et al., Fluorescent imaging of single nanoparticles and viruses on a smart phone, ACS Nano 7 (2013) 9147–9155.
[10] E.Y. Daraghmi, C.-.F. Lin, S.M. Yuan, Mobile Phone Enabled Barcode Recognition for Preferences Monitoring, Advances in Computer Science and Education Applications, Springer, Berlin, Heidelberg, 2011, pp. 297–302.
[11] F. Mandracchia, E. Llauradó, L. Tarro, R.M. Valls, R. Solà, Mobile phone apps for food allergies or intolerances in app stores: systematic search and quality assessment using the mobile app rating scale (MARS), JMIR Mhealth Uhealth 8 (9) (2020) e18339.
[12] A. Arens, N. Rösch, F. Feidert, P. Harpes, R. Herbst, R. Mösges, Mobile electronic patient diaries with barcode based food identification for the treatment of food allergies, GMS Med. Inform. Biom. Epidemiol. 4 (3) (2008) 1–5.
[13] T. Khan, A Cloud-Based Smart Expiry System Using QR Code, 2018 IEEE International Conference on Electro/Information Technology (EIT), IEEE, 2018.

[14] M.R. Hyder, T. Khan, Automatic expiry date notification system interfaced with smart speaker, Int. J. Eng. Sci. Invention 9 (7) (2020) 14–20.

[15] M. Maringer, N. Wisse-Voorwinden, P. van 't Veer, A. Geelen, Food identification by barcode scanning in the Netherlands: a quality assessment of labeled food product databases underlying popular nutrition applications, Public Health Nutr. 22 (7) (2019) 1215–1222.

[16] M. Ramalingam, R. Puviarasi, N.D.A.B. Zakaria, E. Chinnavan, Developing mobile application for medicine expiry date detection, Int. J. Pure and App. Math. 119 (16) (2018) 3895–3901.

[17] J.X.H. Wong, X. Li, F.S.F. Liu, H.-.Z. Yu, Direct reading of bona fide barcode assays for diagnostics with smartphone apps, Sci. Rep. 5 (2015) 1–11.

[18] Y. Zhang, J. Sun, Y. Zou, W. Chen, W. Zhang, J.J. Xi, et al., Barcoded microchips for biomolecular assays, Anal. Chem. 87 (2) (2015) 900–906.

[19] E.-.Y. Kim, J. Stanton, B.T.M. Korber, K. Krebs, D. Bogdan, K. Kunstman, et al., Detection of HIV-1 p24 Gag in plasma by a nanoparticle-based biobarcode- amplification method, Nanomedicine (Lond) 3 (3) (2008) 293–303.

[20] J. Guo, J.X.H. Wong, C. Cui, X. Li, H.-.Z. Yu, Smartphone-readable barcode assay for the detection and quantitation of pesticide residues, Analyst 140 (16) (2015) 5518–5525.

[21] L. Guan, J. Tian, R. Cao, M. Li, Z. Cai, W. Shen, Barcode-like paper sensor for smartphone diagnostics: an application of blood typing, Anal. Chem. 86 (22) (2014) 11362–11367.

[22] D.-.J. Cho, M.-.Y. Bae, A study on development of OTIP system using QR code based on smartphone, Int. J. Multimedia and Ubiquitous Eng. 9 (10) (2014) 261–270.

[23] Y. Cheng, Z. Fu, B. Yu, Improved Visual Secret Sharing Scheme for QR Code Applications, IEEE Trans. Inf. Forensics Secur. 13 (9) (2018) 2393–2403.

[24] J.-.K. Lee, I.-.S. Lee, Y.-.J. Kwon, Scan &learn! use of quick response codes & smartphones in a biology field study, Am. Biol. Teach. 73 (8) (2011) 485–492.

[25] L. Benedict, H.E. Pence, Teaching chemistry using student-created videos and photo blogs accessed with smartphones and two-dimensional barcodes, J. Chem. Educ. 89 (4) (2012) 492–496.

[26] J.P. Devadhasan, S. Kim, An ultrasensitive method of real time pH monitoring with complementary metal oxide semiconductor image sensor, Anal. Chim. Acta 858 (2015) 55–59.

[27] M. Baharfar, Y. Yamini, S. Seidi, M.B. Arain, Approach for downscaling of electromembrane extraction as a lab on-a-Chip device followed by sensitive red-green-blue detection, Anal. Chem. 90 (14) (2018) 8478–8486.

[28] D. Zhang, J. Jiang, J. Chen, Q. Zhang, Y. Lu, Y. Yao, et al., Smartphone-based portable biosensing system using impedance measurement with printed electrodes for 2,4,6-trinitrotoluene (TNT) detection, Biosens. Bioelectron. 70 (2015) 81–88.

[29] C.Shad Thaxtona, R. Elghanian, A.D. Thomas, S.I. Stoeva, J.-.S. Lee, Norm D. Smitha, et al., Nanoparticle-based bio-barcode assay redefines "undetectable" PSA and biochemical recurrence after radical prostatectomy, Proc. Natl Acad. Sci. 106 (44) (2009) 18437–18442.

[30] J. Zhang, I. Khan, Q. Zhang, X. Liu, J. Dostalek, B. Liedberg, et al., Lipopolysaccharides detection on a grating-coupled surface plasmon resonance smartphone biosensor, Biosens. Bioelectron. 99 (2018) 312–317.

[31] D. Gallegos, K.D. Long, H. Yu, P.P. Clark, Y. Lin, S. George, et al., Label-free biodetection using a smartphone, Lab Chip 13 (11) (2013) 2124–2132.

[32] F. Giavazzi, M. Salina, E. Ceccarello, A. Ilacqua, F. Damin, L. Sola, et al., A fast and simple label-free immunoassay based on a smartphone, Biosens. Bioelectron. 58 (2014) 395–402.

[33] A. Roda, E. Michelini, L. Cevenini, D. Calabria, M.M. Calabretta, P. Simoni, Integrating biochemiluminescence detection on smartphones: mobile chemistry platform for point-of-need analysis, Anal. Chem. 86 (15) (2014) 7299–7304.

[34] Q. Mei, H. Jing, Y. Li, W. Yisibashaer, J. Chen, B.Nan Li, et al., Smartphone based visual and quantitative assays on upconversional paper sensor, Biosens. Bioelectron. 75 (2016) 427–432.

[35] M. Rezazadeh, S. Seidi, M. Lid, S. Pedersen-Bjergaard, Y. Yamini, The modern role of smartphones in analytical chemistry, Trends Anal. Chem. 118 (2019) 548–555.

[36] J. Il Hong, B.Y. Chang, Development of the smartphone-based colorimetry for multi-analyte sensing arrays, Lab Chip 14 (10) (2014) 1725–1732.

[37] N. Lopez-Ruiz, V.F. Curto, M.M. Erenas, F. Benito-Lopez, D. Diamond, A.J. Palma, et al., Smartphone-based simultaneous pH and nitrite colorimetric determination for paper microfluidic devices, Anal. Chem. 86 (19) (2014) 9554–9562.

[38] A. Orth, E.R. Wilson, J.G. Thompson, B.C. Gibson, A dual-mode mobile phone microscope using the onboard camera flash and ambient light, Sci. Rep. 8 (1) (2018) 1–8.

[39] P. Edwards, C. Zhang, B. Zhang, X. Hong, V.K. Nagarajan, B. Yu, et al., Smartphone based optical spectrometer for diffusive reflectance spectroscopic measurement of hemoglobin, Sci. Rep. 7 (1) (2017) 1–7.

[40] S.-J. Yeo, K. Choi, B.T. Cuc, N.N. Hong, D.T. Bao, N.M. Ngoc, et al., Smartphone-based fluorescent diagnostic system for highly pathogenic H5N1 viruses, Theranostics 6 (2) (2016) 231.

[41] D. Jung, J.-.H. Choi, S. Kim, S. Ryu, W. Lee, J.-.S. Lee, et al., Smartphone-based multi-contrast microscope using color-multiplexed illumination, Sci. Rep. 7 (1) (2017) 1–10.

[42] S. Lee, G. Kim, J. Moon, Performance improvement of the one-dot lateral flow immunoassay for aflatoxin b1 by using a smartphone-based reading system, Sensors 13 (2013) 5109–5116.

[43] L. Bueno, G.N. Meloni, S.M. Reddy, T.R.L.C. Paix~ao, Use of plastic-based analytical device, smartphone and chemometric tools to discriminate amines, RSC Adv. 5 (2015) 20148–20154.

[44] M.O. Salles, G.N. Meloni, W.R. De Araujo, T.R.L.C. Paix~ao, Explosive colorimetric discrimination using a smartphone, paper device and chemometrical approach, Anal. Methods 6 (7) (2014) 2047–2052.

[45] A.S.M. Mosa, I. Yoo, L. Sheets, A Systematic Review of Healthcare Applications for Smartphones, BMC Med. Inform. Decis. Mak. 12 (1) (2012) 67.

[46] T. Alawsi, Z. Al-Bawi, A review of smartphone point-of-care adapter design, Engineering Reports 1 (2) (2019) e12039.

[47] E. Tekin, D. Vásquez, J.M. Coughlan, Smartphone Barcode Reader for the Blind, Journal on technology and persons with disabilities, Annual International Technology and Persons with Disabilities Conference, NIH Public Access, Vol. 28 2013.

[48] K. Ming, J. Kim, M.J. Biondi, A. Syed, K. Chen, A. Lam, et al., An integrated quantum dot barcode smartphone optical device for wireless multiplexed diagnosis of infected patients, ACS Nano 9 (3) (2015) 3060–3074.

[49] C. Ding, Q. Zhang, S. Zhang, An electrochemical immunoassay for protein based on bio bar code method, Biosens. Bioelectron. 24 (8) (2009) 2434–2440.

Current applications of colourimetric microfluidic devices (smart phone based) for soil nutrient determination

Ying Cheng[a,c], Reuben Mah Han Yang[b,c], Fernando Maya Alejandro[b,c], Feng Li[b,c], Sepideh Keshan Balavandy[b,c], Liang Wang[a,c], Michael Breadmore[b,c], Richard Doyle[c,d], Ravi Naidu[a,c]

[a]Global Centre for Environmental Remediation, College of Engineering, Science and Environment, University of Newcastle, Callaghan, NSW, Australia
[b]The School of Nature Science, University of Tasmania, Sandy Bay, TAS, Australia
[c]CRC for High Performance Soils, University Drive, Callaghan, NSW, Australia
[d]Tasmanian Institute of Agriculture, Private Bag 98, Hobart, TAS, Australia

7.1 Introduction

Soil nutrient concentration is one of the indicators of soil health and a key factor of fertilizer placement in agricultural soil. The inputs of fertilizers, especially the inorganic fertilizers, are crucial to global agriculture and food security [1]. Nitrogen (N) and phosphorus (P) are major essential nutrients in the soil since they are both significant components of nucleic acids that support the growth and reproduction of plant cells. These two nutrients (N and P) are also involved in several key functions in plants such as energy transfer, photosynthesis, and nutrient movement. Soil pH is also a crucial parameter of soil function, as it affects the availability of nutrients from the soil to plant [2]. An adequate net amount of cations such as calcium (Ca^{2+}), magnesium (Mg^{2+}), and potassium (K^+) in the soil is indispensable for plant growth as well. To gain a sustainable and consistent yield with high quality, modern agriculture focusses on soil health management, which regularly analyses soil parameters and makes an informed decision based on the analysis results. Therefore, the efficiency and accuracy of monitoring methods are of paramount importance.

Traditional methods of soil chemistry analysis are based on the 'grab-to-lab' approach of collecting a sample and sending it to a laboratory for analysis by gas and liquid chromatography and mass spectrometry. These are expensive lab instruments and required well trained technical personnel to operate, making this approach both financially and time expensive [3]. Although these traditional methods can detect low concentrations with high accuracy, it does not meet the requirements for regular monitoring of soil nutrients. To address this gap, several rapid testing kits have been developed, but the results are subjective and based on the individual judgment of the person doing the test on-site. Colourimetric methods are feasible, low-cost, and straightforward and are one of the best options for the in situ test [3]. The colourimetric detection relies on a color change given by interaction or reaction between the analyte and reagents to get a rapid and readable result [4]. The accuracy of the test is affected by color perception of the user and awareness in sample preparation procedures, both of which can cause variabilities in the results. Besides, these commercial test kits can also only give an approximate range of each nutrient rather than a quantitative amount [5].

One way in which traditional assays can be improved is through the use of the latest technology, such as digital data processing techniques and microfluidics, to improve the quality and reliability with which highly accurate results can be obtained at point-of-collection. In particular, smartphones and digital cameras offer a more accurate measurement platform with higher color sensitivity to use the colourimetric methods for soil nutrients detection. As a vital gadget in modern life, the smartphone, in particular, is swiftly developing new technology features, including both hardware and applications, and importantly also features telemetry and GPS positioning capability for easy spatial mapping of the field. These new features and the widespread use of smartphones make it a preferred platform to integrate conventional approaches with creative techniques [6].

While this removes and improves the subjective visual components of using the kits, it does not improve the accuracy and reliability of the chemistry. This can be done via microfluidics through the creation of a 'lab-on-a-chip'. Microfluidic devices, especially three-dimensional (3D) printed devices, provide a cost-effective system for the achievement of lab-on-chip technology [7]. The in-field detection could be achieved by using 3D print microfluidic devices for the colourimetric test because of the ease with which complicated laboratory functions can be seamlessly integrated into a simple device that can be operated by anyone. When the output is then read and recorded with a mobile phone, this will ensure high quality and reliable data is obtained quickly and at a low cost. With good analytical data, it is then possible through an app to analyse, interpret, and summarise the testing data, and ultimately provide suggestions for fertilizing based for each specific site based on the monitored, and potentially long-term, soil nutrient levels.

In this chapter, we focus on the reviewing of chemical reagents for colourimetric detection as well as the possible interference in both colourimetric reaction and digital data processing using mobile phones. The recent progress of microfluidic devices used in colourimetric testing is also reviewed.

7.2 Colourimetric methods applied to soil chemistry detection

7.2.1 Colourimetric reagents

7.2.1.1 pH

pH, as an important chemical indicator of soil health, is the mainstay to soil fertility which can affect the biodiversity, the availability of different soil nutrition, and the run-off of fertilizers [8]. However, soil pH values can be significantly different within very short distances, which stimulate the development of colourimetric approaches as a rapid measurement. Colourimetric methods have been applied widely to determine pH in several studies [8–12]. Soil pH is mostly in the range of 4 to 8, but for most crop production systems, a neutral soil pH is considered between 6.0 to 7.0 which is favored for most crop types [75].

There are various color reagents that provide different colors to indicate pH values Fig. 7.1. Methyl red (pK_a = 5.10) is red at low pH and yellow at high pH [12], bromocresol violet (pK_a = 6.17) is coloured yellow below pH 5.2 and violet above pH 6.8 [13], bromothymol blue (pK_a = 7.07) is yellow in acid, blue in alkaline and green in neutral solution, while phenol red (pK_a = 8.00) has a color change from yellow to red over a pH range of 6.8 to 8.2 and becomes bright pink above pH 8.2 [11]. Phenolphthalein is pink in a pH range from 8.2 to 10.0 but becomes colorless in both acidic solutions (pH from 0 to 8.2) and strong basic above pH 10.0 [14]. Thymol blue is red when pH below 1.2, and changes to yellow between pH 2.8 and 8.0, then becomes blue when pH above 9.6 [15]. Most of the formulations use a mixture of several color reagents and are referred to as universal indicators, to provide several smooth color changes over a broader range of pH values [8–10]. The universal pH indicator, which is a typical combination includes thymol blue, methyl red, bromothymol blue, and phenolphthalein is versatile and displays a variety of colors over the whole pH range. However, it requires new digital color processing methods to correct and modify the resolution and color intensities of images, and to improve the credibility of pH detecting results. For instance, a new smartphone tutor application has been developed to quantify the pink color in the titration solution of phenolphthalein-based acid-base titrations with the smartphone camera [14].

7.2.1.2 Nitrogen

The common forms of nitrogen in natural soil exist as nitrate and ammonium. The Griess diazotization reaction is a well-known colourimetric method used extensively to determine nitrites and nitrates for various sample types, such as body fluids and soil samples [16–18].

The Griess reagent is comprised of an aniline compound, i.e. sulfanilimide or sulfanilic acid, and a color dye N-(1-naphthyl) ethylenediamine (NED), which detect nitrite ions in an acidic medium usually with phosphoric acid or hydrochloric acid. The mechanism of the Griess reagent begins with binding of nitrite ions with the aniline compound in an acidic medium to form a diazonium salt. The diazonium salt is then coupled with N-(1-naphthyl) ethylenediamine (NED) to form a so-called "azo-dye" that is pink/purple [19] (Fig. 7.2). The color intensity formed is proportional to the concentration of nitrite ions in the sample. As the Griess reagent's detection is specific to nitrites, the determination of nitrates involves a chemical reduction process that turns nitrate into nitrite before detection by the Griess reagents [20,21]. This reduction process can be done through a reduction by cadmium. However, due to the hazardous nature of cadmium, zinc is used as a safer alternative [21]. Interferences in the quantification of nitrite and nitrate with the modified Griess assay by ions, such as Ag^+, Fe^{3+}, Hg^{2+}, F^- at a low concentration (<10 mg/L) and SO_4^{2-} (<20 mg/L) and Cl^- (<50 mg/L) could cause high error percentages [22].

106 Smartphone-Based Detection Devices

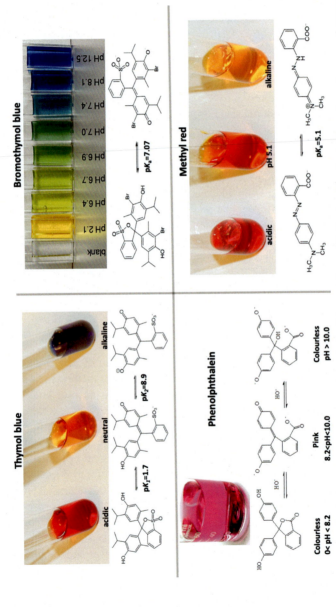

Fig. 7.1 Colour reagents in universal indicator that provide different colors to indicate pH values. The colour change and structure change of colour reagents at different pH (thymol blue, bromoiliymol blue, phenolphthalein, and methyl red).

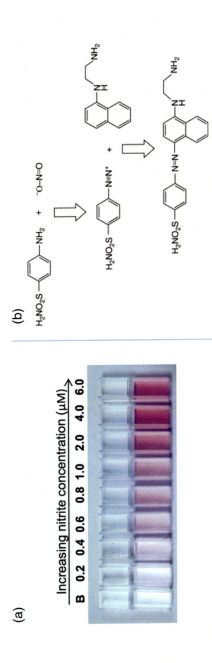

Fig. 7.2 Griess reaction for nitrite detection. (a) The colour change with increasing nitrite concentration for Griess reaction [79] (b) Schematic diagram representing the Griess reaction principle [80].

Other nitrite methods that include s-tetrazine-based chromogenic chemistry are listed by Wang et al. [23]. Nitrite has also been detected using s-dihydrotetrazine as an alternative to the Griess method. The direct measurement of nitrates with the diphenylamine reagent, the ferrous sulfate heptahydrate (iron(III)) reagent have also been reported. However, the use of concentrated sulfuric acid makes these reagents only suitable under strict laboratory conditions.

Ammonium/ammonia is commonly detected using Nessler's reagent [24,25]. Nessler's reagent is an alkali solution, which uses a 90 mmol/L potassium tetraiodomercurate (II)($K_2[HgI_4]$) solution in 2.5 mol/L potassium hydroxide. The method deprotonates ammonium to ammonia by rising solution pH with potassium hydroxide. Then ammonia reacts with potassium tetraiodomercurate (II) and potassium hydroxide to form a colloid (Iodide of million's base) which changes from yellow to brown with the increasing ammonia concentration. Besides, zinc sulfate or ethylenediaminetetraacetic acid (EDTA) can be used to eliminate the turbidity-causing substances while the excess amount of Nessler reagent needs to be added to ensure the reaction between ammonium and Nessler reagent [26]. Another ammonium detection method is the phenol hypochlorite method, in which phenol and hypochlorous acid react with ammonium to create a blue-coloured compound [25]. However, the toxic nature of phenol limits the application in-field.

7.2.1.3 Phosphorus

Determination of phosphate by the molybdenum blue method (MB) is the most frequently performed colourimetric detection assay for both water and soil samples [5,27,28]. The mechanisms involve two parts: first, the formation of orthophosphate and molybdate under acidic conditions, commonly with strong acids since the acid and molybdate concentration are key factors in the reaction, then the reaction with reductants, such as ascorbic acid and Sb(III), or $SnCl_2$ and $N_2H_6SO_4$, to form phosphomolybdenum blue (PMB) species [27,29] Fig. 7.3. The blue color arises because the near-colorless phosphor-molybdate anion, $PMo_{12}O^{3-}_{40}$, can accept more electrons to form an intensely coloured mixed-valence complex, such as $Sb_2PMB(4e^-)$ and $SnPMB(4e^-)$ [30]. Nagul et al. [27] comprehensively reviewed and listed all of the known chemical interferences for phosphate colourimetric determination with the MB method. For instance, a major source of interference is chloride, known as 'salt error' [31], which has been found to cause up t 20 percent error when applied to marine waters [32]. It is worth noting there are other chemicals, including AsO_4^{3-}, $[SiO_{4-x}^{(4-2x)-}]_n$, Fe^{3+}, might also affect the MB reaction [27]. Fortunately, the concentrations of these chemicals usually are not high enough to cause significant interferences when measuring a natural water sample without contamination. Thus, sulfuric acid is generally applied in the MB reaction to maintain the measuring solution at a very low pH to avoid excessive direct reduction of Mo (VI).

7.2.1.4 Sulphide

A typical method for the determination of hydrogen sulphide is the methylene blue method. In an acid condition (hydrochloric acid), sulphides react with

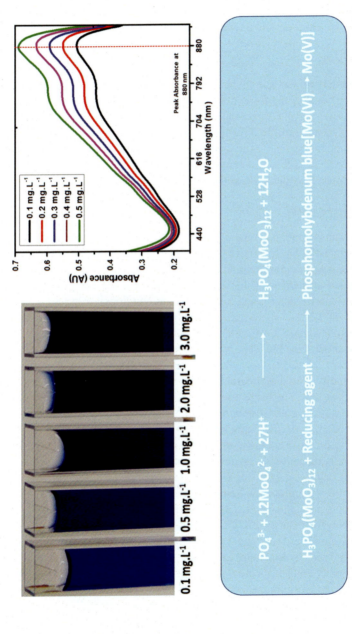

Fig. 7.3 Images and absorption spectra of the reagent treated phosphate solution with molybdenum blue method [61]. Phosphate standard solutions with different concentrations and the absorption spectra of each concentration were shown on the top, with the mechanisms of the molybdenum blue method listed below.

N,N-dimethyl-p-phenylenediamine (DMPD) oxalate to from methylene blue, which proportional to the sulphide concentration [25,33]. However, the limitations have been reported as the method lacks sensitivity at low sulphide concentrations and does not obey Beer's law [34] possibly due to a loss of volatile hydrogen sulphide. It also suffers interference from protein and acid-labile sulphides [34].

7.2.1.5 Chloride

Chloride can be determined using ferric ammonium sulfate and mercuric thiocyanate [35]. The orange color of the ferric thiocyanate cation complex was determined colourimetrically. Alternatively, mercuric nitrate with s-diphenylcarbazide/diphenylcarbazone can also be applied [36,37]. In acid solution pH approx. 3.3, the chloride is found to react with the mercuric ion, thus preventing any color from being developed; however, as soon as an excess of mercuric ions occurred and reacted with s- diphenylcarbazide/diphenylcarbazone, blue-violet color developed.

7.2.1.6 Exchangeable cations

Uranyl zinc acetate is applied to quantify sodium, as sodium is precipitated as sodium uranyl zinc acetate $(UO_2)_3ZnNa(CH_3CO_2)_9 \cdot 6H_2O$, in natural or acid pH [38]. The uranium is determined colourimetrically with potassium ferrocyanide. However, the color developed is unstable and the reaction involves a tedious precipitation step [38]. Potassium can be determined under alkaline conditions using sodium tetraphenylborate, $(C_6H_5)_4BNa$, which reacts with the potassium ions to form potassium tetraphenylborate $(C_6H_5)_4BK$ which is white and has a poor solubility in water thus forming colloidal particles [39]. The turbidity is monitored and is proportional to the potassium ion concentration, thus making it limited to samples that do not have any appreciable turbidity. The dipicrylamine reagent has also been used for rapid in situ measuring of potassium. The reagent involves the chelation of potassium ions to form orange/red crystals. The color formation can be stabilized with 0.7 percent nitric acid for improved readability (Ofite potassium ion strip kit; part NO. 147–90).

The colourimetric measurement of calcium and magnesium are based on two mechanisms, firstly chelation of the desired ions followed by a colourimeric reaction. The chelation of calcium is possible with ethylene-diamine-tetra-acetic acid(EDTA), ethylene glycol-bis(β-aminoethyl ether)-N,N,N′,N′-tetraacetic acid(EGTA), and 1,2-bis(o-aminophenoxy)ethane-N,N,N′,N′-tetraacetic acid(BAPTA) for calcium and 8-hydroxyquinone for magnesium. Upon chelation of the desired ions, the binding with a color dye provides the colourimetric reaction. The reagents of methylthymol blue and glyoxal bis(2-hydroxyanil) class were used to chelate calcium and yield a blue and a red-violet color respectively in a high pH (above 11) that aviods interfereces [40,41]. For magnesium, magon or xylidyl blue forms a red complex with magnesium in alkaline solution [42]. Calcium and magnesium can both be determined by chelation with chlorophosphonazo-III at neutral pH, but only calcium was complexed at pH 2.2 with absorption maximum at 667.5 nm [43]. Therefore, calcium concentration is determined at pH 2.2 while the magnesium concentration need be calculated by the difference between pH 7.0 and 2.2 at 669 nm. Chelating agents for calcium and magnesium can be interchangeable, for instance using a calcium chelator to mask interferences by calcium when measuring magnesium vice versa. Nevertheless, the direct analysis of these exchangeable cations is possible.

7.2.2 Reaction temperature and time

The external environment is another challenge to maintain the accuracy of point of care testing method. Natural effects, such as temperature and luminous intensity, could severely affect the reaction rate, sensitivity, and reliability of the colourimetric reaction. Merely providing a waiting time for measurement without considering the thermal energy condition might cause a wide range of errors. For instance, the universal indicator, which used in pH measurement, is optimized to 20 °C in lab-based predictions. In methylthymol blue method for calcium detection, the signal intensity drops 1.6 times from 10 to 45 °C since the color development is sensitive to temperature changes [44]. However, in natural water, the temperature can range from zero (below zero for seawater) up to 50 °C [45]. Therefore, the conditions for reaction kinetics, such as temperature and reaction time, should also be considered as the causative factors for prediction errors. However, the effects of reaction conditions mainly depend on the fundamental law of each chemical reaction. It is also necessary to follow the sequence of reagent addition order and amount in colourimetric determination.

Another example, the Griess reaction assay-based methods are widely used in nitrate and nitrite determination. It has been reported that the high concentration of hydrogen ion increases the diazotization rate and lower the coupling rate, since the hydrogen ion act first as a reactant and then as a product in principle color reaction [46]. Hence, the best reaction time is determined by the acidities in the whole reaction system. In a modified Griess reaction, the signal response gets higher with the increase in the concentration of the acidic solution and nitrosating agent, and the signal of the azo dye band peaks at 15 min after mixing then stays stable for 30 min. Although the increase of temperature can accelerate the reaction rate in the Griess reaction, the temperature is not shown to be a determining factor in the modified Griess reaction [22].

In molybdenum blue (MB) method for phosphate determination, the blue color arises because the near-colorless phosphor-molybdate anion, $PMo_{12}O^{3-}_{40}$, can accept more electrons, i.e., be reduced, to form an intensely coloured mixed-valence complex [27]. Based on the reaction rate, after a specific time, yellow precipitations, ammonium phosphor-molybdate, will be formed, and MB blue color will change. The time for the yellow precipitation can be varied based on the reaction thermal energy. Molecules at a higher temperature have a higher speed of interaction and so more rapid reaction rates occur. The 'rule of thumb' in chemistry is that the rate of chemical reactions doubles for every 10 °C [47]. Usually, the commercial product, which based on the molybdenum blue method, suggests that before any measurements at least five minutes are needed for most reactions to occur, but all reaction equilibria need to be adjusted to the natural conditions. Therefore, a field-based quantification system with calibration curves is necessary.

7.3 Digital image capture and image processing for smartphone analysis

7.3.1 Digital color models

A digital camera is a powerful tool for image acquisition. Color cameras capture images in RGB (red, green, and blue), and models based on these have been widely used in image processing and color communications. Researchers have applied the intensities of the RGB colors to determine various analytes, including pH [12,48], glucose [18], and trinitrotoluene (TNT) in soil [49]. However, the RGB signals generated by a digital camera are device-dependent (G. [50]). The color elements and their response to the individual R, G, and B levels vary from manufacturer to manufacturer, or even in the same device over time. Hence, the calibration using the RGB model usually is device-dependent. To overcome the issue, Jia et al. [18] developed a self-background correction model for calibration, and the digital RGB values were corrected by setting up black and white sports as background values to limit the errors caused by ambient light and camera differences with smartphone brands. Another shortcoming of the RGB model is the composition of RGB - one digitized color does not always change monotonically with spectral wavelength and intensity [51]. To obtain monotonic variations, the ratios between the three RGB intensities can be applied. For instance, the ratio of G/R and R/B were applied to determine pH values by different research teams [12] (Fig. 7.4).

Given the limitations of using RGB, two popular color models have been developed that convert these into other parameters used in color. These are HSV (hue, saturation, and value) and the Commission Internationale d'Eclairage (CIE). HSV, also known as HSL (hue, saturation, lightness), is derived from RGB with easy mathematical transformations [52]. Hue (H) and saturation (S) are the parameters from HSV that have been predominantly applied colourimetric applications. The CIE color model uses three variables created from the RGB values using the formulat listed in [50,53]: X represents the linear combination of the cone response non-negative curves created, Y represents luminance/lightness, and Z is defined as almost equal to blue. XYZ does not incorporate negative numbers and instead uses the values to define the color space that can be seen by the human eye. CIE XYZ is a perceptually uniform color space where the luminance/grayscale (Y) is mainly applied in colourimetric applications rather than the other two values [8,21].

With potential applicability to solid chemistry measurement, a hue-based pH determination method was reported, and it aimed to do rapid analysis for a large number of samples, such as to analyse pH results of a 384-well plate in one digital image with an error smaller than 0.23 pH units [9]. Another team reported that the hue value could be used in potassium detection with a disposable membrane and optical sensor [54]. RGB data provided monotonic changes to determine phosphorus using the 'molybdenum blue' reaction [27]. Moonrungsee et al. [5] applied a blue intensity to determine phosphorus in their application.

Current applications of colourimetric microfluidic devices (smart phone based) 113

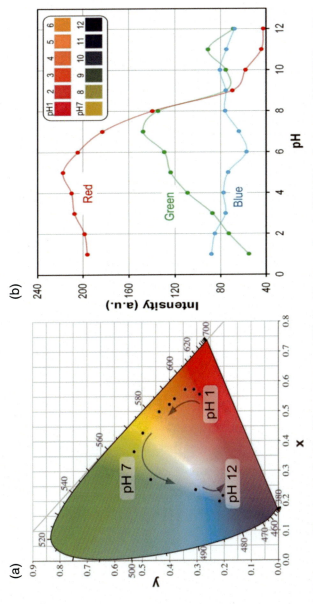

Fig. 7.4 Colour qualification method in pH measurement [8]. (a) pH paper colors in CIE1931 colour space (b) RGB intensities of pH paper color at each pH value.

7.3.2 Background illuminance interference

Although nearly all smartphones' cameras use complementary metal-oxide-semiconductor (CMOS) sensors [55], the vast range of variations in-camera function and sensitivity across different smartphones could be the primary source of error during the point of care detections. In photography, the sensitivity of a camera refers to the sensitivity of the digital camera sensor to light. Therefore, ambient light is one of the main physical deviation factors for densitometry. According to the Beer-Lambert law for absorption, the ambient light is one type of stray light that can interfere with the light with the correct wavelength to reach the detector [56]. It has been proved that the densitometric analysis can be impacted dramatically by different ambient light, especially when the illuminance values are beyond 500 lx [57]. The intensity of ambient light could seriously affect the pixel density of the picture obtained from a mobile phone camera by lowering the precision and accuracy of the measurement [5].

Accordingly, well-controlled constant illumination will produce the best results with the highest precision. Several solutions have been developed to reduce the ambient light impact (Fig. 7.5). Most of the studies avoid ambient light effects by

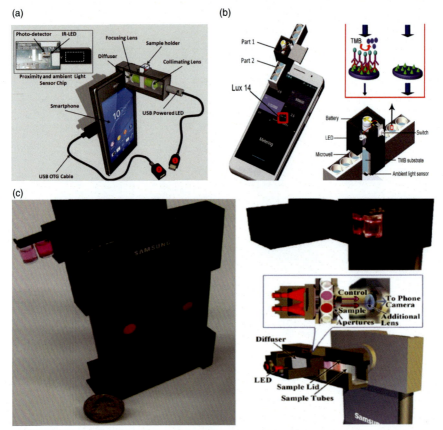

Fig. 7.5 Smartphone accessory for colorimetric analysis. (a) Schematic of the smartphone-based optical device [61], (b) Mechanism of smartphone colourimetric reader [3], (c) The picture and schematic diagram of iTube platform [77].

introducing a light-tight box with LED flash to control the background illumination [58]. To improve the measurement precision and accuracy, an ambient light sensor was designed to measure the transmitted light intensity then adjust the phone screen brightness [3,59]. The fluoride sensor for fluoride concentration detection in drinking water was developed by using the built-in smartphone LED flashlight with an ambient light intensity sensor [60]. Based on the previous study, Hussain et al. [61] designed a photodetector by combining the ambient light sensor with an external LED, this detector can analyse both iron (II) and phosphate concentrations simultaneously with the detection limits of 0.053 mg/L and 0.069 mg/L respectively.

There are other solutions to reduce or compensate for the interference from high ambient light that does not involve the use of a specific lightbox and light meter. The RGB color model can be converted into other absolute color spaces to analyse small changes in color, such as HSV and CIE (XYZ) [62]. For instant, García et al. [52] converted the RGB color model into hue (H) values for measuring potassium to enhance the precision for quantitative analysis and added two additional illumination lamps to avoid ambient light interferences. They reported that when the illumination conditions are homogeneous and well maintained, the influence of other physical factors, such as the distance between target and phone, image focusing, and centering, is negligible at a distance from 5 cm to 50 cm. Choodum et al. [49] converted the RGB color values into a CIE (XYZ) coordinate system for soil color measurement. The built-in flash was used as the only light source with a black PDMS diffuser, which was added to provide uniform illumination, and could avoid the harsh gradients in light intensity [6].

7.4 Microfluidic devices for colourimetric nutrient analysis

The challenge in performing chemical measurements in the field is the implementation of the myriad of processes that occur within the laboratory. This begins with the many laborious and manual sample handling steps and culminates in the use of bulky and complex instruments – making the analysis very time-consuming and costly. Emerging in the late '80s from the MEMS sector was the concept of a 'lab-on-a-chip' in which chemical processes are combined within micron-sized chambers within a single 'chip' to create a sample-in/answer-out platform for chemical (and biochemical) measurement. This is now more broadly known as the field of microfluidics, and this is an ideal approach for soil analysis, because of the ability to integrate all of the steps required for soil analysis in a simple, low-cost and rapid system for point-of-collection measurement. There are many different ways and materials with which microfluidic devices can be made. A few of the more common, and with relevance to soil chemistry analysis, are discussed below.

7.4.1 Polymer-based microfluidic devices for soil analysis

Polymer-based microfluidic devices are attractive because of the diversity of polymer materials that can be selected and their compatibility with mass replication techniques such as injection molding and hot embossing – which theoretically allows them to be produced for a fraction of a cent each. Over the past few years, researchers have explored various polymer materials with polydimethylsiloxane (PDMS) being the most widely used. Dudala et al. [63] presented an integrated, low-cost PDMS-based microfluidic device for the detection of nitrite, pH, and electrical conductivity(EC) within a single device. Nitrite was detected via the Griess reaction by photometric detection using a LED and photodiode, and pH detection was achieved using a commercially available pH probe. EC was detected by using a conductivity cell with copper electrodes powered by an oscillating power source [63]. PDMS offers good optical transparency making it suitable for colorimetric detection, while the fabrication process does not require a clean-room environment. Polymer-based centrifugal microfluidic devices have also been used for colorimetric analysis. Centrifugal microfluidics uses centripetal fluid movement as a result of rapid rotation to perform a comprehensive set of fluidic unit operations such as liquid transport, metering, mixing, and valving. Hwang et al. [64] showed a disk for the colorimetric detection of various nutrients in water. The overall analytical procedure was automatically carried out on a rotating disc-shaped device (Fig. 7.6). The sample and reagents were pumped by centrifugal force in the rotating disc, and their positions and movement controlled through a programmable light from a laser diode. The advantages of these devices are the low sample and reagent consumption and the possibility to run the measurement of the different analytes in parallel. The same device was also used for measuring nutrients in soil [64]. Simultaneous measurement of different analytes is an advantage of centrifugal microfluidics due to its multiple channel design, but it usually requires the external device providing the centrifugation force to drive the device, which compromises its portability. The idea proposed by Bhamla et al. [65] using hand-powered spinning inspired by fundamental mechanics of an ancient whirligig could be an alternative to increase the portability of the centrifugal microfluidics, but controlling the speed, and therefore the forces, is challenging with this approach.

7.4.2 Microfluidic paper-based analytical devices (µPADs) for soil analysis

Paper is the basis of many field test kits and is ideal for field analysis because of its ability to wick fluid. The advent of microfluidics has seen a renasenace in paper as a substrate for chemistry because of this ability, and the possibility to make specific 2 and 3D fluid pathways through the deposition, e.g. printing, of hydrophobic regions as well as being able to deposit reagents in specific locations. Recent colourimetric applications with microfluidic devices have been summarized in Table 7.1 and paper-based microfluidic devices for nutrient analysis in water samples have been recently reviewed [20].

Cardoso et al. [16] reported the use of a disposable microfluidic paper-based device for colourimetric detection of nitrite in environmental samples. The device uses the Griess reaction with a change of color from colorless to strong pink in the

Current applications of colourimetric microfluidic devices (smart phone based)   117

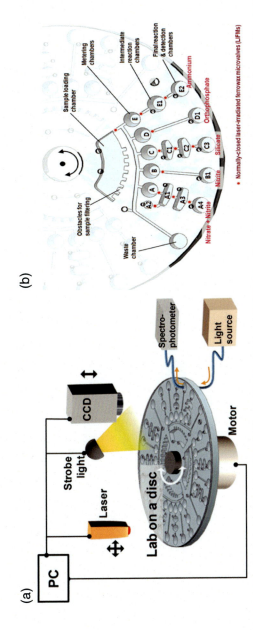

Fig. 7.6 Centrifugal microfluidic platform [64]. (a) Device for the simultaneous determination of five nutrients on a lab-on-a-disc device (b) Schematic illustration of lab-on-a-disc device.

Table 7.1 Recent colourimetric applications using digital colour models.

Analyte	Reagents	Colour model	Regression model	Camera	Device (sensor)	Light Corrections	Ref.
Phosphorous	MB blue	RGB	Blue intensity	Nokia X	Quartz Cuvette	LED light	[5]
Phosphorous	MB blue	RGB	Red intensity	Canoscan™ Lide 700f	Paper microfluidic	NA	[67]
Phosphorous	MB blue	RGB	Red intensity	Canon P4700	Paper microfluidic	UV resistant laminating	[76]
Nitrate	Griess	CIE	Illuminance	Canon 60D	3D printed microfluidic	CIE illuminance	[78]
Nitrate	Griess	CIE	Illuminance	Scanner Hewlett-Packard, G4050	Paper microfluidic	NA	[16]
Nitrate	Griess	HSV	Saturation	iPhone 4	Paper microfluidic	LED light	[66]
Nitrate	Griess	HSV	Hue and Saturation	Samsung Galaxy SII	Paper microfluidic	Camera flash	[11]
Nitrite & Nitrate	s-dihydrotetrazine	RGB	Absorbannce Ave (RGB)	Scanner Canoscan Lide 700f	Paper microfluidic	Inferences, Colour development time	[17]
pH	Universal pH indicator	CIE	X and Y	HTC and BlackBerry	Paper Strip	X & Y (CIE) calibration	[8]
pH	Universal pH indicator	HSV	Hue	Samsung Galaxy	Paper Strip	White balance	[10]
pH	Phenol red, cholorophenol red	HSV	Hue and Saturation	Samsung Galaxy SII	Paper microfluidic	Camera flash	[11]
pH	Universal pH indicator	HSV	Hue	iPhones	3D printed strip	Camera flash and White balance	[6]
Nitrite	Griess	HSV	Saturation	iPhone 4	Paper microfluidic	LED light	[66]
Nitrite	Griess	HSV	Hue and Saturation	Samsung Galaxy SII	Paper microfluidic	Camera flash	[11]

Current applications of colourimetric microfluidic devices (smart phone based) 119

Analyte	Reagent	Colour space	Colour value	Phone/Camera	Platform	Inferences, Colour development time	Ref
Nitrite & Nitrate	s-dihydrotetrazine	RGB	Absorbance Ave(RGB)	Scanner Canoscan Lide 700f	Paper microfluidic	Inferences, Colour development time	[17]
pH	Universal pH indicator	CIE	X and Y	HTC and Blackberry	Paper strip	X & Y (CIE) calibration	[8]
pH	Universal pH indicator	HSV	Hue	Samsung Galaxy	Paper strip	White balance	[10]
pH	Phenol red, chlorophenol red	HSV	Hue and Saturation	Samsung Galaxy SII	Paper microfluidic	Camera flash	[11]
pH	Universal pH indicator	HSV	Hue	iPhone	3D printed strip	Camera flash and White balance	[6]
pH	UCPs and Neutral red	RGB	G/R ration	Canon EOS 50D	Quartz Cuvette	Intensive auto-fluorescent background	[12]
pH	Universal pH indicator	HSV	Hue	Canon EOS Rebel T1	Isoelectric	Unknown	[9]
pH & O$_2$	FITC for pH, Pt-TPFPP for pO$_2$	RGB	R/B and G/B ratios	Canon EOS 50 D	UV light	Unknown	[48]
Ammonia	Nessler, NTP	RGB	Blue intensity	Canon IXUS 125 HS	Paper microfluidic	Unknown	[25]
Sulfide	Methylene blue	RGB	Red intensity	Canon IXUS 125 HS	Paper microfluidic	Unknown	[25]
Urine	Urine test kit	HSV	Hue and Saturation	Samsung Galaxy Note 3	Paper Strip	Black and white background correction	[51]
Glucose	PBA, TMOS, MTMS, CTAB	RGB	Corrected RGB	iPhone 4; MEIZU MX2; Samsung Galaxy SII	Paper Sensor array	Black and white background correction	[18]

3-nitrophenol (NTP), Phenylboronic acid (PBA), tetramethylorthosilicate (TMOS), methyltrimethoxysilane (MTMS), Hexadecyl trimethyl ammonium bromide (CTAB), 4-aminoantipyrine (AAP) and 3,5-dichloro-2-hydroxy-benzenesulfonic acid (DHBS), Fluorescein isothiocyanate (FITC), platinum(II)-5,10,15,20-tetrakis-(2,3,4,5,6-pentafluorophenyl) porphyrin (Pt-TPFPP), N,N-dimethyl-p-phenylenediamine (DMPD) and Ferric chloride hexahydrate (FCH).

presence of nitrite. In this report, the device consists of eight star-like paths, where three of them are used as controls and the other five as analyte detection areas for standards or samples. They used a common office scanner and digital photo software for colourimetric calibration of nitrite quantification. A linear response between 0 and 100 μM with a limit of detection of 11.3 μM was obtained. The paper-based microfluidic device was applied for nitrite determination in river water. The paper-based method, when compared with a batch spectrophotometric method, showed no significant difference at a confidence level of 95 percent.

A number of paper-based devices for the colourimetric analysis of a single analyte such as ammonia [25], nitrite [66], and phosphate [67] have been developed. Ammonia changes to volatile ammonia gas by adding 2.5 μL NaOH(1mol L^{-1}), then reacts with the Nessler reagent to form a yellow color on the acceptor layer of the μPAD. Nitrite has been detected using s-dihydrotetrazine as an alternative to the Griess Method with μPADs, in which nitrite firstly changes to nitrous acid and makes the detection zone of s-dihydrotetrazine turning into pink(absorption maximum at 472 nm) by oxidation [66]. Phosphate analysis in paper-based microfluidic devices has been based on the classic molybdenum blue reaction method, typically used in batch or automated flow-analysis of this nutrient [68]. Shibata et al. [69] reported distance-based μPADs for the pure naked-eye colourimetric determination of calcium ions (Ca^{2+}) in drinking water and tap water without any external equipment. The ion-selective optode nanospheres (nano-optodes) have been deposited onto wax-patterned paper substrates using desktop thermal inkjet printing technology for the simple quantification of Ca^{2+} relying on the visual read-out of the length of a color-changed detection channel. Notably, the μPADs provided improved lowest naked-eye detectable concentrations of Ca^{2+} (0.05 mmol L^{-1}), compared to the value achieved with previously reported paper-based devices. Similar instrument-free μPADs for the determination of cobalt in waters using distance-based read-out were fully integrated for the first time with accurate geometrical alignment between the individual steps [70]. A combination of several fabrication steps of μPADs simply achieved by accurately aligned wax printing in order to make fluidic barriers using a digital craft cutter followed by reagent deposition via plotting through the technical pens. The novel alignment provides a low-cost, simple, rapid, reproducible techniques for μPADs prototyping especially those who involve multiple steps, e.g., wax printing, cutting, and deposition.

However, the practical requirement is an essencial factor to evaluate whether the method can be applied in field test. For example, pre-treatment process, especially the prefiltering of soil samples, is a big obstacle in colourimetric detection. The paper-based microfluidic devices may lack enough selectivity for samples containing a high load of suspended solids. It is reported that even using quartz cuvettes, the filtering of each sample may take about 10 to 15 min [5]. The opaque property of the paper matrixs also can affect the detection sensitivety that makes μPAD only suitable for high concentration analysis if without sample enrichment step.

7.4.3 Three-dimensional (3D) printed microfluidic devices for soil analysis

The main advantage of 3D printing for the design of portable analytical devices is the simplicity and reproducibility of the fabrication procedure. Previously, other manufacturing techniques have been used to develop analytical microfluidic devices for nutrient analysis. While these devices enable the robust analysis of nitrate and nitrate in the field, their construction and use requirements of specialized skills due to the complexity of the system [71]. 3D printing is emerging as an advantageous manufacturing technique for rapid prototyping during development through to small-volume manufacturing of several thousands of units a week. A demonstration of this is the reported multi-material 3D printed device based on fused filament deposition which incorporated an integrated polymeric porous membrane (Fig. 7.7). The membrane was made using a composite filament containing a water-soluble polymer PLA. After 3D printing of the microfluidic device, the water-soluble polymer is removed by flushing with water, leaving a porous structure inside the microfluidic device, which can be used as a membrane to filter soil samples before approaching to color reacting reagents [21]. This device was used with the Griess reagent to detect nitrate in soil.

Similar microfluidic systems have also been applied for the high-performance microfluidic analysis of phosphate in marine waters using the vanadomolybdate method [72]. The 3D printed platform not only proposed manually actuated designed for entirely on-site water quality testing but also multiplexed detecting chambers pre-load with enzymatically targeted colourimetric reagents enable detection of multiple pathogens. Gaal et al. [73] demonstrated the potential of 3D printing techniques to fabricate electrodes of the e-tongue sensor used for soil analysis as an alternative way to simplify the complexity of traditional techniques. Using a home-made FDM 3D printer, nanostructured thin films (sensing units) were deposited into interdigitated electrodes (IDEs) via dip-coating layer-by-layer technique [74]. The frequency response of different macronutrients (N, P, K, S, Mg, and Ca) enriched soil samples verified by electrical impedance spectroscopy was in good agreement with principal component analysis.

New progress has been made on the development of microfluidic devices for colourimetric nutrient analysis based on paper, micro-milled polymers, or 3D printed polymers. 3D printing currently offers new possibilities in the design of such analytical devices. With the various benefits of 3D printing, the potential of 3D printing technology on colourimetric analysis devices remains high. It provides a stable platform of mass-fabrication at a low cost for which can enable immediate real-time and on-site measurements for quick corrective actions and interventions. Prototype design iterations using computer-aided designs (CAD) has been found to be both efficient and cost-effective. 3D printing is however still in its infancy and has a number of challenges that must be overcome. Primary amongst these is the availability of materials with different properties and chemical resistance, particularly enhanced stability against acids that will be needed to facilitate the applicability of 3D printed devices for in-field colourimetric measurements.

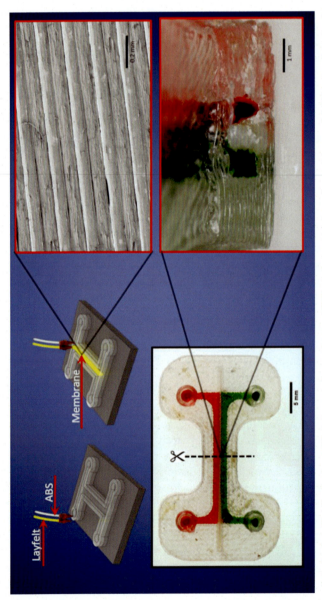

Fig. 7.7 3D printed microfluidic device with integrated membranes [21]. A demonstration of a multi-material 3D printed device based on fused filament deposition which incorporated an integrated polymeric porous membrane.

7.5 Conclusion

Colourimetric analysis is convenient for the detection of soil nutrient concentrations, especially in developing new portable in field test tools. The recent trend of in situ test is to link the data read and analysis to the smartphone platform. Therefore, microfluidic devices are investigated to take full advantage of traditional colourimetric methods with smartphone applications. Additive manufacturing (3D printing) and paper-based microfluidics are becoming increasingly popular in the development of miniaturized devices for portable colourimetric measurements. In both instances, the combination of these supports for analytical method development requires the use of smartphones for the quantification of the analytes, ensuring the portability of the overall method, from sample preparation to analyte detection.

Soil nutrient concentrations are highly related to soil health in farmland, which makes the long-term monitoring of soil nutrient level becoming one of the fundamental requirements in modern agricultural. In the case of nutrient analysis, the smartphone applications read, and analyse image data then store all the digitalized test results to form a database. The continually monitored results can be kept in the database for the modeling development to predict the runoff of soil nutrients. Meanwhile, the fertilization management can also base on the information in database to maintain the soil supply in a sustainable way for optimum yields.

Acknowledgement

We would like to thank Soil CRC for funding this project and the support from industrial partners of Herbert Cane Productivity Services Ltd and Burdekin Productivity Services Ltd, Australia.

References

[1] Karlen, D.L., Veum, K.S., Sudduth, K.A., Obrycki, J.F., & Nunes, M.R. (2019). Soil health assessment: past accomplishments, current activities, and future opportunities. Soil Tillage Res., 195. https://doi.org/10.1016/j.still.2019.104365.

[2] A. Mccauley, C. Jones, K. Olson-Rutz, Soil pH and organic matter, Nutr. Manag. Module 16 (2017) 1–12.

[3] Y. Chen, Q. Fu, D. Li, J. Xie, D. Ke, Q. Song, Y. Tang, H. Wang, A smartphone colorimetric reader integrated with an ambient light sensor and a 3D printed attachment for on-site detection of zearalenone, Anal. Bioanal. Chem. 409 (28) (2017) 6567–6574. https://doi.org/10.1007/s00216-017-0605-2.

[4] S.D. Kim, Y. Koo, Y. Yun, A smartphone-based automatic measurement method for colorimetric pH detection using a color adaptation algorithm, Sensors (Switzerland) 17 (7) (2017). https://doi.org/10.3390/s17071604.

[5] N. Moonrungsee, S. Pencharee, J. Jakmunee, Colorimetric analyzer based on mobile phone camera for determination of available phosphorus in soil, Talanta 136 (2015) 204–209. https://doi.org/10.1016/j.talanta.2015.01.024.

[6] V. Oncescu, D. O'Dell, D. Erickson, Smartphone based health accessory for colorimetric detection of biomarkers in sweat and saliva, Lab Chip 13 (16) (2013) 3232–3238. https://doi.org/10.1039/c3lc50431j.

[7] A.A. Yazdi, A. Popma, W. Wong, T. Nguyen, Y. Pan, J. Xu, 3D printing: an emerging tool for novel microfluidics and lab-on-a-chip applications, Microfluid Nanofluidics 20 (3) (2016) 1–18. https://doi.org/10.1007/s10404-016-1715-4.

[8] L. Shen, J.A. Hagen, I. Papautsky, Point-of-care colorimetric detection with a smartphone, Lab Chip 12 (21) (2012) 4240–4243. https://doi.org/10.1039/c2lc40741h.

[9] J.D. Brown, N. Bell, V. Li, K. Cantrell, Quantitative pH assessment of small-volume samples using a universal pH indicator, Anal. Biochem. 462 (2014) 29–31. https://doi.org/10.1016/j.ab.2014.06.001.

[10] B.Y. Chang, Smartphone-based chemistry instrumentation: digitization of colorimetric measurements, Bull. Korean Chem. Soc. 33 (2) (2012) 549–552. https://doi.org/10.5012/bkcs.2012.33.2.549.

[11] N. Lopez-Ruiz, V.F. Curto, M.M. Erenas, F. Benito-Lopez, D. Diamond, A.J. Palma, L.F. Capitan-Vallvey, Smartphone-based simultaneous pH and nitrite colorimetric determination for paper microfluidic devices, Anal. Chem. 86 (19) (2014) 9554–9562. https://doi.org/10.1021/ac5019205.

[12] R.J. Meier, J.M.B. Simbuürger, T. Soukka, M Schaüferling, Background-free referenced luminescence sensing and imaging of pH using upconverting phosphors and color camera read-out, Anal. Chem. 86 (11) (2014) 5535–5540. https://doi.org/10.1021/ac5009207.

[13] E. Fornasier, F. Fornasier, V. Di Marco, Spectrophotometric methods for the measurement of soil pH: a reappraisal, Spectrochimica Acta - Part A: Molecular and Biomolecular Spectroscopy 204 (2018) 113–118. https://doi.org/10.1016/j.saa.2018.06.029.

[14] B.B. Rathod, S. Murthy, S. Bandyopadhyay, Is this solution pink enough? A smartphone tutor to resolve the eternal question in phenolphthalein-based titration, J. Chem. Educ. 96 (3) (2019) 486–494. https://doi.org/10.1021/acs.jchemed.8b00708.

[15] T. De Meyer, K. Hemelsoet, V. Van Speybroeck, K. De Clerck, Substituent effects on absorption spectra of pH indicators: an experimental and computational study of sulfonphthaleine dyes, Dyes Pigm. 102 (2014) 241–250. https://doi.org/10.1016/j.dyepig.2013.10.048.

[16] T.M.G. Cardoso, P.T. Garcia, W.K.T Coltro, Colorimetric determination of nitrite in clinical, food and environmental samples using microfluidic devices stamped in paper platforms, Anal. Methods 7 (17) (2015) 7311–7317. https://doi.org/10.1039/c5ay00466g.

[17] B.M. Jayawardane, S. Wei, I.D. McKelvie, S.D. Kolev, Microfluidic paper-based analytical device for the determination of nitrite and nitrate, Anal. Chem. 86 (15) (2014) 7274–7279. https://doi.org/10.1021/ac5013249.

[18] M.Y. Jia, Q.S. Wu, H. Li, Y. Zhang, Y.F. Guan, L. Feng, The calibration of cellphone camera-based colorimetric sensor array and its application in the determination of glucose in urine, Biosens. Bioelectron. 74 (2015) 1029–1037. https://doi.org/10.1016/j.bios.2015.07.072.

[19] J. Almog, S. Zitrin, Colorimetric detection of explosives. In: Aspects of Explosives Detection, Elsevier, 2009, pp. 41–58. https://doi.org/10.1016/B978-0-12-374533-0.00004-0.

[20] M I G S Almeida, B.M. Jayawardane, S.D. Kolev, I.D McKelvie, Developments of microfluidic paper-based analytical devices (μPADs) for water analysis: a review, Talanta 177 (2018) 176–190. https://doi.org/10.1016/j.talanta.2017.08.072.

[21] F. Li, P. Smejkal, N.P. Macdonald, R.M. Guijt, M.C. Breadmore, One-step fabrication of a microfluidic device with an integrated membrane and embedded reagents by multimaterial 3D printing, Anal. Chem. 89 (8) (2017) 4701–4707. https://doi.org/10.1021/acs.analchem.7b00409.

[22] M. Irandoust, M. Shariati-Rad, M. Haghighi, Nitrite determination in water samples based on a modified Griess reaction and central composite design, Anal. Methods 5 (21) (2013) 5977–5982. https://doi.org/10.1039/c3ay40913a.

[23] Q.H. Wang, L.J. Yu, Y. Liu, L. Lin, R g Lu, J p Zhu, L. He, Z.L. Lu, Methods for the detection and determination of nitrite and nitrate: a review, Talanta 165 (2017) 709–720. https://doi.org/10.1016/j.talanta.2016.12.044.

[24] B.M. Jayawardane, I.D. McKelvie, S.D. Kolev, Development of a gas-diffusion microfluidic paper-based analytical device (μPAD) for the determination of ammonia in wastewater samples, Anal. Chem. 87 (9) (2015) 4621–4626. https://doi.org/10.1021/acs.analchem.5b00125.

[25] P. Phansi, S. Sumantakul, T. Wongpakdee, N. Fukana, N. Ratanawimarnwong, J. Sitanurak, D. Nacapricha, Membraneless gas-separation microfluidic paper-based analytical devices for direct quantitation of volatile and nonvolatile compounds, Anal. Chem. 88 (17) (2016) 8749–8756. https://doi.org/10.1021/acs.analchem.6b02103.

[26] H. Jeong, J. Park, H. Kim, Determination of NH+ in environmental water with interfering substances using the modified nessler method, J Chem (2013). https://doi.org/10.1155/2013/359217.

[27] E.A. Nagul, I.D. McKelvie, P. Worsfold, S.D. Kolev, The molybdenum blue reaction for the determination of orthophosphate revisited: opening the black box, Anal. Chim. Acta 890 (2015) 60–82. https://doi.org/10.1016/j.aca.2015.07.030.

[28] P.J. Worsfold, L.J. Gimbert, U. Mankasingh, O.N. Omaka, G. Hanrahan, P C F C Gardolinski, P.M. Haygarth, B.L. Turner, M.J. Keith-Roach, I.D McKelvie, Sampling, sample treatment and quality assurance issues for the determination of phosphorus species in natural waters and soils, Talanta 66 (2) (2005) 273–293. https://doi.org/10.1016/j.talanta.2004.09.006.

[29] Pamela, D. (2014). Ascorbic acid method for phosphorus determination.

[30] The Radiochemistry of Polonium (1961). https://doi.org/10.17226/20247.

[31] H. Kobayashi, E. Nakamura, Mechanism of salt error on the color development of phosphomolybdenum blue, Bunseki Kagaku 53 (2) (2004) 119–122. https://doi.org/10.2116/bunsekikagaku.53.119.

[32] J. Murphy, J.P. Riley, A single-solution method for the determination of soluble phosphate in sea water, J. Mar. Biolo. Assoc. U. K. 37 (1) (1958) 9–14. https://doi.org/10.1017/S0025315400014776.

[33] S. Sugahara, M. Suzuki, H. Kamiya, M. Yamamuro, H. Semura, Y. Senga, M. Egawa, Y. Seike, Colorimetric determination of sulfide in microsamples, Anal. Sci. 32 (10) (2016) 1129–1131. https://doi.org/10.2116/analsci.32.1129.

[34] Olson, K.R. (2012). A practical look at the chemistry and biology of hydrogen sulfide. Antioxid. Redox Signaling, 17(1), 32–44. https://doi.org/10.1089/ars.2011.4401.

[35] R.A. CROCKSON, A rapid colorimetric method for the estimation of urinary chlorides, J. Clin. Pathol. 16 (1963) 473–475. https://doi.org/10.1136/jcp.16.5.473.

[36] F.E. Clarke, Determination of chloride in water improved colorimetric and titrimetric methods, Anal. Chem. 22 (4) (1950) 553–555. https://doi.org/10.1021/ac60040a011.

[37] K. Yokoi, Colorimetric determination of chloride in biological samples by using mercuric nitrate and diphenylcarbazone, Biol. Trace Elem. Res. 85 (1) (2002) 87–94. https://doi.org/10.1385/BTER:85:1:87.

[38] H.L.S. McCance, R Alexander, The colorimetric determination of sodium, Biochem. J 25 (1931) 449.
[39] V. Schmidt, A quick method for colorimetric determination of potassium, Scand. J. Clin. Lab. Invest. 3 (4) (1951) 252–254. https://doi.org/10.3109/00365515109060611.
[40] P. Gosling, Analytical reviews in clinical biochemistry: calcium measurement, Ann. Clin. Biochem. 23 (2) (1986) 146–156. https://doi.org/10.1177/000456328602300203.
[41] C.W. Miliigan, F. Lindstrom, Colorimetric determination of calcium using reagents of the Glyoxal Bis(2-Hydroxyanil) Class., Anal. Chem. 44 (11) (1972) 1822–1829. https://doi.org/10.1021/ac60319a019.
[42] M.F. Ryan, H. Barbour, Magnesium measurement in routine clinical practice, Ann. Clin. Biochem. 35 (4) (1998) 449–459. https://doi.org/10.1177/000456329803500401.
[43] J.W. Ferguson, J.J. Richard, J.W. O'laughlin, C.V. Banks, Simultaneous Spectrophotometric Determination of Calcium and Magnesium with Chlorophosphonazo III, Anal. Chem. 36 (4) (1964) 796–799. https://doi.org/10.1021/ac60210a028.
[44] E.M. Gindler, J.D. King, Rapid colorimetric determination of calcium in biologic fluids with methylthymol blue, Am. J. Clin. Pathol. 58 (4) (1972) 376–382. https://doi.org/10.1093/ajcp/58.5.376.
[45] United States Environmental Protection Agency, Temperature. In: Water: Monitoring and Assessment, 2012. https://archive.epa.gov/water/archive/web/html/vms53.html.
[46] D. Giustarini, R. Rossi, A. Milzani, I. Dalle-Donne, Nitrite and nitrate measurement by griess reagent in human plasma: evaluation of interferences and standardization, Meth. Enzymol. 440 (2008) 361–380. https://doi.org/10.1016/S0076-6879(07)00823-3.
[47] K.J. Laidler, Chemical Kinetics, 3rd Edition, Pearson India, 2003 International Economy Edition.
[48] R.J. Meier, S. Schreml, X.D. Wang, M. Landthaler, P. Babilas, O.S. Wolfbeis, Simultaneous photographing of oxygen and pH in vivo using sensor films, Angewandte Chemie - Int. Ed. 50 (46) (2011) 10893–10896. https://doi.org/10.1002/anie.201104530.
[49] A. Choodum, P. Kanatharana, W. Wongniramaikul, N. Nic Daeid, Using the iPhone as a device for a rapid quantitative analysis of trinitrotoluene in soil, Talanta 115 (2013) 143–149. https://doi.org/10.1016/j.talanta.2013.04.037.
[50] G. Hong, M.R. Luo, P.A. Rhodes, A study of digital camera colorimetric characterization based on polynomial modeling, Color Res. Appl 26 (2001). https://doi.org/10.1002/1520-6378.
[51] J.I. Hong, B.Y. Chang, Development of the smartphone-based colorimetry for multi-analyte sensing arrays, Lab Chip 14 (10) (2014) 1725–1732. https://doi.org/10.1039/c3lc51451j.
[52] A. García, M.M. Erenas, E.D. Marinetto, C.A. Abad, I. De Orbe-Paya, A.J. Palma, L.F. Capitán-Vallvey, Mobile phone platform as portable chemical analyzer, Sens. Actuators B 156 (1) (2011) 350–359. https://doi.org/10.1016/j.snb.2011.04.045.
[53] Martínez-Verdú, F., Pujol, J., Vilaseca, M., & Capilla, P. (2003). Characterization of a digital camera as an absolute tristimulus colorimeter. In *Proceedings of SPIE - The International Society for Optical Engineering* (Vol. 5008, pp. 197–208). https://doi.org/10.1117/12.474876.
[54] M.M. Erenas, K. Cantrell, J. Ballesta-Claver, I. De Orbe-Payá, L.F. Capitán-Vallvey, Use of digital reflection devices for measurement using hue-based optical sensors, Sens. Actuators B 174 (2012) 10–17. https://doi.org/10.1016/j.snb.2012.07.100.
[55] B. Purohit, A. Kumar, K. Mahato, P. Chandra, Smartphone-assisted personalized diagnostic devices and wearable sensors, Curr. Opin. Biomed. Eng. 13 (2020) 42–50 https://doi.org/10.1016/j.cobme.2019.08.015.

[56] E. Spectroscopy, Spectrosc. Electron (1999) 2383–2389.
[57] Russell, S.M., Doménech-Sánchez, A., & De La Rica, R. (2017). Augmented reality for real-time detection and interpretation of colorimetric signals generated by Paper-Based Biosensors. ACS Sensors, 2(6), 848–853. https://doi.org/10.1021/acssensors.7b00259.
[58] A. Roda, E. Michelini, M. Zangheri, M. Di Fusco, D. Calabria, P. Simoni, Smartphone-based biosensors: a critical review and perspectives, TrAC - Trends. Anal. Chem. 79 (2016) 317–325. https://doi.org/10.1016/j.trac.2015.10.019.
[59] S. Dutta, Point of care sensing and biosensing using ambient light sensor of smartphone: critical review, TrAC - Trends. Anal. Chem. 110 (2019) 393–400. https://doi.org/10.1016/j.trac.2018.11.014.
[60] I. Hussain, K.U. Ahamad, P. Nath, Low-cost, robust, and field portable smartphone platform photometric sensor for fluoride level detection in drinking water, Anal. Chem. 89 (1) (2017) 767–775. https://doi.org/10.1021/acs.analchem.6b03424.
[61] I. Hussain, A.J. Bora, D. Sarma, K.U. Ahamad, P. Nath, Design of a smartphone platform compact optical system operational both in visible and near infrared spectral regime, IEEE Sens. J. 18 (12) (2018) 4933–4939. https://doi.org/10.1109/JSEN.2018.2832848.
[62] Chodorow, N. (1979). The reproduction of colour.
[63] S. Dudala, S.K. Dubey, S. Goel, Microfluidic soil nutrient detection system: integrating Nitrite, pH, and electrical conductivity detection, IEEE Sens. J. 20 (8) (2020) 4504–4511. https://doi.org/10.1109/JSEN.2020.2964174.
[64] H. Hwang, Y. Kim, J. Cho, J.Y. Lee, M.S. Choi, Y.K. Cho, Lab-on-a-disc for simultaneous determination of nutrients in water, Anal. Chem. 85 (5) (2013) 2954–2960. https://doi.org/10.1021/ac3036734.
[65] M.S. Bhamla, B. Benson, C. Chai, G. Katsikis, A. Johri, M. Prakash, Hand-powered ultralow-cost paper centrifuge, Nature Biomedical Engineering (1) (2017) 1. https://doi.org/10.1038/s41551-016-0009.
[66] I. Ortiz-Gomez, M. Ortega-Muñoz, A. Salinas-Castillo, J.A. Álvarez-Bermejo, M. Ariza-Avidad, I. de Orbe-Payá, F. Santoyo-Gonzalez, L.F. Capitan-Vallvey, Tetrazine-based chemistry for nitrite determination in a paper microfluidic device, Talanta 160 (2016) 721–728. https://doi.org/10.1016/j.talanta.2016.08.021.
[67] B.M. Jayawardane, I.D. McKelvie, S.D. Kolev, A paper-based device for measurement of reactive phosphate in water, Talanta 100 (2012) 454–460. https://doi.org/10.1016/j.talanta.2012.08.021.
[68] P.J. Worsfold, J. Richard Clinch, H. Casey, Spectrophotometric field monitor for water quality parameters. The Determination of Phosphate, Anal. Chim. Acta 197 (C) (1987) 43–50. https://doi.org/10.1016/S0003-2670(00)84711-X.
[69] N. Shahrubudin, T.C. Lee, R. Ramlan, An overview on 3D printing technology: technological, materials, and applications, Procedia Manufacturing 35 (2019) 1286–1296. Elsevier B.V. https://doi.org/10.1016/j.promfg.2019.06.089.
[70] M. Rahbar, P.N. Nesterenko, B. Paull, M. Macka, Geometrical alignment of multiple fabrication steps for rapid prototyping of microfluidic paper-based analytical devices, Anal. Chem. 89 (22) (2017) 11918–11923. https://doi.org/10.1021/acs.analchem.7b03796.
[71] A.D. Beaton, C.L. Cardwell, R.S. Thomas, V.J. Sieben, F.E. Legiret, E.M. Waugh, P.J. Statham, M.C. Mowlem, H. Morgan, Lab-on-chip measurement of nitrate and nitrite for in situ analysis of natural waters, Environ. Sci. Technol. 46 (17) (2012) 9548–9556. https://doi.org/10.1021/es300419u.
[72] F.E. Legiret, V.J. Sieben, E.M.S. Woodward, S.K. Abi Kaed Bey, M.C. Mowlem, D.P. Connelly, E.P Achterberg, A high performance microfluidic analyser for phosphate measurements in marine waters using the vanadomolybdate method, Talanta 116 (2013) 382–387. https://doi.org/10.1016/j.talanta.2013.05.004.

[73] G. Gaál, T.A. da Silva, V. Gaál, R.C. Hensel, L.R. Amaral, V. Rodrigues, A. Riul, 3D printed e-tongue, Front Chem 6 (2018). May. https://doi.org/10.3389/fchem.2018.00151.

[74] A. Riul, A.M.G. Soto, S.V. Mello, S. Bone, D.M. Taylor, L.H.C Mattoso, An electronic tongue using polypyrrole and polyaniline, Synth. Met. 132 (2) (2003) 109–116. https://doi.org/10.1016/S0379-6779(02)00107-8.

[75] W.E., Soil Health for Farming in Tasmania, 2009.

[76] B.M. Jayawardane, W. Wongwilai, K. Grudpan, S.D. Kolev, M.W. Heaven, D.M. Nash, I.D. McKelvie, Evaluation and application of a paper-based device for the determination of reactive phosphate in soil solution, J. Environ. Qual. 43 (3) (2014) 1081–1085. doi:10.2134/jeq2013.08.0336 25602837.

[77] A.F. Coskun, J. Wong, D. Khodadadi, R. Nagi, A. Tey, A. Ozcan, A personalized food allergen testing platform on a cellphone, Lab. Chip. 13 (4) (2013) 636–640. doi:10.1039/c2lc41152k 23254910.

[78] F. Li, P. Smejkal, N.P. Macdonald, R.M. Guijt, M.C. Breadmore, One-Step Fabrication of a Microfluidic Device with an Integrated Membrane and Embedded Reagents by Multimaterial 3D Printing, Anal. Chem. 89 (8) (2017) 4701–4707. doi:10.1021/acs.analchem.7b00409 28322552.

[79] A. Wany, P.K. Pathak, K.J. Gupta, Methods for Measuring Nitrate Reductase, Nitrite Levels, and Nitric Oxide from Plant Tissues, Methods Mol. Biol. 2057 (2020) 15–26. doi:10.1007/978-1-4939-9790-9_2 31595466.

[80] P. Singh, M.K. Singh, Y.R. Beg, G.R. Nishad, A review on spectroscopic methods for determination of nitrite and nitrate in environmental samples, Talanta 191 (2019) 364–381. doi:10.1016/j.talanta.2018.08.028 30262072.

Smartphones as Chemometric applications

Taniya Arora[a], Rohini Chauhan[a], Vishal Sharma[a], Raj Kumar[b]
[a]Institute of Forensic Science & Criminology, Panjab University, Chandigarh
[b]State Forensic Science Laboratory, Madhuban, Karnal, Haryana

8.1 Introduction

Mobile phones or smartphones first appear in the world around the 1990s. Such devices were made to make and receive calls, messages, and facsimiles. However, with the evolution of smartphone technology over the many years, they have become so widespread that no human can expect their lives without them. Based on the previous trends, it is predicted that the expected rate of mobile phone and smartphone users will be 7.1 billion and 3.8 billion in the upcoming year as compared to the 6.95 billion and 3.5 billion in the present year respectively [1–2].

Further, with the evolution of the electrical component's technology, their size has been abridged to such an extent that they can be easily handled with a single hand as compared to older models this may weigh up to 3 Kg. This evolution has not taken place in terms of size only but in software technology as well. Now, mobile devices are not only used to make calls and receive messages but also with the advent of accessing the internet over the mobile devices, they can be used in almost every field of daily requirement and hence, is their popularity. Modern mobile devices offer the ability to make and receive calls and messages, accessing the internet, underwent banking transactions, clicking pictures, making videos, and so on.

Modern mobile devices also provide the facility to store additional applications and information which supplements and intensifies their requirement in every person's life as they made them being able to connect with the world every moment. Further, this provides the people with ease by the use of IoT (Internet of Things) which supports their routine activities with just a simple click or swipe on their smartphone devices. Further, the VoIP applications on mobile devices make it easy for people to interact and share information and being socialized at no cost expected.

All these positive aspects of such devices made them a hot area of scientific research as they can be easily carried to any place and most of the things can be done single-handedly and no trained specialized personnel are required for their operation. Further, their ability to store additional applications and information has made them emerged as an alternative to the primitive analytical instruments. The analytical instruments used so far are expensive and heavy; require trained personnel for their operations. But, the availability of high-quality cameras, low cost, ease of operation of smartphones is making them the easily accessible detection tool. This aspect of smartphone analysis has been taken into a good advantage and varied work has been published in this context, e.g. architectural photogrammetry [3], bacterial detection [4], determination of glucose, uric acid, and urea [5], iris recognition [6], food adulteration [7], real-time monitoring of athlete's performance [8], sports monitoring [9], digital colorimetry [10], automation and modeling [11,12], medical sciences [13,14] and so on. Further, due to their small size, they can be carried to the scene as well and is thus, also helpful in forensic cases as on-site analysis can be simply performed without any contamination and alteration of the evidence.

8.2 Application of smartphones in chemical sciences

As explained, smartphones have various uses in various fields. Smartphones are used in the field of analytical chemistry as they possess the capability to replace analytical instruments [15-17]. Modern smartphones come up with good camera quality which can be used as a detector. Further, they are gaining popularity in analytical chemistry due to their low cost. The analytical instruments used in chemistry are varied and different instruments are required for different applications. But a single mobile phone can act as a detector over all such instruments. In addition to cost-cutting, it also reduces the workforce required as any person can handle the smartphone, and thus, no trained personnel are required for the operation of the instrument. It also reduces the time required for the analysis as only a single picture can serve the purpose of repeatability and reproducibility tests.

Further, the mobile devices possess easy connectivity modes like the USB port, Bluetooth, and Wi-Fi which makes the data transfer and thus, analysis on the laptop and computer handy. They are portable and can easily be carried to any place and are thus, helpful in on-scene investigation and analysis in forensic cases. The basic principle involved in the analysis of chemical compounds using smartphones is the analysis of the picture or video captured by the mobile device. The properties like color intensity, pixel count in the specific area, luminescence, etc. can help in the detection and quantification of the substance in question.

8.2.1 Colorimetric applications

This is the most basic approach to analysis using smartphones. This involves the analysis of the color intensity of the picture captured using a smartphone. The components of the colors captured are resolved into their constituent colors and thus, their analysis serves the purpose of detection and quantification of the substance. The most commonly used color systems are RGB, HSV, and CieLAB [18-21].

a. RGB color system

The RGB system involved the analysis by comparing the model with the cube. The red, green, and blue colors constituents are represented by the x, z, and y-axis respectively.

b. HSV color system

HSV stands for Hue Saturation Value. In this color system, hue represents pure colors that are changed from angle 0° to 360° in the form of a cone. Saturation indicates the range of grey color along the horizontal axis ranging from 0 percent to 100 percent. The value represents the color brightness and is represented along the vertical axis ranging from 0 percent to 100 percent.

c. CieLAB color system

This color system is based upon the spherical model. The L or vertical axis represents lightness from black (0 percent) to white (100 percent). The B or perpendicular axis represents green to red and blue to yellow colors which run from range −128 to 127 respectively. The opposing colors that cannot be combined are represented by axes. By this system, all the colors that the human eye can perceive are represented. The examples utilizing smartphones in colorimetry are as follows:

Tao et al. investigated the color of rice plant leaves using digital colorimetry or smartphones [22]. They have developed the color chart such that the leaves are arranged in the chart according to the different colors specified according to their quality. The setup consisted of the leaf gripper in which the leaf was attached and is photographed. The images are then compared to the color level using a standard color chart. Fig. 8.1(a) shows the experimental procedure and designed a color chart for comparison. Smartphone application gives 92 percent, 95 percent, and 95 percent accuracies for color levels 2, 3, and 4 respectively. The developed app is compatible with other smartphones as well with a processing time of 400 ms.

Silva et al. detected milk adulteration based upon the protein content of milk using the smartphone digital colorimetry [7]. The proposed methodology involves the precipitation of protein contents using copper sulfate and a proportional concentration of Cu^{2+} ions. The precipitates are then photographed using the smartphone and the color of the precipitates defined the protein content. The proposed method detects adulteration above 10 percent v/v water.

Monogarova et al. investigated the drug samples to identify and quantify their active contents by employing digital colorimetry [23]. Transparent 96 celled polypropylene plates are selected as the testing base or matrix of the chip. The results of the test were recorded by smartphones which shows the barcode type color positions referenced to the specific drug (as shown in Fig. 8.1(b)). The proposed methodology was applied to the 40 drugs but can be extended to pharmaceutical drugs and heterogeneous substances as well.

Other work published about colorimetry includes the analysis of pharmaceutical drugs (carbamazepine, ciprofloxacin, and norfloxacin) [24,25], urine sample [26], pH determination from nitrite ions [27], chlorpromazine hydrochloride tablets [28], milk adulterants [29], water [30], food adulteration [31-33] and food dyes [34].

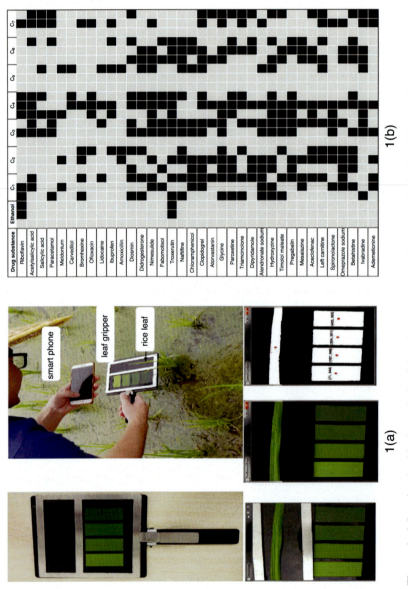

Fig. 8.1 (a) The methodology involved in the analysis of rice plant leaves (reproduced from Reference [22]); (b) The results of digital colorimetry in the form of partial barcodes of different drugs (reproduced from Reference [23]).

8.2.2 Fluorescence microscopy

Fluorescence microscopy has gained great interest in the field of biology (especially cell and molecular biology) in the previous years. The main reason behind this is its ability to reveal only the objects of interest in the otherwise black background; therefore, many such probes have been developed to label the biological systems. Fluorescence is the emission of light within nanoseconds after the absorption of a shorter resonant wavelength. Thus, this type of microscopy can be used for the substances that fluoresce [35].

Recently, the approach of using a smartphone as the fluorescence microscope is prevailing because it analyzes fluorescent materials easily enough as no extra equipment is required and whole imaging and processing can be done on the smartphone screen only. The basis of using a smartphone as the fluorescence microscopy is that the inbuilt flashlight of the smartphone can be used as the excitation source. The examples of such methodology are described below.

Fronczeck et al. investigated Salmonella Typhimurium [36]. The methodology involved the detection of fluorescence created by the nucleic acids of Salmonella Typhimurium. They used the iPhone4 for the analysis of the clinical samples of the salmonella. The designed miniature fluorescent microscope was used as an attachment to record the fluorescence produced by the patterned paper chips. The blue LED light was used as a lighting source while taking pictures on the smartphone. The microscope also uses two objective lenses for better focusing of the fluorescence. The arranged optics is as shown in Fig. 8.2(a). Further, the results obtained by the smartphone microscopy were confirmed with the traditional PCR (Polymerase Chain Reaction).

On a similar basis, Koydemir et al. investigated the waterborne pathogens and also quantified the Giardia Limblia cysts [37]. They have also developed the mechanic algorithm for the automatic counting of the number of fluorescent cysts present for the quantification purpose. The captured images were transferred to this designed application, results interpreted and within no time the results were shown back on the smartphone display. The outlay of customized application's working is as shown in Fig. 8.2(b).

Nguyen et al. have also analyzed water contamination [38]. They have developed a smartphone-based micro-colorimetry detector that can work as both fluorescents as well as the dark-field microscope. They have quantified the lead ions (Pb^{2+}) by the detection of fluorescence produced by PbCrO4 (lead chromate) nanoparticles by the addition chromate (CrO_4^{2-}) solution to the detecting sample. The images captured using smartphone-based fluorescence microscopes are shown in Fig. 8.2(c).

Other publications include the analysis of pathogenic bacteria [39]. Others have also used a smartphone as a compact lens microscope [40].

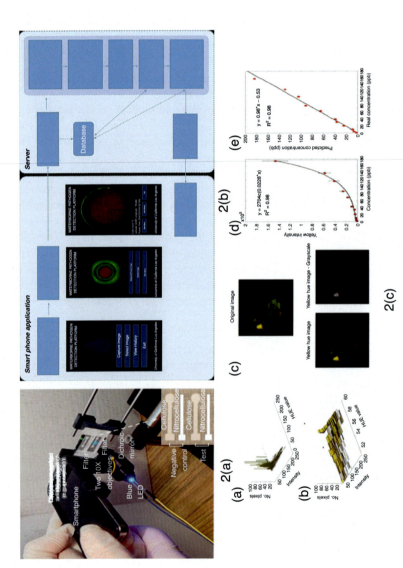

Fig. 8.2 (a) Schematic showing the arrangement of a smartphone with other optical pieces (reproduced from Reference [36]); (b) Outlay of the application designed for the calculation of the no. of cysts (reproduced from Reference [37]); (c) Results of the images captures using a smartphone as the fluorescence microscope (reproduced from Reference [38]).

8.2.3 Barcode reading

Barcodes are the pattern of dark and light lines. The thickness of the lines can be varied. Based upon the thickness and area of occupancy of lines, different letters, alphabets, numerals, and special characters are encoded [41]. Barcodes allow real-time identification and are rapid and accurate in their processing. These codes can be designed for varied things in varied coding protocols such that each product gets a different identification code. Such codes are helpful in product selling in supermarkets, attendance system in different institutions as they possess the information that can be decoded by a simple barcode scanner.

This similar principle is used in chemical sciences for assigning a unique barcode to different chemicals and thus, utilizing these codes for their identification. The usage of these barcodes and scanning those using smartphones is a hot topic as these systems then became portable and can be used for the on-spot analysis of different drugs and other materials conditioned database of their barcodes is available. The basic methodology involves the scanning of these barcodes using the phone's CMOS camera or another way of developing the barcode-based paper sensor for quantification purposes as well. The examples of such work are listed below.

Guan et al. used the paper-based analytical device (PAD) for blood typing [42]. They have developed a PAD that uses the barcode reading concept. The fabricated PAD has three channels, one each for A, B, and D. The concept of agglutination of RBCs and antibodies was employed. The bar channels were loaded with the NaAlg solution and the antibody was poured over it. The length eluted in the bar channel confirms the blood group and Rh factor. The results of detecting blood group from elution length are depicted in Fig. 8.3(a). The agglutinated antibodies travel up to the short bar length. To analyze the blood typing results and accurately predicting the bar length traveled, a smartphone-based application using Java codes was designed. Further, the results were confirmed by using the developed PAD for 98 volunteers.

Guo et al. used the barcode-based PAD that can be read with any freely available barcode reading application [43]. The barcode assay used was developed using the Code 39 encoding technique consisting of '+' and '−'. This barcode assay was used to detect and quantify the pesticide residues. The assay sensor was developed using the irreversible activity of the enzyme acetylcholinesterase by methyl parathion. This reaction gives the yellow color whose intensity is inversely proportional to the quantity of pesticide. The '+' barcode was thus, printed in yellow color to detect the actual intensity of the methyl parathion. The barcode- formatted assay uses the PDMS channel plate as the reaction reservoir below the printed barcode to complete the barcode. This developed barcode was read with the Quickmark app. The whole methodology is shown in Fig. 8.3(b).

Others like Zhang et al. used barcoded microchips for biological assays [44] and Lin et al. used the codes for the management system [45].

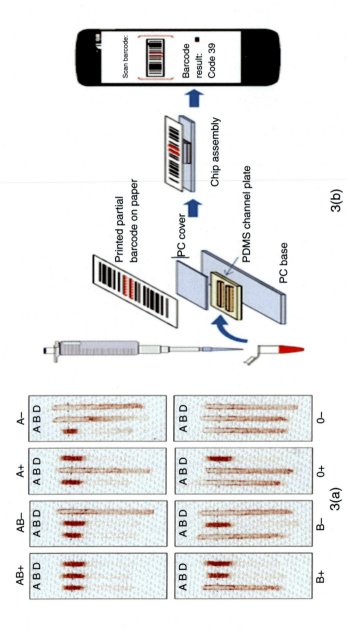

Fig. 8.3 (a) The actual blood typing results using barcoded channels (reproduced from Reference [42]); (b) The micro-pipette analysis to detect and quantify the pesticide residues. The yellow color's intensity indicates the quantity of methyl parathion (reproduced from Reference [43]).

8.2.4 Electrochemistry

The study of inter-conversion of chemical energy and electrical energy is known as electrochemistry [46]. Electrochemistry works upon the principle of electromotive force, i.e. that is whenever there is a potential difference across two ends; the current will flow corresponding to the potential difference. Thus, whenever there is an electrochemical cell, a redox reaction will take place and the current will flow from cathode to anode (the electrodes between which the current is flowing).

Electrochemistry has various applications in developing sensors. Thus, it became necessary to develop procedures that can detect the results in real-time. Therefore, the best detector became the smartphone which requires the external electrochemical cell for processing. The method involved may use a smartphone as the potentiometer, impedance reader, ammeter, or voltammeter. The examples of research in this respect are summarized below.

Zhang et al. investigated 2,4,6-trinitrophenol by employing impedance monitoring [47]. They used the screen-printed paper electrodes specifically coated with TNT active peptides; the electrochemical cell was designed to use methanol as the blank solution. The peptide immobilization and impedance measurement are recorded using the smartphone-based developed application. The impedance analyzer was developed using Arduino control as shown in Fig. 8.4(a).

Guo et al. used chronoamperometry for the detection and characterization of uric acid from the whole blood sample [48]. They have developed an electrochemical based sensor printed on the screen. The electrodes of carbon compounds were used. This screen-printed electrode act as an electrochemical module that directly characterizes the uric acid. They have also used a similar approach to be able to test the uric acid and glucose simultaneously. For this, they developed the two-channel system which is driven by the capillary action. A single drop of blood can run through the two channels and give independent results. They have also prepared the testing strip that permits the analysis of uric acid and glucose on the same substrate [49].

Ji et al. investigated the glucose from the whole blood sample using cyclic voltammetry. They used the graphene-modified screen-printed electrodes as the electrochemical module. Electric information was transmitted to the smartphone using the handheld CV detector. The app was also developed that connects with the CV detector using Bluetooth mode. The app controls the system and produces real-time cyclic voltammograms as shown in Fig. 8.4(c) [50].

Others have used electrochemical based assays for the analysis of uric acid as would analytics [51], potentiostat for wireless detection [52,53].

8.2.5 Pixelation

Pixelation is the term used to describe the blurred sections in the single-color area of the image due to the individual pixels [54]. It is used as an alternative technique to colorimetry for optical detection. It involves the counting of the number of pixels

138   Smartphone-Based Detection Devices

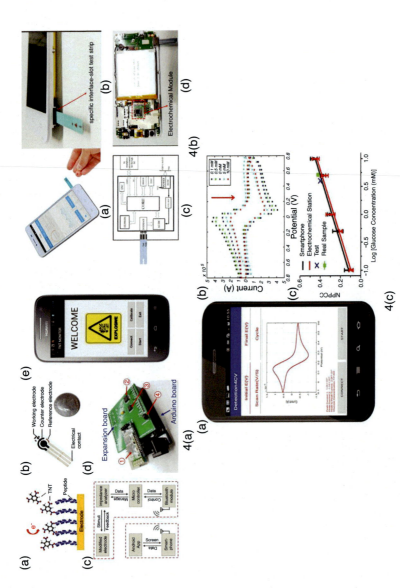

Fig. 8.4 (a) Schematic of the methodology involved for the impedance reading of paper-based sensor using Arduino control (reproduced from Reference [47]); (b) Screen-printed electrode as a testing attachment to the smartphone (reproduced from Reference [48]); (c) Cyclic voltammetry analysis using a customized application to produce real-time curves (reproduced from Reference [50]).

in the standard area as a representative of its concentration. Thus, pixilation can be helpful where quantitative analysis is required. The examples of pixelation-based research are as follows.

Yu et al. investigated the drugs using the smartphone photography of thin-layer chromatography [55]. They prepared the customized assembly that arranges the TLC plates and smartphone's camera in line as shown in Fig. 8.5(a). They developed an application that works on smartphones such that it works on every pixel of the image and characterizes it as spot and non-spot. The retention factor was calculated as follows:

$$Rf = \frac{\text{Distance traveled by the principal spot}}{\text{Distance traveled by the solvent front}}$$
$$= \frac{\text{No. of pixels between the spots center and the origin line}}{\text{No. of pixels between the solvent front line and the origin line}}$$

95 percent of the tested drugs fall into the range of references, i.e., the correct calculation of R_f and thus, the correct drug identification.

Lee et al. analyzed Aflatoxin B1 by improving the one-dot lateral flow immunoassay (LFIA) using a smartphone-based reading system [56]. The developed LFIA reader was attached to the camera lens (as shown in Fig. 8.5(b)) and thus the limit was quantified by taking the peak and the area under the curve. The detection limit is calculated as 5µg/kg.

Other work done in this context includes the quantitative detection of alkaline phosphatase in milk [57] and for bioanalytical applications [58].

8.2.6 Label-free detection

Label-free detection is the method that utilizes molecular biophysical properties such as molecular weight, refractive index, and molecular charge to monitor the presence of molecules or their activity. They are used for real-time monitoring. They provide direct information as they use only native proteins and ligands. The most common types of label-free detection are micro-cantilever, mass spectrometry, quartz crystal microbalance, surface plasmon resonance, and localized surface plasmon resonance, and anomalous reflections of the gold surface [59,60]. Label-free detection can be used coupled with smartphone analysis as the customized smartphone applications can be used for the determination of refractive index and molecular mass. The examples are enlisted as follows.

Gallegos et al. developed a smartphone attachment that can be worked as a spectrometer detector for label-free photonic crystal biosensor [61]. Their method of detection was based upon the measurement of shifts in the resonant wavelength of the sensor. A cradle or holder was developed in such a way that it aligns the camera of the smartphone and the designed optics in alignment with the photonic crystal.

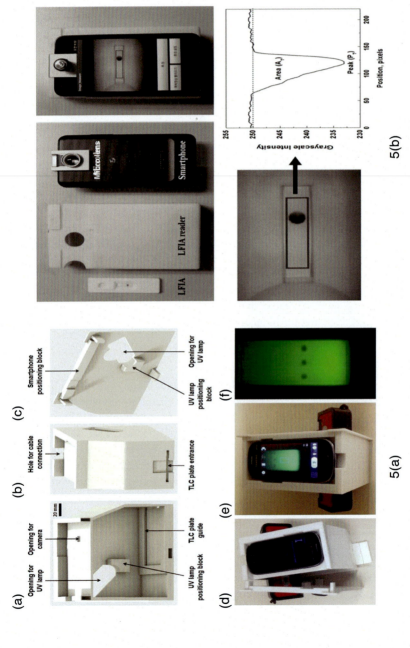

Fig. 8.5 (a) The developed assembly for thin chromatographic results using smartphone (reproduced from Reference [55]); (b) The LFIA assembly attached to a smartphone to give the results and limit of detection by calculating the area under peak (reproduced from Reference [56]).

The methodology involved the collimation and linear polarization of the wavelength that resonantly reflects the narrow band of wavelengths. Further, they developed an application that converts the images captured by the camera into the photonic crystal transmission spectrum in the visible range. The photonic crystal used in the study was porcine immunoglobulin G using an immobilized layer of Protein A. The results confirm that they are in the accuracy range of 0.009 nm.

Dalstein et al. demonstrated the use of a smartphone camera as a detector to mark the changes in diffraction efficiency of photonic metal-organic frameworks (MOF) patterns which are induced by variations in the refractive index [62]. The captured images using CCD cameras were converted to the RGB color space which was further taken to grayscale to calculate the luminance intensity. Further, the methodology of sensor fabrication was confirmed by analyzing styrene in air, which resulted in detection lower to 57 ppm.

Zhang et al. used smartphone-based label-free detection of lipo-polysaccharides on the grating coupled surface plasmon resonance (GC-SPR) [63]. The developed biosensor system relies on the inbuilt flash and camera of the smartphone. It also uses the Au grating and CD as spectra dispersive unit. The methodology reported the detection limit of 32.5 ng/mL in water.

Fig. 8.6 shows the arrangement of optics and smartphones as used in the above different applications.

The other work includes the analysis of salmonella based on paper microfluidics [64] and immunoassays for bio-molecular detection [65].

8.2.7 Fluorescence

Whenever the electron absorbs energy and gets excited to a higher energy level, it emits energy while coming back to its ground state or the lowest energy level. This emission of light is known as luminescence. Fluorescence is the one in which the electron emits energy and falls back to the ground state within nanoseconds of its excitation [66]. This emitted fluorescence can be recorded by the smartphone as the images and thus work as the detector. Various researchers have utilized this fact in their research methodology as described below.

Canning et al. used the smartphone as the optical hardware for the analysis of Rhodamine-123 doped self-assembled giant mesostructured silica sphere by measuring the fluorescence produced by them [67]. The results were confirmed by the SEM images of the hexagonal type structures.

Rajendran et al. analyzed the food-borne bacterial pathogens based on fluorescent detection using smartphones [68]. They used the immune chromatographic test strips for the detection of biosynthesized fluorescent nanoparticles. The produced fluorescence was recorded by the nitrocellulose strip that can be inserted into the smartphone fluorimeter. The pathogens, e.g. Salmonella spp. and Escherichia coli O157 were detected by the present methodology which allows the detection down to the limit of 10^5 cfu/mL.

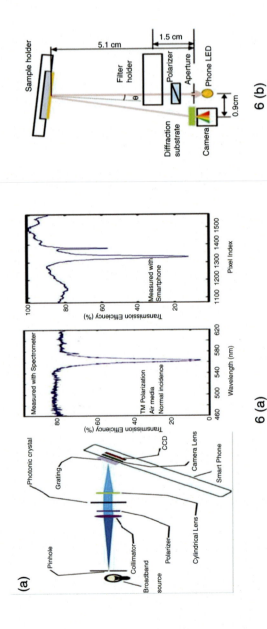

Fig. 8.6 (a) Comparison of results obtained using spectrometer and smartphone (reproduced from Reference [61]); (b) Optical arrangement of smartphone spectrometer showing Au sensor chip (reproduced from Reference [63]).

Shrivastava et al. also analyzed the bacterial pathogen Staphylococcus Aureus using smartphone-based fluorescence [69]. They analyzed the samples in liquid form using aptamer-functionalized magnetic nanoparticles. The detection was done in the form of a magnetic cassette. The proposed method can detect even 10 cfu/mL of concentration. The results were further confirmed for the peanut butter which can be tested within 10 min. Fig. 8.7 shows the comparison of results with smartphone microscopy with that of confocal microscopy. Further, the single pathogen is analyzed by capturing its image and converting it to a gray value graph.

Other work includes fluorescence imaging to analyze the droplet-based assays for the surface and detection of bacteria [70], ascorbic acid assay [71], glucose, and H_2O_2 [72].

8.2.8 Bioluminescence and chemiluminescence

Whenever the light emitted induces the chemical reaction, the luminescence produced is known as chemiluminescence. Chemiluminescence occurring naturally in living beings is known as bioluminescence [66]. Similar to fluorescence, chemiluminescence can also be detected using smartphones. Such examples are described below.

Roda et al. integrated the bio-chemiluminescence for point of need analysis [73]. They used the smartphone for imaging and quantifying bio-chemiluminescence coupled bio-specific enzymatic reactions to detect analytes in biological fluids. They also fabricated the mini-assembly as a reacting and image capturing area. For confirmation of the system, they used two assays viz. SmartChol and SmartBA for testing cholesterol and total Bile Acid in serum (whole blood) and oral fluid respectively. Both the assays can be easily done within 3 min.

Song et al. used a smartphone as the detection platform for rapid molecular diagnostics and spatiotemporal disease mapping [74]. They coupled the smartphone technology with the Bioluminescent Assay in Real-Time and Loop-mediated isothermal AMPlification (BART-LAMP). They also developed a customized app that can monitor spatiotemporal diseases and do target quantification. The schematic of the proposed methodology is as shown in Fig. 8.8.

8.2.9 Photoluminescence

Photoluminescence is the emission of energy released due to the excitation of the electron [66]. The combination of smartphones and photoluminescence can be used as follows.

Petryayeva et al. used smartphones and 3D printing for excitation and detection of semiconductor-based Quantum Dots (QD) [75]. The 3D printed accessory created a dark environment and helps indirect excitation using an in-built flashlight of the smartphone. Multiple colors and compositions of QD were used for the evaluation of the photoluminescence using the standard as fluorescein and R-phycoerythrin (R-PE). Further, the results were confirmed by the assay for avidin. They also used similar procedures for the analysis of thrombin.

144 Smartphone-Based Detection Devices

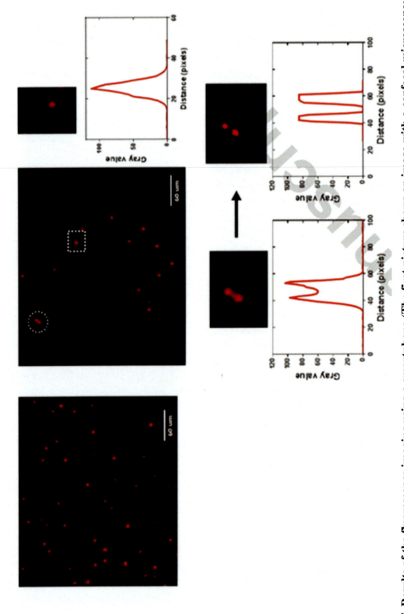

Fig. 8.7 Results of the fluorescence imaging using a smartphone (The first picture shows an image with a confocal microscope; reproduced from Reference [69]).

Smartphones as Chemometric applications 145

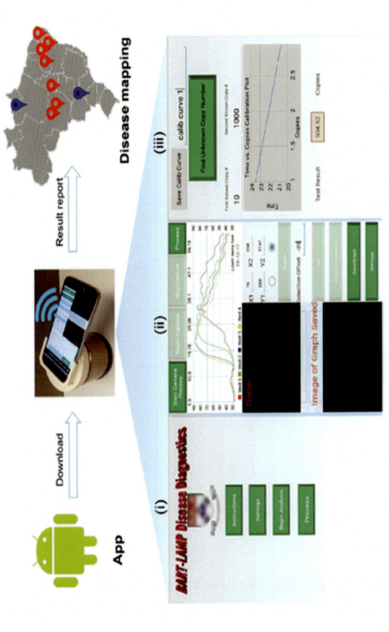

Fig. 8.8 Schematic showing working of application designed for molecular detection and disease (reproduced from Reference [74]).

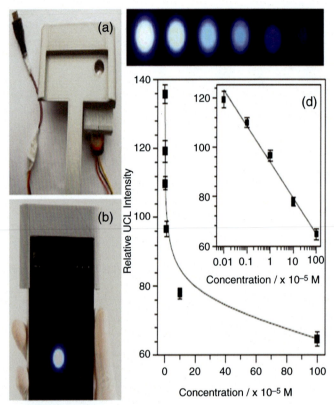

Fig. 8.9 Optical results showing quenching of thiram with decreasing concentration (reproduced from Reference [76]).

Mei et al. analyzed the upconversion paper sensors using smartphone detection of photoluminescence [76]. They analyzed the pesticide thiram based on the principle of quenching as shown in Fig. 8.9 and laser excitation of $NaYF_4$: Yb/Tm. They prepared the paper sensor using 3D printing and the methodology gives concentration down to 0.1 µM.

He et al. used a similar methodology using upconversion nanoparticles for the roadside analysis of cocaine [77]. Schafer et al. used the approach for biosensing of metal ions [78].

8.3 Chemometric analysis using smartphones

Chemometrics is the field of application of mathematical and statistical principles to extract chemical information [79]. Thus, chemometrics has tremendous applications in analytical chemistry. The recent approach of using chemometrics coupled with instrumental techniques is the hot area. This combination makes the data analysis simple and also, provides statistical confidence in hand. This approach has already been used in the pharmaceutical tablets and medicine [80], Viagra [81], counterfeiting [82], fireworks [83], soft drinks [84,85], paint [86], textile [87], cosmetics [88], soil [89], blood [90], ink analysis [91] and so on.

All the aforementioned methods were applied using heavy analytical instruments. But nowadays, rapid and real-time analysis is in demand as they help in reaching the results without any chances of contamination. Thus, smartphones can very easily be taken as an alternative to heavy analytical instruments. Rather they can just be used as a real-time detector connected to any portable analytical instrument. Further, the facility of third-party applications on smart devices makes them even popular as the analysis can be formed without the need to take the sample and results to the laboratories for confirmation. Further, smartphones can be used to develop the customized apps that can very simply make all the analyses including the statistical portion at just a go of click. A similar approach has already been applied as explained in the below-given examples.

a. Valderrama et al. investigated the blue pen inks in documentoscopy using digital image analysis [92].

Experimental procedure: 60 blue pen inks were collected from the local markets of Brazil. The samples were characterized as ballpoint, rollerball, gel point, and felt-tipped. The standard sample in the form of a circle was collected upon different grammage papers. The digital images were collected after 5 days of drying of ink using the iPhone 5S under standard conditions. The collected images were resolved into their RGB constituents using MATLAB. The involved method is as shown in Fig. 8.10. Partial least squares for discriminant analysis (PLS-DA) was employed as the chemometric tool.

Findings: The samples were recovered back to their brands and type with a 100 percent recovery rate within the confidence limit of 95 percent. Fig. 8.10

b. Song et al. developed a sensor system that analyzes smartphone videos and pattern recognition for authentication [93].

Experimental procedure: They have developed a Computer vision system (CVS). The basic methodology involved relies upon the color variations of the adulterated and unadulterated samples. A 5s video is recorded from the camera of the smartphone with a color sequence from purple to red. A region of interest (ROI) is selected in a red box. A succession of frames is extracted and the image was resolved into RGB components with each color channel having 256 color levels. Further, the Local weighted PLS classification (LW-PLSC) on video data is compared with baseline methods including k-nearest neighbors (k-NN), PLS-DA, SVM, and RF

Findings: For milk adulteration, the model can give inaccurate results whereas, for oil adulteration, the PLS-based techniques achieve the best results. The detected accuracy of 95.3 and 96.2 was achieved for training and testing samples respectively. The results are further shown in Fig. 8.11.

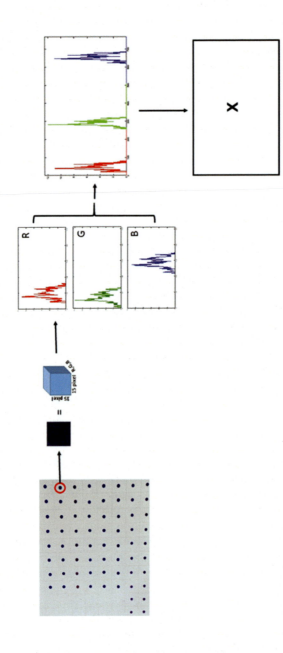

Fig. 8.10 Colorimetric disintegration of blue pen ink using MATLAB (X represents the further chemometric analysis using PLS-DA; reproduced from Reference [92]).

Smartphones as Chemometric applications 149

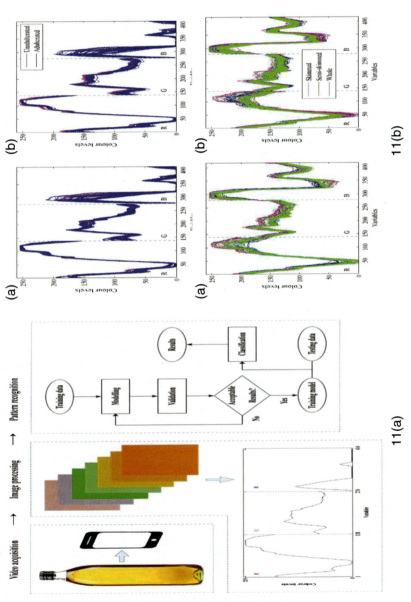

Fig. 8.11 (a) The procedure of image acquisition from a smartphone; (b) UP The spectroscopic results of olive oil; Down Spectroscopic results of milk (reproduced from Reference [93]).

They have also analyzed the extra virgin olive oil using the videos captured from smartphones [94]. 5s video is recorded which was split into 175 frames each in ROI. The 50th frame of the video was used as the representative to differentiate extra virgin olive oil from the vegetable oil. The PLS models are constructed using the UV–Vis, NIR, and image data sets.

Findings: UV–Vis and NIR outperform the CVS approach.

c. Parastar et al. developed a customized android based MVS application that can be used for MLR, PCR, and PLSR [95].

The developed app is based on the Java programming language. The data can be transferred to the smartphone using Bluetooth, Wi-fi, or a USB connection. The spectral, electrochemical, or chromatographic data is fed to the MVS application. After the data is imported, the calibration model is selected. The results are displayed back on the screen. The methodology of the workflow using this app is as shown in Fig. 8.12.

Thus, this mobile phone-based application makes the chemometrics in the reach of every person as the whole software and methodology are present within their pockets. This application is a boon for research groups who are not trained in statistical procedures.

d. Böck et al. also gave the briefing and working of the app PhotoMetrix using smartphones [96].

The app was specially designed for chemical analysis. Further, the app allows chemometric tools like PCA, PLS, and HCA. Various papers have been published using this application for the smartphone analysis of chemical compounds.

8.4 Conclusion and future trends

The present chapter reviews all the applications where a smartphone-based approach can be utilized. Further, the rate at which smartphone use is increasing and its technology is advancing, it has a huge potential to replace the analytical instruments. The smartphones' low-cost, highly advanced technologies used in their manufacturing and programming, availability of good quality camera, large screen display, touch screen programming, and above all the portability made it ideal in all the situations. They can be used in the analysis of drugs and explosives at airports, can be used for the analysis of ink and paper in documents related queries. Application in adulteration in food made them even more practical as they can be used by the common people to protect them from frauds. Further, they can be used in the real-time detection of almost all types of forensic exhibits.

Thus, in conclusion, smartphones possess a great potential to be used in every field, not only in chemical analysis but also in physical and biological assays as well.

Smartphones as Chemometric applications 151

Fig. 8.12 The brief layout of the procedure of the working of the MVC application (reproduced from Reference [95]).

Acknowledgments

This publication did not receive any specific grant from funding agencies in the public and commercial sectors.

References

[1] https://www.statista.com/statistics/330695/number-of-smartphone-users-worldwide ≤accessed on 13 July 2020>.

[2] https://www.statista.com/statistics/218984/number-of-global-mobile-users-since-2010 <accessed on 13 July 2020>.

[3] A.S. Hernán-Pérez, M.G. Domínguez, C.R. González, A.P. Martín, Using iPhone camera in photomodeler for the 3D survey of a sculpture as practice for architecture's students, Procedia Comput. Sci. 25 (2013) 345–347.

[4] S.C.B. Gopinath, T.H. Tang, Y. Chen, M. Citartan, T. Lakshmipriya, Bacterial detection: from microscope to smartphone, Biosens. Bioelectron. 60 (2014) 332–342.

[5] Y. Wu, A. Boonloed, N. Sleszynski, M. Koesdjojo, C. Armstrong, S. Bracha, V.T. Remcho, Clinical chemistry measurements with commercially available test slides on a smartphone platform: colorimetric determination of glucose and urea, Clin. Chim. Acta 448 (2015) 133–138.

[6] K.B. Raja, R. Raghavendra, V.K. Vemuri, C. Busch, Smartphone based visible iris recognition using deep sparse filtering, Pattern Recogn. Lett. 57 (2015) 33–42.

[7] A.F.S. Silva, F.R.P. Rocha, A novel approach to detect milk adulteration based on the determination of protein content by smartphone-based digital image colorimetry, Food Control 115 (2020) 107299.

[8] D. Rowlands, D. James, Real time data streaming from smart phones, Procedia Eng 13 (2011) 464–469.

[9] T. McNab, D.A. James, D. Rowlands, iPhone sensor platforms: applications to sports monitoring, Procedia Eng 13 (2011) 507–512.

[10] P. Thanakiatkrai, A. Yaodam, T. Kitpipit, Age estimation of bloodstains using smartphones and digital image analysis, Forensic Sci. Int. 233 (2013) 288–297.

[11] A. Carrio, C. Sampedro, J.L. Sanchez-Lopez, M. Pimienta, P. Campoy, Automated low-cost smartphone-based lateral flow saliva test reader for drugs-of-abuse detection, Sensors 15 (2015) e29569–e29593.

[12] M. Shoaib, S. Bosch, O.D. Incel, H. Scholten, P.J.M. Havinga, Complex human activity recognition using smartphone and wrist-worn motion sensors, Sensors 16 (2016) e426–e450.

[13] M.O. Baerlocher, R. Talanow, A.F. Baerlocher, Radiation passport: an iPhone and iPod touch application to track radiation dose and estimate associated cancer risks, J. Am. Coll. Radiol. 7 (2010) 277–280.

[14] A.I. Barbosa, P. Gehlot, K. Sidapra, A.D. Edwards, N.M. Reis, Portable smartphone quantitation of prostate specific antigen (PSA) in a fluoropolymer microfluidic device, Biosens. Bioelectron. 70 (2015) 5–14.

[15] A. Niemirich, O. Petrusha, O. Vasheka, L. Trofymchuk, N. Myndrul, Exploring the color of plant powders using computer colorimetry, East. J. Enterp. Technol. 4 (2016) 15–20.

[16] N. Shehzadi, K. Hussain, M.T. Khan, M. Salman, M. Islam, H.M. Khan, Development of a validated colorimetric assay method for estimation of amikacin sulphate in bulk and injectable dosage form, J. Chem. Soc. Pak 38 (2016) 63.

[17] P.G. Sunitha, R. Karthikeyan, B. Ranjith Kumar, S. Muniyappan, Validated colorimetric methods for the estimation of teneligliptin in tablets, J. Drug Deliv. Ther. 7 (2017) 38–40.

[18] T. Smith, J. Guild, C.I.E The, Colorimetric Standards and Their Use, Trans. Opt. Soc. 33 (3) (1931) 73–134.

[19] J. Melorose, R. Perroy, S. Careas, Color Spaces for Computer Graphics 1978, Statew. Agric. L. Use Baseline 2015 (1) (2015) 20–25.

[20] BillmeyerW. Fred, M. Saltzman, Principles of Color Technology, J. Wiley & Sons, Ed.; Wiley-Interscience, New York, 1981 Joost Laurus Dinant.

[21] Y.Zhao Nelis, L. Bura, K. Rafferty, C.T. Elliott, K. Campbell, A randomised combined channel approach for the quantification of colour and intensity based assays with smartphones, Anal. Chem. 92 (11) (2020) 7852–7860.

[22] M. Tao, X. Ma, X. Huang, C. Liu, R. Deng, K. Liang, L. Qia, Smartphone-based detection of leaf color levels in rice plants, Comput. Electron. Agric. 173 (2020) 105431.

[23] O.V. Monogarova, A.A. Chaplenko, K.V. Oskolok, Multisensory digital colorimetry to identify and determination of active substances in drugs, Sens. Actuators: B. Chem. 299 (2019) 126909.

[24] P.C.Feitosa R.S.Lamarca, L. Gomes, A low cost method for carbamazepine, ciprofloxacin and norfloxacin determination in pharmaceutical formulations based on spot-test and smartphone images, Microchem. J. MICROC 152 (2020) 104297.

[25] L. Shen, J.A. Hagen, I. Papautsky, Point-of-care colorimetric detection with a smartphone, Lab Chip 12 (2012) 4240–4243.

[26] I.P. Alvesa, N.M. Reis, Microfluidic smartphone quantitation of Escherichia coli in synthetic urine, Biosens. Bioelectron. 145 (2019) 111624.

[27] N.L. Ruiz, V. F.Curto, M.M. Erenas, F. Benito-López, D. Diamond, J.P. Alberto, L.F. López, Capitan-Vallvey, Smartphone-based simultaneous pH and nitrite colorimetric determination for paper microfluidic devices, Anal. Chem 86 (2014) 9554–9562.

[28] N. Phadungcharoen, N. Pengwanput, A. Nakapan, U. Sutitaphan, P. Thanomklom, N. Jongudomsombut, A. Chinsriwongkul, T. Rojanarata, Ion pair extraction coupled with digital image colorimetry as a rapid and green platform for pharmaceutical analysis: an example of chlorpromazine hydrochloride tablet assay, Talanta 219 (2020) 121271.

[29] J.P. Devadhasan, S. Kim, An ultrasensitive method of real time pH monitoring with complementary metal oxide semiconductor image sensor, Anal. Chim. Acta 858 (2015) e55–e59.

[30] S. Seidi, M. Rezazadeh, Y. Yamini, N. Zamanic, S. Esmaili, Low voltage electrically stimulated lab-on-a-chip device followed by red-green-blue analysis: a simple and efficient design for complicated matrices, Analyst 139 (2014) 5531–5537.

[31] R.A. Costa, C.L.M Morais, T.R. Rosa, P.R. Filgueiras, M.S. Mendonca, I.E.S Pereira, B.V. Vittorazzid, M.B. Lyrad, K.M.G. Lima, W. Romao, Quantification of milk adulterants (starch, H2O2, and NaClO) using colorimetric assays coupled to smartphone image analysis, Microchem. J. 156 (2020) 104968.

[32] J.L.D. Nelis, A.S. Tsgkaris, M.J. Dillon, J. Hajslova, C.T. Elliott, Smartphone based optical assays in the food safety field, Trends Anal. Chem., https://doi.org/10.1016/j.trac.2020.115934.

[33] A.W. Waller, M. Toc, D.J. Rigsby, J.E.Andrade M.Gaytán-Martínez, Development of a Paper-Based Sensor Compatible with a Mobile Phone for the Detection of Common Iron Formulas Used in Fortified Foods within Resource-Limited Settings, Nutrients 11 (2019) 1673, doi:10.3390/nu11071673.

[34] F. Soponar, A. Catalin Mot, C. Sarbu, Quantitative determination of some food dyes using digital processing of images obtained by thin-layer chromatography, J. Chromatogr. A 1188 (2008) 295–300.

[35] J.W. Lichtman, J. Conchello, Fluorescence Microscopy, Nature Publishing Group, 2, 12.

[36] C.F. Fronczek, T.S. Park, D.K. Harshman, A.M. Nicolini, J.Y. Yoon, Paper microfluidic extraction and direct smartphone-based identification of pathogenic nucleic acids from field and clinical samples, RSC Adv. 4 (2014) e11103–e11110.

[37] H.C. Koydemir, Z. Gorocs, D. Tseng, B. Cortazar, S. Feng, R.Y.L. Chan, J. Burbano, E. McLeod, A. Ozcan, Rapid imaging, detection and quantification of Giardia lamblia cysts using mobile-phone based fluorescent microscopy and machine learning, Lab Chip 15 (2015) e1284–e1293.

[38] H. Nguyen, Y. Sung, K. O'Shaughnessy, X. Shan, W.C. Shih, Smartphone nanocolorimetry for on-demand lead detection and quantitation in drinking water, Anal. Chem. 90 (2018) e11517–e11522.

[39] J.M.Sousa V.M¨uller, H. C.Koydemir, M. Veli, D. Tseng, L. Cerqueira, A.Ozcan, N.F. Azevedo, F. Westerlund, Identification of pathogenic bacteria in complex samples using a smartphone based fluorescence microscope, RSC Adv. 8 (2018) 36493.

[40] S.A. Lee, C. Yang, A smartphone-based chip-scale microscope using ambient illumination, Lab Chip 14 (2014) e3056–e3063.

[41] G. Singh, M. Sharma, Barcode technology and its application in libraries and Information centers, Int. J. Next Gen. Library and Technol. 1 (1) (2015) 1–8.

[42] L. Guan, J. Tian, R. Cao, M. Li, Z. Cai, W. Shen, Barcode-like paper sensor for smartphone diagnostics: an application of blood typing, Anal. Chem. 86 (2014) 11362–11367.

[43] J. Guo, J.X.H. Wong, C. Cui, X. Li, H.Z. Yu, A smartphone-readable barcode assay for the detection and quantitation of pesticide residues, Analyst 140 (2015) e5518–e5525.

[44] Y. Zhang, J. Sun, Y. Zou, W. Chen, W. Zhang, J.J. Xi, X. Jiang, Barcoded microchips for biomolecular assays, Anal. Chem. 87 (2015) e900–e906.

[45] Y. Lin, W. Cheung, F. Siao, Developing mobile 2D barcode/RFID-based maintenance management system, Autom. Constr. 37 (2014) 110–121.

[46] W. Clarke, P. D'Orazio, Chapter 9 - Electrochemistry, Editor(s): William Clarke, Mark A. Marzinke, Contemporary Practice in Clinical Chemistry (Fourth Edition), Academic Press, 2020, pp. 159–170. ISBN 9780128154991, https://doi.org/10.1016/B978-0-12-815499-1.00009-0.

[47] D. Zhang, J. Jiang, J. Chen, Q. Zhang, Y. Lu, Y. Yao, S. Li, G.Logan Liu, Q. Liu, Smartphone-based portable biosensing system using impedance measurement with printed electrodes for 2,4,6-trinitrotoluene (TNT) detection, Biosens. Bioelectron. 70 (2015) 81–88.

[48] J. Guo, Uric acid monitoring with a smartphone as the electrochemical analyzer, Anal. Chem. 88 (2016) e11986–e11989.

[49] J. Guo, X. Ma, Simultaneous monitoring of glucose and uric acid on a single test strip with dual channels, Biosensors and Bioelectronic 94 (2017) 415–419 BIOS9614.

[50] D. Ji, L. Liu, S. Li, C. Chen, Y. Lu, J. Wu, et al., Smartphone-based cyclic voltammetry system with graphene modified screen printed electrodes for glucose detection, Biosens. Bioelectron. 98 (2017) e449–e456.

[51] P. Kassal, J. Kim, R. Kumar, W.R. De Araujo, I.M. Steinberg, M.D. Steinberg, J. Wang, Smart bandage with wireless connectivity for uric acid biosensing as an indicator of wound status, Electrochem. Commun. 56 (6) (2015) e10.

[52] A. Ainla, M.P.S. Mousavi, M.N. Tsaloglou, J. Redston, J.G. Bell, M.T. Fern_andez-Abedul, G.M. Whitesides, Open-source potentiostat for wireless electrochemical detection with smartphones, Anal. Chem. 90 (2018) e6240–e6246.

[53] G.F. Giordano, M.B.R. Vicentinia, R.C. Murera, F. Augustob, M.F. Ferrãoc, G.A. Helferd, A.B. da Costad, A.L. Gobbia, L.W. Hantaoa, R.S. Lima, Point-of-use electroanalytical platform based on homemade potentiostat and smartphone for multivariate data processing, Electrochim. Acta 219 (2016) 170–177.

[54] M. Rezazadeh, M.L. S.Seidi, S. Pedersen-Bjergaard, Yadollah Yamini, The modern role of smartphones in analytical chemistry, Trends Anal. Chem. 118 (2019) e548–e555.

[55] H. Yu, H.M. Le, E. Kaale, K.D. Long, T. Layloff, S.S. Lumetta, B.T. Cunningham, Characterization of drug authenticity using thin-layer chromatography imaging with a mobile phone, J. Pharm. Biomed. Anal. 125 (2016) e85–e93.

[56] S. Lee, G. Kim, J. Moon, Performance improvement of the one-dot lateral flow immunoassay for aflatoxin b1 by using a smartphone-based reading system, Sensors 13 (2013) e5109–e5116.

[57] L. Yu, Z. Shi, C. Fang, Y. Zhang, Y. Liu, C. Li, Disposable lateral flow-through strip for smartphone-camera to quantitatively detect alkaline phosphatase activity in milk, Biosens. Bioelectron. 69 (2015) 307–315. http://dx.doi.org/10.1016/j.bios.2015.02.035.

[58] S.K. Vashist, T. Oordt, E.M. Schneider, R. Zengerle, F. Stetten, J.H.T. Luong, A smartphone-based colorimetric reader for bioanalytical applications using the screen-based bottom illumination provided by gadgets, Biosens. Bioelectron.

[59] Label-free detection: technologies, key considerations, and applications, (2019). Fetched from Label-free detection: technologies, key considerations, and applications Molecular Devices (ohsu.edu).

[60] A. Syahir, K. Usui, K. Tomizaki, K. Kajikawa, H. Mihara, Label and Label-Free Detection Techniques for Protein Microarrays, Microarrays(Basel) 4 (2) (2015) 228–244.

[61] D. Gallegos, K.D. Long, H. Yu, P.P. Clark, Y. Lin, S. George, P. Nath, B.T. Cunningham, Label-free biodetection using a smartphone, Lab Chip 13 (2013) e2124–e2132.

[62] O. Dalstein, D.R. Ceratti, C. Boissi_ere, D. Grosso, A. Cattoni, M. Faustini, Submicrometric Nanoimprinted, MOF-based 2D photonic structures: toward easy selective vapors sensing by a smartphone camera, Adv. Funct. Mater. 26 (2016) e81–e90.

[63] J. Zhang, I. Khan, Q. Zhang, X. Liu, J. Dostalek, B. Liedberg, Y. Wang, Lipopolysaccharides detection on a grating-coupled surface plasmon resonance smartphone biosensor, Biosens. Bioelectron. 99 (2018) e312–e317.

[64] T.S. Park, W. Li, K.E. McCracken, J.Y. Yoon, Smartphone quantifies Salmonella from paper microfluidics, Lab Chip 13 (2013) e4832–e4840.

[65] F. Giavazzi, M. Salina, E. Ceccarello, A. Ilacqua, F. Damin, L. Sola, M. Chiari, B. Chini, T. Bellini, R. Cerbino, M. Buscaglia, A fast and simple label-free immunoassay based on a smartphone, Biosens. Bioelectron. 58 (2014) e395–e402.

[66] P.S. Francis, C.F. Hogan, Chapter 13 - Luminescence. S.D. Kolev, I.D. Mckelvie (Eds.), Comprehensive Analytical Chemistry, Elsevier, vol. 54, 2008, pp. 343–373, ISSN 0166-526X, ISBN 9780444530943. https://doi.org/10.1016/S0166-526X(08)00613-2.

[67] J. Canning, A. Lau, M. Naqshbandi, I. Petermann, M.J. Crossley, Measurement of fluorescence in a rhodamine-123 doped self-assembled "giant" mesostructured silica sphere using a smartphone as optical hardware, Sensors 11 (2011) e7055–e7062.

[68] V.K. Rajendran, P. Bakthavathsalam, B.M. Jaffar Ali, Smartphone based bacterial detection using biofunctionalized fluorescent nanoparticles, Microchim. Acta. 181 (2014) e1815–e1821.

[69] S. Shrivastava, W. IlLee, N. Lee, Culture free, highly sensitive, quantitative detection of bacteria from minimally processed samples using fluorescence imaging by Smartphone, Biosens. Bioelectron. 109 30 June (2018) 90–97.

[70] A.M. Nicolini, C.F. Fronczek, J.Y. Yoon, Droplet-based immunoassay on a "sticky" nanofibrous surface for multiplexed and dual detection of bacteria using smartphones, Biosens. Bioelectron. 67 (2015) e560–e569.

[71] N. Li, Y.Q. Zhong, S.G. Liu, et al., Smartphone assisted colorimetric and fluorescent triple channel signal sensor for ascorbic acid assay based on oxidase-like CoOOH nanoflakes, Spectrochim. Acta Part A (2020). https://doi.org/10.1016/j.saa.2020.118412.

[72] T. Liu, S. Zhang, W. Liu, S. Zhao, Z. Lu, Y. Wang, G Wang, P Zou, X Wang, Q Zhao, H Rao, Smartphone based platform for ratiometric fluorometric and colorimetric determination H2O2 and glucose, Sensors and Actuators: B. Chem. (2019). https://doi.org/10.1016/j.snb.2019.127524.

[73] A. Roda, E. Michelini, L. Cevenini, D. Calabria, M.M. Calabretta, P. Simoni, Integrating biochemiluminescence detection on smartphones: mobile chemistry platform for point-of-need analysis, Anal. Chem. 86 (2014) e7299–e7304.

[74] J. Song, V. Pandian, M.G. Mauk, H.H. Bau, S. Cherry, L.C. Tisi, C. Liu, Smartphone-based mobile detection platform for molecular diagnostics and spatiotemporal disease mapping, Anal. Chem. 90 (2018) e4823–e4831.

[75] E. Petryayeva, W.R. Algar, A job for quantum dots: use of a smartphone and 3D-printed accessory for all-in-one excitation and imaging of photoluminescence, Anal. Bioanal. Chem. 408 (2016) e2913–e2925.

[76] Q. Mei, H. Jing, Y. Li, W. Yisibashaer, J. Chen, B. Nan Li, Y. Zhang, Smartphone based visual and quantitative assays on upconversional paper sensor, Biosens. Bioelectron. 75 (2016) e427–e432.

[77] M. He, Z. Li, Y. Ge, Z. Liu, Portable upconversion nanoparticles-based paper device for field testing of drug abuse, Anal. Chem. 88 (2016) e1530–e1534.

[78] M. Schäfer, V. Bräuler, R. Ulber, Bio-sensing of metal ions by a novel 3D-printable smartphonespectrometer, Sens. Actuators B (2017).

[79] Statistical Analysis Methods for Chemists: A Software-based Approach, William P. Gardiner Department of Mathematics, Glasgow Caledonian University, Glasgow, UK.

[80] Hui Chen, Chao Tan, Zan Lin, Express detection of expired drugs based on near-infrared spectroscopy and chemometrics: a feasibility study, Spectrochim. Acta Part A (2019) Volume 22, 05 September 117153.

[81] M.J. Vredenbregt, L. Blok-Tip, R. Hoogerbrugge, D.M. Barends, D. de Kaste, Screening suspected counterfeit Viagra® and imitations of Viagra® with near-infrared spectroscopy, J. Pharm. Biomed. Anal. 40 (2006) 840–849 March.

[82] J. Omar, A.B. B.Slowikowski, Chemometric approach for discriminating tobacco trademarks by near infrared spectroscopy, Forensic Sci. Int. 29 4 January (2019) 15–20.

[83] C. Martín-Alberca, F. Zapata, H. Carrascosa, F.E. Ortega-Ojeda, C. García-Ruiz, Study of consumer fireworks post-blast residues by ATR-FTIR, Talanta 149 (2016) 257–265.

[84] M. Mosleh, S.M. Ghoreishi, S. Masoum, A. Khoobi, Determination of quercetin in the presence of tannic acid in softdrinks based on carbon nanotubes modified electrode using chemometric approaches, Sens. Actuators B 27 21 November (2018) 605–611.

[85] P.K. Sarswat, M.L. Free, Light emitting diodes based on carbon dots derived from food, beverage, and combustion wastes, Phys. Chem. Chem. Phys 17 (2015) 27642–27652.

[86] D. Lambert, C. Muehlethaler, G.Massonnet P.Esseiva, Combining spectroscopic data in the forensic analysis of paint: application of a multiblock technique as chemometric tool, Forensic Sci. Int. 26 (3) June (2016) 39–47.

[87] P. Peets, I. Leito, J. Pelt, Signe Vahur, Identification and classification of textile fibres using ATR-FT-IR spectroscopy with chemometric methods, Spectrochim. Acta Part A 173 (2017) 175–181.

[88] V. Sharma, A. Bharti, R. Kumar, On the spectroscopic investigation of lipstick stains: forensic trace evidence, Spectrochim. Acta Part A 215 (2019) 48–57.

[89] R. Chauhan, R. Kumar, V. Sharma, Soil forensics: a spectroscopic examination of trace evidence, Microchem. J. 139 (2018) 74–84.

[90] Raj Kumar, Kajal Sharma, Vishal Sharma, Bloodstain age estimation through Infrared spectroscopy and chemometric models, Sci. Justice (2020) In press, journal pre-proof Available online 16 July.

[91] V. Sharma, R. Kumar, Fourier transform infrared spectroscopy and high performance thin layer chromatography for characterization and multivariate discrimination of blue ballpoint pen ink for forensic applications, Vib. Spectrosc. 92 (2017) 96–104.

[92] L. Valderrama, P. Valderrama, Nondestructive identification of blue pen inks for documentoscopy purpose using iPhone and digital image analysis including an approach for interval confidence estimation in PLS-DA models validation, Chemom. Intell. Lab. Syst. 156 (2016) 188–195.

[93] W. Song, N. Jiang, H. Wang, J. Vincent, Use of smartphone videos and pattern recognition for food authentication, Sens. Actuators B 304 1 February (2020) 127247.

[94] W. Song, Z. Song, J. Vincent, H. Wang, Z. Wang, Quantification of extra virgin olive oil adulteration using smartphone videos, Talanta 216 (2020) 120920.

[95] H. Parastar, H. Shaye, M.V.C. App: A smartphone application for performing chemometric methods chemometrics and intelligent laboratory systems, 147, 15 October 2015, pp. 105–110.

[96] F.C. Böck, G.A. Helfer, A.B. da Costa, M.B. Dessuy, M.F. Ferrão, PhotoMetrix and colorimetric image analysis using smartphones, J. Chemom. (2020) e3251.

Reconstruction of human movement from large-scale mobile phone data

Haoran Zhang[a], Yuhao Yao[b]
[a]LocationMind Inc, Japan
[b]Center for Spatial Information Science, The University of Tokyo, Tokyo, Japan

9.1 Introduction

9.1.1 Background

Human mobility estimation provides sufficient information for urban planning and transportation management [1,2], which plays an important role in improving urban mobility, accessibility, and quality of residents' life. On the one hand, individual trajectory estimation helps to track specific target, which is important in criminal investigation and infectious disease tracking. The spread of Coronavirus at the beginning of 2020 alarms us that being able to collect individual's trajectory is extremely significant. If the government could collect the trajectory information of each resident, the spreading of the virus could be controlled very swiftly. On the other hand, statistic movement data such as Origin Destination (OD) matrices, Traffic Volume, etc. helps to optimize transportation networks to adapt residents' requirement not only at present but also fit future needs.

In the past, this was usually accomplished by household questionnaires, road surveys or some sensor-based methods in small sample sizes and low frequencies. The shortcomings are obvious: they are limited in both spatial and temporal scale. With the development of positioning technologies, trajectory reconstruction based on GPS data becomes general [3–7]. Comparing with traditional methods, it has extremely high accuracy profited by high precision of GPS. However, it still cannot solve the problems of limited sample scale and strong bias since GPS device is usually set on vehicles.

Call Detail Record (CDR) is a kind of mobile phone data that generated whenever a mobile terminator device has data interaction with a cell tower and it records the related information of the data interaction [8,9], including device's ID and cell tower's position. To be more specific, the data interaction event includes: the beginning, procedure, and end of a placed or received call; Sending or receiving a message; Running application interacting with the Internet. With the development of mobile communication technologies, intelligent terminals are strong enough to support applications running in background to continuously interact with the Internet. Meanwhile, the popularization of SNS applications such as Line, Facebook, Twitter, etc. make them almost always online in most utilizers' smartphones, which means CDR could achieve significant short sampling interval and high coverage time period. Therefore,

Smartphone-Based Detection Devices. DOI: https://doi.org/10.1016/B978-0-12-823696-3.00015-5
Copyright © 2021 Elsevier Inc. All rights reserved.

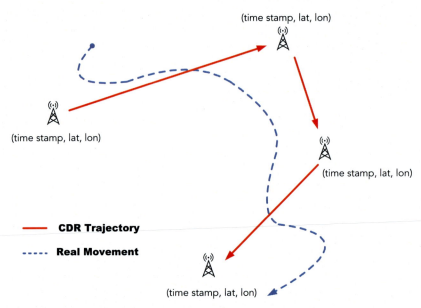

Fig. 9.1 The Comparison between CDR trajectory and real movement.

CDR becomes extremely general and could be utilized as a great data source for many human mobility related studies and applications [10–13].

However, the position of cell tower only indicates a course range of the mobile terminator device's position, which results in the massive difference between CDR trajectory and the real movement as Fig. 9.1 shows. How to extract the ground truth position of the device's holder from connected cell tower's position becomes the main challenge.

To address this challenge, trajectory reconstruction methods for CDR could be roughly divided into three types:

- Map-matching-based methods: In all methods of this type, there is a basic assumption that people always move on the road network [14–16]. Then each location point in the trajectory is corresponding with a road segment in the road network and these road segments form the complete trajectory. To find the corresponding road segment, Jagadeesh et al. [17] utilized a Hidden Markov Model to generate coarse map-matched trajectories and identified it based on a pre-training route selecting model to get more accurate result. Xiao et al. [18] reconstruct trajectories from CDR based on a conditional random field model, in which the contextual relationships between different trajectory points are utilized as indicator features.
- Interpolation-based methods: Methods in this type focus on interpolating the missing location information due to long time interval based on spatial-temporal correlations among data. They assume that the relationship between the missing points and the given points could be described as a simple mathematical model such as Gaussian function. For interpolation, features like distances and time spans between missing point and its contextual points is utilized to measure weights and estimate missing point's position [19–22].
- Pattern-learning-based methods: With the development of machine learning, utilizing artificial intelligence to learn the latent pattern of human mobility becomes efficient. historical

trajectories are utilized to explore the characteristics of human movement behavior, which greatly improve various models. For instance, Fan et al. [23] proposed a collaborative filtering method based on LDA topic model to infer the movement of people. Liu et al. [24] utilized a back propagation neural network based on the gyration radius as indicator to estimate mobile phone user's position. Li et al. [25] proposed a multi-criteria data partitioning technique to measure the similarity among individuals for reconstruction model.

A map-matching-based method will be introduced to demonstrate a standard framework of trajectory reconstruction for CDR, which could be divided into several steps: data filtering, mode detection and trip segmentation, map-matching and interpolation.

9.1.2 Dataset

In this research, two datasets are utilized:

- The field research CDR dataset collected in Hongkong, China: Data is collected by a programed smartphone application, which could record the CDR in time interval of 5 min and corresponding GPS position in time interval of 30 s. Six researchers holding signal receivers and smartphones in both Android and IOS systems with application running randomly moved in Hongkong, China. Researchers randomly changed moving mode (by foot, car, bus and metro) and manually recorded the mode and change time. The field research lasted for 5 days and every day the program ran at least 6 h. More than 2,500 CDR are recorded.
- The simulated CDR dataset based on GPS data in Tokyo, Japan: The GPS data utilized is a part of *Konzatu-Tokei(R)* provided by *Zenrin DataCom INC*. *Konzatsu-Tokei (R)* Data refers to people flow data collected by individual's location data sent from mobile phone under utilizers' consent, through Applications provided by *NTT DOCOMO, INC*. These data are processed collectively and statistically in order to conceal the private information. Original location data is GPS data (latitude, longitude) sent in about every 5 min and does not include the information to specify individual. CDR data is simulated by a simulation model from GPS data.

Several other spatial information is also utilized:

- Road network: We utilize road network to apply interpolation and map-matching. The road network is extracted from *OpenStreetMap* (*OSM*). *OSM* is an online map collaboration project, which aims to create a world map with free content that can be edited by everyone.
- Cell tower distribution maps: We utilize cell tower distribution maps to simulate CDR data. The cell tower distribution map is constructed from *Opencellid*. *Opencellid* is the largest Open Database of Cell Towers and their locations.
- Tokyo land use map: We estimate the land type based on the Tokyo land use map for CDR simulation. The Tokyo land use map is an open data which contains the information of land use of whole Japan, provided by Ministry of Land, Infrastructure, Transport and Tourism of Japan government.

9.2 Data filtering

By analyzing the field research CDR data, we mainly detected two harmful phenomena: *Pendulum Phenomenon* and *Flash Phenomenon*.

Selecting which cell tower to connect is a complex process for mobile terminator device, which depends not only on the distance, but also on the environment and the load state of cell towers. Therefore, sometimes even the device holder does not move, the change of the signal environment may cause the selected cell tower change, which we call it *Pendulum Phenomenon*. Moreover, due to some error of the system, sometimes one CDR may lose the cell tower ID, and for default it will contain an incorrect cell tower id which was connected in the past, we call it *Flash Phenomenon*.

9.2.1 Pendulum phenomenon

This phenomenon usually happens when the mobile terminator device holder stays at the junction of several cell towers' coverage area. Because the signal is not stable, the connected cell tower may change rapidly between different cell towers though the device does not move, which will mislead the mode detection step.

Notice that when this phenomenon happens, in a short period the mobile terminator device only connects with specific cell towers around, which is different from the device holder is really moving and the device continuously connect with different cell towers. Therefore, we detect and filter it as follows:

- Assume that the set of CDR of an individual u in date d is $R(u, d) = \{r_0, r_1, ..., r_n\}$, where r is one CDR that contains at least the timestamp t and the position of the cell tower p. We apply a sliding window $W = \{r_k, r_{k+1}, ..., r_{k+l}\}$ with length $l + 1$ where $k \geq 0, k + l \leq n$, which contains $l + 1$ CDR
- We start W with $k = 0$ and sliding step s. Every time we check whether the number of unique cell tower in the sliding window is smaller than a threshold m. If so, it means during the period of records in this sliding window, the device holder has repeated some routes. Then we start to temporally expand the sliding window until the number of unique cell tower will also raise, the length is l'.
- We define a maximum cluster distance threshold ΔS and the Eq. (9.1).

$$maxdistance(p_{r_i}, p_{r_j}) < \Delta k < i, j < k + l' \qquad (9.1)$$

If the records in the sliding window fit the formula, it means records in the window forms a pendulum phenomenon. Then we utilize the most frequent cell tower position to replace the others in the window.

In this step, parameter l, s and m is related with time interval of records and ΔS is related with the density of cell tower.

9.2.2 Flash phenomenon

This phenomenon happens randomly, sometimes when the device holder keeps moving in one direction, one record jumps back to a past cell tower, which may dramatically interfere with the speed estimation and trip segmentation.

Notice that when this phenomenon happens, the noise record usually appears out of the moving tendency and has high probability to be far away from the real position, which will cause the cell tower transferring speed extremely high. Therefore, we apply two rules to check incorrect CDR.

9.2.2.1 Triangle rule

Assume there are three points A, B and C. If point A, B and C are in the same moving trend, the angle they form ∠ABC usually has a large degree unless the moving trend just make a turn between point A and C. Thus we could determine whether point B is out of the trend or not by ∠ABC. For any CDR series $R_k = \{r_k, r_{k+1}, r_{k+2}\}$ of an individual u in a date d, where $k \geq 0$ and $k + 2 \leq m$, set an angle threshold $\Delta\theta$ as the minimum angle for normal situation, based on the law of cosines we have the Eq. (9.2):

$$\cos^{-1} \frac{distance(p_{r_k}, p_{r_{k+1}})^2 + distance(p_{r_{k+1}}, p_{r_{k+2}})^2 - distance(p_{r_k}, p_{r_{k+2}})^2}{2 distance(p_{r_k}, p_{r_{k+1}}) distance(p_{r_{k+1}}, p_{r_{k+2}})^2} < \theta \quad (9.2)$$

If the CDR series fit the formula (2.2), then the middle CDR may be a noise record and we come to the second rule.

9.2.2.2 Speed rule

When the device holder moves in a low speed like go for a walk, the mobile terminator device may connect to different cell towers in the same small area, which may also generate lots of acute triangles that fit the formula of Triangle Rule.

To avoid this occasion, we need to check the connected cell tower transferring speed. We define a minimum cell tower transfer speed threshold ΔV and Eq. (9.3):

$$\frac{distance(p_{r_k}, p_{r_{k+1}})}{t_{r_{k+1}} - t_{r_k}} > \Delta V, \frac{distance(p_{r_{k+1}}, p_{r_{k+2}})}{t_{r_{k+2}} - t_{r_{k+1}}} > \Delta V \quad (9.3)$$

If the CDR series also fit the Eq. (9.3), then the middle CDR is a noise record generated by error and we just remove it.

9.3 Mode detection and trip segmentation

9.3.1 Problem description

We divide moving mode into four types: *Stay*, *Non-motorized*, *Motor* and *Metro*. *Stay* type means the device holder stops moving and stays in a small region. *Non-motorized* type means the device holder moves without motor vehicle, usually by foot or bicycle. *Motor* type means the device holder moves by motor vehicle, usually by own car or bus. *Metro* means the device holder moves by metro, including underground and ground. We correspondingly utilize {1, 2, 3, 4} to code them.

For a CDR set of an individual u in date d as $R(u, d) = \{r_0, r_1, ..., r_n\}$, where r is one CDR that contains timestamp t and the position of the cell tower p, we could have the moving mode sequence $S(u, d) = \{s_1, s_2, ..., s_n\}$, $s_i \in \{1, 2, 3, 4\}$, where m_i is the moving mode between r_{i-1} and r_i. Since moving mode sequence could not be observed directly, we need to estimate $S(u, d)$ based on $R(u, d)$.

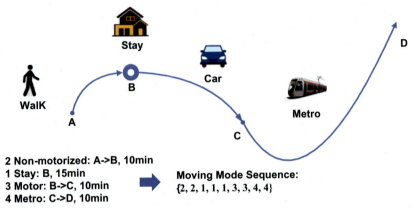

Fig. 9.2 Concept of mode detection.

For example, as Fig. 9.2 shows, if one mobile terminator device holder walks at first 10 min to go back home, stay at home for 15 min then take a taxi for 10 min, finally take the metro for 10 min, a 5 min interval moving mode sequence should be {2, 2, 1, 1, 1, 3, 3, 4, 4}.

Based on moving mode sequence, we could divide the whole day trajectory into different segments by moving mode changing.

9.3.2 Hidden markov model

If we describe moving mode changing as a Markov process, the moving mode sequence could be regarded as hidden states sequence and we need to estimate it based on the observable information. We apply a Hidden Markov Model to solve it.

Hidden Markov Model (HMM) is a statistical model that describes a Markov process with unknown parameters. Its states cannot be observed directly but could be indirectly observed through the observation vector sequence. Each observation vector is expressed as various states through some probability density distribution, and each observation vector is generated by a state sequence with corresponding probability density distribution. Therefore, HMM is a dual stochastic process, i.e. a set of hidden Markov chains with a certain number of states and a set of dominance random functions. As an important direction of signal processing, it has been successfully utilized in speech recognition, behavior recognition, character recognition, fault diagnosis and other fields.

An HMM could be described as 5 elements: hidden states S, observable states O, initial state probability matrix π, state transfer probability matrix A and emission probability matrix B. In our model, we create two layers v and t for observable states, therefore we have two observable states O_v and O_t to form O and two emission probability matrix B_v and B_t to form B.

Hidden states S in our model represents moving mode of device holders, i.e. *Stay*, *Non-motorized*, *Motor* and *Metro*.

9.3.2.1 Transferring speed state

Observable states O_v relate to the connected cell tower transferring speed. We divide transferring speed into four types: *Stop, Low Speed, Medium Speed* and *High Speed* and correspondingly utilize {0, 1, 2, 3} to code them. We create a connected cell tower transferring speed distribution boxplot of different moving modes from the field research as Fig. 9.3 shows.

From the graph, we can see that the transferring speed of moving by foot is usually lower than 6 km/h, sometimes equals to 0 or more than 6 km/h, but never more than 15 km/h. The transferring speed of moving by motor vehicle is usually more than 15 km/h, sometimes between 6 and 15 km/h. Metro is similar with motor type, but with higher average speed and could achieve lower than 6 km/h due to unstable signal. Although the sampling conditions may affect the speed distribution, the influence is very small, so we just base on this boxplot to build the observable state. For the connected cell tower transferring speed mode sequence of an individual u in date d as $V(u, d) = \{v_1, v_2, \ldots, v_n\}$, $v_i \in \{0, 1, 2, 3\}$, where v_i is the connected cell tower transferring speed mode between r_{i-1} and r_i, we have:

$$v_i = \begin{cases} 0, \dfrac{distance(p_{r_{i-1}}, p_{r_i})}{t_{r_i} - t_{r_{i-1}}} = 0 \\ 1, 0 < \dfrac{distance(p_{r_{i-1}}, p_{r_i})}{t_{r_i} - t_{r_{i-1}}} \leq 6 \\ 2, 6 < \dfrac{distance(p_{r_{i-1}}, p_{r_i})}{t_{r_i} - t_{r_{i-1}}} \leq 15 \\ 3, \dfrac{distance(p_{r_{i-1}}, p_{r_i})}{t_{r_i} - t_{r_{i-1}}} > 15 \end{cases}, 1 \leq i \leq n \quad (9.4)$$

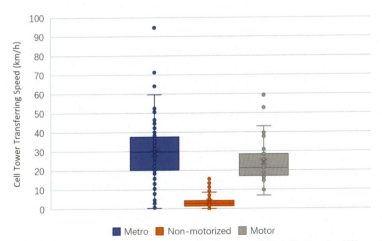

Fig. 9.3 Boxplot of connected cell tower transferring speed distribution in different moving modes.

9.3.2.2 Cell tower type state

Observable states O_t depend on the type of the connected cell tower. Basically, cell tower could be classified based on the position where they were set. In addition, where they were set could affect their coverage area. For example, cell tower set on the roof of tall building is flexible, usually could be connected to both inside that building and outside; cell tower set inside the metro station could only be connected to inside the station; cell tower set in a tunnel could just be connected to while passing the tunnel and etc. Based on this, we divide the type of connected cell tower into five types: *Outdoor, Indoor, Mix, Tunnel* and *MT* and correspondingly utilize {0, 1, 2, 3, 4} to code them.

- *Outdoor* means the cell tower is set at some open area and could only be connected to outside.
- *Indoor* means the cell tower is set inside some tall buildings, usually shopping mall, that could only be connected to inside the building.
- *Mix* means the cell tower is set on a tall building's roof and could be connected to both outdoor and indoor.
- *Tunnel* means the cell tower is set inside the tunnel and could only be connected to inside the tunnel.
- *MT* means the cell tower is set inside the metro station and could only be connected to inside the station.

Usually, the cell tower type could be found in CDR. If CDR does not cover it, we could just focus on type 4 (*Metro*) by checking whether the cell tower is near the metro station. If not, we just set the cell tower type a default value 2 (*Mix*).

9.3.3 Model training

Basically, the model training method could be divided into supervised learning and unsupervised learning two ways. Because the ground truth data is not sufficient to calculate the accurate parameters, we need to utilize unsupervised method.

Usually, *Baum-Welch Algorithm* is utilized to train HMM parameters as unsupervised learning method. Because the parameter estimation of hidden Markov model is a maximum likelihood estimation of hidden variables, *Baum-Welch Algorithm* utilize *Expectation-Maximization (EM) algorithm* to solve the above-mentioned parameter estimation problem. From *EM algorithm*, it gets Q function, from which the partial derivative of Q function and the extreme value of maximum likelihood function could be calculated. The properties of node graph help to calculate the joint probabilities to reduce the complexity of node calculation with intermediate variables.

However, due to the loss of information, the result of *Baum-Welch Algorithm* is not good enough. For our case, its max estimation accuracy is no more than 60 percent. Meanwhile, indeed we have sufficient priori knowledge about the relationship between two observable states and hidden states. Therefore, we generate several rules based on prior knowledge and manually construct each probability matrix based on those rules.

For example:

1. *It tends to maintain its hidden state rather than change it*
2. *It tends to start with "Stay"*

3. *"Stay"* state only act as *"Stop"*
4. *"Non-motorized"* state can also act as *"Stop"* but no more than 10 min
5. *"Low Speed"* tends to be *"Non-motorized"*, but sometimes *"Car"* or *"Metro"*
6. *"MT"* cell tower to *"MT"* cell tower must be *"Metro"*

After constructing the probability matrices, we manually do some simple adjust to correct them based on ground truth.

9.3.4 Decoding

We utilize *Viterbi Algorithm* to decode the hidden states sequence. *Viterbi algorithm* is proposed by *Andrew Viterbi* in 1967 for deconvolution in digital communication links to eliminate noise. It is a dynamic programming algorithm utilized to find the most likely hidden states sequence that may generate the observable states sequence, especially in the context of Markov information source and Hidden Markov Model. Nowadays, it is widely utilized in speech recognition, keyword recognition, computational linguistics, and bioinformatics.

For a given Hidden Markov Model S, assume that the probability of initial state i is π_i, the probability of state i transferring into j is $a_{i,j}$, the emission probability of hidden state i to observable state j is $b_{i,j}$. If the observable sequence is $[y_1, y_2, ..., y_T]$, and the corresponding most likely hidden states sequence is $[x_1, x_2, ..., x_T]$, based on *Viterbi algorithm* we could have:

$$V_{1,k} = b_{k,y1} * \pi_k \tag{9.5}$$

$$V_{t,k} = B_{k,yt} * max_{x \in x}(a_{x,k} * V_{t-1,x}) \tag{9.6}$$

Where $V_{t,k}$ is the probability of the first t events' most likely hidden states sequence of which the final state is k. We could save the states x in Eq. (9.6) to get the Viterbi path. Define a function $Path(k, t)$, when $t = 1$, it returns k; else it returns x which is utilized in calculating $V_{t,k}$. Then we can get the hidden states sequence by:

$$x_T = argmax_{x \in s}(V_{T,x}) \tag{9.7}$$

$$x_{t-1} = Path(x_t, t) \tag{9.8}$$

9.4 Map-matching and interpolation

9.4.1 Problem description

Map-matching is a process in which the orderly GPS positions sequence of vehicles are associated with the road network of electronic map, and the sampling sequence under the GPS coordinate is converted into sequence under the coordinate of road network.

For CDR, map-matching means to estimate the ground truth position on the road network based on the corresponding connected cell tower's position. Different from

GPS map-matching, the distance between the known position and ground truth position is usually very large, which makes it very complex to estimate the real position.

Assume that the road network is G, the time interval of the CDR is T, a CDR trip segment in moving mode M of an individual u starting at time t as $R(u, t, M) = \{r_0, r_1, \ldots, r_n\}$, the corresponding ground truth position sequence is $P(u, t) = \{p_0, p_1, \ldots, p_n\}$, p_i is the position of u at $t + T*i$, where $p_i \in G$, $0 \le i \le n$. We need to estimate $P(u, t)$ based on $R(u, t, M)$.

After map-matching, a sparse trajectory with original CDR time interval T has been obtained. However, the original CDR time interval is usually relatively large, which makes the route between every two points still unclear. Therefore, we need to interpolate the sparse trajectory into a fine trajectory. That is, for a coarse position sequence $P(u, t) = \{(p_0, t), (p_1, t+T), \ldots, (p_n, t+T*n)\}$, p_i is the position of u at $t+T*i$, where $p_i \in G$, $0 \le i \le n$ we need to estimate a fine position sequence as $P'(u, t) = \{(p_0', t_0), (p_1', t_0), \ldots, (p_m', t_m)\}$, p_i' is the position of u at t_i, where $p_i' \in G, (p_i', p_{i+1}') \in G, 0 \le i \le m$.

9.4.2 Multi-Steps least cost algorithm

For a given cell tower position Y, assume that the nodes of road network in its coverage area with radius R is nodes set $S = \{node_0, node_1, \ldots, node_n\}$. In consideration of that due to the information loss, restore the precise position of the mobile terminator holder as what GPS can is impossible, we assume that every time when the CDR is generated, the device holder is at one node of the road network. Therefore, it is possible for all the nodes in S to be a candidate ground truth position of the mobile terminator holder who connect to the given cell tower. Since observable attributions of CDR sequences are too limited, the probability of each node to be the ground truth is mainly indicated by two factors:

- The probability for the node to choose the cell tower *celltower$_y$*:

Assume that the cell towers set $C = \{celltower_1, celltower_2, \ldots, celltower_y, \ldots, celltower_n\}$ is the set of which the coverage area includes the node Q, the probability for the node to choose cell tower P means the probability for Q to connect to *celltower$_y$* instead of other cell towers in C.

It could be calculated by *Maximum Likelihood Estimation* when data is sufficient, i.e. using the times of utilizer staying at node Q and mobile terminator device connecting to cell tower Y to divide the total times of utilizer staying at node Q.

- The least impedance of the paths between two continuous points.

Same with GPS map-matching, the least impedance (usually the least time cost) of the paths between two continuous points is usually a significant indicator. Two nodes that are easy to move from one to the other is obviously more reasonable than two nodes that are nearly impassable.

To calculate the impedance, traditional GPS map-matching methods usually utilize the length of the path. Some methods also assign average speed to each road segment based on its road type and calculate the time cost, which is more reasonable. *Dijkstra Algorithm* is applied to find the least impedance path between two points.

However, those methods ignore an important feature, that is the moving mode of device holder. For GPS data, using the length of the road to divide the assigned average speed as impedance may be reasonable, because GPS data is usually collected by vehicles, which need to obey the speed limitation of the road and have different speed in different roads. When it comes to CDR data, things are different.

While the mobile terminator device holder is driving or taking motor vehicles, of course time cost of the trip could be calculated by the road length and assigned speed. But when the device holder is taking non-motorized way, the type of the road almost does not affect the holder's moving speed. For instance, if the device holder needs to move from A to B. There are two routes for the holder to choose: one is a direct small path with length of 0.2 km, and the other is a main road but need to take a big turn, of which the length is 0.5 km. We assume the average speed for vehicles in small path is 20 km/h and in main road is 60 km/h. If the moving mode of the device holder is *motor* type, obviously he will choose the main road because the time cost is 0.5 km/ (60 km/h), nearly 30 s while small path will take 0.2 km/ (20 km/h) nearly 36 s. But if his moving mode is *non-motorized* type, of which the moving speed is a constant value, such as 5 km/h, he will choose the small path which is shorter with no doubt. In addition, if the mobile terminator device holder is in *metro* mode, he must move on metro line, which will be far away from the route of shortest path.

Therefore, we apply different strategy for different moving mode:

- *Non-motorized* Mode: For this moving mode, we utilize the whole road network, and just utilize the length of the road as the road impedance.
- *Metro* Mode: For this moving mode, we extract metro lines from the road network to compose the metro road network. We just utilize the length of the road as the road impedance. When the time interval is not large, the cost will not affect too much because usually there is just one route could be selected.
- *Motor* Mode: For this moving mode, we first remove all roads with road types that cannot find the corresponding speed limitation, such as *path*, *pedestrian*, to form the motor road network. Because these road types are not for vehicles but just for pedestrian. Then we assign the speed limitation as the average speed to each road type. The speed limitation of each road type varies in different country.

Assume that the length of the road is L, the speed limitation of the road is V, the basic impedance I of the road is:

$$I = \frac{L}{V} \tag{9.9}$$

However, the impedance of the road should not be constant. This kind of calculation method ignore the load of the road. When the number of vehicles on the road is smaller than the capacity of the road, which means there is no traffic jam on this road, vehicles could move in average speed and the impedance make sense. Once on some core roads, especially during the commute time, the vehicles are too much that exceed the capacity of the road, the average speed will dramatically drop down. If we simply utilize this impedance as the indicator for map-matching and interpolation, the result will be that all vehicles prefer to select several core roads ignoring the number of vehicles on this road, and the number of vehicles on these roads will seriously exceed the ground truth number.

To solve it, we combine the impedance with *The Bureau of Public Roads Function*. *The Bureau of Public Roads (BPR) Function* is utilized to calculate the free travel time of road sections. In the traffic assignment stage of the stages for traffic planning, the time impedance of a certain road section should be considered to allocate the traffic flow. The correction of road travel time can be determined according to the relationship between travel time and road traffic volume, i.e. the road impedance function.

The BPR function could be described as:

$$t_i = t_{i0} * (1 + \alpha(\frac{Q}{C})\beta) \tag{9.10}$$

Where t_i is the actual time cost required to pass through the road section. t_{i0} is the free time cost of the road section, i.e. assume that there is no other vehicle on the road. Q is the traffic volume passing through the road section at that time. C is the actual traffic capacity of the road section. α and β are the undetermined parameters of the model, of which the recommended values are respectively 0.15 and 4. However, since the traffic environment varies in different country, the value should be determined according to the actual situation.

In consideration of that the actual capacity of each road section is unknown from the road network, but the number of lines in each road is known, which is related to the capacity, we utilize the number of lines multiply a coefficient to replace the capacity. The coefficient could be merged into α. To save calculating time, we save the temporal traffic volume in $Q(road, timewindow)$, every time when a new trajectory is determined, the traffic volume will be updated. In this case, at first, Q for each road is empty, route will be chosen for free. With more and more trips interpolated at that time, route choosing will be seriously affected by traffic volume. In this way, we do not need calculate it twice to get the traffic volume first and re-allocate the volume then. Although in this way the trajectory for individual may not be close to the ground truth because which trip is earlier is random, the aggregated statistic data will be close to the ground truth like traffic volume.

The new impedance function could be constructed as:

$$I = \frac{L}{V} * (1 + \alpha(\frac{Q}{C})\beta) \tag{9.11}$$

Parameters α and β is determined by dataset scale.

The problem of map-matching and interpolation could be transferred into a kind of multi-steps shortest path problem as Fig. 9.4 shows:

Then, it could be solved together by a multi-step Dijkstra algorithm as follow:

Assume that $G = (V, E)$ is the graph, where V is the set of vertices, E is the set of edges, $\Delta t (v_i, v_j)$ is the cost between vertex v_i and vertex v_j. Start vertex set is $C_0 = \{v_{start}\}$, End vertex set is $C_n = \{v_{end}\}$, middle nodes set $C_i = \{v_0, v_1, ..., v_k\}$ from a given nodes set sequence $C = \{C_1, C_2, ..., C_{n-1}\}$ need to be passed. Utilize C_{last} to present the last nodes cluster need to pass and C_{next} to present the next nodes cluster need to pass. Prepare a set S to represent the vertices that have been checked, then $V - S$ is the vertices that have not been checked. We utilize a vector *dist* to store the

Reconstruction of human movement from large-scale mobile phone data

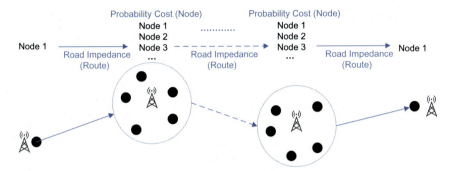

Fig. 9.4 Multi-steps shortest path problem of map-matching and interpolation.

cost between the points, a vector *path* to store the passed by vertices, which interprets what the shortest path is, and a vector *time* to store the arriving time for vertices.

- Step 1: Initialize S as $\{v_{start}\}$. $C_{last} = C_0$, $C_{next} = C_2$. Construct the sub-graph $G' \in G$ between C_{last} and C_{next} as the temporal graph. For vertex v_i in $V - S$, if $\Delta t (v_0, v_i)$ exists, set $dist(v_i) = \Delta t(v_0, v_i)$ and record v_0 into $path(v_i)$; else set it *NAN*
- Step 2: For any point in S, obtain the distance between it and any point in $V - S$ from *dist*. Select the point with the least cost and record it, which is the new point for shortest path that could be determined. Assume the last vertex as min, put min into S.
- Step 3: Set min in Step 2 as the intermediate node, find all the vertices in $V - S$, set it as K to check whether $dist(min) + \Delta t(min, K) < dist(k)$ If so, update $dist(K)$ into the value of $dist(min) + E(min, K)$, and $path(K)$ records min.
- Step 4: Repeat Step 2 and Step 3 until $C_{next} \in S$ or there is no new route.
- Step 5: If $C_{next} = C_n$, Finish. For v_i in C_{next} and corresponding v_j in C_{last}, for v_k in $path(v_i) - path(v_j)$, record $t_k = t_j + (t_i - t_j) * \dfrac{dist(v_k) - dist(v_j)}{dist(v_i) dist(v_j)}$ into $time(v_k)$. Update

$dist(v_i) = dist(v_i) + \dfrac{\alpha}{1 + \beta * p_i}$. For v_i in $S - C_{last}$, initialize $dist(v_i) = \{\}$ and $path(v_i) = \{\}$.

Initialize S as C_{last}. $C_{last} = C_{next}$, $C_{next} = C_{next+1}$, Go back to Step 2.
- The map-matched and interpolated path is $(path(v_{end}), time(v_{end}))$.

9.5 Validation and analysis

In this section, we validate the estimated human movement products in different CDR sampling conditions.

We mainly change two conditions: the cell tower density and the sampling time interval. For cell tower density, we fix the sampling time interval as 5 min and change the cell tower density in [120 /km², 40 /km², 10 /km², 2.5 /km², 0.5 /km²], which means the number of cell tower per kilometers. For sampling time interval, we fix the cell tower density as 40 /km² and change the sampling time interval in [5 min, 10 min, 15 min, 30 min, 1 h].

9.5.1 Mode detection result

In this section, we validate the accuracy of mode detection on the real field research dataset with data filtering. Then we compare the accuracy of mode detection in different CDR sampling conditions on the simulation dataset Assume that there are totally N records pair in GPS time interval, ground truth moving mode is $M_{i,real}$, $i \in N$, the corresponding estimated moving mode is $M_{i,estimated}$, $i \in N$, the total accuracy A is calculated by:

$$Y_{i,j} = \begin{cases} 0, i \neq j \\ 1, i = j \end{cases} \quad (9.12)$$

$$A = \frac{\sum_{i=0}^{N} Y_{M_{i,real}, M_{i,estimated}}}{N} \quad (9.13)$$

The accuracy A_m for a specific mode m is calculated by:

$$A_m = \frac{\sum_{i=0}^{N} Y_{M_{i,estimate},m} * Y_{M_{i,real},m}}{\sum_{i=0}^{N} Y_{M_{i-real},m}} \quad (9.14)$$

9.5.1.1 Mode detection validation

We compare the accuracy of mode detection result of our Hidden Markov Model with a simple method on our field research data.

In consideration of that the real moving mode change is always happened between two records' time stamp, we assume that estimating it the previous mode or the current mode between two records when mode change happen are both correct.

The simple method is defined as:

- Set it 0 (*Stay*) if the cell tower transferring speed is 0.
- Set it 1 (*Non-motorized*) if the transferring speed is between 0 and 15 km/h.
- Set it 2 (*Motor*) if the transferring speed is more than 15 km/h.
- After trip segmentation, check all the motor trip and set it 3 (*Metro*) if any cell tower of this *Motor* trip is *MT* type.

The accuracy result is shown as Table 9.1:

Table 9.1 Mode detection accuracy comparison.

	Total	Stay	Non-motorized	Motor	Metro
HMM	92.85 percent	94.13 percent	93.59 percent	91.19 percent	88.04 percent
Simple Method	77.00 percent	98.82 percent	53.81 percent	81.41 percent	97.82 percent

Reconstruction of human movement from large-scale mobile phone data   173

Table 9.2 Accuracy of mode detection in different conditions.

	Total	Stay	Non-Motorized	Motor	Metro
5 min, 120/km^2	99.59 percent	99.87 percent	98.07 percent	96.98 percent	100.0 percent
5 min, 40/km^2	99.46 percent	99.88 percent	97.14 percent	96.27 percent	100.0 percent
5 min, 10 /km^2	99.14 percent	99.88 percent	94.92 percent	95.04 percent	100.0 percent
5 min, 2.5 /km^2	98.32 percent	99.90 percent	88.90 percent	93.51 percent	100.0 percent
5 min, 0.5 /km^2	95.84 percent	99.94 percent	70.66 percent	91.24 percent	100.0 percent
5 min, 40 /km^2	99.46 percent	99.88 percent	97.14 percent	96.27 percent	100.0 percent
10 min, 40 /km^2	97.07 percent	97.85 percent	97.29 percent	77.77 percent	80.90 percent
15 min, 40 /km^2	95.03 percent	96.01 percent	96.86 percent	67.06 percent	68.73 percent
30 min, 40 /km^2	89.03 percent	90.14 percent	96.53 percent	44.90 percent	39.53 percent
1 h, 40 /km^2	78.85 percent	79.12 percent	97.44 percent	16.37 percent	13.70 percent

9.5.1.2 Mode detection analysis

We calculate the accuracy of mode detection by (9.12), (9.13) and (9.14) on simulation dataset in different sampling conditions. The result is shown as Table 9.2.

9.5.2 Stay point extraction result

In this section, we compare the accuracy of stay point extraction in different CDR sampling conditions on the simulation dataset, which is very significant for trip extraction.

We define that when a device holder stays at a point for more than 15 min, it is a *stay point*. Stay point is a very significant feature for extracting workplace, home, etc. It is also a very important feature for trip segmentation. Here the concept of trip is different from the trip in map-matching step. In map-matching, trip is defined as the same moving mode trajectory, but here trip means the trajectory from one stay point into another, usually consists of different moving mode.

Here we define the accuracy as the number of matched stay point to divide the total number of stay point. The result is shown in Table 9.3 and Fig. 9.5.

9.5.3 Home location extraction result

In this section, we compare the accuracy of home location extraction in different CDR sampling conditions on the simulation dataset. The result could not only explain the accuracy of home location extraction but also explain the accuracy of other important stay point like workplace, which is very helpful for human mobility pattern extraction.

We utilize the follow process to extract the home location:

- For an individual, do mode detection and then extract all his stay points during period from 10:00 PM to 6:00 AM to form the set S.
- For S, utilize *DBSCAN* Algorithm with minimum points number N and scan radius R to find the best biggest cluster as home location.

Table 9.3 Accuracy of stay point detection in different conditions.

	Stay Point Detection Accuracy
5 min, 120 /km²	99.28 percent
5 min, 40 /km²	99.00 percent
5 min, 10 /km²	98.42 percent
5 min, 2.5 /km²	96.79 percent
5 min, 0.5 /km²	91.10 percent
5 min, 40 /km²	99.00 percent
10 min, 40 /km²	91.48 percent
15 min, 40 /km²	85.42 percent
30 min, 40 /km²	70.65 percent
1 h, 40 /km²	52.72 percent

DBSCAN (Density-Based Spatial Clustering of Applications with Noise) is a representative density-based clustering algorithm. Different from the partition and hierarchical clustering methods, it defines a cluster as the largest set of density connected points. It can divide the region with high density into clusters and can find clusters of arbitrary shape in the spatial database of noise. The process of *DBSCAN* could be described as:

Start with any point and find all nearby points within radius R (including R).

- If the number of nearby points is greater than or equal to minimum number N, the current point and its nearby points form a cluster, and the start point is marked as visited. Then recursively process all the points in the cluster that are not marked as visited in the same way, so as to expand the cluster.
- If the number of nearby points is smaller than N, the point is temporarily marked as a noise point.

If the cluster is fully extended, i.e. all the points in the cluster are marked as visited, then utilize the same way to deal with the unvisited points.

We measure the accuracy of home location by calculate the average distance between ground truth home location (extracted from GPS) and the home location extracted from CDR as average error. The result is shown in Table 9.4 and Fig. 9.6.

9.5.4 OD matrices estimation result

In this section, we compare the accuracy of Origin-Destination (OD) Matrices extraction in different CDR sampling conditions on the simulation dataset.

Origin-Destination (OD) matrices play a significant role in transportation networks planning and management. If OD matrices could be precisely calculated, the government can optimize the transportation networks to best fit the need of each aspects.

To calculate OD matrices, we first divide geographical area under analysis into grids: $grid_i$, $i = 1, \ldots, n$. Then we extract origin position o and destination position d with starting time t for each trip of each utilizer u as $trip(u, o, d, t)$. Finally, we group trips with the same origin grids i and destination grids j in different time window tw to form the daily OD table $M(tw, i, j)$:

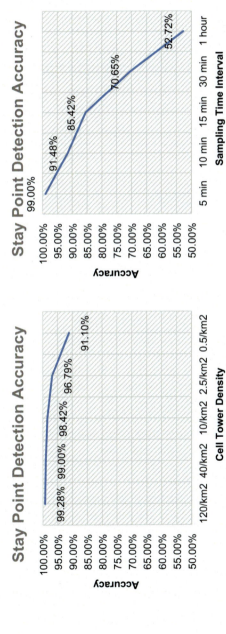

Fig. 9.5 Accuracy of stay point detection versus cell tower density and sampling time interval.

Table 9.4 Accuracy of home location extraction in different conditions.

	Home Location Average Error
5 min, 120 /km²	135.27 m
5 min, 40 /km²	157.20 m
5 min, 10 /km²	175.01 m
5 min, 2.5 /km²	244.97 m
5 min, 0.5 /km²	453.62 m
5 min, 40 /km²	157.20 m
10 min, 40 /km²	167.16 m
15 min, 40 /km²	159.19 m
30 min, 40 /km²	151.30 m
1 h, 40 /km²	148.18 m

Then we measure the accuracy by:

- Error rate:

$$Error\,Rate = \frac{\sum\sum\sum(|M_t(tw,i,j) - M_e(tw,i,j)|)}{\sum\sum\sum(M_t(tw,i,j))} \quad (9.15)$$

- Coefficient of determination:

$$r^2 = 1 - \frac{\sum\sum\sum(M_t(tw,i,j) - M_e(tw,i,j))^2}{\sum\sum\sum(M_t(tw,i,j) - \overline{M}_t)^2} \quad (9.16)$$

Where M_t is the OD table matrix of GPS, and M_e is the OD table matrix of CDR

Here error rate indicates the reliability of the estimated result, if it equals 0, it means the estimated result is just the ground truth. It is possible for it to be more than 100 percent, because the estimated trips number is not equal to the real trips number. The coefficient of determination which in the range of 0 to 1 represents how relative the estimated result is to the ground truth, while close to 0 means no relationship and close to 1 means it is strictly linear correlation.

We utilize two time windows: hourly and daily, which correspondingly means gather the trip by hour and by day.

The number of trips extracted from CDR and accuracy of OD Table is shown as Table 9.5. Distribution graphs of the comparison between estimated matrix and ground truth matrix for several conditions are also provided in Fig. 9.7.

9.5.5 Trajectory estimation result

In this section, we first validate the similarity between the estimated whole day trajectory from CDR data and the ground truth trajectory on the field research dataset comparing with three methods as baseline. Then we compare the similarity between estimated trajectory and ground truth in different CDR sampling conditions on simulated CDR dataset.

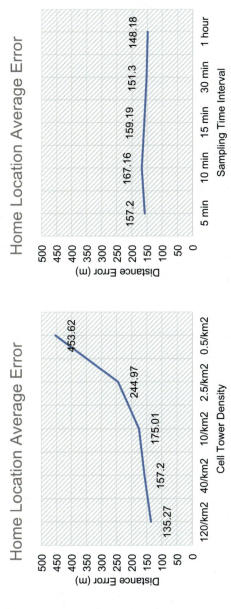

Fig. 9.6 Average distance error of home location estimation versus cell tower density and sampling time interval.

Table 9.5 Accuracy of OD matrix estimation in different conditions.

| | | Number | \multicolumn{8}{c|}{Grids Number (Side Length)} |
| | | | 5 × 5 (8 km) | | 10×10 (4 km) | | 20×20 (2 km) | | 40×40 (1 km) | |
		Real: 184,345	Error rate	r^2	Error rate	r^2	Error rate	r^2	Error rate	r^2
5 min, 120 /km²	hourly	183,988	3.68 percent	0.9995	9.94 percent	0.9957	21.70 percent	0.9744	45.11 percent	0.8662
	daily		2.66 percent	0.9997	7.72 percent	0.9979	18.48 percent	0.9885	41.08 percent	0.9271
5 min, 40 /km²	hourly	183,593	4.70 percent	0.9992	12.49 percent	0.9936	27.01 percent	0.9646	56.36 percent	0.8205
	daily		3.50 percent	0.9996	9.73 percent	0.9968	22.76 percent	0.9833	51.06 percent	0.9016
5 min, 10 /km²	hourly	182,318	6.59 percent	0.9986	16.45 percent	0.9902	36.59 percent	0.9438	75.97 percent	0.7285
	daily		5.01 percent	0.9991	12.62 percent	0.9950	30.67 percent	0.9724	68.54 percent	0.8461
5 min, 2.5 /km²	hourly	178,303	11.02 percent	0.9963	24.35 percent	0.9774	51.93 percent	0.8924	99.95 percent	0.5706
	daily		8.92 percent	0.9975	19.36 percent	0.9852	44.02 percent	0.9364	90.75 percent	0.7320
5 min, 0.5 /km²	hourly	164,049	22.18 percent	0.9800	42.49 percent	0.9176	79.26 percent	0.7453	130.28 percent	0.3300
	daily		18.85 percent	0.9825	34.16 percent	0.9350	69.35 percent	0.8289	123.03 percent	0.5072
5 min, 40 /km²	hourly	183,593	4.70 percent	0.9992	12.49 percent	0.9936	27.01 percent	0.9646	56.36 percent	0.8205
	daily		3.50 percent	0.9996	9.73 percent	0.9968	22.76 percent	0.9833	51.06 percent	0.9016
10 min, 40 /km²	hourly	170,206	11.26 percent	0.9971	19.40 percent	0.9888	35.01 percent	0.9470	63.53 percent	0.7819
	daily		9.10 percent	0.9991	14.37 percent	0.9961	27.26 percent	0.9794	54.76 percent	0.8886
15 min, 40 /km²	hourly	159,098	17.52 percent	0.9914	26.12 percent	0.9780	42.33 percent	0.9203	70.13 percent	0.7343
	daily		14.70 percent	0.9982	19.64 percent	0.9936	32.21 percent	0.9721	58.73 percent	0.8696
30 min, 40 /km²	hourly	134,439	31.15 percent	0.9620	40.38 percent	0.9321	56.93 percent	0.8384	82.05 percent	0.6144
	daily		27.87 percent	0.9942	31.89 percent	0.9825	43.37 percent	0.9474	66.77 percent	0.8121
1 h, 40 /km²	hourly	106,157	46.58 percent	0.8641	56.17 percent	0.7939	72.19 percent	0.6610	93.56 percent	0.4307
	daily		42.89 percent	0.9817	45.92 percent	0.9405	55.71 percent	0.8808	75.92 percent	0.6879

Reconstruction of human movement from large-scale mobile phone data

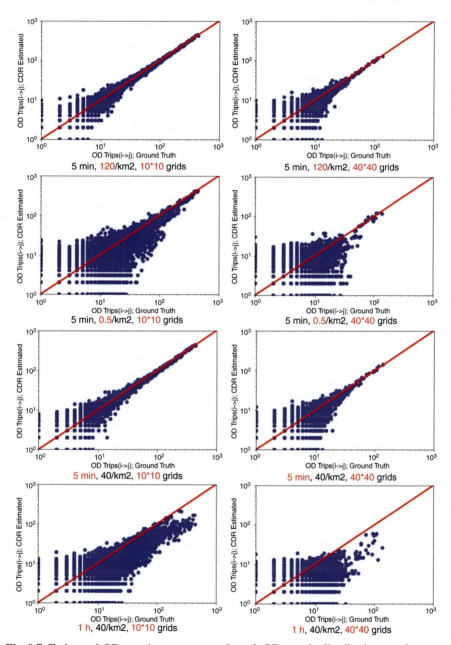

Fig. 9.7 Estimated OD matrix versus ground truth OD matrix distribution graph.

To measure the similarity, we utilize three values:

- *Non-Coincidence Ratio*: We compare the estimated trajectory with the ground truth trajectory and record the length of non-coincidence part in estimated trajectory to divide the total length of the estimated trajectory as *Non-Coincidence Ratio*.
- *Average Dynamic Time Warping (DTW) Distance*: *DTW Distance* is a value to measure the similarity of two time series with different length. To measure average *DTW Distance* between estimated trajectory A and ground truth trajectory B, we map points in A to points in B in time sequence by dynamic programming to ensure the total distance is the shortest one, then calculate the average distance between each node pair.
- *Fréchet Distance*: *Fréchet Distance* is a value to measure the space similarity between two paths, proposed by French mathematician *Maurice René Fréchet* in 1906. To measure *Fréchet Distance* between estimated trajectory A and ground truth trajectory B, we first find the shortest distance for each point in A with B, then find the maximum one in these shortest distances as *Fréchet Distance*.

9.5.5.1 Trajectory estimation validation

To validate the estimated result, the baseline method we use is as follows:

- *Nearest Shortest*: In this method, we assume that the ground truth position of each CDR is the closest node in the road network to the cell tower. Then we directly apply *Dijkstra Algorithm* to interpolate the trajectory, of which the road impedance is set as the time cost.
- *OD Shortest*: In this method, we assume that the device holder always moves on the shortest path between origin and destination. Therefore, we assume the closest node in the road network to the cell tower is the ground truth position of origin and destination, then for each trip segment we ignore the middle CDR and directly apply *Dijkstra Algorithm* to interpolate the trajectory just for origin and destination, of which the road impedance is set as the time cost.
- *Map-matching of Hidden Markov Model*: In this method, we utilize a classical HMM based map-matching method [26] to map-match the CDR and utilize Dijkstra Algorithm to interpolate the trajectory, of which the road impedance is set as the time cost.

In the field research, the cell tower density is nearly 40/km^2. The average comparison result is shown in Table 9.6.

9.5.5.2 Trajectory estimation analysis

We compare the similarity between estimated result and ground truth trajectory in different conditions, the result is shown in Table 9.7.

Table 9.6 Trajectory similarity in different methods.

	Non-Coincidence Ratio	Average DTW Distance	Fréchet Distance
Proposed method	**47.89 percent**	75.36 m	258.12 m
Nearest shortest	55.61 percent	113.29 m	435.13 m
OD shortest	59.23 percent	259.89 m	732.84 m
MM with HMM	**53.37 percent**	102.91 m	416.58 m

Reconstruction of human movement from large-scale mobile phone data

Table 9.7 Trajectory similarity in different conditions.

	Non-Coincidence Ratio	Average DTW Distance	Fréchet Distance
5 min, 120 /km²	13.74 percent	45.58 m	296.8 m
5 min, 40 /km²	19.02 percent	52.89 m	425.2 m
5 min, 10 /km²	24.40 percent	101.64 m	547.72 m
5 min, 2.5 /km²	34.70 percent	173.73 m	785.82 m
5 min, 0.5 /km²	45.45 percent	217.88 m	1071.89 m
5 min, 40 /km²	19.02 percent	52.89 m	425.2 m
10 min, 40 /km²	20.34 percent	83.44 m	675.04 m
15 min, 40 /km²	24.39 percent	122.08 m	814.34 m
30 min, 40/km²	31.76 percent	189.82m	1039.67m
1 h, 40 /km²	38.45 percent	250.65 m	1285.36 m

9.5.6 Traffic volume and average speed estimation result

In this section, we validate the accuracy of traffic volume and average speed estimation on the simulation dataset. Then we compare the accuracy of traffic volume and average speed estimation in different CDR sampling conditions on the simulation dataset.

We aggregate the data in one month into one day in order to expand the traffic volume to avoid the traffic volume being too small. We set the time window as hourly. Assume that the ground truth traffic volume of road i in time window tw is $V_t(tw, i)$, while the estimated traffic volume is $V_e(tw, i)$, the ground truth average speed of road i in time window tw is $S_t(tw, i)$, while the estimated average speed is $S_e(tw, i)$, we measure the accuracy by:

- Error rate of traffic volume:

$$Error\ Rate = \frac{\sum\sum\sum (|V_t(tw,i) - V_e(tw,i)|)}{\sum\sum\sum (V_t(tw,i))} \qquad (9.17)$$

- Coefficient of determination of traffic volume:

$$r^2 = 1 - \frac{\sum\sum\sum (V_t(tw,i) - V_e(tw,i))^2}{\sum\sum\sum (V_t(tw,i) - \overline{V}_t)^2} \qquad (9.18)$$

- Error rate of average speed:

$$Error\ Rate = \frac{\sum\sum\sum (|S_t(tw,i) - S_e(tw,i)|)}{\sum\sum\sum (S_t(tw,i))} \qquad (9.19)$$

- Coefficient of determination of average speed:

$$r^2 = 1 - \frac{\sum\sum\sum (S_t(tw,i) - S_e(tw,i))^2}{\sum\sum\sum (S_t(tw,i) - \overline{S}_t)^2} \qquad (9.20)$$

Table 9.8 Accuracy of traffic volume estimation comparison.

	Traffic Volume		Average Speed	
	Error Rate	r^2	Error Rate	r^2
Proposed method	**24.64 percent**	0.9207	**26.02 percent**	**0.5282**
Nearest shortest	33.80 percent	**0.9298**	28.40 percent	0.5098
OD shortest	43.67 percent	0.7843	37.22 percent	0.4396
MM with HMM	31.34 percent	0.9223	26.96 percent	0.5119

Table 9.9 Accuracy of traffic volume estimation in different conditions.

	Traffic Volume		Average Speed	
	Error Rate	r^2	Error Rate	r^2
5 min, 120 /km^2	23.82 percent	0.9216	24.30 percent	0.5801
5 min, 40 /km^2	24.64 percent	0.9207	26.02 percent	0.5282
5 min, 10 /km^2	26.67 percent	0.9080	28.40 percent	0.5098
5 min, 2.5 /km^2	34.28 percent	0.8651	33.51 percent	0.4142
5 min, 0.5 /km^2	45.76 percent	0.7517	39.55 percent	0.4096
5 min, 40 /km^2	24.64 percent	0.9207	26.02 percent	0.5282
10 min, 40 /km^2	28.56 percent	0.8933	28.63 percent	0.4716
15 min, 40 /km^2	37.80 percent	0.8705	31.46 percent	0.4474
30 min, 40 /km^2	55.28 percent	0.8413	36.48 percent	0.4114
1 h, 40 /km^2	79.71 percent	0.8136	46.85 percent	0.3544

9.5.6.1 Traffic volume and average speed estimation validation

We validate the result under the sampling condition of 40 /km^2 cell tower density and 5 min time interval, the baseline methods are the same with Section 9.5. The result is shown as Table 9.8.

9.5.6.2 Traffic volume and average speed estimation analysis

We calculate the accuracy by Eq. (9.17), (9.18), (9.19) and (9.20) in different sampling conditions. The result is shown as Table 9.9.

References

[1] K. Liu, S. Gao, F. Lu, Identifying spatial interaction patterns of vehicle movements on urban road networks by topic modelling, Comput Environ Urban Syst 74 (2019) 50–61.
[2] Y. Yue, T. Lan, A.G.O. Yeh, Q.-.Q Li, Zooming into individuals to understand the collective: a review of trajectory-based travel behavior studies, Travel Behavior and Society 1 (2014) 69–78.
[3] M. Ahmed, C. Wenk, Probabilistic street-intersection reconstruction from GPS trajectories: approaches and challenges, Proceedings of the Third ACM SIGSPATIAL

International Workshop on Querying and Mining Uncertain Spatio-Temporal Data, 2012, 2012 34–37.
[4] N. Ivchenko, Y. Yuan, E. Linden, Post-flight trajectory reconstruction of suborbital free-flyers using GPS raw data, J. of Geodetic Sci. 7 (2017) 94–104.
[5] E. Ozdemir, E. Ozdemir, A.E. Topcu, A.E. Topcu, M.K. Ozdemir, M.K. Ozdemir, A hybrid HMM model for travel path inference with sparse GPS samples, Transp. (Amst) 45 (2018) 233–246.
[6] C. Spagnol, R. Muradore, M. Assom, A. Beghi, R. Frezza, Trajectory reconstruction by integration of GPS and a swarm of MEMS accelerometers: model and analysis of observability, Proceedings. The 7th International IEEE Conference on Intelligent Transportation Systems (IEEE Cat. No.04TH8749), 2004, 2004 64–69.
[7] S. Miyazawa, X. Song, T. Xia, R. Shibasaki, H. Kaneda, Integrating GPS trajectory and topics from twitter stream for human mobility estimation, Frontiers of Comput. Sci. 13 (3) (2018) 460–470.
[8] R. Ahas, A. Aasa, S. Silm, M. Tiru, Daily rhythms of suburban commuters' movements in the Tallinn metropolitan area: case study with mobile positioning data, Transp. Res. Part C: Emerging Technol. 18 (2010) 45–54.
[9] Y. Yuan, M. Raubal, Y. Liu, Correlating mobile phone usage and travel behavior - A case study of Harbin, China, Comput Environ Urban Syst 36 (2012) 118–130.
[10] T. Pei, S. Sobolevsky, C. Ratti, S.L. Shaw, T. Li, C. Zhou, A new insight into land utilize classification based on aggregated mobile phone data, Int. J. of Geographical Information Sci. 28 (9) (2014) 1988–2007.
[11] D. Shin, D. Aliaga, B. Tunçer, S.M. Arisona, S. Kim, D. Zünd, G. Schmitt, Urban sensing: using smartphones for transportation mode classification, Comput Environ Urban Syst 53 (2015) 76–86.
[12] Y. Xu, A. Belyi, I. Bojic, C. Ratti, Human mobility and socioeconomic status: analysis of Singapore and Boston, Comput Environ Urban Syst 72 (2018) 51–67.
[13] K.M. Gurumurthy, K.M. Kockelman, Analyzing the dynamic ride-sharing potential for shared autonomous vehicle fleets using cellphone data from Orlando, Florida, Comput Environ Urban Syst 71 (2018) 177–185.
[14] G.R. Jagadeesh, T. Srikanthan, Probabilistic map-matching of sparse and noisy smartphone location data, 2015 IEEE 18th International Conference on Intelligent Transportation Systems (ITSC 2015), Los Alamitos, IEEE, 2015, pp. 812–817.
[15] J. Chen, M. Bierlaire, Probabilistic multimodal map-matching with rich smartphone data, J. of Intell. Transp. Syst. 19 (2015) 134–148.
[16] E. Algizawy, T. Ogawa, A. El-Mahdy, Real-time large-scale map-matching using mobile phone data, ACM Trans Knowl Discov Data 11 (2017) 52.
[17] G.R. Jagadeesh, T. Srikanthan, Online map-matching of noisy and sparse location data with hidden Markov and route choice models, IEEE Trans. Intell. Transp. Syst. 18 (2017) 2423–2434.
[18] Z. Xiao, H. Wen, A. Markham, N. Trigoni, Lightweight map-matching for indoor localization using conditional random fields, IPSN'14: Proceedings of the 13th International Symposium on Information Processing in Sensor Networks, Piscataway, IEEE, 2014, pp. 131–142.
[19] M. Ficek, L. Kencl, Inter-Call mobility model: a spatio-temporal refinement of call data records using a Gaussian mixture model, 2012 Proceedings IEEE INFOCOM, IEEE, 2012, pp. 469–477.
[20] S. Hoteit, S. Secci, S. Sobolevsky, G. Pujolle, C. Ratti, Estimating real human trajectories through mobile phone data, 2013 IEEE 14th International Conference on

Mobile Data Management (MDM 2013). 2. 2013 IEEE 14th International Conference on Mobile Data Management (MDM 2013), Los Alamitos, IEEE, 2013, pp. 148–153.

[21] S. Hoteit, S. Secci, S. Sobolevsky, C. Ratti, G. Pujolle, Estimating human trajectories and hotspots through mobile phone data, Comput. Networks 64 (2014) 296–307.

[22] H. Yu, A. Russell, J. Mulholland, Z. Huang, Using cell phone location to assess misclassification errors in air pollution exposure estimation, Environ. Pollut. 233 (2018) 261–266.

[23] Z. Fan, A. Arai, X. Song, A. Witayangkurn, H. Kanasugi, R. Shibasaki, A collaborative filtering approach to citywide human mobility completion from sparse call records, in: S. Kambhampati (Ed.), Proceedings of the Twenty-Fifth International Joint Conference on Artificial Intelligence (IJCAI-16), Palo Alto, AAAI Press, 2016, pp. 2500–2506.

[24] Z. Liu, T. Ma, Y. Du, T. Pei, J. Yi, H. Peng, Mapping hourly dynamics of urban population using trajectories reconstructed from mobile phone records, Transactions in GIS 22 (2018) 494–513.

[25] M. Li, S. Gao, F. Lu, H. Zhang, Reconstruction of human movement trajectories from large-scale low-frequency mobile phone data, Comput Environ Urban Syst 77 (2019) 101346.

[26] P. Newson, J. Krumm, Hidden Markov map matching through noise and sparseness, Proceedings of the 17th ACM SIGSPATIAL International Conference on Advances in Geographic Information Systems, ACM, 2009, pp. 336–343.

Chemical analysis

10

Eman I. El-Kimary, Marwa A.A. Ragab
Faculty of Pharmacy, Department of Pharmaceutical Analytical Chemistry,
University of Alexandria, El-Messalah, Alexandria, Egypt

10.1 Introduction

The wide spread of smartphone around the world with its computing power that has grown to the same degree as desktop computers has attracted attention to its possible use for chemical analysis. Its graphic processing and central processing units have nearly the same computing abilities as desktop computers. Smartphone is used for performing analysis in various fields depending on its processing capabilities and its built-in camera quality. It is widely used in analytical sensing as a detector, illumination source and power supply [1]. Regarding chemical analysis, different smartphone based colorimetric, fluorimetric and electrochemical assays were proposed. Moreover, smartphone could control and adjust experiments steps, acquire, analyze and display data. The use of smartphone in these applications provided a simple, inexpensive and portable sensing system for sample analysis. Different analytes were quantified in different matrices allowing the performance of the analysis by non-trained personal outside the well-equipped labs. The different discussed reports utilized smartphone with its built-in camera alone or with optical accessories that are important in cases of optical conditions are needed to be adjusted when capturing images. In most of these reports, smartphone applications (app) are designed to allow image processing and analyte quantification [1]. Miniaturized extraction, derivatization or pre-concentration procedures were conducted in most reports before image capturing and analyzing.

10.2 Colorimetric-based techniques

Smartphone camera could be used for capturing the image of colorimetric sensors or color obtained after certain chemical reactions and then the obtained images are processed using certain software application to calculate the Hue-Saturation-Lightness (HSL) value or Red-Green-Blue (RGB) value and correlate it to the corresponding analyte concentration. The camera should be a high resolution camera (> 3 MPixel) with digital zoom and autofocus properties [2]. Different smartphone-based colorimetric assays have been reported for the analysis of different analytes.

In the first discussed work [3], two smartphone image analysis apps (Color Lab and ON Color Measure) were used for extracting RGB and HSV color data of water soluble organic dyes. These dyes were extracted by magnetic textile solid phase extraction. These two apps were tested either to be directly applied after image capturing using smartphone camera or indirectly applied after photographs image analysis.

The dyes concentrations were calculated after capturing images under normal daylight in standard illuminating room using smartphone camera (Samsung Galaxy SIII). The obtained data were compared with those obtained by computer based software "ImageJ software" and they were very close to the standard color pictures. Following to HSV image analysis, S (saturation) value of the image was correlated to the dye concentration. The S value relative standard deviation (RSD) obtained directly upon app analysis of captured photos or indirectly after photograph image analysis ranged from 2.48 to 7.96 percent. This indicated that both methods could be efficiently used for image analysis.

In another report [4], phenol as a model of phenolic compounds was analyzed using a combined technique of microextraction (sample preparation) and a smartphone (sensing device). The phenol index was estimated which is an important indicator for water quality. In order to minimize extraction steps and contamination due solution transfer, pycnometer was used as extraction vial and detection cell. Phenol was derivatized using 4-aminoantipyrine to produce a colored compound which was extracted and transferred to the detection cell followed by phenol quantitation using smartphone (LG-G4 with 16 Mpixels camera) which acts as detector and data processing unit. Image of the color produced was captured and opened by RGB analyzer software (Android 6.0) followed by extraction of the red, green and blue values of the image selected area. The analytical signal was the difference between blank and sample blue values. The method sensitivity allowed quantification of phenol in range of 5–100 µg/mL and detection limit (DL) of 2 µg/mL. The combined emulsification microextraction and smartphone based technique was superior when compared to reported conventional spectrophotometric methods utilizing the same reagent. The method is suitable for in-field analysis, doesn't need skilled operator, and consumes less sample and reagent volumes.

Another analytical strategy [5] was proposed by Fashi et al. involving miniaturized liquid phase microextraction procedure (ultrasonic assisted flat membrane) combined with smartphone for image analysis. This strategy was investigated for the analysis of malondialdehyde as a model analyte found in biological fluids. Malondialdehyde derivatization is crucial due its chemical instability. The extraction procedure was designed to provide efficient mass transfer of the analyte in a relatively comparable time when compared to other reported techniques (32 min). The collected organic phase was then transferred into a vial and an image was captured by smartphone camera (iPhone 7 plus with 16 Mpixels camera) followed by its RGB analysis adjusting the illumination environment and the distance for capturing images as mentioned in paper [5]. The G value of the image was inversely proportional to malondialdehyde concentration and the extraction procedure parameters were optimized to achieve the lowest G value for the derivatized product. In order to validate RGB image analyzing process, the organic acceptor phase was analyzed side by side using the proposed smartphone image analyzer and HPLC-VIS/UV. The results of the two analysis methods showed good agreement reaching linearity range of 8 to 100 ng/mL and DL of 1.2 ng/mL for malondialdehyde. Moreover, the image analyzer method is fast, accurate, cheap and sensitive to analyze malondialdehyde in its low concentrations.

Costa et al. proposed an analytical procedure demonstrating the effectiveness of using digital images and smartphone based applications with the aid of chemometric methods [6]. Three milk adulterant analytes (starch, sodium hypochlorite and hydrogen peroxide) were analyzed after specific derivatization of each and capturing of their color images (using Samsung S7 with 12 Mpixels camera) followed by image analyzing using two smartphone apps. These apps (RedGIM® and PhotoMetrix®) were based on partial least square regressions with RGB images histograms. The proposed procedure allowed analytes identification and quantification with low cost, good robustness and high speed. The coefficient of correlation mean for the image processing demonstrated values higher than 0.9929 and 0.9653 by PhotoMetrix® and RedGIM®, respectively. However, a limitation was noticed upon using RedGIM® app represented by unsatisfactory accuracy results of a number of sample solutions. This inaccuracy may be due to the variation in reaction time and the fast precipitation of the colored products. PhotoMetrix® which can capture the images immediately could minimize such inaccuracy.

Photo editing using smartphone (Samsung) was important for adjusting images of the color reaction product distance in microfluidic paper device (MPD). The device enabled the cheap, selective, sensitive and in field analysis of quaternary ammonium compounds [7]. The principle of the device could be used for analytes that need in field analysis. The distance based detection was conducted by localizing the chromogenic agent (polydiacetylene liposome) in microchannels on the paper substrate. Also the paper had a sample reservoir for sample addition. The analyte can react specifically with the blue chromogenic agent yielding a red color bar with a specific length that varies with the analyte concentration. This sensing distance is the red colored distance (reacted analyte) and it should be differentiated from the completely blue distance (un-reacted reagent). The adjustment of this distance is very important to allow accurate and precise quantification of the analyte and is done via smartphone photo editor. Enhancement of images color transition and thus improvement of the readouts of the distance based sensing system was achieved by adjusting color saturation to 100. The proposed MPD combined with image color adjustment using smartphone yielded coefficient of correlation values for the three analyzed quaternary ammonium compounds above 0.997 and it was compared with standard microplate reader method which proved the absence of any significant difference among their results.

Ascorbic acid has been determined quantitatively using smartphone-based CD-spectrometer as a simple, cost effective and sensitive colorimetric method with minimum requirement of devices for data processing [8]. A home-made spectrometer has been made using a part of CD for light diffraction instead of gratings, a light source, a cuvette and a slit and a smartphone as a detector. The distance between these parts has been optimized and the optimized spectrometer was applied for the determination of ascorbic acid based on its reaction with 3,3′,5,5′-tetramethylbenzidine-manganese dioxide nano-sheets. The platform used the camera of Vivo X7 smartphone interfaced on an opaque box that contains the light source, cuvette, a slit and CD fragment with a cover for adding the sample. The photos are obtained using the smartphone, HSV color model modified from RGB model is employed and the obtained V-values are used to compute the intensity in image analysis. A set of standard solutions are used to correlate the measured color intensity and the analyte concentration. A wide linearity range (0.6250–40 µM) was obtained with great sensitivity (0.4946 µM as DL) making this platform suitable for various point of care applications.

Detection of cyanide as a hazardous chemical has been done by both fluorogenic and colorimetric sensors using smartphone-based technique [9]. The assay involves synthesis of two selective diaminomalonitrile-based fluorescent and colorimetric sensors or receptors to detect anions that undergo deprotonating or hydrogen bonding interaction. The developed receptors allowed selective determination of cyanide without interference from F^- or acetate ions. Phenolic hydroxyl and diaminomalonitrile amino groups are the anionic receptors where the deprotonating of these groups promoted a noticeable color change for both receptors for cyanide detection. Moreover, the presence of cyanide increased the fluorescence of one receptor and decreased the fluorescence intensity of the second receptor. The two receptors were coated separately on test paper on which the sample is added, the color of the paper changed from yellow to dark or somon pink in presence of cyanide ion only. The RGB value of the developed colors are read using Apple iPhone 6s Plus smartphone through an installed color assist application. The R/G ratios were found in linear relation with cyanide concentration wit good sensitivity (3.25–4.12 µM). This smartphone-base test paper strip technique could be applied for the qualitative and quantitative determination of cyanide ion in different samples.

Another colorimetric method was described for the multiplex determination of six heavy metals simultaneously using three dimensional multiplex paper-based colorimetric sensor [10]. The paper contains the chromogenic reaction sensor for each target ion in different detection zones and different layers and the sample is allowed to flow on the paper in a circuitous route. Both a LED lamp and the flash built in the smartphone (iPhone 7) have been tried as light sources to avoid interference from ambient light. The LED lamp was found optimum as light source for obtaining the images of the paper by the smartphone as a color detector. The color intensity of the images (ΔGray and ΔE values) exhibited linear relationship with the studied metals. This smartphone-based technique allowed great sensitivity and wider linearity range for the rapid determination of heavy metals.

A unique smartphone-based application has been reported for the determination of volatile organic compounds in plant leaves as a tool for pathogen detection [11]. A paper colorimetric multiplex sensor was used which contained gold nanoparticles and certain organic dyes for the determination of 10 volatile leaf compounds or phytochromones the main of which is (E)–2-hexenal within 1 min analysis time. The colorimetric sensor is monitored by 3D-printed smartphone reader that is calibrated with standard known concentrations of plant volatile compounds to quantify volatile compounds released by healthy or diseased plants at ppm detection level. A small attachment box is mounted or fixed on the smartphone (LG V10) camera in which the sensor is inserted and imaged by the phone camera. Also, a micro pump is fixed on the phone for active gas sampling. Results were compared with GC–MS method and were found comparable with better DL in the smartphone-based method for most of the analyzed volatile compounds. This smartphone-based detection of volatiles in leaf is a useful portable low cost rapid tool for early detection of a number of pathogens affecting the volatile compounds content in leafs.

Smartphone app used for image analysis allowed the accurate color measurement using high resolution smartphone camera (Honor 9). This idea enabled extending the applicability of some color tests that were used originally as qualitative and semi-quantitative assays into accurate and precise quantitative assays. One of these tests is the H_2O_2 detection strips which their color was basically read with naked eyes [12]. Smartphone camera is used to capture the strips color images after incident light from LED is applied to the strips located in dark box. The RGB factors were converted to HSL using matrix transformation algorithm. The S value was calculated in order to calculate the H_2O_2 concentration. A portable, sensitive and rapid biosensor was introduced for quantifying *Salmonella Typhimurium* bacteria using magnetic nanoparticle immunoseparation followed by nanocluster signal amplification using polyclonal antibodies and glucose oxidase modified nanoclusters [12]. After the specific capturing of bacteria, it was allowed to catalyze glucose to produce hydrogen peroxide which was detected by peroxide test strip and the smartphone app was used for colored strips image analysis. Accordingly, the bacteria could be quantified based on amount of H_2O_2 found. The results of the proposed biosensor with smartphone app were compared with those of other biosensors and other smartphone apps and the proposed methodology showed favorable results concerning linearity range and DL. The achieved linearity range was 1.0×10^1–1.0×10^5 and DL was 1.6×10^1 CFU/mL.

A colorimetric immunoassay [13] has been reported for the determination of Alternariol monomethyl ether mycotoxin. The assay is performed on a microfluidic chip utilizing Alternariol monomethyl ether monoclonal antibodies modified gold nanoparticles as a probe for detection of mycotoxin and the color change of the nanoparticles upon binding with the mycotoxin is monitored by either UV spectrophotometry or using a smartphone with 8 Mpixels camera on which an imaging application has been loaded and the HSL color space has been utilized instead of the RGB color space for the image processing. The saturation increases directly with increasing the mycotoxin concentration in the studied calibration range. The method has been applied for detection of this mycotoxin in spiked fruit samples. Although the sensitivity obtained using the smartphone imaging was lower than that obtained using UV-spectrophotometric method (200 pg/mL versus 12.5 pg/mL, respectively), it still adds to the advantages of the method including simplicity and the need for smaller device with lower cost for detection and data processing with minimal user efforts.

Smartphones have been utilized to obtain process and organize data in various analytical methodologies as lateral flow immunoassays in which antibodies and/or oligonucleotides are immobilized at preset positions on a membrane which is the capture zone followed by visual reading or imaging the obtained lines on the test strip. The phone may act only as the capturing device that only transmit the images to another server without processing the data or the color intensity in the photos are converted to analyte concentration using specific software applications [2]. Different lateral-flow and flow-through immunoassays have been developed [14] for the detection of total peanut and hazelnut proteins as allergens and smartphone has been used as readout system by capturing images using open Camera under normal light conditions and the images are analyzed by two different smartphones and using two software applications (freely downloaded from Google Play Store) without the need for a specific software application. The red, green, blue (RGB) values were acquired for test zones by RGB color detector and converted to luminosity by Nix Pro-Color, a device-independent color space that represents how humans read color intensity. The obtained luminosity is correlated to the concentration of the allergen in the sample. The optimized assays were used to calculate the allergen concentrations in spiked buffer (DL 0.1–0.5 ppm total proteins) or spiked biscuit matrix (DL 1–5 ppm total proteins) with good sensitivity in short analysis time (1 min) and comparable results were obtained using two different models of smartphone.

A label-free aptasensor-based simple colorimetric assay with smartphone imaging for determination of chroramphenicol has been developed [15]. Aptamer-modified gold nanoparticles were used and their aggregation was mediated by lanthanum ion in absence of chloramphenicol leading to the change of color of nanoparticles (red to blue). In presence of chromaphenicol, the aptamer combined to the drug forming complex preventing the binding of the aptamer to nanoparticles and in this case the color of the nanoparticles changed to violet-red. The UV–Vis absorbance of the developed colors are measured in a microplate reader or by smartphone imaging using Android Huawei STFAL10 smartphone on which the Touch Color imaging app has been installed to analyze the color changes. The camera was put at 90° to the color solution and the digital images were obtained followed by their RGB analysis by the software application and the data obtained were utilized to correlate to analyte concentration. Linear plot between the B/R values and chloramphenicol concentrations was obtained. The proposed smartphone-based platform showed comparable sensitivity to the UV spectrophotometric method (5.88 nM for smartphone-based method versus 7.65 nM for UV method) within a wide linearity range (0–450 nM) for both methods indicating reliability of using smartphone as detector instead of normal UV–Vis spectrophotometry minimizing the required instrumentation for performing the analysis.

An aptamer-based colorimetric assay with smartphone readout has been developed for the quantitative determination of cadmium ions [16]. Competitive binding of an aptamer with a cationic polymer was utilized in presence of gold nanoparticles as signal element. In presence of the cadmium ions, they bind to the aptamer and prevent its binding with the polymer. The free polymer will aggregate with the nanoparticles changing the color of the nanoparticle solution from red to blue. The color of the nanoparticle solution could be monitored by a smartphone on which a ColorAssist app is installed. To avoid effects from ambient light, a designed black box having fixed light source. RGB model is operated and the R value was recorded by the smartphone and was found to exhibit linear relation with the cadmium concentration in the range (1–400 ng/mL). The results of this simple cost effective smartphone-based technique is compared to the more complicated atomic absorption technique and they were found comparable indicating the easier practical application of the developed platform as an alternative method for the quantitative determination of cadmium in various samples using any smartphone.

Again, smartphone as an image acquisition and data processor has been combined in an enzyme-based colorimetric detection of epoxides [17]. Epichlorohydrin as an example of epoxides has been determined using Halohydrin dehalogenase as recognition enzyme. Bromothymol blue was used as a pH indicator to detect color change caused by the ring opening of the epoxide catalyzed by the enzyme. This induced color change in the sample solution (present in 96-well plate) is monitored by smartphone camera (at 30 cm from the solution) without requiring a complicated instrument or computer based color analysis. The reaction also can be performed in test tube or eppendorf tubes and the analysis time is only 8 min. The Color Lab android free application is installed on the smartphone and it utilizes the Eye dropper function to compute arithmetical value for different color channels that show linear response to the pH induced color change. The hue channel of Hue Saturation Value, R and B in RGB color space and Lightness Chroma Hue color spaces showed linear relation with the change of pH but the first one was found optimal as a quantitative parameter to determination of Epichlorohydrin in the range 0.13–2 mM and DL of 0.07 mM. Results of this smartphone-based simple platform were compared with a more complicated gas chromatographic analysis and the results were comparable confirming the applicability of this simple non-instrumental platform as an efficient alternative for more sophisticated techniques.

Another enzyme-based colorimetric multiplex assay has been reported for the determination of glucose, hydrogen peroxide and catechol (as example of poly phenolic compound) with the aid of smartphone as color detector [18]. A composite polymer film was used on which different enzymes (whose substrates are the analytes) have been immobilized by adsorption. In presence of the substrates (analytes), the enzymes catalyze a reversible redox reaction for the polymer film causing color change from green to blue. The color change in the wells or strips was monitored either by a microplate absorbance reader or a smartphone camera followed by processing the images by a free android operating ColorLab software application. The Hue values of the strips or wells are calculated applying the HSV model and were found to exhibit linear relation with the studied analytes concentrations. The smartphone is held at 6 cm from the wells or strips and images were captures under flashlight conditions to minimize errors from ambient light. The smartphone results were comparable to the spectrophotometric measurements. Thus, the developed polymer film platform acted as disposable, simple, low cost and portable biosensor for the determination of the selected analytes within short analysis time (2 min for hydrogen peroxide to 8 min for glucose) with good analytical performance.

10.3 Fluorescence-based techniques

In addition, some fluorescent smartphone-based platforms have been reported. A fluorescent smartphone-based assay for glucose have been reported [19]. Modified carbon nanoparticles were used as fluorescent probe for the determination of glucose where its fluorescence increased in presence of glucose. This increase in fluorescence could be monitored by smartphone camera operating RGB model where the G channel of the digital images obtained showed linear relation with the concentration of glucose in the range 32 µM-2 mM with DL (8 µM). The reaction is performed in a cuvette and the mobile (HTC ONE SV) is held a certain distance from the cuvette for capturing the images. The images were analyzed by RGB model installed on the smartphone. The reported platform allowed determination of glucose in various samples in single step with low cost and ease of portability for further application at the point of care settings or in chemical or environmental fields.

Another fluorescent assay was described for the sensitive determination of spermidine at fM levels [20]. The assay utilized gold decorated hybrid material and surface plasmon-coupled emission platform. The change in visible color of emission of fluorescent molecules from red to yellow as the spermidine concentration decreases was captured by smartphone camera, then further processed using color grab android application and depicted in CIE chromaticity. Being a natural polyamine whose concentration can indicate different implications, spermidine concentration monitoring is important. Thus, the developed smartphone-based platform allowed rapid simple detection of this bioamine for point of care settings with the ease of data acquisition and handling.

Another fluorescent platform was found in the literature for the determination of picric acid explosives with portable smartphone-interfaced fluorimeter module [21]. The assay utilized a pyrene-based fluorescent sensor for picric acid in which fluorescence quenching occurred in presence of picric acid. The fluorescence measurements are made by portable fluorimeter operating LED-photodetector and the fluorescence data obtained by the detector is sent wirelessly to a smartphone with Bluetooth modem (HC-05) and the phone operating open-source Bluetooth android application. This platform used smartphone not as detector but only for data collection and processing for rapid real-time detection of picric acid explosives with great sensitivity (LD is 99 nM).

A portable kit (D. [22])with Cu nanoparticles embedded in agarose hydrogel was constructed to quantify and detect dimethoate (pesticide). The Cu nanoparticles could fluoresce (green emission) upon the control addition of ammonia through urease-catalyzed urea reaction. This hydrolysis reaction could be suppressed by the analyte addition consequently the fluorescence response was changed. This change could be monitored and the analyte could be quantified after capturing the fluorescent kit image, under UV, using smartphone camera (Huawei Nova 3e). The images were analyzed by ImageJ software. The change in images color could be represented by Hue histograms. These digital data obtained by the software showed good linearity ranges (0.001–5.0 mg/L) for dimethoate. The analysis time (55 min) was comparable to other portable kits for quantifying the analyte. The strategy of smartphone-based nanocolorimetry achieved precise, accurate and in-field analysis of dimethoate.

Again smartphone based platform was used for fluorescence sensing. This platform [23] based on 3D-printing technology which allowed rapid and sensitive detection by the aid of fluorescent paper strip. Internal reference of red-emitting CdTe quantum dots and signal report unit of blue-emitting carbon dots linked to silica nanoparticles outer surface were incorporated on the paper strip. The blue fluorescence could be quenched by gold nanoparticles and then recovered with the analyte (thiram as sulphur containing pesticide which can combine with gold). Though the fluorescence of the paper strip could be detected by naked eye, it is impossible to quantify the analytes unless using the proposed smartphone based platform. When fluorescence is detected on the paper strips due to analyte presence, images were captured and RGB image analysis was done using Color Recognizer App installed in smartphone. The R/B values could be used for analyte quantification with a sensitive DL (59 nM). The smartphone base technology allowed thiram quantitation through its fluorescence activity on the paper strip.

Immunoassay fluorescence sensing strategy was proposed for trace analysis of drugs using a compact smartphone 3D printed device that contains the source of light excitation and the produced fluorescence response was captured by smartphone camera [24]. Quantum dots hybrid and redox reaction allowed multiple color responses to display with different drug concentrations. Distinguishing distinct tonality was possible with naked eye (providing DL 0.37 ng/mL) and the quantitative assay was enabled using HSV analysis with smartphone app (providing DL 0.057 ng/mL). The work achieved designing of a complete smartphone based device which is miniaturized, cheap and compact for detecting fluoresce signal sensing. Moreover, it showed higher sensitivity than classical complicated immunoassays methods as ELIZA in analyzing model drug (amantadine) in complex matrices.

A smartphone optical fiber endoscopic spectrofluorimeter is designed for diagnosis of olive oil quality, freshness and edibility [25]. This portable miniaturized instrument with internet connection is made of smartphone camera attached to 3D printable attached unite. Each component alignment and position was adjusted by optical design and the fluorescence measurements were run by a designed smartphone app. Extra virgin oil fluoresce at 670 nm upon UV excitation at 370 nm. These specific wavelengths are unique to qualify and test extra virgin olive oil purity and check the presence of degradation products or impurities. The data were analyzed and the measurement report could be displayed on smartphone screen or auto archived to the web using a designed smartphone app. The app was used for imaging the light and for plotting spectrum (intensity versus wavelength).

10.4 Foam measurement technique

One of the unique applications of smartphone is the determination of total saponins content in quinoa using foam measurement technique with macro-lens coupled smartphone [26]. The assay is performed in 96-well microplate having black bottom found at 12.1° from a smartphone (Huawei P8 Lite) supported on selfie and coupled to mobile phone macro lens (universal FG) connected to PC. The contrast between the formed foam and the black bottom of the plate increased the quality of the images taken and the stability of the foam is enhanced using some additives like EDTA and protein. The Gray values profile was measured on a rectangular area of the well containing foam (as a free sensor) utilizing ImageJ software. The linearity range was $9.0–180.7 \times 10^{-4}$ mg/mL saponins showing 50 times enhancement in sensitivity and a wider linearity range compared to the afrosimetric assay that relied on the foam height measurement.

10.5 Electrochemical-based techniques

Some smartphone-based techniques with electrochemical detection have been reported. Smartphones have open-source operating system, allowing its combination with various sensing systems reducing cost and improving portability of the systems. Mobile health applications are now available on most smartphones helping the users to self-monitor various health settings including blood glucose concentration and blood pressure. Combing smartphones with electrochemical detection for detecting biomolecules of interest is great advantageous in supporting health services and point of care settings. One report [27] involved dopamine detection (as an important emotional biomarker) using smartphone-based potentiostat system with electrochemical detection. It involved the use of screen-printed electrode modified with poly (3,4-ethylenedioxythiophene, chitosan and graphene as sensor, a potentiostat and a smart phone with a particularly designed software application. The modified electrodes converted and amplified the electrochemical signals whereas the potentiostat generated excitation signals and gather the sensor's converted signals. Smartphone was connected through Bluetooth to the detector to perform commands to the detector to choose the detection method or control the detection parameters, collect the data, process it for calculations, and plot charts in real time. The developed Android application (in Android Studio 3.3.1) is integrated with different electrochemical detection methods including differential pulse voltammetry, cyclic voltammetry, chronoamperometry and square wave voltammetry allowing different operation modes and the processing of different types of electrochemical data. The developed platform allowed detection of dopamine within wide linearity range (0.05–70 μM.) and with acceptable sensitivity (DL 0.29 μM) and could be used for the determination of various biomolecules using different sensors selective for each analyte. The same research group previously reported a similar platform for the determination of α-amylase in saliva as a stress biomarker [28] using different electrochemical sensor. It allowed the determination of α-amylase in the range 50–1200 U/mL with a D of 1.6 U/mL.

Another smartphone-based platform with electrochemical detection [29] has been reported for the determination of different heavy metals (Hg^{2+}, Cd^{2+} and Pb^{2+}) with great sensitivity (2 μg/L ion). For resource-limited users, a smartphone is used as the information terminal allowing the performance of electrochemical analyses with the phone and sharing the data on the internet. This platform is similar to previously discussed platforms where it contains an electrochemical sensor, a wireless potentiometric controller connected by Wi-Fi to a smartphone (Xiao Mi Note, operating Android version 7.0) with an installed application software written with Java language (Sun Microsystems, USA) to collect data and process the obtained electrochemical data.

Also, an electrochemical tag [30] for detecting metals as: lead and cadmium was developed. The tag could be applied for detecting the metals in situ with no wires by attaching it inside water or food containers for long time. The tag was connected to square wave anodic stripping voltammetry circuit and electrode array. The tag showed high sensitivity and selectivity. Moreover, the tags were connected to NFC modules which allowed the wireless connection with smartphone for data transmission and power delivery. The smartphone connection was important as it allowed the proposed system to function without needing external equipment wired connections or on-board batteries. This augments the tag miniaturization and flexibility.

Another application of smartphone in the electroanalytical field was its control of a minipotentiostat coupled to an electronic micropipette via Bluetooth [31]. By the aid of adaptor, three electrodes were inserted into the tip of micropipette to allow carrying out the electrochemical measurement of small liquid volumes without affecting accuracy and precision. A model analyte, H_2O_2, was analyzed successfully by the designed device which suggests its possible future application in clinical and enzymatic biosensing fields. The smartphone connection to this device enabled the full control of the experiment steps and its monitoring via Bluetooth. In addition, smartphone permitted the portability of the integrated device to allow in-field analysis.

10.6 Conclusion

In all the discussed applications, the use of smartphone in the analysis method greatly added to the advantages of the developed platform. In all the discussed colorimetric-based applications, the smartphone is used as a detector minimizing the need for other bulk instrumental detectors. If a colorimetric derivatization is required, the reactions are performed on a paper strip, in a microtiter plate or even in eppindorf or test tubes which are then imaged by a smartphone built-in camera or a macro-lens attached to the smartphone. In fluorimetric-based applications, the smartphone either acted as the detector if the visible color of the fluorescence is measured but if the fluorescence itself is measured then the smartphone is connected wirelessly to the detector to collect and handle data only. Meanwhile, in electrochemical discussed applications, the smartphone was not the detector but only is connected wirelessly to the detector by Wi Fi or Bluetooth to control the commands to the detector (adjusting measuring conditions) and to collect and process data easily like a personal computer but with better portability. For both colorimetric and fluorimetric applications, the imaging conditions are always specified in each study either under ambient light or using the built-in flash light. In some cases, the mobile may be mounted on a black box for controlling the lighting conditions and minimizing the effect of normal light. Regarding the software applications required to be installed on the phone for image analysis, some of them are specially designed while others are free downloadable applications adding to the advantages of the method where no paid special applications are required. In most of the discussed applications, the results of the smartphone based methods were compared with another instrumental techniques as spectrophotometry or chromatography and the results were in good agreement proofing the validation and reliability of the data obtained by smartphone-based techniques as an excellent alternative to other expensive instruments.

References

[1] A.J.S. McGonigle, T.C. Wilkes, T.D. Pering, J.R. Willmott, J.M. Cook, F.M. Mims, A.V. Parisi, Smartphone Spectrometers, 18, Sensors, Basel, Switzerland, 2018. https://doi.org/10.3390/s18010223.

[2] A. García, M.M. Erenas, E.D. Marinetto, C.A. Abad, I. de Orbe-Paya, A.J. Palma, L.F. Capitán-Vallvey, Mobile phone platform as portable chemical analyzer, Sens. Actuators B 156 (1) (2011) 350–359. https://doi.org/10.1016/j.snb.2011.04.045.

[3] I. Safarik, E. Baldikova, J. Prochazkova, K. Pospiskova, Smartphone-based image analysis for evaluation of magnetic textile solid phase extraction of colored compounds, Heliyon 5 (12) (2019) e02995. https://doi.org/10.1016/j.heliyon.2019.e02995.

[4] A. Shahvar, M. Saraji, D. Shamsaei, Smartphone-based on-cell detection in combination with emulsification microextraction for the trace level determination of phenol index, Microchem. J. 154 (2020) 104611. https://doi.org/10.1016/j.microc.2020.104611.

[5] A. Fashi, M. Cheraghi, H. Badiee, A. Zamani, An analytical strategy based on the combination of ultrasound assisted flat membrane liquid phase microextraction and a smartphone reader for trace determination of malondialdehyde, Talanta 209 (2020) 120618. https://doi.org/10.1016/j.talanta.2019.120618.

[6] R. Costa, C. Morais, T. Rosa, P. Filgueiras, M. Mendonça, I. Pereira, B. Vittorazzi, M. Lyra, K. Lima, W. Romão, Quantification of milk adulterants (starch, H2O2, and NaClO) using colorimetric assays coupled to smartphone image analysis, Microchem. J. 156 (2020) 104968. https://doi.org/10.1016/j.microc.2020.104968.

[7] B. Chutvirasakul, N. Nuchtavorn, M. Macka, L. Suntornsuk, Distance-based paper device using polydiacetylene liposome as a chromogenic substance for rapid and in-field analysis of quaternary ammonium compounds, Anal. Bioanal. Chem. 412 (13) (2020) 3221–3230. https://doi.org/10.1007/s00216-020-02583-y.

[8] L. Kong, Y. Gan, T. Liang, L. Zhong, Y. Pan, D. Kirsanov, A. Legin, H. Wan, P. Wang, A novel smartphone-based CD-spectrometer for high sensitive and cost-effective colorimetric detection of ascorbic acid, Anal. Chim. Acta 1093 (2020) 150–159. https://doi.org/10.1016/j.aca.2019.09.071.

[9] S. Erdemir, S. Malkondu, On-site and low-cost detection of cyanide by simple colorimetric and fluorogenic sensors: smartphone and test strip applications, Talanta 207 (2020) 120278. https://doi.org/10.1016/j.talanta.2019.120278.

[10] F. Li, Y. Hu, Z. Li, J. Liu, L. Guo, J. He, Three-dimensional microfluidic paper-based device for multiplexed colorimetric detection of six metal ions combined with use of a smartphone, Anal. Bioanal. Chem. 411 (24) (2019) 6497–6508. https://doi.org/10.1007/s00216-019-02032-5.

[11] Z. Li, R. Paul, T. Ba Tis, A.C. Saville, J.C. Hansel, T. Yu, J.B. Ristaino, Q. Wei, Non-invasive plant disease diagnostics enabled by smartphone-based fingerprinting of leaf volatiles, Nat. Plants 5 (8) (2019) 856–866. https://doi.org/10.1038/s41477-019-0476-y.

[12] R. Guo, S. Wang, F. Huang, Q. Chen, Y. Li, M. Liao, J. Lin, Rapid detection of Salmonella Typhimurium using magnetic nanoparticle immunoseparation, nanocluster signal amplification and smartphone image analysis, Sens. Actuators B 284 (2019) 134–139. https://doi.org/10.1016/j.snb.2018.12.110.

[13] Y. Man, A. Li, B. Li, J. Liu, L. Pan, A microfluidic colorimetric immunoassay for sensitive detection of altenariol monomethyl ether by UV spectroscopy and smart phone imaging, Anal. Chim. Acta 1092 (2019) 75–84. https://doi.org/10.1016/j.aca.2019.09.039.

[14] G.M.S. Ross, G.I. Salentijn, M.W.F Nielen, A critical comparison between flow-through and lateral flow immunoassay formats for visual and smartphone-based multiplex allergen detection, Biosensors 9 (4) (2019). https://doi.org/10.3390/bios9040143.

[15] Y.Y. Wu, B.W. Liu, P. Huang, F.Y. Wu, A novel colorimetric aptasensor for detection of chloramphenicol based on lanthanum ion–assisted gold nanoparticle aggregation and smartphone imaging, Anal. Bioanal. Chem. 411 (28) (2019) 7511–7518. https://doi.org/10.1007/s00216-019-02149-7.

[16] L. Xu, J. Liang, Y. Wang, S. Ren, J. Wu, H. Zhou, Z. Gao, Highly Selective, Aptamer-Based, Ultrasensitive Nanogold Colorimetric Smartphone Readout for Detection of Cd(II), Molecules 24 (15) (2019) 2745. https://doi.org/10.3390/molecules24152745.

[17] G. Ijaz, B.T. Fantaye, D. Jiao, C. Yong, F. Ruiqin, F. Juan, T. Lixia, Enzyme-based detection of epoxides using colorimetric assay integrated with smartphone imaging, Biotechnol. Appl. Biochem. (2020). https://doi.org/10.1002/bab.1898.

[18] O. Hosu, M. Lettieri, N. Papara, A. Ravalli, R. Sandulescu, C. Cristea, G. Marrazza, Colorimetric multienzymatic smart sensors for hydrogen peroxide, glucose and catechol screening analysis, Talanta 204 (2019) 525–532. https://doi.org/10.1016/j.talanta.2019.06.041.

[19] N. Alizadeh, A. Salimi, R. Hallaj, A strategy for visual optical determination of glucose based on a smartphone device using fluorescent boron-doped carbon nanoparticles as a light-up probe, Microchim. Acta 187 (2019). https://doi.org/10.1007/s00604-019-3871-1.

[20] S. Bhaskar, N C S S Kowshik, S.P. Chandran, S.S Ramamurthy, Femtomolar detection of spermidine using Au decorated SiO2 nanohybrid on plasmon-coupled extended cavity nanointerface: a smartphone-based fluorescence dequenching approach, Langmuir 36 (11) (2020) 2865–2876. https://doi.org/10.1021/acs.langmuir.9b03869.

[21] A. Kathiravan, G. Ag, T. Khamrang, M. Kumar, N. Dhenadhayalan, K.-.C. Lin, M. Velusamy, J. Madhavan, Pyrene-based chemosensor for picric acid-fundamentals to smartphone device design, Anal. Chem. 2019 (2019). https://doi.org/10.1021/acs.analchem.9b03695.

[22] D. Kong, R. Jin, T. Wang, H. Li, X. Yan, D. Su, C. Wang, F. Liu, P. Sun, X. Liu, Y. Gao, J. Ma, X. Liang, G. Lu, Fluorescent hydrogel test kit coordination with smartphone: robust performance for on-site dimethoate analysis, Biosens. Bioelectron. 145 (2019) 111706. https://doi.org/10.1016/j.bios.2019.111706.

[23] S. Chu, H. Wang, X. Ling, S. Yu, L. Yang, C. Jiang, A portable smartphone platform using a ratiometric fluorescent paper strip for visual quantitative sensing, ACS Appl. Mater. Interfaces 12 (11) (2020) 12962–12971. https://doi.org/10.1021/acsami.9b20458.

[24] W. Yu, C. Jiang, B. Xie, S. Wang, X. Yu, K. Wen, J. Lin, J. Wang, Z. Wang, J. Shen, Ratiometric fluorescent sensing system for drug residue analysis: highly sensitive immunosensor using dual-emission quantum dots hybrid and compact smartphone based-device, Anal. Chim. Acta 1102 (2020) 91–98. https://doi.org/10.1016/j.aca.2019.12.037.

[25] M. Hossain, J. Canning, Z. Yu, Quality using an endoscopic smart mobile spectrofluorimeter, IEEE Sens. J. 20 (8) (2020) 4156. https://doi.org/10.1109/JSEN.2019.2961419.

[26] N. León-Roque, S. Aguilar-Tuesta, J. Quispe-Neyra, W. Mamani-Navarro, S. Alfaro-Cruz, L. Condezo-Hoyos, A green analytical assay for the quantitation of the total saponins in quinoa (Chenopodium quinoa Willd.) based on macro lens-coupled smartphone, Talanta 204 (2019) 576–585. https://doi.org/10.1016/j.talanta.2019.06.014.

[27] X. Shen, F. Ju, G. Li, L. ma, Smartphone-based electrochemical potentiostat detection system using PEDOT: pSS/Chitosan/Graphene modified screen-printed electrodes for dopamine detection, Sensors 20 (2020) 2781. https://doi.org/10.3390/s20102781.

[28] L. ma, F. Ju, C. Tao, X. Shen, Portable, low cost smartphone-based potentiostat system for the salivary α-amylase detection in stress paradigm*, Conference proceedings: ... Annual International Conference of the IEEE Engineering in Medicine and Biology Society. IEEE Engineering in Medicine and Biology Society. Conference, 2019, 2019 1334–1337. https://doi.org/10.1109/EMBC.2019.8856360.

[29] W. Zhang, C. Liu, F. Liu, X. Zou, Y. Xu, X. Xu, A smart-phone-based electrochemical platform with programmable solid-state-microwave flow digestion for determination of heavy metals in liquid food, Food Chem. 303 (2020) 125378. https://doi.org/10.1016/j.foodchem.2019.125378.

[30] G. Xu, X. Li, C. Cheng, J. Yang, Z. Liu, Z. Shi, L. Zhu, Y. Lu, S.S. Low, Q. Liu, Fully integrated battery-free and flexible electrochemical tag for on-demand wireless in situ monitoring of heavy metals, Sens. Actuators B 310 (2020) 127809. https://doi.org/10.1016/j.snb.2020.127809.

[31] J. Barragan, L. Kubota, Minipotentiostat controlled by smartphone on a micropipette: a versatile, portable, agile and accurate tool for electroanalysis, Electrochim. Acta 341 (2020) 136048. https://doi.org/10.1016/j.electacta.2020.136048.

Applications of smartphones in analysis: Challenges and solutions

Jemmyson Romário de Jesus[a], Marco Flôres Ferrão[b], Adilson Ben da Costa[c], Gilson Augusto Helfer[d], Marco Aurélio Zezzi Arruda[e]

[a]Universidade Federal de Viçosa, Deparatmento de Química, Laboratório de Pesquisa em bionanomateriais, Viçosa, Minas Gerais, Brazil
[b]Universidade Federal do Rio Grande do Sul, Departamento de Química Inorgânica, Laboratório de Instrumentação Analítica e Químiometria – LAQIA, Porto Alegre, RS, Brazil
[c]Universidade de Santa Cruz do Sul, Departamento de Ciências da Vida, Santa Cruz do Sul, RS, Brazil
[d]Universidade de Santa Cruz do Sul, Engenharias, Arquitetura e Computação, Santa Cruz do Sul, RS, Brazil
[e]Universidade de Campinas, Instituto de Química, Departamento de Química Analítica, Grupo de Espectrometria, Preparo de amostras e Mecanização – GEPAM, Campinas, São Paulo, Brazil

11.1 Introduction

Hello! This is a usual expression used for attending a phone call, and since the first public call using a cordless phone, in April 1973 by Martin Cooper [1], it is projected a total of 3.5 billion smartphones in the world in 2020 (Statistica.com 2020).

It is curious, but smartphones are used today much more than establishing an oral call. Text and images are now easily sent through messages, photos, and videos using different apps like Short Message Service (SMS), Multimedia Messaging Service (MMS), and others. All this new reality is possible due to the advent not only of miniaturization and different operating systems, but also the improvement of cameras, audio, digital platforms, and component qualities of these devices.

In this way, it is easy to rationalize a revolution in terms of information acquisition through smartphones, once they present simplicity, low-cost and portability, making themselves a proper way not only for acquiring communication information, but also that information from different areas of science through a diversity of analyses. In fact, these devices are used today as microscopes and endoscopes [2], coupled to drones, as lab-on-a-drone for remote sensing [3] inside the context of agriculture 4.0 [4], as a tool for learning in the classrooms [5], among others. Due to their characteristics, smartphones also became a tool for in situ analysis, so that the literature features a multitude of applications focusing on medical and health-care, agronomical, environmental, chemical, biochemical, among others [6].

Due to these new functions of smartphones, they come to be considered as an analytical tool, needing software for their proper application. This necessity imposes the development of a diversity of analyses software, using different computational languages and applied to the most used operating systems like iOS, Android, and Windows [6].

When considering accurate and precision analysis, a diversity of consolidated and well established analytical techniques can be mentioned such as spectrophotometry, mass spectrometry, atomic/emission spectrometry, liquid and gas chromatography, electrochemistry, among others, which are applied to different samples, presenting all of the necessary analytical figures of merits, for accurate and precise results acquisition, being then used as primary tools for clinical, environmental, chemical and biochemical analysis [7,8].

However, although there is a revolution in progress for using smartphones in analyses, and excellent advances already done in a diversity of their components, some challenges still need to be faced before their complete acceptance from the scientific community. For example, and considering that they are currently used for colorimetric purposes through image acquisition, there are current problems with ruggedness, high stray light, low response, lack of accurate wavelength, or energy calibration, among others.

Despite these problems, smartphones platforms for in situ analysis have been gaining more and more attention to the scientific community so that more than 60 k publications are found in the Google Scholar when considered smartphones and chemical analysis, indicating a powerful platform for such task. Then, inside this context, and trying to contextualized the smartphone platforms in analysis, this Chapter was designed to point out the state-of-the-art regarding some important aspects to these applications, trailing issues as development of a mobile colorimetric analysis tool, and applications of smartphones in chemical, environmental, forensic, and clinical analysis, being all of these subjects exemplified through a diversity of applications. As a large avenue is visualized in terms of future developments, possible trends are also pointed out in this Chapter focusing on the analytical scenarios to gain future insight considering the applications of these platforms in the analysis.

11.2 Development of a mobile colorimetric analysis tool: challenges and solutions

A smartphone today, it can be any simpler model, it is much more than it is presented. The market, obviously, focuses on components such as processor, display, memory, and battery to classify and price them, but it is the interior of the phone that we find a range of different components and parts that intervene in its proper functioning. However, there is a group that can sometimes be overlooked, but whose operation is fundamental and will define whether the phone works or not as it should, and the most interesting thing is that they perform measurements. These are called sensors.

a. Gyroscope: the gyroscope is undoubtedly one of the most important sensors in a smartphone and, fortunately, most current devices have one. It is a small electronic device that is responsible for accurately measuring the position of the smartphone. Many applications use the gyroscope to perform some of its functions and, in recent years, with the rise of augmented and virtual reality, we have seen how programmers are increasingly exploring the capabilities of this sensor in their applications and games.

b. Accelerometer: the user changes the position of his smartphone from vertical to horizontal, or vice versa, and the content of the display adapts to the new situation. Since this sensor generates less information about the position of the device than the gyroscope or the combination of gyroscope and accelerometer, normally its functionality is limited to detect the orientation of the phone and programmers can also take advantage, for example, by incorporating gestures in the applications and games. Normally, it will have to be calibrated before use.

c. Proximity sensor: another sensor that smartphone phones have had for years is the proximity sensor. This is the sensor that, in some terminals, allows the display to go out during a call when the user raises the phone to his ear or that the ringing is deactivated while the user takes it in his pocket to avoid accidental touches. Physically, the proximity sensor bases its operation on an infrared LED (light-emitting diode) that, when generating a beam, can measure the distance between the emitting diode and the surface on which it is bounced.

d. Light sensor: if you use the automatic brightness option on your smartphone, you are using the brightness sensor and you may not know it. This sensor, also known as an ambient light sensor, captures the existing light in the environment and it is responsible for sending the information to the operating system in units that it is capable of interpreting (normally, in lux or lumen/m²), to trigger different actions, such as increasing or decreasing brightness if the ambient light is high or low.

e. Compass: the compass sensor on a smartphone can have several different functions. Technically, it works as an electronic compass, measuring the position of the device through magnetic fields or currents. Therefore, some applications and geolocation systems can rely on this sensor to perform its function.

f. Barometer: although it is not an essential sensor since it is not capable of making very accurate measurements, there are more and more smartphones that incorporate a barometer, also known as a pressure sensor. This component is responsible for measuring atmospheric pressure, which allows the operating system to be able to identify the altitude at which the device is located.

g. Thermometer: almost all devices have temperature sensors to monitor the CPU and battery. If there is another temperature sensor to measure the air temperature, it must be implanted far away enough from the battery and the CPU. Even if these measurements are maintained, we cannot rule out the possibility that their reading will be affected by the general temperature of the cell phone. It would be difficult to obtain an accurate ambient temperature, so it stopped being shipped.

h. Microphone: a microphone sound sensor captures variations in ambient sound. The sound variation can also predict health factors and allows you to capture variations in ambient noise. In this case, the sensor is a light diaphragm that is excited by changes in air pressure, responding in a way that can produce an electrical signal.

Camera is also considered as a sensor, and it will be more discussed in the following sections. But how do I know the sensors on my smartphone? There are hundreds of different device models, and in each of them, you can find different types of sensors available. The simplest is to install an application like CPU-Z for Android or Sensors for iOS, which allows you to know all these components. (see https://developer.android.com/guide/topics/sensors)

11.2.1 PhotoMetrix®: the case study using images

Digital images of a given scene are a great source of data, through the extraction of the figure. elements, called pixels, where each pixel is characterized by a series of variables called channels. Therefore, its use offers an opportunity to propose simple, fast, inexpensive, and non-destructive analytical methods, for example, by using univariate and multivariate analysis techniques.

11.2.1.1 Composing an image

The visible light spectrum occupies a very narrow range of the total spectrum of electromagnetic radiation (Fig. 11.1). For the color to be seen, it is necessary that the electromagnetic energy hits the eye through the light reflected by the object. The theory of chromatic perception by the human eye is based on a hypothesis formulated by Young in 1801, which establishes that the cones (photosensitive cells that make up the retina together with rods) are subdivided into three classes, with different sensitivity maxima located around of the red ("R"), green ("G") and blue ("B"). Thus, all the color sensations perceived by the human eye are combinations of the intensities of the stimuli received by each of these types of cones [9].

For standardization purposes, the CIE (Commission Internationale de l'Eclairage) assigned, in 1931, the following wavelengths to these primary colors: blue = 435.8 nm, green = 546.1 nm, red = 700 nm. RGB colors are called additive primary colors because it is possible to obtain any other color from an additive combination of one or more of them, in different proportions. The mixture of the primary colors, two by two, produces the so-called secondary colors, which are: magenta ("R + B"), yellow ("R + G"), and cyan ("G + B"). The mixture of the three primary colors or a secondary one with its "opposite" primary color produces white light, and on the contrary, in the subtractive the union of the three primary colors or of a secondary color with its opposite primary produces black [10].

The RGB model is based on a Cartesian coordinate system, which can be seen as a cube where three of its vertices are the primary colors, the other three are the secondary colors, the vertex next to the origin is black, and the furthest from the origin corresponds to white, as shown in Fig. 11.2. In this model, the grayscale extends across

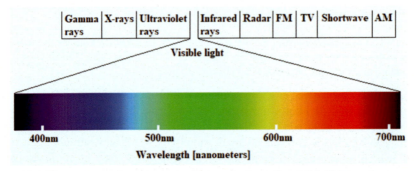

Fig. 11.1 Electromagnetic spectrum, with emphasis on the visible light region.

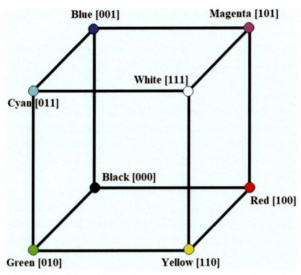

Fig. 11.2 RGB color model.

a line (the diagonal of the cube) that runs from the origin (black) to the vertex furthest from it (white). For convenience, it is generally assumed that the maximum values of R, G, and B are normalized in the range of 0 to 1 [9].

The RGB histogram can be generated from the RGB colors. The histogram of an image corresponds to the distribution of the gray levels of the same, which can be represented by a graph indicating the number of pixels in the image for each gray level. These values are usually represented by a bar graph or frequency distribution. By viewing the histogram of an image, an indication of its quality is obtained in terms of the contrast level and its average brightness (if the image is predominantly light or dark) by the number of times that the gray level occurs in the image [11].

Fig. 11.3 shows a scaled image and its histogram. In this case, there is a low exposure of white light and the histogram graph tends to the left (darker intensities) [12].

From the RGB system, new color models such as HSV (Hue-Saturation-Value), HSL (Hue-Saturation-Lightness), HIS (Hue-Saturation-Intensity), among others, can

Fig. 11.3 Grayscale image and its histogram.

be generated. The HSV model allows us to separate the hue, saturation, and value (luminance) components of the color information in an image, the way people perceive it. Its utilization is more intense in artificial vision systems strongly based on the color perception model by the human being, such as, for example, an automated fruit harvesting system, in which it is necessary to determine if the fruit is ripe enough to be harvested from of its external coloring [13].

Geometrically, the HSV model is a solid whose horizontal cuts produce triangles, in which the vertices contain the primary colors and the center corresponds to the combination of these colors in equal proportions. This combination will be closer to black or white, depending on the height at which the cut was made [12], illustrated and calculated in Fig. 11.4.

Hue (H) is an attribute that describes the pure color as yellow, orange, or red, between a normalized value of 0 and 1, relative to 360°. Saturation (S) is a measure of how much a pure color is diluted by white light, normalized between 0 and 1 [9].

Another important component in computer vision is represented by the brightness of a color. In the HSV model, brightness is indicated by the value of "V" (value) and is defined as the largest component of an RGB color. This put all three primary colors, as well as all "secondary colors" - cyan, yellow, and magenta - in the same plane with white, forming a hexagonal pyramid outside the RGB cube [14].

In the HSL model, the lighting related to the brightness is defined as the average of the largest and smallest RGB color components. This definition also places the primary and secondary colors on the same plane, but on a plane that passes halfway between white and black [15]. However, the simplest definition of brightness is found in the HSI model, where "I" represents intensity, applying only the average of the three RGB components. In theory, it consists of the projection of a point on the neutral axis or the vertical height of a point on the inclined RGB cube (Hanbury, 2007).

It is worth mentioning that three factors have a direct influence on the pixelization process: lighting, the distance between the object and the lens, and, finally, the nonlinearities of the CMOS sensors. The first case can be activated by using the flash, when available, or the second use of the support (tripod, selfie-stick). The third indicator of better-quality cameras (sensors of CCD, charge-coupled device) estimates more accurate values. Even so, we do not guarantee the reproducibility of results between devices [16].

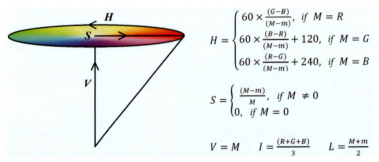

$$H = \begin{cases} 60 \times \frac{(G-B)}{(M-m)}, & \text{if } M = R \\ 60 \times \frac{(B-R)}{(M-m)} + 120, & \text{if } M = G \\ 60 \times \frac{(R-G)}{(M-m)} + 240, & \text{if } M = B \end{cases}$$

$$S = \begin{cases} \frac{(M-m)}{M}, & \text{if } M \neq 0 \\ 0, & \text{if } M = 0 \end{cases}$$

$$V = M \quad I = \frac{(R+G+B)}{3} \quad L = \frac{M+m}{2}$$

Fig. 11.4 HSV model with horizontal section and conversion equations via RGB.

11.2.1.2 Univariate analysis methods

As the name itself defines, they are analyzing a variable represented by a point, only, where a set of samples were a vector. Its greatest applicability is performed in univariate calibration. The univariate concentration of a sample is not an observable physical quantity. In any analytical process, the concentration is always obtained indirectly, from measurements of other quantities such as absorption or emission of light, conductivity, and even weights or volumes.

In this sense, in order to arrive at the concentration value, it is necessary to find a function that relates the measures actually performed with the concentration sought, that is, it is necessary to make a calibration. [17].

The calibration process normally consists of two steps. In the descriptive stage, measurements performed on a series of analytical standards of known concentrations are used to build a mathematical model that relates the measured quantity with the concentration of the species of interest. The second, predictive stage, uses this model to predict concentrations of new samples, according to the analytical signals measured by them [17].

Regression models are mathematical systems that relate the behavior of one variable Y with another X. When the function F that relates two variables is of the type $F(x) = a * x + b$ we have the simple regression model. The variable X is the variable independent of the equation while $y = F(x)$ is the variable dependent on the variations of X, where the estimator "a", called slope, represents the slope of the curve, and the estimator "b", called a linear coefficient, represents the intersection of the line on the Y-axis [18]. A linear equation is defined as perfect when the linear regression coefficient tends to one (1) unit, positive or negative, depending on the concentration patterns (increasing or decreasing) [18].

11.2.1.3 Multivariate analysis method

Unlike univariate analysis, multivariate features are the use of many variables in a sample, whose representation is defined by a vector and the set of samples by a matrix. Among the multivariate exploratory analysis methods, Principal Component Analysis, Hierarchical Cluster Analysis, among others, stand out.

Principal Component Analysis (PCA), or Karhunen-Loeve expansion, is a classic method for dimensionality reduction or exploratory data analysis [19]. PCA is the fundamental basis of most modern unsupervised methods for the treatment of multivariate data and consists of manipulating the data matrix in order to represent the variations present in many variables through less "factors". A new system of axes (routinely called factors, main components, latent variables, or even eigenvectors) is constructed to represent the samples, in which the multivariate nature of the data can be visualized in a few dimensions [20].

The factor analysis is performed on a data matrix that relates a set of variables to different experiments (samples). A data matrix "A" can be centered on the mean or scaled, being decomposed into the product of three matrices through the singular value decomposition algorithm (SVD) where $A = U * S * V^t$ [19]. The matrix "V" (transposed) is called the loadings matrix, the columns of matrix V correspond to eigenvectors and "S" is a diagonal matrix. The "U" and "V" matrices are orthogonal to each other and the product between "U" and S matrices are called scores.

As a result of the analysis of main components, the original data set is grouped according to the existing correlation between the variables, generating a new set of axes (main components) orthogonal to each other and of simpler mathematical manipulation [21].

Geladi et al. [22] observed that multivariate images can be used both for classification in the variable space and for exploratory analysis. In many cases, the transformations for univariate images don't allow a complete interpretation because the underlying phenomena are of multiple nature.

Strategies must be adopted to perform data prediction when there are many variables involved, such as the generation of analytical models through Machine Learning methods. These algorithms eliminate variables that do not correlate to the property of interest, such as those that add noise, nonlinearities, or irrelevant information [23].

After the variables are selected, Machine Learning methods such as Partial Least Squares Regression can be used. Once ready and adjusted through figures of merit as linearity, sensitivity, and selectivity, prediction can be made [24].

PLS are generally applied in situations where process variables have high levels of correlation, noise, missing observations, and imbalance in the proportion of variables and observation. So, a small number of linear combinations, independent of process variables, are generated. These new variables, called components, account for most of the divergence that is presented in the process of dimensionality reduction. Thus, a few components or leverages are retained to represent tens or even hundreds of processing variables [25].

11.2.1.4 Data pre-processing

Pre-processing the data is an operation oriented between the variables for a set of samples of two or more samples, allowing to compare them in different dimensions. In a matrix, it can be described as a column schedule [26]. Among the most used pre-processing in multivariate analysis, the process of self-scaling the data and centering it on the mean stands out [27]. Centering the data on the mean consists of subtracting each intensity by the respective average value for each variable. Already self-scaling the data, it means centering the data on the mean and dividing them by the respective standard deviation [28].

11.2.1.5 Proposed solution

The purpose of this work is to develop a mobile solution for univariate quantitative and qualitative multivariate analysis, both from the decomposition of digital images acquired using the main camera of a mobile device.

The use of this application can contribute to the development of classroom experiences or extra-class activities, allowing the teacher and students to determine elements and chemical compounds present in liquid solutions by the colorimetric technique. Thus, for example, practices for determining the concentration of chlorophyll in plant extracts, as well as determining the concentration of chlorine, metals, and nutrients in water can be easily developed.

So, from images that refer to a colorimetric calibration, they are decomposed in the RGB color channels, giving rise to three data matrices. The data are processed by generating graphs of calibration curves. The predictions of results of new unknown concentration samples are made from simple linear regression (univariate analysis) or multiple linear regression (multivariate calibration). Furthermore, the data also processed can generate classification graphs using the singular value decomposition algorithm and dendrograms, as shown in Fig. 11.5.

11.2.1.6 Development the methodologies

Test-driven Development (TDD) is a technique for building software that guides its development through the writing of tests. Its essence is in its mantra: test, code, and refactor. This technique emerged from refactoring, providing the basis for guiding the coding.

TDD generates learning because it helps to learn good programming practices and create automated tests. First, the test code is written and then (only after) the production code. In this way, the coding will be guided by tests and the production code will be covered by them. The tests will guide the development [29].

In general, the following tasks must be performed: requirements gathering, development, software testing, delivery, and maintenance [30].

11.2.1.7 Planning

To identify, analyze, specify, and define the needs that the application should provide to solve the proposed problem, the requirements gathering activity was carried out. Right after an interface design, it was illustrated to make it practical and use the visualization modeling task. This project was realized only for univariate analysis interfaces. For multivariate analysis, only established design standards were followed to always use intuitive icons and images, which easily describe functions in the application as well as objective and clear messages with comfortable viewing for the user.

The application was developed for the Windows Phone, Android, and iOS platform, using C#, Java, and Swift programming language, respectively, according to a software design standard known as the Model-View-View Model (MVVM) standard. The MVVM standard is a way of organizing and dividing code in a project into manageable layers, which can be independently developed, tested, and modified. It is an especially effective development standard for Windows Presentation Foundation (WPF) and Silverlight projects and can be adapted to the Android and iOS platforms. As three main classes of standard MVVM, the display mode is displayed (interface), the model (database), and the View Model (classes of access to the local database, image processing algorithms, and other controls), such as illustrated in Fig. 11.6.

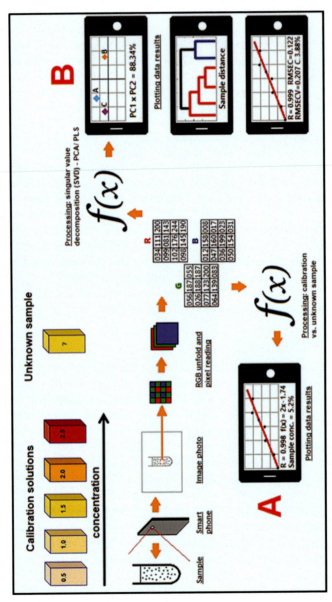

Fig. 11.5 Proposal for univariate (A) and multivariate (B) analysis.

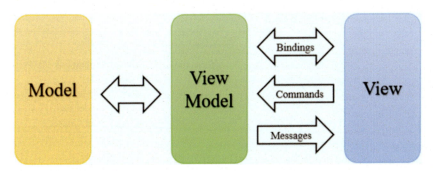

Fig. 11.6 MVVM pattern.

In general, code reuse was adopted as good development practices, such as, for example, the adoption of static methods for conversion between color systems and operations in vectors and matrices, separation of classes by layers, structuring of projects following the same name as variables, routines, and functions, respecting the particularities of C #, Java and Swift programming languages.

11.2.1.8 Development

The applications were generated in an integrated development environment (IDE). The IDEs function is to bring together features and tools to support the software's construction to streamline this process. For this purpose, [31,32] were used, both having a high level of abstraction of controls and classes, resulting from the use of the [33], Android SDK 15 and Swift 2.2 package, respectively.

In addition, both applications have an internal database to store the analyzes performed. In this case, SQL-LINQ DataContext, SQLite, and Realm were used, respectively. For processing the most complex calculations (decomposition of the singular value), open-source libraries such as Alglib, EJML, and the native Accelerate framework were embedded. For graphics generation, the OxyPlot, AFreeChart, and iOS-Chart libraries performed satisfactorily.

It was necessary to use the PHP-Mailer library to the development of an electronic mail web service for sending the exported data according to the email address configured by the user. Good web hosting is necessary to provide all these services. Table 11.1 summarizes the technologies used in the PhotoMetrix project.

Finally, the development demands can be summarized in the following tasks: manage univariate analysis, manage multivariate analysis, model database, define the interface for data entry, manage camera parameters, manage image capture by the camera, process image and decompose in RGB histograms, develop algorithms for converting RGB colors to HSV, HSL and HSI systems, develop algorithms to calculate simple and multiple linear regression, develop graphics, create web services for sending data, manage advanced statistical libraries in matrices, manage graphics, develop algorithms for screen capture, develop and manage application settings.

Table 11.1 Common technologies used.

Tools	Android	Windows Phone	iOS
IDE	Andoid Studio 1.2.1	Visual Studio 2012 Pro	Xcode 7.3
Framework	Android SDK 15	[33]	Swift 2.2
Database	SQLite	SQL-LINQ DataContext	Realm
Statistics library	EJML 0.27	Alglib 3.9	Accelerate
Charts	AFreeChart 0.0.4	Oxyplot v.2014.1.546	iOS-Charts 2.2.3
WebService	PHP 5.5		
UML	Astah 6.9		
Interface mockup	https://wireframe.cc/		
ER Diagrams	MySql Workbench 6.3		
Testing	Microsoft Excel 2015 e ChemoStat 1.0		

11.2.1.9 Testing

To validate the results emitted by the applications, new analyzes of them were made in other software. To validate Univariate Analysis, Microsoft Excel® 2015 was used, as well as to validate Multivariate Analysis, ChemoStat® version 1.0 software was used, in addition to unit tests.

A usual situation that happens in the decomposition of singular values is the mirroring of the graphical results of the PCA in different software. This effect was observed between the PhotoMetrix® application and the ChemoStat® software, that is, opposite signs. This can happen in only one, the first two or more main components. Scott Ramos [34] chief scientist at Infometrix® company, commented, at the Pirouette software algorithms forum on 11/14/2005, reproduced in Appendix H, that this is an expected result and that the distance between the points remains the same, that is, the signal is not critical in the analysis, but the magnitude (which in this case has not changed) is extremely important.

PLS algorithm was tested and compare with ChemoStat® during the interpretation of the thermal stability of raw Milk through the Alizarol test [35].

11.3 Applications of smartphones in analysis

11.3.1 Applications of smartphones in chemical analysis

From the images captured by a smartphone camera, two approaches are most commonly used in the development of alternative methodologies: (1) the first, uses the averages of the color channels (RGB, HSV, among others), performing univariate regression with each channel color and selecting the best performing regression; (2) and, the second, uses histograms (for example RGB histograms) to apply unsupervised (such as PCA and hierarchical cluster analysis, HCA) or supervised (such as PLS and PLS-DA, Partial Least Squares Discriminant Analysis) multivariate methods.

Regardless of the strategy used, another representative difference is whether image capture, decomposition, and processing are carried out on the smartphone, or whether only capture and/or decomposition are carried out, in which case the data is exported and processed outside the smartphone.

11.3.1.1 Applications using smartphone and with external processing

In Table 11.2, representative applications are presented chronologically through the acquisition of digital images by smartphones and the processing of data carried out externally, with the majority developing colorimetric methodologies alternative to the classic methodologies by spectrophotometry in the visible region using benchtop spectrophotometers.

Colorimetric detection by digital images has been the most common approach for detecting and quantifying through smartphones. The use of RGB color histograms has been the most used due to its simplicity. However, one of the first applications in the chemical analysis used the CIE 1931 color space, which can be used to encode colorimetric images. The CIE is another color-coding system in which the color is represented by the parameters x and y to determine the chromaticity of a color (x) and the brightness of the color (y). The CIE 1931 was used for pH measurements in the approach using pH indicator strips which were dipped into a range of pH buffer solutions and then imaged with a smartphone camera, presenting a linear response in the range of 1 to 12 [36].

Using a plastic device, which consisted of a series of membranes with five pH indicators (namely alizarin, bromophenol blue, chlorophenol red, methyl red, and thymol blue), a methodology was proposed to discriminate the different types of amines (triethylamine, isobutylamine, isopentylamine) [37]. To this end, the authors used an iPhone® to obtain the images and extract the RGB histograms, which were treated by unsupervised pattern recognition methods (HCA and PCA), performed using Statistica 12.0 software. Based on this analysis, it was possible to clearly discriminate the studied amines without incorrect classification, being possible to quantify the concentrations of amines up to 1 ppm.

Table 11.2 Chronology of applications using smartphone and with external processing.

Detection Target	Chemometric Aproach	Reference
pH measurements	Calibration Curve using chromaticity values	[36]
Amine-based volatile compounds	PCA and HCA	[37]
Analysis of catechols	PCA, HCA PLS-DA and PLS Regression	[40]
Bisphenol A from water	Univariate Calibration	[41]
Furfural in sugarcane spirits	Univariate Calibration	[38]
Methanol in sugarcane spirits	Univariate Calibration	[39]
Allura red dye in hard candies	N-PLS	[42]
Anionic surfactants in milk	Univariate Calibration	[44]
Oxidative degradation in lubricating oil	PCA and PLS-DA	[45]
dye degradation by the Fenton process	PLS Regression	[46]

Analytical procedures using univariate calibration were proposed for the determination of furfuraldehyde and methanol in situ in samples of sugarcane spirit (cachaça), based on the combination of a local colorimetry test and a digital imaging-based method (DIB). In both procedures, a portable system with cheap materials as well as a system to standardize the lighting were employed. The first method proposed is based on the reaction of furfuraldehyde with aniline in an acidic medium, resulting in a colored product acquired by means of a spot test; a digital image was captured with a smartphone camera and extracted the RGB values [38]. In order to identify which RGB channel presented the best response for the colorimetric product, different furfural concentrations were evaluated using the Hewitt method, and it was found that the G channel showed greater sensitivity compared to other channels. The dynamic linear range for determining furfural was linear from 1.67 to 10.0 mg/100 mL (anhydrous ethanol; $R^2 = 0.996$) with detection and quantification limits of 0.34 and 1.15 mg/100 mL, respectively. For the second application, methanol was oxidized to methanal and consequent formation of a violet chromophore in the presence of chromotropic acid [39]. From the optimization of the system, analytical curves were constructed that showed good linearity for the G channel (from RGB), with a regression coefficient (R^2) of 0.998. Both methods were compared with the reference method by spectrophotometry with a 95 percent confidence level ($n = 3$) with no significant differences.

Wang et al. proposed a smartphone-based colorimetric system coupled to a remote server for quick on-site analysis of 13 catechols. For this, sets of smaller scale 2 × 2 colorimetric sensors composed of pH indicators and phenylboronic acid were configured using the RGB system for chemometric strategies [40]. A schematic diagram representing data acquisition and processing by the smartphone/colorimetric sensor array is shown in Fig. 11.7. Initially, the data were evaluated by unsupervised methods of exploratory analysis (PCA and HCA), followed by the development of models supervised by PLS-DA and PLS regression. The array data (ΔRGB) for the 13 catechols in 6 concentration levels (0.5, 5, 12.5, 25, 50, 75, 100, 150 mM) were loaded on the remote server, where the PLS-DA and PLS regression models were archived

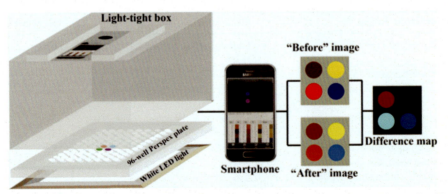

Fig. 11.7 Schematic diagram of smartphone-based colorimetric reader. Reproduced with permissions from Ref. [40]; Elsevier.

to provide qualitative discrimination and quantitative calculation, respectively, of the 13 catechols in real-time. To validate the applicability of the method in a real-life scenario, the analysis of a water sample of the Yangtze River was carried out in loco. The results of the feedback on the smartphone showed that the method was able to identify the catechols with 100 percent accuracy and predict the concentrations within the standard deviation range of 0.706–2.240 mM.

One of the few applications that employ fluorescence detection was presented by McCracken et al. [41] to estimate bisphenol A (BPA) in water samples using capture by a smartphone camera (iPhone 5s or Nexus 5X). For this purpose, 8-hydroxypyrene-1,3,6-trisulfonic acid (HPTS) was used as a fluorescent probe with specificity for BPA, with a detection limit of 4.4 mM. These results demonstrate the potential for application in smartphone platforms.

In the work presented by Botelho et al. [42] was the development of an optical sensor device using a smartphone in a homemade darkroom (built with recycled materials) for the determination of the azo allura red dye in hard candies. The images were captured with a smartphone and the RGB histograms extracted and processed using MATLAB 7.13 [43]orks, Natick, USA], PLS Toolbox 6.5 (Eigenvector Technologies, Manson, USA), Image Processing Toolbox 8.0 (Mathworks) and The N-way Toolbox for Matlab 3.30 (R. Bro & C. Andersson, University of Copenhagen, Denmark). For the construction of the PLS and N-way PLS regression models, 238 samples of sweets of four flavors and different brands and lots were used. The PLS model showed high forecasting errors, attributed to the presence of texture variations in the images, and the histograms are then pre-processed by a fast two-dimensional Fourier transform followed by multi-path calibration with N-way PLS, resulting in mean square errors calibration (RMSEC) and prediction (RMSEP) of 4.8 and 6.1 mg kg^{-1}, respectively. After a complete multivariate analytical validation, the analytical range was established between 22.9 and 78.8 mg kg^{-1} of red-brown. Were estimated as advantages we can highlight that the method developed is simple, fast, and non-destructive.

A new procedure is proposed using RGB values based on simultaneous protein precipitation and liquid-liquid microextraction (LLME) of the ion-pair formed between the surfactants and the methylene blue [44]. Sample treatment, ion-pair extraction, and photometric detection with a smartphone camera were performed in a single vessel, significantly simplifying the procedure and minimizing the risk of contamination and loss of analytes. Sodium dodecyl sulfate was used as a model surfactant and the values of channel B extracted from digital images, using only 100 μL of the sample, 25 μg of methylene blue and 2.9 mg of EDTA, a linear response of 10 to 50 mg L^{-1}, $R^2 = 0.998$ ($n = 5$), with a detection limit (99.7 percent confidence level) and variation coefficient of 2.2 mg L^{-1} and 1.9 percent ($n = 10$), respectively. To assess accuracy, the proposed procedures and spectrophotometry were applied to the determination of anionic surfactant in samples of whole milk from different suppliers, presenting concordant results at the 95 percent confidence level. The proposed procedure is a simple, fast, and economical approach and based on green analytical chemistry for the detection of milk adulterations by surfactants.

In recent work, an approach was proposed that combines analytical methods based on computer vision and chemometric tools to detect the oxidation of lubricating oils

[45]. Samples of non-oxidized lubricating oil purchased in the local market were analyzed by ATR-FTIR (Attenuated Total Reflectance Fourier Transform Infrared) and digital images acquired by the camera of a smartphone with a resolution of 13 MP. After the samples were subjected to heating and radiation exposure to force degradation, they were analyzed by ATR-FTIR that confirmed the degradation. The histograms of the digital images were evaluated using the PCA, which revealed the possibility of differentiating non-oxidized samples from those that suffered degradation at some level. In the sequence, a supervised PLS-DA model was built which proved to be effective in classifying these samples, presenting a correct classification of 100 percent. This new strategy is in line with green analytical chemistry, as it does not use sample handling and does not generate waste, allowing obtaining quick and in situ results of the degree of oxidation of lubricating oils.

In a recent study, it was proposed to monitor the degradation of methylene blue by the Fenton reaction using RGB histograms and multivariate regression by PLS [46]. In building the PLS regression model, 31 methylene blue solutions with different concentrations were prepared, 20 samples for calibration (from 0.0000 to 0.06253 mmol L^{-1}), and 11 for validation (from 0.00469 to 0.05471 mmol L^{-1}). For the acquisition of the images, an iPhone was used, and to standardize the lighting conditions, the photo box was illuminated by two emergency lights of 30 LEDs with a tilt angle of 45° The results of the method employing digital and spectrophotometric images (UV–VIS) with no significant difference being observed. Thus, it is concluded that the combination of digital imaging devices (universally available) with multivariate regression can be a fast and low-cost alternative for studies of laboratory and field monitoring of dye degradation.

11.3.1.2 Applications using smartphone with processing embedded

Initially, applications were restricted to the development of alternative methodologies using color channels independently in the construction of analytical curves to estimate components of interest [47–49], as well as the application of exploratory analysis (for example PCA) using color histograms to assess similarities between different samples [50,51]. From the availability of the new version called Photometrix PRO, which incorporated a routine for regression by partial least squares (PLS), methodologies using RGB histograms or sets of averages of the color variables (R, G, B, H, S, V, L and I) in the development of PLS models that are successful applied for several applications [52,53,35]. These and other applications are shown chronologically in Table 11.3.

One of the first works that use digital images obtained and processed on the smartphone sought to separate six of the most commonly used commercial tanning extracts from different sources [50]. Thirty (30) samples of commercial tanning extracts were used, with five different lots of each, including black wattle (*Acacia mearnsii*), quebracho (*Schinopsis lorentzii*), tara (*Caesalpinia spinosa*), chestnut (*Castanea sativa*), myrobalan (*Terminalia chebula*) and valonea (*Quercus aegilops*). In this work, the researchers used the PCA, available in the PhotoMetrix® app. to separate the varieties

Applications of smartphones in analysis: Challenges and solutions

Table 11.3 Chronology of applications using smartphone with processing embedded.

Detection Target	Chemometric Approach	Reference
Commercial Tannin Extracts	PCA	[50]
Iron in vitamin tablets	Univariate Calibration	[51]
Iodine value in biodiesel	Univariate Calibration	[49]
Acetic Acid in vinegar samples	Univariate Calibration	[48]
pH determination in raw milk	Multivariate Regression (PLS)	[35]
Ethanol in Sugar Cane Spirit	Multivariate Regression (PLS)	[52]
Determination of methanol in biodiesel	Univariate Calibration	[54]
Determination of methanol in biodiesel	Univariate Calibration	[54]
Quantification of biodiesel in diesel blends	Univariate Calibration	[56]
Estimation of endoglucanase activity	Multivariate Regression (PLS)	[53]

of tannins through digital images. These images (64 × 64 pixels) were captured using a cardboard box (Fig. 11.8), with a hole for the phone's camera, and the lighting was performed with an ultraviolet lamp (25 W). The graphs of PCA scores and loadings were obtained by means of the color channels H, S, V, L, and I. The results obtained by the PCA of the digital images agreed with the results obtained by infrared and ultraviolet evaluated in previous works of the authors, indicating the ability to separate the different extracts using a cheap and easily accessible tool, such as a camera, and an application available free of charge for smartphones.

In the work developed by Nogueira et al. it is proposed to monitor acid-base titration reactions through univariate calibration of digital images using the PhotoMetrix® [48]. The authors carried out titrations in microzones of wax printed paper (5 mm in diameter each), where the extract of jaboticaba bark was used as a natural pH indicator. First, buffer solutions from pH 2 to pH 12 were used to construct the analytical

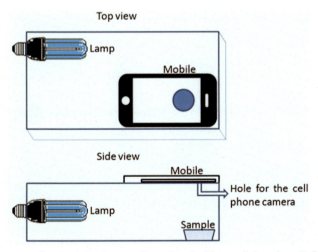

Fig. 11.8 Apparatus for obtaining images. Reproduced with permissions from Ref. [50]; JBCS.

curve; then, three vinegar samples were used to quantify acetic acid. The images (64 × 64 pixels) were captured in ambient light with controlled capture distances (10 cm) and the acid-base titration took about 5 min, using volumes of 20 μL. The analytical best response was obtained by using color information from channel S (saturation), and the achieved data were linearly fitted and presented a coefficient of determination equal to 0.99. The results obtained with the digital imaging method were compared with those of the conventional titration method (reference method) and the difference between them was less than 8 percent. When applying the Student t-test, it was possible to prove that there were no significant differences between the two methodologies for a 95 percent confidence level. The required volume of samples and reagents was 1000 times less than the amount normally used in standard volumetric methods. Consequently, it is possible to conclude that the generation of waste is minimal. The instrumental and economic advantages obtained to enable greater dissemination of knowledge of analytical chemistry at all levels, without the need for sophisticated infrastructure for volumetric analysis.

Several methodologies have been developed seeking to evaluate the quality of biodiesel using image capture and univariate calibration using the PhotoMetrix app. In a first study, Soares et al. developed a test to indirectly determine the value of iodine in biodiesel using digital images [49]. The analysis is based on the residual iodine from the halogenation reaction of unsaturated esters of biodiesel. The steps involved are shown in Fig. 11.9, from the addition of an iodine solution (24 μL) in a biodiesel sample (40 μL), and then 5 μL of this mixture is transferred to a paper, in which the iodine, not reacting, is then developed by the starch solution. The images of the blue complex are captured in an environment with controlled lighting and the calibration curve is obtained using PhotoMetrix®.

Two methodologies for the determination of methanol in biodiesel by digital images have recently been proposed by Soares et al. In the first, the procedure was based on the reaction of oxidation of methanol to formaldehyde and subsequent

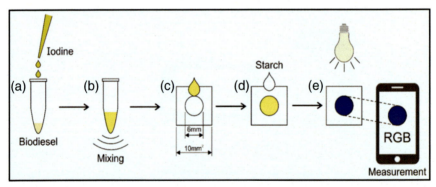

Fig. 11.9 Schematic diagram of the colorimetric spot test. (a) Reaction of biodiesel with iodine; (b) Mixing and subsequent reaction time; (c) Transfer of 5 μL of the mixture to the paper; (d) Addition of 5 μL of starch (1 percent m/v) and (e) acquisition of the RGB value by employing the software PhotoMetrix®. Reproduced with permissions from Ref. [49]; Elsevier.

reaction with Schiff's reagent [54]. The quantification of methanol, through images acquired with lighting and controlled distance, was performed directly in the 1.5 mL polypropylene microtubes with PhotoMetrix®. In the second, the authors present a micro distillation methodology followed by the determination of methanol in biodiesel [55]. In this procedure, the evaporated methanol from the biodiesel matrix is subsequently oxidized by a solution of potassium permanganate in an acidic medium, the analyte being indirectly determined by decolorizing the permanganate. The procedures for obtaining the calibration curve were similar to the previous one. In both methodologies, the results found were similar to those obtained by reference methodologies, generating low-cost alternatives, and reagent consumption.

The last methodology for determining biodiesel in diesel/biodiesel blends was proposed by Soares et al. In this, the determination was based on the colorimetric reaction of the formation of the violet complex between Fe (III) and hydroxamate ion, generated by the reaction of alkyl esters characteristic of biodiesel with hydroxylamine and subsequent quantification by image capture and univariate calibration using the application PhotoMetrix. Recovery tests were performed and the results obtained were in accordance with the reference methodology [56].

An alternative methodology to assess the thermal stability of raw milk by colorimetric reaction with alizarin was proposed by Helfer et al. through multivariate regression by partial least squares (PLS) of digital images captured and processed by PhotoMetrix PRO® [35]. The image of each solution was captured by the smartphone camera (64 × 64 pixels), in a polyethylene black box (150 × 70 and 95 mm in height) with brightness controlled by a 6W LED lamp, with the RGB histograms extracted by PhotoMetrix PRO for PLS regression (see Fig. 11.10). Thirty samples (three for each pH value, from 3 to 12) were used to build the PLS model, and 7 samples of raw milk obtained from local producers were evaluated and compared with the reference method (Potentiometric). The RMSEP and R2Pred values were 0.301 and 0.994, respectively. The pH values obtained with the proposed method for raw milk samples agreed with those obtained by potentiometry in the range of 95.0 percent to 100.9 percent, with no statistical difference ($p < 0.05$). This methodology is fast, accurate and of low cost, allowing its use in any place by the quality control and inspection organs of the commercialized raw milk.

In a recent application of multivariate digital image regression, Böck et al. quantified the ethanol content in sugarcane spirits using RGB histograms and PLS regression, captured and modeled by PhotoMetrix PRO [52]. The colorimetric reaction between ethanol and cerium and ammonium nitrate in nitric acid was initially optimized to form a red complex, using a Doehlert matrix. The authors used a black polyethylene box (10 × 4 × 6 cm) with luminosity controlled by two LED strips with 6 lamps each, resulting in illumination of 9 W/m^2. With the best experimental condition, a PLS model was built, obtaining RMSEP values of 0.0677 percent (v/v), considered satisfactory, since the ethanol concentration range in the cachaça is between 38 and 48 percent (v/v), R2pred of 0.97, LOD 0.19 percent (v/v) and LOQ 0.62 percent (v/v). There was no statistically significant difference ($p < 0.05$; t-test) between the proposed method and the reference method by UV–Vis spectrophotometry. These results showed the efficiency of digital imaging techniques

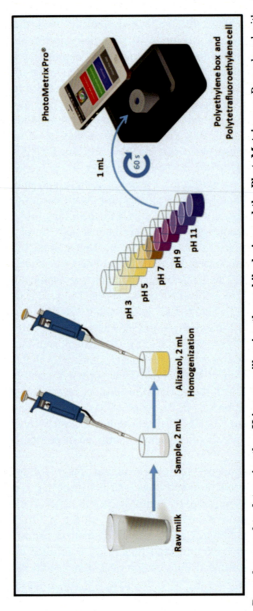

Fig. 11.10 Procedure used to determine the pH in raw milk using the mobile device and the PhotoMetrix app. Reproduced with permissions from Ref. [35]; Springer.

using mobile devices to optimize experiments and/or quantify chemical species, in addition to resulting in a fast and low-cost methodology.

Finally, a recent study describes the estimation of the enzymatic activity of endoglucanase based on the 3,5-dinitrosalicylic acid (DNS) method and on the substrate of carboxymethylcellulose (CMC) using glucose as an analyte, based on the colorimetric analysis of the colored product obtained by the reaction between glucose and DNS. In this work, PhotoMetrix PRO was also used to capture the images and to build the PLS multivariate regression model using RGB histograms. The model obtained showed a correlation coefficient of 0.983 for the linear range between 0.60 and 2.60 mg mL^{-1} of glucose, and the standard cross-validation error (SECV) was 0.219 mg mL^{-1}. The proposed method was applied to samples of cellulase enzymes from *Aspergillus niger* and *Trichoderma reesei*, presenting precise results comparable to those of the reference method, revealing an interesting alternative to estimate the enzymatic activity of endoglucanase.

11.3.1.3 Advantages, limitations, and perspectives

Certainly, among the advantages described by the works that propose analytical methodologies using smartphones, deserve mention: the portability, the cost reduction, and the great availability of smartphones in the current moment of civilization.

Regarding the comparison with the most widespread spectrometric technique in colorimetry (spectrometry in the visible region), it is worth mentioning the greater flexibility in the use of smartphones. While the vast majority of methodologies and spectrophotometers use only the samples in solution, digital images captured by the cameras of the smartphones can be obtained from solutions, solids, colored products supported on paper, among others.

Most of the methodologies that have been developed seek to be in tune with green analytical chemistry, greatly reducing the amount of reagents and samples, as well as a minimal waste generation in relation to traditional and reference methodologies [44,52,42,46,53,35,48,49].

Fig. 11.11 shows three different ways to acquire images related to the articles previously presented that used the smartphone with processing incorporated in their analysis, where to some work cameras they were made to capture images, and, to others, the acquisition of images was performed without control of light. Images of the reaction product were also captured from samples packaged in paper microzones. These examples demonstrate the versatility of digital image analysis for various types of matrices.

In contrast, difficulties related to the interference of lighting are also discussed in the works described above, which should be optimized to improve both the precision and the accuracy of the proposed alternative methodologies.

Additionally, the diversity of smartphone models and the different performances of the cameras can be a limiting factor in the standardization of analytical responses. In this sense, a new version of Photometrix called Photometrix UVC that uses an external camera (USB) and distributed free of charge can be an alternative to be tried.

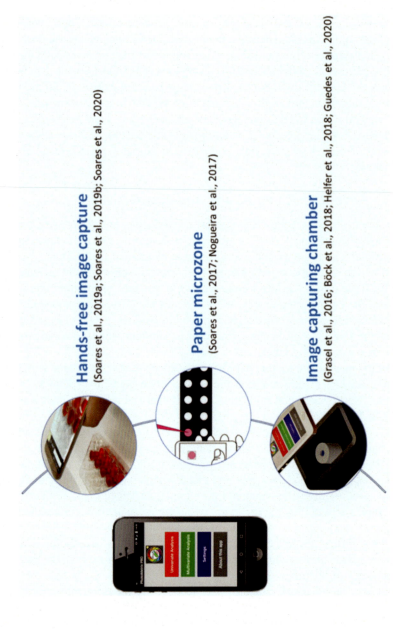

Fig. 11.11 Flexibility in the acquisition of digital images inherent to smartphones.

11.3.2 Applications of smartphones in environmental analysis

The growth of smartphone use drives the improvements of the hardware, software, and high-end imaging and sensing technologies embedded, transforming the mobile phone into a cost-effective and yet extremely powerful platform to run scientific measurements that would normally require advanced analytical instruments [57,6].

Currently, the role of smartphones in environmental analytical chemistry has been highlighted, due to the growing concern with the environment and the analytical characteristics that the use of these devices confers for analysis (Fig. 11.12). Several studies have investigated the determination of chemical species of environmental interest using digital image processing [58,59], embedded ambient light sensor (ALS) [60,61], microfluidic systems [62] and sensors [63–66].

11.3.2.1 Applications

Smartphones have been used a lot in analytical methods to conduct optical detection, as a colorimetric analyzer. This occurs because these mobile devices incorporate the two functional units for this type of operation, namely: source (white LED-emitting light from flash) and detector of radiation (digital camera). In addition, its technical characteristics of usability, portability, connectivity, and high energy autonomy and storage capacity of the data have promoted its use as a platform for field analysis.

The analytical methods using the smartphone-embedded camera are typically easy-to-use, low-cost, and approachable (Fig. 11.13). For this reason, they have been widely used in environmental education and science teaching activities [67,68].

However, other studies have investigated the development of analytical devices using additive manufacturing (3D printing), auxiliary lighting and camera, to minimize interference in the acquisition of images caused by ambient light and variations in the distance between the camera lens and the sample surface [67].

Colorimetry detections methods are the most used in the environmental analysis by smartphones. These spectrophotometric methods are typically based on official compendiums of analytical methods to be used on the smartphone platform. The digital images captured are converted into a numerical form using several methods or color spaces. RGB is the most common color space used in computer vision-based analytical chemistry [69].

In the last few years, smartphone applications in the environmental analysis have evolved a lot in liquid sample tests, and in a smaller quantity in solid or gaseous samples. This preference occurs because in many cases the liquid samples can be analyzed directly, or after simplified sample pre-treatment (filtration for example). On the other hand, samples in the solid or gaseous state may require additional steps, such as extraction and preconcentration of the analyte. Furthermore, many works are engaged with proving the proposed analytical concept, so using samples of easy preparation. However, these devices and analytical procedures can be employed to in a variety of environmental matrices, using an appropriate sample preparation protocol.

In monitoring natural waters, Levin et al. [70] developed a compact sampling chamber to make measuring sample water simpler, automate the dispensing of reagent, and provide a standard container with a fixed geometry for the image acquisition

Fig. 11.12 Advantage of smartphone-based analytical platforms in environmental analysis.

Fig. 11.13 Experimental procedure to image acquisition using the smartphone-based analytical platform in an environmental education activity (iron determination in water samples using *o*-phenanthroline method).

by the main camera of a smartphone. The test chamber prevents outside light from interfering with the tests, the only light being provided from the phone flash. An app was developed for recording and analyzing the RGB color of the picture. The app uses Euclidian's distance to compares the color extract from the sample image against pre-calibrated colors to calculate the result value. As a proof of concept, this device was used to fluoride determination (0 - 2 mg L^{-1}) using the commercially available zirconium xylenol orange reagent (SPADNS reagent).

PhotoMetrix® app to determinate the fluoride and phosphorus in water, using an image acquisition chamber made in high-density polyethylene (HDPE), containing polytetrafluoroethylene (PTFE) cell sample [71]. The analytical procedure was adapted from SPADNS and molybdenum blue reference methods for fluoride and phosphorus measurements, respectively. The samples are illuminated indirectly, by diffusing light from a 6 W LED installed on the PTFE wall. The digital images were acquired (64 × 64 pixels) by the main camera from the smartphone through the hole on the top cover, and the RGB norm vector standard as the analytical answer was used in the univariate analysis module from the app. The same app was used monitoring of iron and hydrogen peroxide concentrations in advanced oxidation processes (AOPs) for the degradation of pollutants [59]. However, a PLS regression model was performed using RGB histogram values from digital images. The results evidencing that the app offers significant advantages in terms of simplicity, cost-effectiveness, smaller sample volumes, and less environmental impact.

A similar device has been proposed for the fluoride determination in water [61]. However, for the development of the proposed sensor was used the ambient light sensor (ALS) of the smartphone as a detector and its flashlight as an optical source. To improve the sensitivity and selectivity, a narrow-band optical filter (peak transmission wavelength of 568 nm) was used. With optical fiber, the output light signal is transmitting through a pair of collimating and focusing lens arrangement up to quartz sample cell installed before the ALS of the smartphone. Thus, the principle of the method is based directly on the Beer-Lambert absorption law, and not on the processing of digital images, as in other works. This feature enhances the application of this device to determine other analytes as well.

In previous work, the authors had used the infrared proximity sensor embedded in the smartphone to develop water turbidity sensing. This device was based on the Mie-scattering principle where suspended microparticles in water medium scatter the light signal that can be detected by sensor [60].

A home-made portable smartphone-based spectrophotometer with the purpose of measuring trace levels of two important anions, chlorine and nitrite ions in water samples, was recently proposed by Sargazi and Kaykahaii [72]. The results evinced values of limits of detections of 5.00×10^{-2} mgL^{-1} and 8.60×10^{-3} mgL^{-1} were obtained for chlorine and nitrite, respectively.

A smartphone platform for water quality detection based on the use of a built-in camera for capturing a single-use reference image was proposed to nitrite, phosphate, hexavalent chromium, and phenol determination, according to the classical colorimetric methods [58]. The image acquisition was performed using a portable photo studio made in white cardboard. White LED were placed inside on the box ceiling 15 cm over the samples for illumination purposes. The smartphone was located 30 cm away from the samples. An app was used to digital image processing and to compare two color matching algorithms of low computational complexity. The first one calculates correlation coefficients between the test image and trained images, white the second one is based on measuring color differences between two images using distance metric obtained from the CIE 1931 color space.

PCA was also applied for soil type classification, using a smartphone-based soil color sensor [73]. The smartphone-embedded camera was directly used as the sensor and the flashlight was used as the light source. The peripheral components included external lens, shading devices, and color calibration card, which were assembled on the phone directly. The colors of soil and proofread cards were acquired by the smartphone and converted into RGB signals. With RGB signals, after simple processing, rapid soil classification could be achieved using multivariate analysis.

An application integrating functions including calibration, signal processing, GPS real-time positioning, data storage and transmission, was recently reported to control a dual-functional smartphone-based sensor for colorimetric and chemiluminescent detection [74]. The optical system was designed and fabricated from the acrylonitrile butadiene styrene (ABS) polymer using a 3D printer. In colorimetric detection, due to the use of white LED and diffraction grating, this sensor could determinate five contaminants (fluoride, ammonia nitrogen, ortho-phosphate, turbidity, and chlorine). The images were captured in RGB color format, and the weighted averaged gray value (WAG) was calculated according to [75] (WAG = 0.299R + 0.587G + 0.114B). To convert the WAG into absorbance value using the Beer-Lambert law, where WAG_0, WAG_S are the intensity of the blank and sample cuvette, respectively (Absorbance = log10 (WAG_0/WAG_S)). In the chemiluminescence procedure to hydrogen peroxide determination, the RGB values had converted them into chemiluminescence intensity.

The paper-based devices have emerged with progress towards the expansion of smartphone applications in environmental studies. The use of smartphones to captures and processing of digital images from paper-based microfluidic analysis devices (μPADs) has allowed sensitive and selective determination of chemical species of ecotoxicological interest, such as mercury [62], cadmium [76], nitrite [77], fluoride [77].

The procedures for acquiring and processing digital images on μPADs are the same as for liquid samples, described earlier. However, the μPADs have many advantages, including a high sampling rate and are more environmentally friendly, because they reduce the consumption of reagents and the generation of waste [78].

For the determination of mercury (II) in water, for example, Jarujamrus et al. [62] propose the design of the μPAD witch a hydrophilic circular area with 4 mm inner diameter. 2 μL sample solutions were applied to the detection zone of double-layer doped with 2 μL of unmodified silver nanoparticles (AgNPs). A smartphone application was used to monitoring the gray intensity in the blue channel as a colorimetric analyzer. Under the optimized conditions, the developed approach showed high sensitivity and low limit of detection (0.003 mg L^{-1}). The proposed technique allows for a rapid, instant report of the final mercury concentration via smartphone display.

An optical paper sensor for in situ measurement of hydrogen sulfide in waters and atmospheres, based on the immobilization of the reagent N,N-Dimethyl-p-phenylenediamine and $FeCl_3$ in paper support, was proposed by Pla-Tolós et al. [79]. Two sampling strategies for H_2S caption from the air have been assayed: active and passive sampling. The H_2S is adsorbed to give rise to the formation of methylene blue as a reaction product. However, a free image editor software was employed to image analysis, while smartphones were used only for image acquisition in de field.

The integration of smartphones with paper sensors has been gain increasing attention also for organic pollutants. Luminescence images from upconversion nanosensors were used to pesticide thiram ($C_6H_{12}N_2S_4$) determination by point-of-care analyzer on a paper-based platform [80]. Other authors develop a screening tool smartphone-based app that can be paired with low-cost colorimetric badges for the detection of formaldehyde [81].

The electroanalytical sensor combined smartphones are particularly interesting for the development of point-of-use platforms by exhibiting high analytical performance, portability, low-cost and operational simplicity.

A smartphone-based analytical platform for multivariate analyses by home-made potentiostat and smartphone was proposed by Giordano et al. [63] to pattern recognition of honey samples according to their botanical and geographic origins. This system combined high-performance detection of the linear sweep, cyclic, and square wave voltammetry with great simplicity, low-cost, portability, energy autonomy, and cable-free (wireless communication) device. The method relied on the unsupervised technique of PCA and the assays were performed by cyclic voltammetry using a single and non-modified working electrode of gold.

Microbiological analysis can also be performed using a smartphone-based sensor. A sensitive microfluidic bacteria pre-concentrator and sensor based on a smartphone through wireless connection was made to detect the density of as low as 10 *E. coli* cells per milliliter of water [82]. For this, a specifically-designed impedance (EIS) network analyzer chip with a microcontroller together performs electrochemical impedance spectroscopy measurement and analysis. A smartphone application app has been developed to enable recording and visualization of testing results as well as control of the sensor electronics. The real-time measurement data will be transmitted to a smartphone by a Bluetooth circuit module.

More recently, a device wireless water quality monitoring and spatial mapping were proposed [65]. A disposable whole-copper electrochemical sensor (WCES) and a handheld detector, controlled by a smartphone, were proposed to determine the chemical demand for oxygen (COD) and lead in the water. The smartphone was used to control the system, process data, and display results in real-time through a customized App. The sample-related information, including the measured results, dates, and Global Positioning System (GPS) coordinates of the testing sites, are upload to the Cloud map website.

The miniaturization of many instruments has allowed the development of new portable equipment. The standalone miniature spectrophotometer, of open source, was proposed recently by [64]. The assembled device can measure absorption in the wavelength range from 450 to 750 nm with a resolution of 15 nm and is housed in a 90 × 85 × 58 mm casing, made by a 3D printer (Fig. 11.14). To test the open-source miniature spectrophotometer (OSMS) in practical applications, cyanocobalamin (vitamin B12), and phosphate successfully determined in water, in a comparative study with other commercial spectrophotometers. Open-source hardware based on Arduino platforms, also used in the development of miniaturized analytical systems, connected to smartphones to monitoring environmental parameters in air, soil, and water samples [83,66]. Other authors [84] develop an app for the treatment and multivariate analysis of spectral data obtained by a low-cost near-infrared spectrophotometer connected to a smartphone, and to develop a portable analytical methodology for the determination of nitrogen in plant tissue. Although they have been tested in some analytes, the characteristics of these devices allow the development of methodologies for application in determining various analytes of environmental interest.

11.3.2.2 Challenges and perspectives of the smartphone applications s in environmental analysis

The improvements of the hardware, software, and high-end imaging and sensing technologies embedded in the smartphones in few years, transform the smartphones into a low-cost and powerful platform to perform scientific measurements that would normally require advanced laboratory instruments. Moreover, the ease of development and availability of applications has contributed a lot to the customization of the smartphone-based analytical platforms. In the most recent works, the GPS of smartphones has been very useful for mapping results in real-time [65,74]. In addition, combined with high data storage capacity, connectivity, energy autonomy, and the ability to run

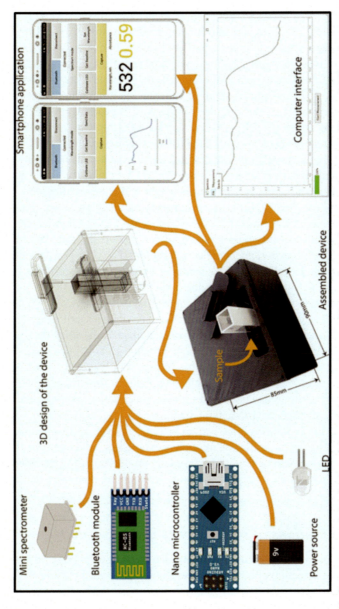

Fig. 11.14 Components of the device of the mini-spectrometer, a Bluetooth module, nano microcontroller, an LED and a power source are featured on the left side [64].

multiple applications simultaneously, the smartphone-based platforms have been the attention of the scientific community in environmental monitoring studies. Indeed, the use in field analysis is the main vocation of the smartphone-based analytical platform.

Therefore, nowadays the role of the smartphone in environmental analytical extend from the simple digital images acquisition to external processing [79] up to the complete analytical platform, suitable for the acquisition of analytical information, processing, storage, sharing and presentation of results in real-time [61,74].

However, we must recognize that the embedded sensors for smartphones available were not designed for analytical purposes, at least at the level of laboratory instruments. Thus, many of the studies in environmental analyzes smartphones-based are home-made adaptations of the original functions of these sensors, with the aid of customized applications [61,58,71]. Thus, despite the promising results for environmental analysis, some obstacles must be addressed before for greater acceptance and general use by the scientific community and environmental agencies. To illustrate, the characteristics of current smartphone cameras do not comply with the demands of precision spectrophotometry. Accurate wavelength and energy calibration, low stray light, rapid response, and ruggedness are currently challenging issues. This problem has been overcome, at least in part, by developing apps that use multivariate methods for the development of calibration or classification models [67,40]. The smartphone showed the ability for processing complex and multivariate data, and currently, smartphone applications based on machine learning classification algorithms have been used successfully to minimize interference in lighting conditions [85].

Most current smartphones are able to wireless, Bluetooth, and/or USB connection in other smart devices. Due to this, smartphones coupled to small external detection devices are an important trend in-field environmental analysis, allow the use of based sensors according to some other analytical principles. Thus, spectrophotometers, potentiostats, electrodes, and other portable sensors, commercially available or produced in laboratories, can be used in a complete platform for environmental monitoring. In this way, a single device would produce information from several analytes, allowing a more accurate environmental diagnosis.

To conclude, the evolution of these environmental analysis platforms smartphones-based converges to the connection of multiple technologies, which allow environmental monitoring in the field or remote, information processing using machine learning principles and expression of results in real-time, what we know by the Internet of Things (IoT).

11.3.3 Applications of smartphones in clinical analysis

Developments in the fields of clinical chemistry, molecular biology, and biosensors have enabled the detection of human disease biomarkers with a significant resolution, precision, and clinical sensitivities [86,87]. Various methods that employ the cell phone camera for clinical analysis have been developed and proposed [88–90]. In this sense, the smartphone can be considered a low-cost system with the potential for clinical analysis and for diagnosis of pathologies in remote areas using samples, such as saliva, blood, urine, stools, and tissues [88,57]. These applications are possible because the smartphones present the two functional components necessary for accomplishing optical detection: source LED (white light from flash) and radiation detector (digital camera) [91,90]. The cameras exhibit good resolution and sensitivity. Indeed, the smartphone is a powerful output toward deploying point-of-care (POC) technologies by providing simple analyses, portable hardware, continuous operation without external power supply (autonomy) during hours and the ability for remote data transmission [91]. Although the smartphones is not designed and developed for clinical applications, it can be adapted for this purpose using compatible accessories that include the necessary hardware to perform microscopic imaging [92], and interface with diagnostic tests integrated in lab-on-a-chip devices [93]. Parallel advancements within fields of molecular analysis, biosensors, mathematical algorithms, microfabrication, 3D-printing and microfluidics have made possible to adapt smartphones as portable, versatile and highly connected read-out platforms with the capability of capturing the microscopic world ranging from cells and tissues to individual DNA molecules [91]. Here, we provide a brief overview of the principles behind common diagnostic tests using smartphones technology.

11.3.3.1 Identifying biomarker of human disease using smartphone technology as portable detector

11.3.3.1.1 Smartphones as optical microscopy

Optical microscopy has been established as a widely used central diagnostic tool because it allows identifying pathological changes directly on clinical samples [88]. In the last decade, the development of optical microscopy has been focused on simplifying the hardware to allow for portable and battery-powered systems with robust performance. The adaptation of CCD or CMOS (Complementary Metal Oxide Semiconductor)-based cameras and machine-learning algorithms has been essential developments to achieve this goal [91]. Furthermore, with the help of histological and microbiological staining techniques, differential observation of cellular components is then possible, allowing study and evaluate their integrities and identities at morphological and anatomical scales [89,91].

Breslauer et al. reported the development of a microscope attached to a smartphone, demonstrating the applicability of this device for clinical diagnostics of *P. Falciparum* and *M. tuberculosis*, providing an important tool for disease diagnosis and screening, particularly in remote areas, such as rural areas and developing world where laboratory facilities are scarce, but the mobile-phone infrastructure is extensive [94].

The use of smartphones has been also employed in complex situations, such as performing minimally invasive surgery. For example, [2], investigated the usefulness of smartphone-endoscope integration in performing different types of minimally invasive neurosurgery. The authors present a new surgical tool that integrates a smartphone with an endoscope by use of a specially designed adapter (Fig. 11.15). This eliminates the need for the video system, currently used for endoscopy. The authors used this novel combined system to perform minimally invasive surgery on patients with various neuropathological disorders, including cavernomas, cerebral aneurysms,

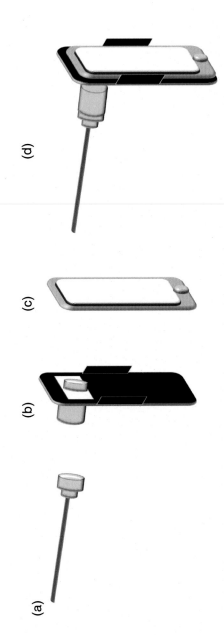

Fig. 11.15 Scheme showing use of the smartphone and neuroendoscope. (a) Endoscopes with different lengths (4, 6, 9, and 12 cm), thicknesses (2.7–4 mm in diameter), and angles (0°, 30°, and 45°); (b) The ClearScope Smart Phone Adaptor allows the connection of a mobile device to almost any endoscope. The unit secures the smartphone to a rigid platform that connects to an adjustable telescope over the camera lens; (c) Smartphone with good resolution camera; (d) The Scope Adaptor is an older version with the same functions.

hydrocephalus, subdural hematomas, contusion hematomas, and spontaneous intracerebral hematomas. As a result, it was observed that all procedures were successfully performed, and no complications related to the use of the new method were observed. As smartphone cameras can record images at high definition or 4K resolution, the quality of the images obtained with the smartphone was high enough to provide adequate information to the neurosurgeons. Moreover, because the smartphone screen moves along with the endoscope, surgical mobility was enhanced with the use of this method, facilitating more intuitive use. In fact, this increased mobility was identified as the greatest benefit of the use of the smartphone-endoscope system compared with the use of the neuro endoscope with the standard video set. Therefore, for the authors, the use of smartphones with endoscopes is a safe and efficient new method of performing endoscope-assisted neurosurgery that may increase surgeon mobility and reduce equipment costs [2].

11.3.3.1.2 Smartphones as colorimetric detector

Measuring the concentration of proteins, small molecules, metabolites, hormones, drugs, and lipids as well as the activity of enzymes is essential for the prognosis and diagnosis of diseases such as diabetes, renal failure, and heart attack amongst others [91]. The concentration of the biomarker of interest can be determined directly or by means of a specific chemical reaction using either end-point or rate measurements [86]. Generally, these analyses are performed via optical or electrochemical measurements mainly in blood and urine samples. Optical methods include absorbance, fluorescence, chemiluminescence, turbidimetry, and nephelometry, whilst electrochemical methods are mainly based on potentiometric measurements. For diseases such as pancreatitis, the lipase activity is quantified by measuring the rate reaction. This is done by quantifying the absorbance of glycerol molecules generated over time upon lipase action on diglycerides.

Smartphone-based spectroscopy has received much more attention as the quality of the image sensor continues to improve [87,114]. The high precision requirements of spectroscopic methods have been prohibitive in the past, but as smartphones continue to get more sophisticated, their qualifications increase to perform many clinical analyzes. Smartphone cameras use low-cost CMOS technology, which is able to detect RGB light and therefore ideal for optical quantification in the visible wavelength range [91]. In addition, this device shows the computational capability and operation interface better than typical cell phones. Another crucial advantage is the high number of subscriptions worldwide [94]. The optical design to achieve spectrophotometric measurements using smartphones is relatively simple (Fig. 11.16).

For diabetes diagnosis, for example, the colorimetric detection of glucose in the blood is based on two catalytically induced reactions: first, glucose oxidase (GOx) reacts with glucose producing H_2O_2, and second horseradish peroxidase (HRP) in the presence of the generated H_2O_2 acts upon achromogenic substrate producing an apparent color change. This is measured by quantifying the absorbance of the sample after reaction completion by means of an optical detector and correlated to the initial concentration of glucose [95].

Fig. 11.16 Illustration of the prototype for colorimetric detection used in the disease's diagnosis. (a) Microtube with biological sample; (b) Microplate used in colorimetric analysis; (c) mobile device for colorimetric detection using smartphone.

Recently, the construction of a smartphone-based spectrometer was reported using a PhotoMetrix app [51] for colorimetric quantification of proteins in serum samples [96]. In this study, it was employed the combination of gel electrophoresis and smartphone technology. Good accuracy and precision were obtained using 32 × 32 pixels for images acquiring in the region of the proteins band. The device was able to sample the entire visible spectrum. Another study reported a powerful point-of-care platform to conduct the Bradford assay to quantify total protein in human plasma [97]. In this study, the authors showed a method based on smartphones for colorimetric quantification of total protein in human blood plasma, presenting as an advantage low cost, simplicity, portability, autonomy, and the ability for remote transmission of the data. The authors suggest these features to contribute the application of the method by non-specialist people.

To fluorescent microscopy applications, smartphones have proven themselves to be more than capable of delivering diagnostically relevant results when equipped with the proper attachments. A typical smartphone fluorescence microscope consists of an excitation light source (LED or laser diode) and an emission filter, in addition to the above-mentioned external lenses with the brightfield modality [98]. How the samples are illuminated is the key consideration when designing a smartphone-based fluorescence microscope. This design enables the capture of fluorescent signals from a few hundred fluorophores within a diffraction-limited spot using the CMOS sensor, offering opportunities to visualize single 100 nm nanoparticles, individual virus particles, and single DNA molecules [99]. DNA amplification uses the convective polymerase chain reaction technique, and the detection is carried out with the variation in the fluorescence. The fluorescence increment used the brightness of the image before and after the DNA amplification, and if there is a difference before and after the DNA amplification, the test is positive. This process can be used for screening hepatitis B virus plasmid samples [89].

Other optical smartphone biosensing platforms based on the surface plasmon resonance (SPR) properties of gold nanoparticles (AuNPs) have been proposed as diagnostic tools [89,91]. This is possible because actual smartphone screens provide more than sufficient wide-angle illumination to perform SPR measurements [100]. For example, the concentration of vitamins such as B12 and D in blood and serum samples has been measured by performing an AuNP-based immunoassay along with a smartphone-based colorimetric reader [89]. Other applications including the SPR detection system based on a smartphone was proposed by Preechaburana et al. where the authors demonstrated that the resolution of the device employing the smartphone is comparable with conventional analytical SPR. The assays were made for the detection of β2 microglobulin, biomarker for cancer, and were achieved a limit of detection of 0.1 mg mL^{-1} in urine [101].

11.3.3.1.3 Smartphones as electrochemical detector

Electrochemical biosensors have also been used in combination with smartphones [91]. According to the employed electrochemical apparatus, smartphone-based electrochemical sensors have been classified into two types: (i) amperometric biosensor, (ii) potentiometric biosensor [102].

The amperometric biosensor relies on the variation of current resulting from the oxidation or reduction of the electrical mediating species when the potential is

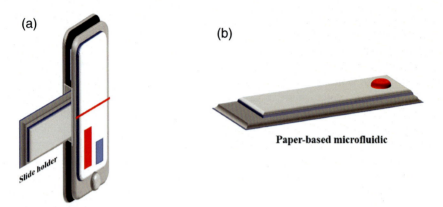

Fig. 11.17 Illustration of smartphone-based electrochemical for diseases diagnostic; (a) Smartphone-based electrochemical; (b) Microfluidics on paper for electrochemical analysis using a smartphone.

constant, which is proportional to the concentration of analytes. Amperometry is the most widely used technique combined with smartphone due to its different methods to detect different analytes [102].

Wang and Lin et al. first reported a paper-based microfluidic smartphone electrochemical biosensor for label-free white blood cell (WBC) counting (Fig. 11.17). WBCs separated from whole blood were physically captured on the microporous paper with interdigitated gold microelectrodes, which was used to determine the cell concentration via the method of differential pulse voltammeter. This platform could be applied to rapidly quantify the concentration of WBC in a few samples with high repeatability. The smartphone-based amperometric biosensor provides the opportunity for fast cell quantification to achieve POC diagnosis of WBC for patients with leukemia or other immune diseases [103].

Another example of the electrochemical biosensor using smartphones includes a potentiostat for monitoring lung infections in cystic fibrosis patients by tracking the concentration of secretory leukocyte protease inhibitor. This was achieved using an electrochemical ELISA (Enzyme-Linked Immunosorbent Assay) platform that is plugged into the audio port of the smartphone [104].

Potentiometry is suitable for sensing applications in which the cumulative electrical charges result in the differences in electrical potential on top of the dielectric layer [102]. Garcia et al. reported a smartphone-based potentiometric biosensor for POC testing of salivary α-amylase (sAA) for personal psychological measurement. The sAA reacts with the preloaded reagents when saliva enters into the reaction zone on the sensing chip, resulting in the conversion of $Fe(CN)_6^{3-}$ to $Fe(CN)_6^{4-}$. Then the sensing chip was pressed to transfer the mixture to the detection zone. The potential read by the smartphone-based potentiometric sensor was sent to the sAA-detection application through the USB port and converted to the concentration of sAA according to a calibration curve. This platform provided an emerging noninvasive biomarker for psychological health for POC testing of sAA [105].

11.3.3.2 Smartphone technology limitations for clinical analysis

Although the recent advancement of smartphone-based analytical biosensor has provided people with numerous portable, cost-effective, and easy-to-operate platforms for POC and mobile health applications, many limitations are noted [88,89,57]. For example, the background illumination is always a challenge when designing spectroscopic systems. This problem exacerbates when the system must be portable and easily attached to a smartphone. In addition, due to the small pixel sizes and larger operating temperatures, smartphone devices are in general not as sensitive to light signals as their commercial counterparts [91,90]. To enhance the fluorescence detection the use of glass capillary tubes as sample containers are useful for both reducing sample volumes and improving detection sensitivity. A study using an optical fiber bundle to direct light from the smartphones flash directly onto the sample was reported as a way to solve part of this problem. The light from the sample was then delivered by an endoscopic fiber bundle for a diffraction grating. From this point, the light was focused using a cylindrical lens with a focal distance of 2.0 cm onto the CMOS image sensor of the smartphone. Plasmonics can also be used to enhance the signal intensity of smartphone fluorescence microscopy enabling the detection of single 50 nm fluorescent beads and individual QDs (ca. 20 nm dia.) with a smartphone [91].

Another challenge involving smartphone technology as a sensor for clinical analysis is associated with ensuring analytical sensitivity, precision, and reproducibility of the method. Thus, these parameters must be strictly guaranteed and validated by well-accredited technologies such as the gold standard benchtop instruments so that they can be translated into practical or clinical applications [90].

An important consideration to be considered for the development of smartphone-based diagnostic tools is the availability of ready-to-use components on smartphones that would allow their use in a modular way. Initiatives that defend the scalable manufacture diagnostic devices based on smartphone technology are crucial for the diagnosis of disease. For example, CMOS sensors can be used to build lens-free microscopy systems that hold promise for on-site diagnostics because they allow for wide FOVs, which could be relevant for infectious diseases such as malaria and tuberculosis [91].

11.3.4 Applications of smartphones in forensics analysis

Undoubtedly, the forensics field has quietly become one of the biggest affected by the advance in the technology [106]. Starting with the advent of fingerprint and DNA databases, the forensics innovation has expanded into nursing, toxicology, ballistics, accounting, blood spatter analysis, and digital and media arenas. This technological boom in forensic complements other advances in science, making it easier to solve cold cases and crimes with less and less visible evidence, increasing proficiency helps law enforcement, providing a strong deterrent and fewer crimes [107].

In the field of crime scene investigation, the number of forensic apps available for mobile phones appears to be growing with many current technologies focused on collecting documents and evidence to solve criminal practices or identify people with criminal behavior. Therefore, these apps appear as alternative tools for crime scene investigators to do their job more efficiently, using a pocket-friendly technology [108].

However, some questions arise regarding the application quality, for example (i) what is the importance of these new technologies really? (ii) how are they developed or chosen? (iii) What ensures that the applications are reliable and that the measurements given are correct, accurate, and reproducible? (iv) Are there criteria to evaluate them and strategies to supervise their use? (v) What are the problems, the risks, and the advantages associated with the use of applications on smartphones and tablets? Then, the following are some criteria used to develop applications for forensics investigation are shown, highlighting the challenge and solution during its development.

11.3.4.1 Design and development of application for crime scene

In general, the crime scene is defined as a place where a crime has occurred. Crime scene reporting is an integral part of investigation and reconstruction. An investigating officer usually documents the crime scene with the help of photography, recordings, sketches, and notes [109]. Measures for the overall crime scene and the documentation for permanent records, trial review, and crime reconstruction will be taken during the crime scene sketching. Measurement is usually a tedious and time-consuming task, in addition, it requires a lot of effort and several personals involvements [110].

Recently, some studies have created smartphone applications to be used in crime scenes as alternative forensics tools [109,111]. Assessing the quality of the application involves a set of important and simple parameters that are used for processing crime scenes [112,106]. Fig. 11.18 shows some of these parameters.

- ✓ Evidence accuracy - this parameter is defined as any features of the app that removes disparities in all the various crime scene investigating agencies – any feature that provides uniformity in all aspects;

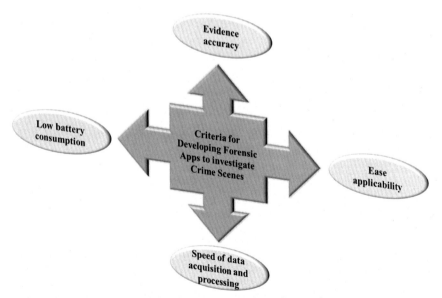

Fig. 11.18 Criteria for developing forensic apps to investigate crime scenes.

- ✓ Ease applicability - the application must be easy to learn and use, allowing users (investigators) to document and record all aspects of a crime scene including access records, notes, evidence, sketches/diagrams, photographs, techniques processing, and interviews. For this, the local investigator (LI) must make a brief exploration of the app;
- ✓ Speed of data acquisition and processing- the app must process the data crime scene quickly. In other words, it must increase the efficiency of the investigation providing answers in a short time, as well as, to reduce data entry time
- ✓ Low battery consumption - the application should not drain the battery too much, allowing it to be used in prolonged or multiple investigations.

In addition, the app should be low cost, that is, the application must not be very costly, and the same can be said for the devices and accessories. The architecture of the application for documenting the crime scene can consist of two main parts: (i) the investigator side and (ii) the server-side (Fig. 11.19) [111]. The server side is responsible for saving the crime scene investigation (CSI) database. The investigator's side is responsible for capturing and documenting the crime scenes. The investigator's side has two forms of application, such as Web Application and Mobile Application which communicate with the server using HTTPS (Hyper Text Transfer Protocol Secure) protocols to prevent attacks on the network [111].

Mobile Application is developed based on the Android or iOS system with JAVA or Swift languages. Users are investigators and investigative officials. The application can be activated both within the office and other locations by a wireless internet connection, such as Wi-Fi/3G/4G to send and receive data between the server and work offline via a system backup database to be used offline on the mobile.

Web application is available through a web browser and can be used on any device anywhere. In this section, users are investigators, investigative officials, service officers, commanders, and administrators. It is suitable for office use but can be used on mobile devices if the mobile device also does not support the Android or iOS operating system.

Five applications developed for investigation of crimes scenes were evaluated by Baechler et al., [112], they are: (i) CrimePad®; (ii) MagicPlan CSI®; CASE®; World Drugs DB®; and Document Compendium®. These applications were installed and tested with an iPad Air® Wi-Fi 32GB MD786CL/B while using the iOS 8 operating system. For more details, see [112]. Table 11.4 shows the evaluation of five applications according to their relevance, reliability, and responses to operational requirements using a scale from +++ to - (see Table 11.5 for symbols meaning).

In general, the app does not help with the note-taking process, unless the user prefers to type his notes instead of writing them by hand. Since the investigator is still typing on a blank page, this is no different than writing on blank paper. However, this can be beneficial for those whose handwriting is difficult to read. For example, if an investigator was retired or is out of town, and someone needs to access his notes, the comprehension of the notes taken would be easier (Davis et al. 2009; [111]).

Even if the app makes the scene documentation take longer, it does not mean that it will not speed up the investigation process of the crime scene as a whole. The reduction, or not, of the total time spent working on a case will depend on the specific protocols of each agency. If a protocol requires investigators to upload their

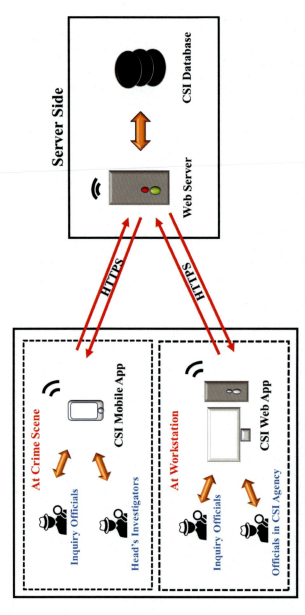

Fig. 11.19 **Illustration of architecture for crime scene investigation (CSI) using smartphone technology.** The CSI system consists of two main parts: the investigator side and the server-side.

Table 11.4 Evaluation of the five applications according to their relevance, reliability, and answers to operational requirements using a scale from +++ to - (see Table 11.5 for symbols meaning).

	Relevance	Reliability	Operational Requirements
CrimePad®	++	++	–
MagicPlan CSI®	+	++	+
CASE®	–	–	+
World Drugs DB®	++	–	++
Documed Compendium®	+++	+++	++

Table 11.5 Meaning of symbols used to describe the evaluation in Table 11.4.

	Relevance
+++	Responds extremely well to the objective—contributes considerably to the crime scene investigation
++	Responds well to the objective—contributes significantly to the crime scene investigation
+	Satisfying response to the objective—contributes somewhat to the crime scene investigation
–	Does not respond or responds in an unsatisfactory manner to the objective/critical needs of crime scene investigation

notes electronically and take them manually, it will take the time that they could be used for something more productive. In this case, using the app could be faster (Davis et al. 2009).

11.3.4.2 Limitations for applications in crime scenes investigation

Although there is an initial enthusiasm for a technological solution, in practice, many applications usable in the crime scene, and their functions, have not been specifically designed to be used for crime scene investigation [112]. Therefore, it is important to verify that the application and its use are indeed appropriate according to the needs of investigators and their organizations. The relevance assessment also makes it necessary to determine whether the use of the application negatively affects the quality of the work on crime scene investigators. The apparent help and ease that the applications bring can guide to automate certain tasks and focus the investigator's attention mainly on the elements and information on the mobile device, to the detriment of the concentration, time and effort put into thinking, establishing of a hypothesis, and the investigation of the crime scene. Therefore, using the application can create a risk of reducing thinking and imagination, lead to removing the responsibilities of an investigator, or even introducing the tunnel vision and bias when examining the crime scene [112].

In addition, data security is essential for the adoption of a mobile crime scene investigation application. Rong et al. explore the idea of cloud computing, discussing related security challenges, and emphasizing technological approaches that can improve security in the cloud [113]. Cloud computing provides convenient, on-demand access, to a shared set of data, allowing access to resources without requiring detailed knowledge of the underlying technologies. Community-based clouds allow members as forensics teams to share resources. However, to avoid leaks of confidential data stored in the cloud, the use of homomorphic or incremental encryption is suggested by Rong et al. [113].

11.4 Conclusions

The popularization of smartphone technology makes possible its application in a vast field of science, as well demonstrate inside this Chapter. Additionally, it points out not only a diversity of applications regarding the use of smartphones for in situ analysis but also shows its main components, describing their importance in the context of smartphone technology.

In view of the number of examples, much more than talk, smartphones open up an avenue of possibilities for carrying out analysis inside and, sometimes most important, outside the laboratories, amplifying the dynamic of work based on smartphones-based-platforms.

Nonetheless, it is of utmost importance that some drawbacks such as ruggedness, high stray light, low response, lack of accurate wavelength, or energy calibration, among others, are solved for figures of merits obtained through smartphone-based-platforms are similar or better than those acquired through consolidated techniques.

Anyway, as possible trends, the utilization of smartphones as a voice detector, indeed, may be used for evaluating some imperfections in the voice propagation, and such imperfections would be the basis for detecting some voice and/or respiratory diseases. Another important trend could be an addition of a thermal camera to smartphones, enlarging their applications for diagnosis of diseases, fever detection, forensic, among others, being all of these applications made through a drone or not, depending on the necessity.

In fact, in our point of view, the use of smartphones will go where our imagination allows. To infinite and beyond!

List of Abbreviation

AgNPs	Silver nanoparticles
ALS	Ambient Light Sensor
AOPs	Advanced Oxidation Processes
ATR-FTIR	Attenuated Total Reflectance Fourier Transform Infrared
AuNPs	Gold Nanoparticles

BPA	Bisphenol A
CCD	Charge-Coupled Device
CIE	Commission Internationale de l'Eclairage
CMC	Carboxymethylcellulose
CMOS	Complementary Metal Oxide Semiconductor
COD	Chemical Demand for Oxygen
DIB	Digital Image-based
CSI	Crime Scene Investigation
DNS	3,5-dinitrosalicylic acid
ELISA	Enzyme-Linked Immunosorbent Assay
GOx	glucose oxidase
GPS	Global Positioning System
HCA	Hierarchical Cluster Analysis
HDPE	High-Density Polyethylene
HIS	Hue-Saturation-Intensity
HPTS	8-hydroxypyrene-1,3,6-trisulfonic acid
HRP	Horseradish Peroxidase
HSL	Hue-Saturation-Lightness
HSV	Hue-Saturation-Value
IDE	Integrated Development Environment
LED	Light-emitting diode
LI	Local Investigator
LLME	Liquid-Liquid Microextraction
µPADs	Paper-Based Microfluidic Analysis Devices
MMS	Multimedia Messaging Service
MVVM	Model-View-View Model
OSMS	Open-Source Miniature Spectrophotometer
PCA	Principal Component Analysis
PLS	Partial Least Squares
PLS-DA	Partial Least Squares Discriminant Analysis
POC	Point-of-Care
PTFE	Polytetrafluoroethylene
RGB	Red, Green and Blue
RMSEC	Root Mean Square Error of Calibration
RMSEP	Root Mean Square Error of Prediction
sAA	Salivary α-amylase
SECV	Standard Error of Cross Validation
SMS	Short Message Service
SPR	Surface Plasmon Resonance
SVD	Singular Value Decomposition algorithm
TDD	Test-Driven Development
WBC	White Blood Cell
WCES	Whole-Copper Electrochemical Sensor
WPF	Windows Presentation Foundation

References

[1] P. Kennedy, Who Made That Cellphone? The New York Times, 2013. Disponible in. https://www.nytimes.com/2013/03/17/magazine/who-made-that-cellphone.html Accessed at May, 25th, 2015.

[2] M. Mandel, C.E. Petito, R. Tutihashi, W. Paiva, S. Abramovicz, F.C.G. Mandel Pinto, A.F. Andrade, M.J. Teixeira, E.G Figueiredo, Smartphone-assisted minimally invasive neurosurgery, J. Neurosurg. 130 (1) (2019) 90–98. https://doi.org/10.3171/2017.6.JNS1712.

[3] K. Anderson, D.G.L. DeBell, S. Hancock, J.P. Duffy, J.D. Shutler, W.J Reinhardt, A Griffiths, A grassroots remote sensing toolkit using live coding, smartphones, kites and lightweight drones, PLoS One 5 (2016) 1–22. https://doi.org/10.1371/journal.pone.0151564.

[4] M. De Clercq, A. Vats, A. Biel, Agriculture 4.0: the Future of Farming Technology, Proceedings of the World Government Summit, Dubai, UAE, 2018 1–30 11–13 February 2018.

[5] A.J. Williams, H.E. Pence, Smart phones, a powerful tool in the chemistry classroom, J. Chem. Educ. 88 (2011) 683–686. https://doi.org/10.1021/ed200029p.

[6] M. Rezazadeh, S. Seidi, M. Lid, S. Pedersen-Bjergaard, Y Yamini, The modern role of smartphones in analytical chemistry, Trends Anal. Chem. 118 (2019) 548–555.

[7] R.J.C. Brown, M.J.T Milton, Analytical techniques for trace element analysis: an overview, Trends Anal. Chem. 24 (3) (2005) 266–274. https://doi:10.1016/j.trac.2004.11.010.

[8] W. Miekisch, J.K. Schubert, From highly sophisticated analytical techniques to lifesaving diagnostics: technical developments in breath analysis, Trends Anal. Chem. 25 (7) (2006) 665–673. https://doi.org/10.1016/j.trac.2006.05.006.

[9] R.C. Gonzalez, R.E. Woods, Digital Image Processing, Pearson Education, New Jersey, 2008.

[10] O. Marques Filho, H. Vieira Neto, Processamento Digital de Imagens, Brasport, Rio de Janeiro, 1999.

[11] H. Pedrini, W.R. Schwartz, Análise de Imagens Digitais - Princípios, Algoritmos e Aplicações, Thomson Learning, São Paulo, 2008.

[12] W. Burger, M. Burge, Principles of Digital Image Processing - Fundamental Techniques, Springer-Verlag, London, 2009.

[13] A. Antonelli, M. Cocchib, P. Favaa, G. Focab, G. Franchinia, D. Manzinib, A. Ulricia, Automated evaluation of food colour by means of multivariate image analysis coupled to a wavelet-based classification algorithm, Anal. Chim. Acta 515 (2004) 3–13. https://doi.org/10.1016/j.aca.2004.01.005.

[14] Alvy Ray Smith, Color gamut transform pairs, Comput Graph (ACM) 12 (3) (1978) 12–19. agoAvailable at:. http://www.icst.pku.edu.cn/course/ImageProcessing/2013/resource/Color78.pdf Accessed at: 25 February. 2015.

[15] M.K. Agoston, Computer Graphics and Geometric Modeling: Implementation and Algorithms, Springer, London, 2005.

[16] A. Theuwissen, How to Measure Non-Linearity? Disponible, 2013. http://harvestimaging.com/blog/?p=1125 Accessed at: May, 25th, 2015.

[17] M.F. Pimentel, B. Barros Neto, Calibração: uma revisão para químicos analíticos, Química Nova 3 (19) (1996) 268–277.

[18] J.M. Miller, J.C. Miller, Statistics and Chemometrics for Analytical Chemistry, Pearson Education, Harlow, 2010.

[19] L.I. Smith, Tutorial on Principal Component Analysis, Disponible, 2002. http://www.sccg.sk/~haladova/principal_components.pdf/ Accessed at February, 25th, 2015.
[20] Ferreira, M.C., Antunes, A., Melgo, M., Volpe, P. 1999. Quimiometria I: calibração multivariada, um tutorial. *Química Nova* 22(5): 724-731. https://doi.org/10.1590/S0100-40421999000500016.
[21] Marco Flôres Ferrão, Aplicação de técnicas espectroscópicas de reflexão no infravermelho no controle de qualidade de farinha de trigo, 2000, 219 f Tese (Doutorado) - Universidade Estadual de Campinas, Campinas, 2000.
[22] P. Geladi, H. Grahn, K. Esbensen, E. Bengtsson, Multivariate Image Analysis, Trends in Analytical Chemistry, 1992. 11, 3. https://doi.org/10.1016/0165-9936(92)85010-3.
[23] M.M.C. Ferreira, Quimiometria: Conceitos, Métodos e Aplicações, Editorada Unicamp, Campinas-SP, 2015.
[24] P. Valderrama, J.W.B Braga, R.J. Poppi, Estado da arte de figuras de mérito em calibração multivariada, Química Nova 32 (5) (2009) 1278–1287. https://doi.org/10.1590/S0100-40422009000500034.
[25] M.J. Anzanello, Seleção de variáveis para classificação de bateladas produtivas com base em múltiplos critérios, Production 23 (4) (2013) 858–865. https://doi.org/10.1590/S0103-65132013005000001.
[26] INFOMETRIX Inc.: Pirouette User Guide. Version 4.5, Bothell, WA, 2011.
[27] R. Bro, A.K. Smilde, Centering and scaling in component analysis, J. Chemom. 17 (2001) 16–33. https://doi.org/10.1002/cem.773.
[28] G. Matos, E. Pereira-Filho, R. Poppi, M. Arruda, Análise exploratória em química analítica com emprego de quimiometria: PCA e PCA de imagens, Revista Analytica (6) (2003) 38–46.
[29] D. Astels, G. Miller, M.M. Novak, Extreme programming: guia prático, Rio de Janeiro: Campus (2002) 342.
[30] I. Sommerville, Engenharia de Software, 6. ed., Pearson Addison Wesley, São Paulo, 2003, pp. 592.
[31] MICROSOFT VISUAL STUDIO 2012®. Microsoft Corporation: Redmond, WA, USA.
[32] MICROSOFT OFFICE EXCEL 2015®. Microsoft Corporation: Redmond, WA, USA.
[33] Souza, C.R. The Accord. NET Framework. 2012. Disponible in http://accord.googlecode.com Accessed in February 25th, 2015.
[34] S. Ramos, Signs of PCA Scores are Inverted in 2 Nearly Identical Sets, 2005. Infometrix Software Forum Index: Pirouette AlgorithmsAvailable in: http://infometrix.biz/phpbb2/viewtopic.php?t=32&sid=c09eaec56c92f19afda7c344404219b2 Access in: 25 set. 2012.
[35] G.A. Helfer, B. Tischer, P.F. Filoda, A.B. Parckert, R.B. dos Santos, L.L. Vinciguerra, M.F. Ferrão, J.S. Barin, A.B. da Costa, A new tool for interpretation of thermal stability of raw milk by means of the alizarol test using a PLS model on a mobile device, Food Anal. Methods 11 (7) (2018) 2022–2028. https://doi.org/10.1007/s12161-018-1190-5.
[36] L. Shen, J.A. Hagen, I. Papautsky, Point-of-care colorimetric detection with a smartphone, Lab Chip 12 (21) (2012) 4240–4243. https://doi.org/10.1039/c2lc40741h.
[37] L. Bueno, G.N. Meloni, S.M. Reddy, T.R.L.C Paixão, Use of plastic-based analytical device, smartphone and chemometric tools to discriminate amines, RSC Adv. 5 (26) (2015) 20148–20154. https://doi.org/10.1039/c5ra01822f.
[38] K.F.M. De Oliveira, W.T. Suarez, V.B dos Santos, Digital image method smartphone-based for furfural determination in sugarcane spirits, Food Anal. Methods 10 (2) (2017) 508–515. https://doi.org/10.1007/s12161-016-0605-4.

[39] K.F.M. De Oliveira, W.T. Suarez, V.B dos Santos, Smartphone application for methanol determination in sugar cane spirits employing digital image-based method, Food Anal. Methods 10 (6) (2017) 2102–2109. https://doi.org/10.1007/s12161-016-0777-y.

[40] Y. Wang, Y. Li, X. Bao, J. Han, J. Xia, X. Tian, L. Ni, A smartphone-based colorimetric reader coupled with a remote server for rapid on-site catechols analysis, Talanta 160 (2016) 194–204. https://doi.org/10.1016/j.talanta.2016.07.012.

[41] K.E. McCracken, T. Tat, V. Paz, J.Y. Yoon, Smartphone-based fluorescence detection of bisphenol A from water samples, RSC Adv. 7 (15) (2017) 9237–9243. https://doi.org/10.1039/c6ra27726h.

[42] B.G. Botelho, K.C.F. Dantas, M.M Sena, Determination of allura red dye in hard candies by using digital images obtained with a mobile phone and N-PLS, Chemom. Intell. Lab. Syst. 167 (2017) 44–49. https://doi.org/10.1016/j.chemolab.2017.05.004.

[43] MATLAB®. The Mathworks, Inc.: Natick, MA, USA.

[44] M.S.M.S.F. Acevedo, M.J.A. Lima, C.F. Nascimento, F.R.P Rocha, A green and cost-effective procedure for determination of anionic surfactants in milk with liquid-liquid microextraction and smartphone-based photometric detection, Microchem. J. 143 (2018) 259–263. https://doi.org/10.1016/j.microc.2018.08.002.

[45] F.C.P. Ribeiro, A.S. Oliveira, A. Araújo, W. Marinho, M.P. Schneider, L. Pinto, A.A Gomes, Detection oxidative degradation in lubricating oil under storage conditions using digital images and chemometrics, Microchem. J. 147 (2019) 622–627. https://doi.org/10.1016/j.microc.2019.03.087.

[46] V.H.J.M. Dos Santos, D. Pontin, G.S. Oliveira, T.A. Siqueira, M Seferin, Multivariate analysis of digital images as an alternative to monitor dye degradation by the Fenton process, Química Nova 43 (2020) 599–606. https://doi.org/10.21577/0100-4042.20170531.

[47] G.A. Helfer, V.S. Magnus, F.C. Böck, A. Teichmann, M.F. Ferrão, A.B. da Costa, PhotoMetrix: an application for univariate calibration and principal components analysis using colorimetry on mobile devices, J. Braz. Chem. Soc. 28 (2) (2017) 328–335. https://doi.org/10.5935/0103-5053.20160182.

[48] Nogueira S.A., Sousa L.R., Silva N.K.L., Rodrigues P.H.F., Coltro W.K.T. (2017). "Monitoring acid – base titrations on wax printed paper microzones using a smartphone." Micromachines 139:1–10. https://doi.org/10.3390/mi8050139.

[49] S. Soares, M.J.A. Lima, F.R.P Rocha, A spot test for iodine value determination in biodiesel based on digital images exploiting a smartphone, Microchem. J. 133 (2017) 195–199. https://doi.org/10.1016/j.microc.2017.03.029.

[50] F.S. Grasel, M.F. Ferrão, G.A. Helfer, A.B. Costa, Principal component analysis of commercial tannin extracts using digital images on mobile devices, J. Braz. Chem. Soc. 27 (12) (2016) 2372–2377. https://doi.org/10.5935/0103-5053.20160135.

[51] G.A. Helfer, V.S. Magnus, F.C. Böck, A. Teichmann, M.F. Ferrão, A.B. Da Costa, PhotoMetrix: an application for univariate calibration and principal components analysis using colorimetry on mobile devices, J. Braz. Chem. Soc. 28 (2) (2017) 336–347. https://doi.org/10.5935/0103-5053.20160182.

[52] F.C. Böck, G.A. Helfer, A.B. da Costa, M.B. Dessuy, M.F. Ferrão, Rapid determination of ethanol in sugarcane spirit using partial least squares regression embedded in smartphone, Food Anal. Methods 11 (7) (2018) 1951–1957. https://doi.org/10.1007/s12161-018-1167-4.

[53] W.N. Guedes, G.N. Lucena, A.V. de Paula, R.F.C. Marques, F.M.V Pereira, Easy Estimation of endoglucanase activity using a free software app for mobile devices, Brazilian J. Anal. Chem. 7 (2020) 27–35. https://doi.org/10.30744/brjac.2179-3425.AR-33-2019.

[54] S. Soares, K.G. Torres, E.L. Pimentel, P.B. Martelli, F.R.P Rocha, A novel spot test based on digital images for determination of methanol in biodiesel, Talanta 195 (2019) 229–235. https://doi.org/10.1016/j.talanta.2018.11.028.

[55] S. Soares, F.R.P Rocha, A simple and low-cost approach for microdistillation: application to methanol determination in biodiesel exploiting smartphone-based digital images, Talanta 199 (2019) 285–289. https://doi.org/10.1016/j.talanta.2019.02.064.

[56] S. Soares, L.C. Nunes, W.R. Melchert, F.R.P Rocha, Spot test exploiting smartphone-based digital images for determination of biodiesel in diesel blends, Microchem. J. 152 (2020) 104273. https://doi.org/10.1016/j.microc.2019.104273.

[57] Aydogan. Ozcan, Mobile phones democratize and cultivate next-generation imaging, diagnostics and measurement tools, Lab Chip 14 (17) (2014) 3187–3194. https://doi.org/10.1039/c4lc00010b.

[58] V. Kilic, G. Alankus, N. Horzum, A.Y. Mutlu, A. Bayram, M.E. Solmaz, Single-image-referenced colorimetric water quality detection using a smartphone, ACS Omega 3 (5) (2018) 5531–5536.

[59] E.C. Lumbaque, B.A. da Silva, F.C. Böck, G.A. Helfer, M.F. Ferrão, C. Sirtori, Total dissolved iron and hydrogen peroxide determination using the PhotoMetrixPRO application: a portable colorimetric analysis tool for controlling important conditions in the solar photo-Fenton process, J. Hazard. Mater. 378 (2019) 120740.

[60] I. Hussain, K. Ahamad, P. Nath, Water turbidity sensing using a smartphone, RSC Adv. 6 (27) (2016) 22374–22382.

[61] I. Hussain, K.U. Ahamad, P Nath, Low-cost, robust, and field portable smartphone platform photometric sensor for fluoride level detection in drinking water, Anal. Chem. 89 (1) (2017) 767–775.

[62] P. Jarujamrus, R. Meelapsom, S. Pencharee, A. Obma, M. Amatatongchai, Ditcharoen, S. Chairam, S. Tamuang, Use of a smartphone as a colorimetric analyzer in paper-based devices for sensitive and selective determination of mercury in water samples, Analytical Science 34 (1) (2018) 75–81.

[63] G.F. Giordano, M.B.R. Vicentini, R.C. Murer, F. Augusto, M.F. Ferrão, G.A. Helfer, A.B. da Costa, A.L. Gobbi, L.W. Hantao, R.S. Lima, Point-of-use electroanalytical platform based on homemade potentiostat and smartphone for multivariate data processing, Electrochim. Acta 219 (2016) 170–177.

[64] K. Laganovska, A. Zolotarjovs, M. Vázquez, K. Mc Donnell, J. Liepins, H. Ben-Yoav, V Karitans, K. Smits, Portable low-cost open-source wireless spectrophotometer for fast and reliable measurements, HardwareX 7 (2020) e00108. https://doi.org/10.1016/j.ohx.2020.e00108.

[65] J. Liao, F. Chang, X. Han, C. Ge, S. Lin, Wireless water quality monitoring and spatial mapping with disposable whole-copper electrochemical sensors and a smartphone, Sens. Actuators B 306 (2020) 127557.

[66] F.J. Mesas-Carrascosa, D. Verdú Santano, J.E. Meroño, M. Sánchez de la Orden, A. García-Ferrer, Open source hardware to monitor environmental parameters in precision agriculture, Biosystems Eng. 137 (2015) 73–83.

[67] F.C. Böck, G.A. Helfer, A.B. Costa, M.B. Dessuy, M.F. Ferrão, PhotoMetrix and colorimetric image analysis using smartphones, J. Chemom. (2020) 1–19. https://doi.org/10.1002/cem.3251.

[68] A. Kahl, Smartphone spectrometers: the intersection of environmental chemistry and engineering, Chemistry and the Environment: Pedagogical Models and Practices, 1214, *American Chemical Society*, 2015, pp. 69–73.

[69] L.F. Capitán-Vallvey, N. López-Ruiz, A. Martínez-Olmos, M.M. Erenas, A.J. Palma, Recent developments in computer vision-based analytical chemistry: a tutorial review, Anal. Chim. Acta 899 (2015) 23–56.

[70] S. Levin, S. Krishnan, S. Rajkumar, Halery, P. Balkunde, Monitoring of fluoride in water samples using a smartphone, Sci. Total Environ. 551-552 (2016) 101–107.

[71] C. Pappis, M. Librelotto, L. Baumann, A.B. Parckert, R.O. Santos, I.D. Teixeira, G.A. Helfer, E.A. Lobo, A.B.d. Costa, Point-of-use determination of fluoride and phosphorus in water through a smartphone using the PhotoMetrix® App, Brazilian J. Anal. Chem. 6 (25) (2019) 9.

[72] M. Sargazi, M. Kaykhaii, Application of a smartphone based spectrophotometer for rapid in-field determination of nitrite and chlorine in environmental water samples, Spectrochimica Acta Part A 227 (2020) 117672.

[73] P. Han, D. Dong, X. Zhao, L. Jiao, Y. Lang, A smartphone-based soil color sensor: for soil type classification, Comput. Electron. Agric. 123 (2016) 232–241.

[74] Y. Xing, Q. Zhu, X. Zhou, P. Qi, A dual-functional smartphone-based sensor for colorimetric and chemiluminescent detection: a case study for fluoride concentration mapping, Sens. Actuators B 319 (2020) 128254.

[75] Z. Liu, Y. Zhang, S. Xu, H. Zhang, Y. Tan, C. Ma, R. Song, L. Jiang, C. Yi, A 3D printed smartphone optosensing platform for point-of-need food safety inspection, Anal. Chim. Acta 966 (2017) 81–89.

[76] H. Wang, L. Da, L. Yang, S. Chu, F. Yang, S. Yu, C. Jiang, Colorimetric fluorescent paper strip with smartphone platform for quantitative detection of cadmium ions in real samples, J. Hazard. Mater. 392 (2020) 122506.

[77] E. Vidal, A.S. Lorenzetti, A.G. Lista, C.E. Domini, Micropaper-based analytical device (μPAD) for the simultaneous determination of nitrite and fluoride using a smartphone, Microchem. J. 143 (2018) 467–473.

[78] Y. He, Y. Wu, J.-Z. Fu, W.-B. Wu, Fabrication of paper-based microfluidic analysis devices: a review, RSC Adv. 5 (95) (2015) 78109–78127.

[79] J. Pla-Tolós, Y. Moliner-Martínez, J. Verdú-Andrés, J. Casanova-Chafer, C. Molins-Legua, P. Campíns-Falcó, New optical paper sensor for in situ measurement of hydrogen sulphide in waters and atmospheres, Talanta 156-157 (2016) 79–86.

[80] Q. Mei, H. Jing, Y. Li, W. Yisibashaer, J. Chen, B. Nan Li, Y. Zhang, Smartphone based visual and quantitative assays on upconversional paper sensor, Biosens. Bioelectron. 75 (2016) 427–432. https://doi.org/10.1016/j.bios.2015.08.054.

[81] S. Zhang, N. Shapiro, G. Gehrke, J. Castner, Z. Liu, B. Guo, J. Prasad, J. Zhang, S.R. Haines, D. Kormos, P. Frey, R. Qin, K.C. Dannemiller, Smartphone app for residential testing of formaldehyde (SmART-Form), Build. Environ. 148 (2019) 567–578.

[82] J. Jiang, X. Wang, R. Chao, Y. Ren, C. Hu, Z. Xu, G.L. Liu, Smartphone based portable bacteria pre-concentrating microfluidic sensor and impedance sensing system, Sens. Actuators B 193 (2014) 653–659.

[83] M. Eskin, M. Torabfam, H. Kurt, E. Psillakis, A. Cincinelli, M. Yüce, Real-time water quality monitoring of an artificial lake using portable, affordable, simple, arduino-based open source sensors, Environ. Eng. - Inženjerstvo okoliša 6 (1) (2019). 2019. https://doi.org/10.37023/ee.6.1.2.

[84] L. Baumann, M. Librelotto, C. Pappis, G.A. Helfer, R.O. Santos, R.B. dos Santos, A.B. da Costa, Nanometrix: an app for chemometric analysis from near infrared spectra, J. Chemometrics n/a(n/a) (2020). https://doi.org/10.1002/cem.3281.

[85] M.E. Solmaz, A.Y. Mutlu, G. Alankus, Kılıç, A. Bayram, N. Horzum, Quantifying colorimetric tests using a smartphone app based on machine learning classifiers, Sens. Actuators B 255 (2018) 1967–1973.

[86] M. Mascini, S Tombelli, Biosensors for biomarkers in medical diagnostics, Biomarkers 13 (7–8) (2008) 637–657. https://doi.org/10.1080/13547500802645905.
[87] S.R. Steinhubl, E.D. Muse, E.J. Topol, The emerging field of mobile health, Sci. Transl. Med. 7 (283) (2015) 1–7. https://doi.org/10.1126/scitranslmed.aaa3487.
[88] R. Buechi, L. Faes, L.M. Bachmann, M.A. Thiel, N.S. Bodmer, M.K. Schmid, O. Job, K.R. Lienhard, Evidence assessing the diagnostic performance of medical smartphone apps: a systematic review and exploratory meta-analysis, BMJ Open 7 (12) (2017) 1–8. https://doi.org/10.1136/bmjopen-2017-018280.
[89] D.B. Hernández, J.L. Marty, R.M. Guerrero, Smartphone as a portable detector, analytical device, or instrument interface, smartphones from an applied research perspective, Nawaz Mohamudally (2017). IntechOpen https://doi.org/10.5772/intechopen.69678.
[90] X. Huang, D. Xu, J. Chen, J. Liu, Y. Li, J. Song, J. Song, X. Ma, J. Guo, Smartphone-based analytical biosensors, Analyst 143 (22) (2018) 5339–5351. https://doi.org/10.1039/c8an01269e.
[91] I. Hernández-Neuta, F. Neumann, J. Brightmeyer, T. Ba Tis, N. Madaboosi, Q. Wei, A. Ozcan, M. Nilsson, Smartphone-based clinical diagnostics: towards democratization of evidence-based health care, J. Intern. Med. 285 (1) (2019) 19–39. https://doi.org/10.1111/joim.12820.
[92] L. Bellina, E. Missoni, Mobile cell-phones (M-Phones) in telemicroscopy: increasing connectivity of isolated laboratories, Diagn Pathol 4 (1) (2009). https://doi.org/10.1186/1746-1596-4-19.
[93] J. Wu, M. Dong, C. Rigatto, Y. Liu, F Lin, Lab-on-chip technology for chronic disease diagnosis, Npj Digital Medicine 1 (1) (2018) 1–11. https://doi.org/10.1038/s41746-017-0014-0.
[94] D.N. Breslauer, R.N. Maamari, N.A. Switz, W.A. Lam, D.A. Fletcher, Mobile phone based clinical microscopy for global health applications, PLoS One 4 (7) (2009) 1–7. https://doi.org/10.1371/journal.pone.0006320.
[95] Y. Xia, Jingjing Ye, K. Tan, J. Wang, G Yang, Colorimetric visualization of glucose at the submicromole level in serum by a homogenous silver nanoprism-glucose oxidase system, Anal. Chem. 85 (13) (2013) 6241–6247. https://doi.org/10.1021/ac303591n.
[96] J.R. Jesus, I.C. Guimarães, M.A.Z Arruda, Quantifying proteins at microgram levels integrating gel electrophoresis and smartphone technology, J. Proteomics 198 (2019) 45–49.
[97] C.L.V. Camargo, B.R. Marcia, A.L. Gobbi, D.S.T. Martinez, R.S. Lima, Smartphone for point-of-care quantification of protein by bradford assay, J. Braz. Chem. Soc. 28 (4) (2017) 689–693.
[98] A. Tapley, N. Switz, C. Reber, J.L. Davis, M. Cecily, J.B. Matovu, W. Worodria, L. Huang, D.A Fletcher, A. Cattamanchi, Mobile digital fluorescence microscopy for diagnosis of tuberculosis, J. Clin. Microbiol. 51 (6) (2013) 1774–1778. https://doi.org/10.1128/JCM.03432-12.
[99] Q. Wei, H. Qi, W. Luo, D. Tseng, S.J. Ki, Z. Wan, Z. Göröcs, Bentolila Laurent A., Wu Ting-Ting, Sun Ren, Ozcan Aydogan, Fluorescent imaging of single nanoparticles and viruses on a smart phone, ACS Nano 7 (10) (2013) 9147–9155. https://doi.org/10.1021/nn4037706.
[100] Y. Liu, Q. Liu, S. Chen, F. Cheng, H. Wang, W. Peng, Surface plasmon resonance biosensor based on smart phone platforms, Sci. Rep. 5 (2015) 1–9. https://doi.org/10.1038/srep12864.
[101] Pakorn Preechaburana, Marcos Collado Gonzalez, Anke Suska, Daniel Filippini, Surface plasmon resonance chemical sensing on cell phones, Angew. Chem. 51 (46) (2012) 11585–11588. https://doi.org/10.1002/anie.201206804.

[102] A.C. Sun, D.A Hall., Point-of-care smartphone-based electrochemical biosensing, Electroanalysis 31 (1) (2019) 2–16. https://doi.org/10.1002/elan.201800474.
[103] Xinhao Wang, Guohong Lin, Guangzhe Cui, Xiangfei Zhou, G.ang Logan Liu, White blood cell counting on smartphone paper electrochemical sensor, Biosens. Bioelectron. 90 (2017) 549–557. https://doi.org/10.1016/j.bios.2016.10.017.
[104] A.C. Sun, Y. Chengyang, Ag Venkatesh, A.H. Drew, An efficient power harvesting mobile phone-based electrochemical biosensor for point-of-care health monitoring, Sens. Actuators B 235 (2016) 126–135. https://doi.org/10.1016/j.snb.2016.05.010.
[105] P.T. Garcia, L.N. Guimarães, A.A. Dias, C.J. Ulhoa, W.K.T Coltro, Amperometric detection of salivary α-amylase on screen-printed carbon electrodes as a simple and inexpensive alternative for point-of-care testing, Sens. Actuators B 258 (2018) 342–348. https://doi.org/10.1016/j.snb.2017.11.068.
[106] M. LoGrande, The utilization of mobile technology for crime scene investigation in the san francisco bay area, Themis: Research Journal of Justice Studies and Forensic Science 4 (1) (2016) 9.
[107] T.I. Kitsaki, A. Angelogianni, C. Ntantogian, C. Xenakis, A forensic investigation of android mobile applications, ACM International (2018) 58–63. https://doi.org/10.1145/3291533.3291573.
[108] J. Fish, L. Miller, M. Braswell, Crime scene investigation. In: Crime Scene Investigation, 2012 1–479. https://doi.org/10.5005/jp/books/12350_28.
[109] C. Baber, P. Smith, M. Butler, J. Cross, J. Hunter, Mobile technology for crime scene examination, Int. J. Human Comput. Studies 67 (5) (2009) 464–474. https://doi.org/10.1016/j.ijhcs.2008.12.004.
[110] P. Margot, Forensic science on trial-what is the law of the land? Australian J. Forensic Sci. 43 (2–3) (2011) 89–103. https://doi.org/10.1080/00450618.2011.555418.
[111] Nutcharee Wichiennit, Khamron Sunat, Sirapat Chiewchanwattana, Boonchai Louchaisa, Boonyarin Onnoom, Design and Development of Application for Crime Scene Notification System, Ubi-Media 2017 - Proceedings of the 10th International Conference on Ubi-Media Computing and Workshops with the 4th International Workshop on Advanced E-Learning and the 1st International Workshop on Multimedia and IoT: Networks, Systems and Applications, 2017. https://doi.org/10.1109/UMEDIA.2017.8074103.
[112] S. Baechler, A. Gélinas, R. Tremblay, K. Lu, F. Crispino, Smartphone and tablet applications for crime scene investigation: state of the art, typology, and assessment criteria, J. Forensic Sci. 62 (4) (2017) 1043–1053. https://doi.org/10.1111/1556-4029.13383.
[113] C. Rong, S.T. Nguyen, M.G. Jaatun, Beyond lightning: a survey on security challenges in cloud computing, Comput. Electr. Eng. 39 (1) (2013) 47–54. https://doi.org/10.1016/j.compeleceng.2012.04.015.
[114] G. Helfer, F. Bock, L. Marder, J.C. Furtado, A.B. da Costa, M.F. Ferrão, Chemostat, um software gratuito para análise exploratória de dados multivariados, Química Nova 38 (4) (2015) 575–579. https://doi.org/10.5935/0100-4042.20150063.

Applications of smartphones in food analysis

Adriana S. Franca, Leandro S. Oliveira
Universidade Federal de Minas Gerais, Av. Pres. Antônio Carlos, Belo Horizonte, MG, Brazil

12.1 Introduction

Food safety and quality are essential to people's survival and lives [32]. These are two different concepts that altogether must ensure to the consumer that each specific food product has met the industry's standards. The concept of food safety is associated with reducing the risk of individuals becoming sick from foodborne illnesses, by handling, preparing and storing food products accordingly. The major goal is to prevent diseases caused by pathogenic microorganisms, misuse of food additives and contaminants such as chemical or biological toxins or adulterants. Food quality, on the other hand, is associated to several attributes that influence the consumers' decisions in terms of acquiring a specific food product. Examples of these attributes include color, flavor, taste, nutritional value, provenance, environmentally friendly production, among others. Given the inherent complexity of food products, most instrumental techniques employed for quality and safety evaluation are time demanding, costly, and require considerable manual labor [1,18]. A wide variety of analytical techniques have been employed for food analysis, including both liquid and gas chromatography [11,16], DNA-based techniques [51], isotope ratio and elemental analysis [17,31], spectroscopic methods [16,18,25,28] and sensor techniques [47], and each method has its advantages and disadvantages. Chromatographic methods, isotope ratio, elemental analysis, and DNA based methods have been shown to be highly accurate and sensitive, but the major drawbacks are associated to the long times required for analyses, and the facts that they are expensive, labor intensive, and usually involve many purification steps. Furthermore, the majority of these traditional analysis methods can only be performed in professional laboratories by trained personnel, thus being unsuitable for portable food evaluation, especially in the case of consumer-oriented detection. Therefore, in the recent past, there has been an increasing interest in simpler, faster, and reliable analytical methods for assessing food quality and safety attributes. Most studies have focused on non-destructive and non-invasive techniques including vibrational, hyperspectral, fluorescence and nuclear magnetic resonance (NMR) spectroscopy, as well as sensor-based techniques including electronic tongues and electronic noses. Even though these techniques are rapid, easy to operate, cost-effective, and involve less or no sample preparation, a major drawback is their lower sensitivity when compared to the more sophisticated techniques. To this end, smartphone-based biosensors have emerged as researchers have attempted to better replicate the conventional analysis tools using powerful and portable devices [32]. Although there is an extensive amount of literature data regarding the development and application of smartphone-based systems for detection of pathogens and chemicals that are of interest for food quality and safety evaluation, this review (not intended to be exhaustive of the literature) focused only on recent studies that tested the proposed methodologies in actual food products. The following reviews that address the application of smartphones in food analysis are indicated as complementary reading material [32,39,40,48,49].

12.2 Food quality and authenticity

The concept of food quality is rather broad and is strictly tied to a set of desired attributes of a specific food product. It encompasses a variety of physical, chemical and sensory properties, such as color, texture, tissue damage, maturity stage, presence or absence of specific compounds, degree of browning, spoilage by microorganisms, freshness, staleness, aroma, flavor, presence of adulterants, among several others, either isolated or combined. Food authenticity is a specific subset of food properties and attributes that define its quality and is related to aspects such as geographical origin, plant or animal species, specific plant or animal parts, presence of adulterants and others. Food quality is regularly monitored and controlled along the entire food supply chain, from production to commercialization, employing costly methodologies, bulky equipment and substantial amounts of chemicals that directly affect the cost of the final product to the consumer. Techniques such as gas and liquid chromatography, UV–Vis and infrared spectroscopies, ultrasound probing and chemical sensing by electronic noses are among the most commonly employed for such endeavor. These techniques, although suitably accurate for the task, are usually cost ineffective and far from user friendly for monitoring food quality in the field (e.g., by the producers) or by the consumers at point of consumption. To tackle these issues, a slow but steady shift has been occurring in the employment of analytical techniques, moving from time-consuming environmentally aggressive and costly methodologies to portable reliable and user-friendly systems with environmentally benign and green characteristics [43]. In this aspect, measurements based on acquisition and adequate processing of digital images have satisfactorily introduced advances to analytical procedures in regard to practicality and productivity, at the same time exploiting low-cost instrumentation and minimizing harmful waste generation [33]. Due to the popularization of smartphones and associated user-friendly software for image analysis, these types of devices have attracted attention from the scientific community in regard to their employment as portable convenient devices for one-site monitoring of food quality by imaging and data analysis [12]. Current smartphones are equipped with high-resolution cameras, high-speed processors, and large storage capacities, which, together with availability of a plethora of image-analysis application software, make them appropriate for use in point-of-care measurements for food quality monitoring. These devices also offer wireless connectivity, real-time geotagging and cloud computing capabilities [12], which not only facilitates transfer but also tracking of information along the food supply chain, from production all the way to the consumer.

Furthermore, different types of miniaturized sensors can be integrated into the device, expanding its capability for sensing specific chemical compounds independently of strictly color measurements. Although the applicability of smartphones as analytical tools is still greatly limited by their low selectivity and sensitivity, these devices have been aptly used in analytical procedures as detection systems, processing units or simply as light sources for digital image acquisition [43]. Furthermore, the possibility of integrating novel sensing technologies into smartphones enables the development of lab-on-smartphone platforms that will facilitate food quality monitoring along the food supply chain, allowing consumers to more easily and expertly decide on the acquisition of a certain food product at the point of consumption. Due to the multiplicity of built-in functionalities, smartphones could be either used independently as data acquisition and analysis system altogether or as an integral part of a larger system performing specific tasks such as digital image acquisition, processing of sensor-acquired data, or simply as an illumination source. A review on smartphone-based technologies for food diagnostic, focused on custom modules to enhance smartphone sensing capabilities, was presented [39], followed by another review [32] that focused on the development of biosensors for smartphones for food evaluation by colorimetry, fluorescence and electrochemistry. Research on applications of such systems to evaluate food quality are thus herein reviewed. Cruz-Fernandez and collaborators [10] developed a methodology in which a Meizu M2 mini smartphone was used to take digital images of cold meat products (e.g., salchichón, chorizo, salami and cured ham) and the images were processed to correlate their pixel color parameters with the fat content of the respective meat product as determined by the Soxhlet method. The built-in LED flash of the smartphone was used as illumination source. Images of the meat products were taken in a cylindrical illumination chamber (24 cm high and 7 cm diameter) with a hole in the top portion to fit in the smartphone camera and flash light at an angle of 16° in regard to the product. The whole setup was devised to ensure reproducible and steady illumination conditions and it weighed 50 g to allow for easy portability. The extracted RGB values of the digitally acquired images were used as input variables for the Partial Least Squares (PLS) and Support Vector Machine (SVM) models constructed for their correlation with the fat content of samples. The best correlations were obtained for salchichón and salami samples using SVM, with relative errors of calibration, cross-validation and prediction of 18, 20 and 16 percent, respectively. A colorimetric method based on color measurements from smartphone digital images of sparkling wine was developed by [37] to study wine browning. Images were taken by Apple iPhone 4S (Apple Inc. USA) smartphone built-in camera with the support of a diffuse light source placed in a suitable black box to eliminate external influences and ensure reproducibility of image acquisitions. The box was made of black foam core board internally covered with matte black velvet paper, with a height large enough to minimize the differences due to distinct optical paths. Wine samples were arranged into the black box in a 96-well plate and the smartphone was placed in a holder at the top of black box at a fixed distance from the 96-well plate. Images of the samples were taken at the highest possible resolution and saved as JPEGs. With this arrangement, all samples could be measured simultaneously to avoid disturbances due to lighting source or smartphone eventual undesirable behaviors. Four sparkling Cava wines were submitted to an accelerated browning process and the samples images acquired along the process were split into the three basic RGB channels and their individual values used to monitor the browning process. Results demonstrated that browning process affected primarily the Blue channel, with its decay being time dependent. The Blue channel decay percentage over time was proposed as a quality marker for wine undergoing browning. The Blue channel decay percentage value presented a high correlation with absorbance at 420 nm and with 5-hydroxymethyl-2-furfural content, the latter being the most common marker of wine browning. [8] proposed a repurposing of food's barcode as a colorimetric sensor array to monitor food conditions, acting as geometric substitute for QR codes. The developed sensor was based on three distinct types of vapor sensitive dyes individually encapsulated in resin microbeads and uniquely drop-casted onto triangle, square and circle patterns in a low cost paper substrate, thus enabling the distinction of the different sensing elements using shape-based encoding. The geometric barcode stamp was used in packages of chicken meat to monitor aging and eventual spoilage under different temperature conditions. An Apple iPhone camera, employing ambient illumination, took images of the developed geometric barcode sensors. The images were processed to extract the R, G, B channel intensities of the distinct sensing dyes at different time intervals. Principal component analysis (PCA) was applied to the colorimetric data arranged as a vector of all the three channels of R, G, B intensities at the square, triangle and circle areas at different times, for discrimination of sensor responses to different degrees of chicken meat spoilage. It was possible to discriminate sensor response using just the red (R) channel information of each sensing dye. Results demonstrated that the maximum change in color profile occurred when the chicken meat was deteriorating at a faster rate. The coupling of colorimetric geometric barcode sensors in food packages with smartphones for food quality diagnostic readouts demonstrated the technology to be amenable for use by consumers at point of consumption. An analytical procedure was developed by [2] for determination of anionic surfactants as milk adulterants, based on a smartphone camera photometric detection of the liquid-liquid micro-extracted ion-pair formed between the analytes and methylene blue. Protein precipitation, ion-pair extraction, and photometric detection were successfully carried out in a single vessel, in an attempt to simplify the procedure and minimize risks of contamination and analyte loss. Sodium dodecylsulfate was used as a model surfactant. Milk samples (100 µL) were processed by mixing with 25 µg of methylene blue and 2.9 mg of EDTA in a 1.5-mL Eppendorf microtube, and the microtube was inserted in a styrofoam box (14 cm-high, 16 cm-wide, and 10 cm-deep), with a support to place a smartphone at a distance of 8 cm from the measurement tubes. Constant illumination of the samples was achieved by a LED-based lamp (6 W) placed at the top of the box. RGB values of the digitally acquired images of treated milk samples were obtained by using the free application Color Grab (Loomatix, version 3.5.2) installed in the Motorola Moto G4 Plus smartphone used to capture the images. Using a standard method for calibration, a linear response was achieved for surfactant concentrations in the range of 10 to 50 mgL^{-1}, $r = 0.999$ ($n = 5$), with detection limit (99.7 percent confidence level) and coefficient of variation of 2.2 mgL^{-1} and 1.9 percent ($n = 10$), respectively. [12] developed a colorimetric sensor array–smartphone–remote server coupled system for rapid on-site testing of saccharides. The affinity interaction between boric acid compounds (boric, phenylboronic and 3-nitrophenylboronic acids) and cis-dihydroxy group in saccharides was tested and 3-nitrophenylboronic acid presented the highest binding capacity, thus being used in the construction of the pH-indicator sensor array. The colorimetric sensor array (2 × 2 cm) was comprised of an organic glass plate with 96 holes, with each individual hole being filled with a fully reacted mixture of 3-nitrophenylboronic acid, a pH indicator (xylenol orange, thymol blue, acid rose red or cresyl violet) and a saccharide solution with a specific concentration. Nineteen different types of saccharides were studied with concentrations ranging from 1 to 400 mM. A white LED light was fixed on the bottom of a light-tight box, a Samsung Galaxy A8 smartphone was placed at the top of the box and the glass plate with the analytes was placed over the white light. The smartphone camera was employed to acquire images of the array before (holes with combined probe: 3-nitrophenylboronic acid and pH indicator) and after reaction (holes with combined probe and saccharide solution) and a self-developed application software was used for image rendering, automatic color recognition and calibration. The obtained RGB color difference data (ΔR, ΔG and ΔB) were sent to a remote server where they were analyzed by principal component analysis (PCA), hierarchical cluster analysis (HCA), linear discriminant analysis (LDA) and partial least squares (PLS) for discrimination of the 19 studied types of saccharides. The PLS models presented discrimination accuracy of 100 percent. The effectiveness of the developed system in rapid on-site detection of saccharides was successfully verified by spiking and recovering experiments with "Oriental leaves" black tea beverage. An image processing program based on Android application 'SmartEYE' was developed and used by [35] to measure image color and texture of tomatoes in three maturity classes (turning, light red, and red). Image processing program was developed using Android Studio in Java language program. The tomatoes were placed in a color comparison cabinet (YL-33 17 A, Taiwan) in D65 mode and jpeg images of the tomatoes were taken with a Samsung DUOS smartphone camera and further processed by the developed Android-based SmartEYE application software. The software generated color values in Lab space as well as image textures in terms of entropy, energy, contrast, and homogeneity based on OpenCV library function and Grey Level Co-occurrence Matrix. Results demonstrated that Lab color values and chroma (C) values increased with maturity advancement stage. There were no significant differences in homogeneity values among the three distinct stages of maturity, whereas image entropy and contrast Tomatoes in three different classes have different image entropy and contrast, while for homogeneity values differed among the three maturity stages. It was successfully demonstrated that the maturity stage of tomatoes can be classified based on Lab space 'a' and 'b' values. [4] developed a methodology for rapid determination of ethanol in Brazilian cachaça spirits using a colorimetric procedure based on the analysis of images taken by a Samsung GALAXY A3 smartphone (Android 5.0.2, 8 MP camera). Ethanol from the spirits were placed in a 10-mm-wide cuvette to react with ceric ammonium nitrate (CAN) and nitric acid (NA) and form a red-colored complex. The cuvette was placed in a black polyethylene box 10 × 4 × 6 cm with controlled luminosity (two led strips with 12 lamps each providing a luminosity of 9 W m^{-2}). The smartphone was suitably placed in the box to take images of the reaction over time. The images were processed by the software PhotoMetrix Pro-via partial least squares (PLS). Results from the smartphone imaging were compared to those from a UV–VIS reference method and presented a root mean square error of prediction (RMSEP) of 0.0677 percent (v/v) with no significant difference between the employed methods ($P < 0.05$). A microtiter macro lens-coupled smartphone assay for the quantitation of the total saponins in quinoa based on foam measurement was developed and validated by [26]. An afrosimetric method was used as reference to measure the content of saponins in quinoa seeds. Quinoa seeds were placed in screw-cap test tube (100 mm long × 7.7 mm diameter) with water and the tube was capped and shaken on vortex for 30s. After 5 to 10s, the mixture was transferred to a 96-well plate and video images were recorded using the smartphone Huawei P8 Lite with an adapted macro lens at an inclination angle of 12.102° and at a distance of 10 mm from the samples. Photograms from the recorded video were extracted at different time intervals and the foam height was measured and analyzed by the free Image J software (https://imagej.nih.gov/ij/), after setting the scale with the tube diameter. Correlation and Bland-Altman analyses and Passing-Bablok regression showed good agreement between total saponin content in quinoa as determined by the macro lens-coupled smartphone and afrosimetric assays. An analytical procedure was developed by [33] for the determination of ethanol in distilled beverages using smartphone-based digital images. The procedure was based on the influence of ethanol concentration on radiation absorption by phenolphthalein in alkaline medium. A support for coupling the reaction microtube to the smartphone was built in acrylonitrile butadiene styrene (ABS) with a 3D printer (GTMAX-3D, model Corel H4). The digital images of the reacting medium were taken by a Samsung Galaxy J7 smartphone camera, with illumination provided by the smartphone LED light. The images were processed by the application software Color Grab 3.6.1 (Loomatix) that measured color intensity and converted to RGB values with G-channel readings providing the desired analytical responses. The reflected radiation, inversely proportional to the color intensity, was quantified. The values were manually transferred to Google Spreadsheets for data handling and calibration curves obtained by least squares regression were used to determine the ethanol concentrations. The procedure presented a linear response from 10 to 70 percent (v/v) ethanol ($r = 0.998$, $n = 7$), with a coefficient of variation of 1.2 percent ($n = 8$) and a limit of detection (99.7 percent confidence level) estimated at 2.1 percent (v/v). The developed procedure was validated by comparison of its results with those of AOAC reference procedures, agreeing at the 95 percent confidence level. A machine vision-based smartphone app was developed and verified by [20] to predict beef tenderness from fresh beef images captured under uncontrolled conditions. An image-processing algorithm was developed to eliminate the effects of uncontrolled imaging conditions in such a way images of beef samples could be taken with more degree of freedom in terms of luminance, rotation, scale, and translation, without worrying about the accuracy of the results. Images of fresh beef were taken with an LG G4 H815 smartphone camera intentionally under different conditions in terms of illumination, rotation, distance, and translation between the camera and the sample planes. The captured images were processed by the developed Android application (App) for extraction of textural features. The performance of the App was assessed by analyzing 30 beef samples. The preprocessed image textural features correlated well with instrumental data obtained using Warner-Bratzler shear force measurement by applying an artificial neural network technique. An average probability of 0.92 of occurrence of 2-D correlation coefficients was obtained from the analyses of all the beef samples subjected to the image processing demonstrating the robustness of the developed algorithm. The best neural network model predicted beef tenderness values with mean absolute percentage error of 3.28 percent and coefficient of determination (R2) of 0.97. A novel paper-based headspace-thin film microextraction in combination with smartphone-based on-cell detection was developed by [43] to detect sulfite in food samples. Sulfite was converted to SO_2 by acidification of the sample solution and the SO_2 headspace micro-extracted by adsorption onto a thin cellulose paper impregnated with Fe(III), 1,10-phenanthroline, and acetate buffer (pH = 5.5). The sulfur dioxide adsorbed on the thin film reduced Fe(III) to Fe(II) forming a red color complex. A wooden box was built using black MDF sheets to serve as a controlled illuminated chamber for image capturing of the colored cellulose film. A Samsung Galaxy J7 smartphone was used for imaging and data processing and a white LED light (60 mW) was used for the illumination of the box inside. Image analysis was performed by an RGB analyzer software for Android and the difference between the blue value of blank and sample was taken as the analytical signal. The method presented three distinct linear ranges (0.1–1.0, 1–50, and 50–700 μgL^{-1}) and a detection limit of 0.04 μgL^{-1}. Recoveries were higher than 94 percent with relative standard deviations lower than 4.8 percent. Chilling injuries on the surfaces of zucchini were analyzed by [36] using a customized image acquisition process image developed for commercial smartphones. A Samsung Galaxy S5 G900F smartphone was used to capture the digital images by video recording in mp4 format. The frames were decoded by the smartphone into a Java ByteBuffer with the Open GL command glReadPixels. Image of each point of the surface of the zucchini is registered from different perspectives as it is rotated, with the number of images depending on the rotation speed and time of exposure. The set of images are stitched together, and the final image is processed to assess the injuries at the surface of the zucchini by calculating the number of pixels inside of areas damaged by cold and using the total number of pixels of damaged surface to compute the damaged ratio in relation to the total surface. [23] employed a smartphone-based spectrometer (740–1070 nm) to study the feasibility of determining salted minced meat composition at industrial scale. Two NIR instruments were used: the smartphone-based SCiO (Consumer Physics, Israel) that can acquire reflection spectra in the range of 740 to 1070 nm controlled by Samsung Galaxy Core Prime smartphone with an Android 5.1.1 operating system; and a diode array Polychromix Spectral Probe (Polychromix Inc., Wilmington, USA) with InGaAs detector, covering the spectral region between 940 and 1700 nm. Both devices were used for acquiring 1312 spectra from meat samples stored at four different temperatures ranging from −14 °C to 25 °C. The obtained spectra for each spectrometer were processed separately to allow for comparison between the two devices. PLS and Random Forest regression models were built for each temperature and global models were created to predict the fat, moisture, and protein contents. The global model was able to predict fat and moisture in a wide range of temperatures by using the smartphone-based spectrometer, which demonstrated an acceptable accuracy for quality control purposes (RPD>7) that was comparable to the accuracy of a benchtop spectrometer. A smartphone was used by [15] to capture and process images of a colorimetric spot test to determine ascorbic acid in Brazilian Amazon native and exotic fruits. The test was based on reduction of Fe(III) by ascorbic acid and subsequent complexation with 1,10-phenanthroline and was carried out on a porcelain plaque with 12 wells. The plaque was placed in a black plastic chamber of 21 × 10 × 15 cm to avoid spurious light and opening was perforated in the chamber through which the camera of a Samsung Galaxy S4 Mini smartphone was accommodated. Internal illumination was provided by four white LEDs. Images of the reaction were captured at 5 min intervals over 30 min, and further processed and decomposed into an RGB matrix to study the colorimetric kinetics of the chemical reaction. Brazilian Amazon native fruits such as bacuri, cupuaçu, muruci, yellow mombin, and others such as cashew, mango, orange and passion fruit were analyzed. The limit of detection for ascorbic acid in the samples was 8.5×10^{-7} mol L^{-1}. Iodometric titration was used as a reference method for determination of ascorbic acid. The results obtained with the smartphone device were in close agreement with those from the reference method, with a confidence level of 95 percent (paired t-test). Moreover, recoveries ranged from 87.1 to 116 percent. A smartphone colorimetric assay was developed by [44] for the detection of milk adulteration by dilution based on the determination of the protein content. The methodology was based on precipitating the milk proteins by salting out with copper sulfate, which proportionally adsorbs on the precipitate, and photometrically determining the remaining Cu(II) concentration in whey after complexation with ethylenediaminetetraacetic acid (EDTA). The intensity of the blue color measured via smartphone-based colorimetry is inversely proportional to the protein content in the sample and was thus used as an indicator of milk adulteration by dilution. The precipitation of the proteins was performed in Falcon tubes, which were centrifuged, and the supernatant subsequently transferred to colorless Eppendorf tubes. The Eppendorf was placed in a styrofoam box (14 cm high, 16 cm wide, and 10.5 cm deep) with a support to keep the smartphone 9 cm apart from the tube and an LED-based lamp (Osram Superstar, R50 40 6W E14 30°) placed on the top to maintain a constant illumination. Photometric measurements were performed directly on the tubes with an LG K10 Pro (equipped with a 13-megapixel camera with resolution 4128 × 2096 pixels and lens aperture f/2.2, equipped with Android 7.1) smartphone, and further processed by the free application Color Grab (Loomatix, version 3.6.1, 2017) to convert the images to RGB values. Photometric measurements were performed near the center of the Eppendorf tubes, with a region of interest of 32 × 32 pixels. The values of the R channel were directly used to quantify the proteins in milk. For the classification of the milk samples, S values were calculated by subtracting each measured R value from the reference value obtained using an Eppendorf tube filled with water (R mean value = 208). The methodology accuracy was verified by comparing its results with those of the NF ISO 21,543 reference method using near infrared spectroscopy, ageing at a 95 percent confidence level. The proposed procedure throughput was 32 assays/h with a coefficient of variation of 3 percent ($n = 20$), and a limit of detection of 1 percent v/v water in adulterated milk. [6] developed a paper-based chemosensor based on the Folin-Ciocalteu (FC) colorimetric assay for rapid quantitative detection of polyphenols in extra virgin olive oil (EVOO) without the need for preliminary sample extraction. N-propanol was used as solvent dilute EVOO samples making them compatible with the alcoholic environment in which the FC reaction takes place. The color change resultant from the reduction of FC reagent in the presence of phenolic compounds was measured using a Samsung S8 smartphone camera. Paper supports loaded with FC reagent were prepared by dispensing the reagent on 1 × 1 cm pieces of chromatographic paper, which were placed on supports and left to dry on air and inserted in a specially constructed microfluidic system. The microfluidic system was then stored under vacuum in the dark and at +4 °C. The chemosensor system was comprised of the microfluidic system (into which the solutions were delivered), a lab-case enclosing the microfluidic system and a dark box (45 × 30 × 55 mm) that connected the lab-case to the smartphone holder avoiding interference from ambient light. It also contained a light diffuser for illumination of the paper supports using the integrated smartphone flash and a 0.4X two-lenses optical element to allow imaging of the paper supports by the smartphone camera. Images of the paper supports were acquired with the smartphone five minutes after delivering the solutions on the activated paper supports. The images were saved in TIFF format and quantitative image analyses were performed by the freeware software ImageJ v.1.46 (National Institutes of Health, Bethesda, MD). The images were converted to grayscale and the regions of interest (ROIs), corresponding to the paper supports, were defined, with the color intensity of each paper support quantified as mean grey intensity (MGI) over the ROI. For each analysis, a calibration curve was obtained by plotting the normalized MGI values (i.e., the actual MGI values divided for that of the blank) measured for the blank and the two gallic acid standard solutions against the gallic acid concentration. To obtain the total polyphenols content of the EVOO sample expressed in gallic acid equivalents (GAE), the MGI value of the sample was normalized and interpolated on the calibration curve. A conventional FC assay was used as a reference method and the results for total polyphenol content of EVOO samples of the developed method were in good agreement with the reference assay, with a limit of detection of 30 µg GAE g^{-1} EVOO. A smartphone coupled with image processing and chemometrics was used by [45] to quantify adulterant contents in extra virgin olive oil. Videos of the samples were recorded by a smartphone and converted into color spectra by image-processing techniques. A Samsung Galaxy S6 smartphone was used to generate a sequence of colors on its screen, varying from purple to red, and illuminate the olive oil samples. The smartphone was placed at a parallel distance of 5 cm from the sample surface, with the center of screen facing the sample surface. A 5-second video of 32-bit and 960 × 720 pixels was recorded by the smartphone front camera (5 MP and f/1.9 aperture). The videos were uploaded to a laptop computer and processed with MATLAB 7.12 R2011a software (The MathWorks Inc., USA). The videos were split into 175 frames (images) that were used in the selection of region of interest (ROI). The ROI images were decomposed into RGB color channels with color levels varying from 0 to 255. The average of the color levels for a ROI image was taken and the color spectral information for each sample was represented as a 525-dimensional vector (175 frames × 3 color channels). Partial least squares regression models were constructed from such video data and compared to the data obtained from near-infrared, ultraviolet–visible and digital imaging to quantify the content of vegetable oil in extra virgin olive oil in the range 5 percent–50 percent (v/v). The video approach (R2 = 0.98 and RMSE = 0.02) presented comparable performance to baseline spectroscopy techniques and outperformed the computer vision system approach. An approach for the detection of hydrogen peroxide as a milk adulterant was proposed by [27] based on a colorimetric spot test and smartphone-based photometry. Hydrogen peroxide is an adulterant added to milk for control of microbial growth. The colorimetric assay is based on analyte oxidation of iron(II) followed by formation of a red iron(III)-thiocyanate complex. The oxidation of Iron(II) and complex formation reactions were carried out in translucent, colorless 1.5-mL polypropylene microtubes with conical bottoms and the digital images were taken directly from these microtubes under controlled illumination conditions without sample pretreatment. The photometric system was comprised of a Styrofoam box (28 cm × 17 cm × 20 cm) with a platform of white LEDs (24 cm × 6 cm × 4 cm) attached to its bottom. A white sheet of paper was glued over the LEDs to disperse the emitted radiation. A square cutout (6 × 6 cm) was made in the front of the box for the smartphone camera and a hole was made in the top of the box for insertion of the microtube support that stood in front of the camera. Images were taken with a Motorola smartphone (Moto X, Android 6.0) camera and the application Color Grab (Loomatix, v3.6.1) was used to convert the digital images to RGB (red, green, and blue) values, which were transferred to a Microsoft Excel spreadsheet for data processing. Photometric measurements were taken with at a 12-cm distance from the microtubes, with a region of interest of 25 × 25 pixels. Based on the absorption spectrum of the Fe(III)-SCN− complex and spectral ranges of the red (R), green (G), and blue (B) channels, higher analytical responses for the B and G channels were expected. The sensitivities of the G and B channels were nearly identical, but better linearity was observed for the G channel ($r > 0.995$), which was selected for further experiments. The results of the proposed method were validated by comparison with the reference method based on direct bioelectrocatalysis of H_2O_2 with horseradish peroxidase (HRP) measured via a disposable microfluidic electrochemical device. A linear response was attained from 2.5 to 25.0 mg L^{-1}, with a limit of detection of 1.7 mg L^{-1} at a 95 percent confidence level. Repeatability and reproducibility, expressed as coefficients of variation, were 4.8 percent and 6.8 percent, respectively [10].

12.3 Food safety

Food products are expected to be rigorously controlled during processing, storage, and retail, but food contamination is still an issue nowadays. Foodborne diseases are responsible for a significant amount of illness and even deaths in both rich and poor countries [24]. A wide variety of diseases can be transmitted to humans through the ingestion of food that is contaminated with microorganisms or chemicals and thus food safety evaluation is of the utmost importance. Traditional methods for safety evaluation are usually restricted, due to high costs, lengthy time demands, bulky instruments and the need for trained personnel. Therefore, rapid and accurate analysis of food products has drawn considerable attention in the past years, and smartphones have emerged as interesting platforms with several applications in food analysis including detection contaminants, toxins, pathogens and allergens, among others. An overview of some of the recent studies that have been developed in this area is given below. Tetracyclines (TC) are used in agricultural practice as antibiotics for suppressing the growth of bacteria, in both plant and animals. Evaluation of TC residues in milk is thus a major concern in terms of human health. [34] developed an iPhone-based digital image colorimeter (DIC) for monitoring tetracycline in bovine milk. TC solutions were extracted from milk using solid phase extraction (SPE) and photographed by an iPhone. The color parameters red (R), green (G), blue (B), hue (H), saturation (S), brightness (V), and gray (Gr) were measured from each picture and compared to those obtained for TC standard solutions. The developed app was able to predict TC concentrations in the range of 0.5–10 µg mL^{-1}, with a limit of detection (LOD) of 0.5 µg mL^{-1} and limit of quantitation (LOQ) of 1.5 µg mL^{-1}. The major limitations of the proposed technique were the need for sample concentration by SPE and the need for ensuring uniformity of the photographic conditions, thus hindering its portable application. Wang and collaborators [50] recently proposed a smartphone-integrated lanthanide-based ratiometric fluorescent sensing system for visual and point-of-care testing of tetracycline with high sensitivity and accuracy. The ratiometric fluorescent sensor, defined as DPA-Ce-GMP-Eu (DPA: dipicolinic acid, GMP: Guanosine 5′-monophosphate disodium salt hydrate), was developed with DPA and GMP as ligands for coordinating Ce^{3+}, and then changed into red emission by further coordinating with Eu^{3+}. Once exposed to a tetracycline environment, the fluorescence of the as-prepared DPA-Ce-GMP-Eu sensing platform changes from red to blue due to the stronger coordination between tetracycline and Eu^{3+}, enabling the visual and quantitative detection of tetracycline under the assistance of UV lamp. Tetracycline detection was accomplished within a wide concentration range (0.01 to 45 µM) and a low limit of detection (6.6 nM). The implementation of the developed fluorescent assay requires a fluorophotometer, which is not available for in-situ determination. Thus, a common smartphone installed Color Picker APP was used as signal reader and analyzer, converting image color signals into digital values representing RGB color channels. A portable detection device was developed, made of black plastic polymer materials to eliminate the interference from external light source. A UV lamp with 302 nm output was fixed at a black strip and set at the bottom of the device to provide excitation light source. A sample stage with a cavity for holding either cuvette or testing paper was set on the device to fix the distance between sample and UV light source. A hole on the top surface of the device was used for capturing the image with the smartphone. The system was tested with spiked milk samples and presented good recovery (97 to 111 percent) and precision (RSD<5.2 percent). Milk is one of the several food products that can be easily adulterated with several substances, with the most common being water, starch, hydrogen peroxide (H_2O_2), and sodium hypochlorite (NaClO). Some of these adulterants can be of concern in terms of human health.

Excessive ingestion of starch can lead to diarrhea, and accumulation of starch in the body may even be fatal for diabetic patients. Preservatives such as H_2O_2 and NaClO can cause gastritis and inflammation of the intestine [3,9]. In view of the aforementioned, [9] proposed a smartphone-based methodology, based on specific reactions for detecting each adulterant, in which the sample color changes. The employed reactants were iodopolyvidone solution (starch detection), hydrochloric acid (HCl) and potassium iodide (KI) (H_2O_2 detection), and potassium iodide (KI) and starch (NaClO detection). Image analysis was performed by lab-made apps (PhotoMetrix®, and RedGIM®) based on partial least squares regression of the red-green-blue images histograms. The mean values of correlation coefficients were 0.9997/0.9785 for H_2O_2, 0.9929/0.9653 for NaClO, and 0.9974/0.9653 for starch, based on PhotoMetrix®/ RedGIM®. Ochratoxin A (OTA) is one of the most important and most commonly occurring mycotoxins, with well documented hepatotoxic, nephrotoxic and teratogenic effects. OTA has been detected in several food products including cereals, wine, coffee, beer, dried fruits, among others. A methodology for quantifying concentrations of ochratoxin A using a smartphone as fluorescence device was proposed by [5] and tested in beer samples. The methodology is based on a system in which the sample is placed between an UV light emitting diode (LED) and a smartphone camera. The fluorescence image obtained from the sample is captured by the smartphone camera and transferred to a computer by a wireless connection. The image is then processed by an automatic data-processing interface (Mathlab) to obtain the RGB model of the image. Given that OTA is naturally fluorescent, when the light is emitted through a solution with OTA, it provides a blue fluorescence. The system was calibrated with OTA solutions of known concentrations in the range 2–20 μg/L and the limit of detection (LOD) was 2 μg/L. In the recent study developed by Liu et al. [29] a smartphone-based quantitative dual detection mode device was proposed, aiming at multiplex detection of several mycotoxins. It was integrated with gold nanoparticles (GNPs) and time-resolved fluorescence microspheres (TRFMs) lateral flow immunoassays (LFIA). The goal was to integrate the most frequently used visible light and fluorescence detection modes into a single device. A user-friendly application was also developed in order to rapidly quantify results. The detected toxins were aflatoxin B1 (AFB1), zearalenone (ZEN), deoxynivalenol (DON), T-2 toxin (T-2), and fumonisin B1 (FB1). Recovery values in spiked samples ranged from 84.0 percent–110.0 percent. A parallel analysis was also performed using 30 naturally contaminated cereal samples and the results were verified by liquid chromatography–tandem mass spectrometry (LC-MS/MS). The proposed methods were shown to be reliable and accurate. In order to meet the requirements of timely and in-field detection for point-of-need food safety inspection, (Zhonggang [30]) developed a 3D-printed smartphone-based platform (SBP) for specific quantitation of food-safety related markers. A solution of aptamer conjugated with gold nanoparticles (AuNPs) was used as the colorimetric indicator whereas the smartphone was employed for light detection and data processing. Aptamers are single-stranded oligonucleotides isolated through a combinatorial biology technique called SELEX; their chemical synthesis is well developed and cost-effective, and they can be selected for any given target. Gold nanoparticles have become ideal platforms for developing easy-to-use and sensitive sensing systems for quantitation of various target analytes in complicated matrices, being efficient colorimetric indicators due to the extremely high extinction coefficient and distance-dependent absorption property. A user-friendly Android application (App) was developed to analyze the images captured by the smartphone rear camera, report and share the detection results. Streptomycin (STR) was used as the proof-of-concept target, with the specific recognition accomplished by an anti-STR aptamer, and its concentration indicated by the ratio of absorbances at 625 and 520 nm, consistent with the color change of the AuNPs solution. The proposed device was evaluated with honey, milk and tap water samples. There was satisfactory agreement between the results obtained using SBP and those of UV–Vis absorption spectrometry and LC-MS, with an LOD of 2.3 nM (8.97 mg kg^{-1}), below the maximum permitted levels of STR in milk. The authors estimated fabrication costs of ~13.0 USD per SBP set and ~5 USD per assay, confirming the cost effectiveness of the proposed device. Melamine, employed as a raw material for the synthesis of plastics, coatings, paints, flame-retarding agents and others, presents a high concentration of nitrogen, and thus can be illegally added to food products to mask and increase apparent protein content [21]. Melamine can react with cyanuric acid in human body, resulting in the formation of insoluble crystals that cause renal failure and may even lead to death [38]. The conventional method (Kjeldahl method) employed for the estimation of protein content in milk and other food products is based on nitrogen content, and thus foods that are tainted with melamine will not be directly detected. In view of the aforementioned, [21] proposed the use of Gold nanoparticles@carbon quantum dots nanocomposites (Au@CQDs) for the analysis of melamine in milk. Fluorescent emission of Au@CQDs was enhanced with the increase of melamine concentration. A smartphone-based fluorescence visualization device was developed and then used for visual fluorescence detection of melamine. The device consisted of a detection cell (365 nm UV light with power of 12W) and a smartphone, used to acquire and display fluorescent images, with the camera being held at a 45-degree angle in front of the detection cell. Images were analyzed in terms of fluorescence intensity against a calibration curve and a fluorescence standard array based on the Au@CQDs. The system was tested for milk adulterated with different concentrations of melamine, with recoveries of 103–106 percent and coefficient of variation smaller than 2 percent. Method accuracy was equivalent to high-performance liquid chromatography (HPLC) determinations. Zearalenone (ZEN) is a mycotoxin produced by several Fusarium species and commonly found in maize and in other crops such as wheat, barley, sorghum, and rye. The Panel on Contaminants in the Food Chain established a tolerable daily intake (TDI) for zearalenone of 0.25 µg/kg (EFSA, 2011). (Yuan [7]) developed a smartphone colorimetric reader integrated with an ambient light sensor and a 3D-printed attachment for the readout of liquid colorimetric assays. The reader used a simplified electronic and light path design and was compatible with different smartphones. As a proof of principle, the utility of this device was demonstrated using it in conjunction with an enzyme-linked immunosorbent assay to detect zearalenone in corn flour samples. Results were consistent with those obtained using a professional microplate reader. The limit of detection was 2.04 ng/mL and the recovery rates (84 – 90 percent) were similar to those obtained using a professional microplate reader. Later, the same research group ([29]) modified the proposed device using solid phase detection in a nitrocellulose filter (NC) membrane instead of colorimetric solutions. The developed method was tested in both cereal and feed samples of corn, bran and wheat. The limits of quantification for ZEN in cereals and feeds were 2.5 and 3.0 µg/kg, respectively, and the limits of detection were 0.08 and 0.18 µg/kg, respectively. Aflatoxin B1 (AFB1) is recognized as one of the most toxic representatives of aflatoxins, carcinogenic substances produced by microscopic fungal species *Aspergillus flavus* and *Aspergillus parasiticus*. Cereals (maize, millet, wheat, and others), spices, nuts, cocoa and coffee beans, as well as some fruits and vegetables are often contaminated with these mycotoxins ([42]). Recent studies have shown that an effective way to detect toxic molecules is the use of chemical sensors based on synthetic mimics of enzymes or antibodies. These sensors can provide fast and selective analysis and are more stable than biosensors based on natural biomolecules ([41]). Thus, a recent study by Sergeyeva and collaborators [42] focused on the development an easy-to-use fluorescent smartphone-based biomimetic sensor system based on free-standing MIP membranes for rapid detection of AFB1. The developed MIP membranes, using 2-acrylamido-2-methyl-1-propansulfonic acid and acrylamide as functional monomers, were capable of selective recognition of the target analyte, generating a fluorimetric sensor response that was registered by the smartphone camera. The obtained images were analyzed using a commercially-available App, Spotxel Reader (Sicasys Software GmbH, Germany). The limit of detection was 20 ng mL^{-1} for the sensor developed with acrylamide as a functional monomer, and the storage stability of the developed sensors was estimated to be one year if kept at 22 °C. Cachaça or sugarcane spirit is an alcoholic beverage typical of Brazil, obtained by distillation of fermented sugarcane, and with an alcohol content of 38–48 percent (v/v) at room temperature. Furfural and hydroxymethylfurfural (HMF) are aromatic aldehydes formed mainly from the pyrolysis of organic matter, and present both toxic and carcinogenic properties. These substances are relevant in terms of contamination of this specific type of spirit, because they may be present in the juice of sugarcane when the harvest is preceded by straw burning or may appear during the aging process of the beverage in wooden vessels. Methanol is also of concern in terms of cachaça contamination, because it is toxic and can cause severe acidosis, affecting the respiratory system and eventually leading to coma. Although there are a wide variety of methods evaluating contaminants in cachaça, most of them involve sophisticated and expensive equipment that cannot be used to monitor quality during the beverage production process. Therefore, [13,14] proposed the combination of a colorimetric spot test and a digital image-based (DIB) method for in situ determination of the previously mentioned contaminants in sugarcane spirit (cachaça) samples. For furfural, the proposed method was based on the reaction of furfural with aniline in acid medium, resulting in a colored product whose digital image was captured with a smartphone camera. In the case of methanol, it was oxidized to methanal in the presence of chromotropic acid, leading to a violet chromophore. The digital images were decomposed by an RGB approach using a free software. The limits of detection and quantification were 0.34 and 1.15 mg/100 mL for HMF, and 0.5 and 5.0 mg/100 mL for methanol. The recovery rates were satisfactory (81 – 110 percent). The proposed method presented a similar performance to that of the reference spectroscopic methods. Foodborne pathogens have become one of the major hazards in food safety and, according to World Health Organization, approximately 10 percent the world's population are stricken by foodborne illnesses every year [19]. It is estimated by the CDC that *Salmonella* alone causes 1 million foodborne illnesses every year in the United States (CDC, 2020). In a recent study, Guo and collaborators [19] developed a portable biosensor for detection of *Salmonella* Typhimurium using magnetic nanoparticle immunoseparation, nanocluster signal amplification and smartphone image analysis. Magnetic nanoparticles were conjugated with monoclonal antibodies to form the magnetic bacteria. The polyclonal antibodies and glucose oxidase (GOx) were incubated with calcium chloride to synthesize the immune GOx-nanoclusters (GNCs); these were reacted with the magnetic bacteria to form the nanocluster bacteria. The GNCs on the nanocluster bacteria were then used to catalyze glucose into hydrogen peroxide, which was measured using a peroxide test strip. The image of the strip was collected and analyzed using a Hue-Saturation-Luminosity color space-based smartphone APP for the determination of the target bacteria. The biosensor was able to detect *Salmonella* Typhimurium within the linear range of 10^1 to 10^5 CFU/mL, with a detection limit of 1.6×10^1 CFU/mL, and recovery values on spiked chicken samples ranged from 85 to 110 percent. Okadaic acid (OA) is a poisonous toxin that is produced by some unicellular algae and accumulates in the digestive glands of shellfish. It can cause several symptoms such as diarrhea, nausea, vomiting and abdominal pain, and has also been identified as a tumor promoter [46], thus being of high concern in terms of shellfish poisoning. Su and co-workers [46] developed a portable smartphone-based system using cell viability biosensor (CVBS) for detection of OA cells. The system was comprised of CVBS, a light source and a smartphone. The CVBS includes living cells, a microtiter plate (MTP) and a CCK-8 kit, responsible for monitoring cell activity. The smartphone installed with a homemade App - iPlate Monitor, is responsible for image acquisition and analysis, data storage and transmission. Results showed that the system could synchronously detect OA in 96 channels. The biosensor presented a good performance to various OA concentrations, with a wide linear detection range (10–800 μg/L). Tests employing OA-spiked shellfish extracts indicated that the proposed system performed similarly to the reference AOAC method, with average percentage recoveries of 99.06 and 100.16, for the proposed and reference methods, respectively. Escherichia coli (*E. coli*) is one of the most dangerous types of bacteria than can cause food poisoning, resulting in serious conditions including hemorrhagic diarrhea, hemolytic uremic, and even death [50]. In the study by Wang and collaborators [50], a smart fluorescent and colorimetric dual-readout sensing system was developed for *E. coli* determination. The sensor is based on the Cu^{2+}-triggered oxidation of *o*-phenylenediamine (OPD) that results in an orange-yellow fluorescence and visible pale-yellow color, and such oxidation is inhibited by *E. coli*, thus reducing both the fluorescence and the UV–vis absorbance signals of the system. A filter paper embedded with OPD-Cu^{2+} solutions was employed as substrate. In the presence of *E. coli*, the color of the paper substrate should change from green to dark-blue under a 302 nm UV lamp. The image is then captured by the smatphone and analyzed using a color-scanning app. The method was tested with spiked degreased milk samples, within the range of 79.8 to 100.3 percent. Another type of smartphone-based system, using lateral flow immunoassay (LFA), was proposed by [22] for *E. coli* detection. LFA is a widely used analyte-detection platform for several applications including pregnancy tests, chemical residue determination, and toxin detection. Two different brands of LFA strips were used: *E. coli* O157:H7 rapid detection kit and Rapidcheck *E. coli* O157:H7 test kit. The smartphone-based imaging system was comprised of major sub-components: (i) sample cartridges for each brand of LFA strip, (ii) interchangeable optical imaging box, and (iii) smartphone cradle. The optical imaging box consists of a reflector to redirect the smartphone LED light onto the strip. An optical diffuser is present in order to avoid specular illumination. The sample LFA is then viewed with a plano-convex lens that relays the image to the smartphone camera lens. Ground beef and spinach artificially inoculated with *E. coli* O157:H7 were tested using the developed apparatus, and contamination was detected in the range of 10^4 to 10^5 CFU/mL.

12.4 Concluding remarks

The scientific community employed smartphones as analytical tools for food diagnostics in a variety of ways, with the majority of applications focusing on some kind of photometric detection. Although the initiative of using smartphones for food diagnostics has been taken for a handful of years now, the technology is still in its infancy and is usually based on an adaptation of existing analytical methodologies. However, several factors point to a significant increase in the use of smartphone technology as analytical tools for food diagnostic in the near future. Firstly, it cannot be ignored that, as of 2017, there are more working smartphones than there are people in the world. This fact alone is a testament of the enormous potential of functionalities these devices can provide and, the more user-friendly application software and miniaturized smartphone-compatible sensors become available, the trendier will be the use of such devices for food diagnostics by producers, processors, merchants and consumers. Secondly, the portability, processing and storage capacity of these devices make them suitable for applications that require real-time and on-site determination of specific food attributes that allows for a quick and reliable evaluation and decision about the quality of the product. Lastly, not only this technology allows for miniaturization of analytical systems and consequent minimization of waste generation, but it allows for a significant reduction in cost and time of the analytical procedure. Current detecting platforms are still dependent on full-scale laboratory equipment, and are labor intensive and demanding of excessive use of chemicals. However, with the fast advancement of micro-electromechanical systems, 3D printing and smartphone technologies, the future of small-scale analytical systems that employ smartphone technology can be easily envisioned.

Acknowledgements

The authors acknowledge financial support from CNPq (Conselho Nacional de Desenvolvimento Científico e Tecnológico, Brazil).

References

[1] O. Abbas, M. Zadravec, V. Baeten, T. Mikuš, T. Lešić, A. Vulić, J. Prpić, L. Jemeršić, J. Pleadin, Analytical methods used for the authentication of food of animal origin, Food Chem. 246 (2018) 6–17.

[2] M S M S F Acevedo, M.J.A. Lima, C.F. Nascimento, F.R.P Rocha, A green and cost-effective procedure for determination of anionic surfactants in milk with liquid-liquid microextraction and smartphone-based photometric detection, Microchem. J. 143 (2018) 259–263.

[3] T. Azad, S. Ahmed, Common milk adulteration and their detection techniques, Int. J. Food Contam. 3 (1) (2016).

[4] F.C. Böck, G.A. Helfer, A.B. da Costa, M.B. Dessuy, M.F. Ferrão, Rapid determination of ethanol in sugarcane spirit using partial least squares regression embedded in smartphone, Food Anal. Methods 11 (7) (2018) 1951–1957.

[5] D. Bueno, R. Muñoz, J.L. Marty, Fluorescence analyzer based on smartphone camera and wireless for detection of Ochratoxin A, Sens. Actuators B 232 (2016) 462–468.

[6] D. Calabria, M. Mirasoli, M. Guardigli, P. Simoni, M. Zangheri, P. Severi, C. Caliceti, A. Roda, Paper-based smartphone chemosensor for reflectometric on-site total polyphenols quantification in olive oil, Sens. Actuators B 305 (2020) 127522.

[7] Yu Chen, G. Fu, Y. Zilberman, W. Ruan, S.K. Ameri, Y.S. Zhang, E. Miller, S.R. Sonkusale, Low cost smart phone diagnostics for food using paper-based colorimetric sensor arrays, Food Control 82 (2017) 227–232.

[8] Yuan Chen, Q. Fu, D. Li, J. Xie, D. Ke, Q. Song, Y. Tang, H. Wang, A smartphone colorimetric reader integrated with an ambient light sensor and a 3D printed attachment for on-site detection of zearalenone, Anal. Bioanal. Chem. 409 (28) (2017) 6567–6574.

[9] R.A. Costa, C.L.M. Morais, T.R. Rosa, P.R. Filgueiras, M.S. Mendonça, I.E.S. Pereira, B.V. Vittorazzi, M.B. Lyra, K.M.G. Lima, W Romão, Quantification of milk adulterants (starch, H2O2, and NaClO) using colorimetric assays coupled to smartphone image analysis, Microchem. J. 156 (2020) 104968.

[10] M. Cruz-Fernández, M.J. Luque-Cobija, M.L. Cervera, A. Morales-Rubio, M de la Guardia, Smartphone determination of fat in cured meat products, Microchem. J. 132 (2017) 8–14.

[11] L. Cuadros-Rodríguez, C. Ruiz-Samblás, L. Valverde-Som, E. Pérez-Castaño, A. González-Casado, Chromatographic fingerprinting: an innovative approach for food \ textquotesingleidentitation\textquotesingle and food authentication – A tutorial, Anal. Chim. Acta 909 (2016) 9–23.

[12] Y. Cui, T. Chen, B. Li, X. Liu, J. Xia, J. Han, Y. Wu, M. Yang, Colorimetric sensor array–smartphone–remote server coupling system for rapid detection of saccharides in beverages, J. Iran. Chem. Soc. 15 (5) (2018) 1085–1095.

[13] M. de Oliveira Krambeck Franco, W.T. Suarez, V.B. dos Santos, Digital image method smartphone-based for furfural determination in sugarcane spirits, Food Anal. Methods 10 (2) (2016) 508–515.

[14] M. de Oliveira Krambeck Franco, W.T. Suarez, M.V. Maia, V.B. dos Santos, Smartphone application for methanol determination in sugar cane spirits employing digital image-based method, Food Anal. Methods 10 (6) (2016) 2102–2109.

[15] V.B. dos Santos, E.K.N. da Silva, L.M.A. de Oliveira, W.T. Suarez, Low cost in situ digital image method, based on spot testing and smartphone images, for determination of ascorbic acid in Brazilian Amazon native and exotic fruits, Food Chem. 285 (2019) 340–346.

[16] M. Esteki, Z. Shahsavari, J Simal-Gandara, Use of spectroscopic methods in combination with linear discriminant analysis for authentication of food products, Food Control 91 (2018) 100–112.

[17] S. Fan, Q. Zhong, H. Gao, D. Wang, G. Li, Z. Huang, Elemental profile and oxygen isotope ratio ($\updelta$ 18 O) for verifying the geographical origin of Chinese wines, J Food Drug Anal 26 (3) (2018) 1033–1044.

[18] A.S. Franca, L.M.L Nollet, Spectroscopic Methods in Food Analysis, CRC Press, 2017.

[19] R. Guo, S. Wang, F. Huang, Q. Chen, Y. Li, M. Liao, J. Lin, Rapid detection of Salmonella Typhimurium using magnetic nanoparticle immunoseparation, nanocluster signal amplification and smartphone image analysis, Sens. Actuators B 284 (2019) 134–139.

[20] S. Hosseinpour, A.H. Ilkhchi, M. Aghbashlo, An intelligent machine vision-based smartphone app for beef quality evaluation, J. Food Eng. 248 (2019) 9–22.

[21] X. Hu, J. Shi, Y. Shi, X. Zou, M. Arslan, W. Zhang, X. Huang, Z. Li, Y. Xu, Use of a smartphone for visual detection of melamine in milk based on Au@Carbon quantum dots nanocomposites, Food Chem. 272 (2019) 58–65.

[22] Y. Jung, Y. Heo, J.J. Lee, A. Deering, E. Bae, Smartphone-based lateral flow imaging system for detection of food-borne bacteria E.coli O157:H7, J. Microbiol. Methods 168 (2020) 105800.

[23] A. Kartakoullis, J. Comaposada, A. Cruz-Carrión, X. Serra, P. Gou, Feasibility study of smartphone-based Near Infrared Spectroscopy (NIRS) for salted minced meat composition diagnostics at different temperatures, Food Chem. 278 (2019) 314–321.

[24] M.D. Kirk, S.M. Pires, R.E. Black, M. Caipo, J.A. Crump, B. Devleesschauwer, D. Döpfer, A. Fazil, C.L. Fischer-Walker, T. Hald, A.J. Hall, K.H. Keddy, R.J. Lake, C.F. Lanata, P.R. Torgerson, A.H. Havelaar, F.J. Angulo, Correction: world health organization estimates of the global and regional disease burden of 22 foodborne bacterial, protozoal, and viral diseases, 2010: a data synthesis, PLoS Med. 12 (12) (2015) e1001940.

[25] K. Kucharska-Ambrożej, J. Karpinska, The application of spectroscopic techniques in combination with chemometrics for detection adulteration of some herbs and spices, Microchem. J. 153 (2020) 104278.

[26] N. León-Roque, S. Aguilar-Tuesta, J. Quispe-Neyra, W. Mamani-Navarro, S. Alfaro-Cruz, L. Condezo-Hoyos, A green analytical assay for the quantitation of the total saponins in quinoa (Chenopodium quinoa Willd.) based on macro lens-coupled smartphone, Talanta 204 (2019) 576–585.

[27] M.J.A. Lima, M.K. Sasaki, O.R. Marinho, T.A. Freitas, R.C. Faria, B.F. Reis, F.R.P Rocha, Spot test for fast determination of hydrogen peroxide as a milk adulterant by smartphone-based digital image colorimetry, Microchem. J. 157 (2020) 105042.

[28] Z. Lin, L. He, Recent advance in SERS techniques for food safety and quality analysis: a brief review, Curr. Opin. Food Sci. 28 (2019) 82–87.

[29] Zhiwei Liu, Q. Hua, J. Wang, Z. Liang, J. Li, J. Wu, X. Shen, H. Lei, X. Li, A smartphone-based dual detection mode device integrated with two lateral flow immunoassays for multiplex mycotoxins in cereals, Biosens. Bioelectron. 158 (2020) 112178.

[30] Zhonggang Liu, Y. Zhang, S. Xu, H. Zhang, Y. Tan, C. Ma, R. Song, L. Jiang, C. Yi, A 3D printed smartphone optosensing platform for point-of-need food safety inspection, Anal. Chim. Acta 966 (2017) 81–89.

[31] F. Longobardi, G. Casiello, A. Ventrella, V. Mazzilli, A. Nardelli, D. Sacco, L. Catucci, A. Agostiano, Electronic nose and isotope ratio mass spectrometry in combination with chemometrics for the characterization of the geographical origin of Italian sweet cherries, Food Chem. 170 (2015) 90–96.

[32] Y. Lu, Z. Shi, Q. Liu, Smartphone-based biosensors for portable food evaluation, Curr. Opin. Food Sci. 28 (2019) 74–81.

[33] O.R. Marinho, M.J.A. Lima, F.R.P. Rocha, B.F. Reis, M.Y Kamogawa, A greener, fast, and cost-effective smartphone-based digital image procedure for quantification of ethanol in distilled beverages, Microchem. J. 147 (2019) 437–443.

[34] P. Masawat, A. Harfield, A. Namwong, An iPhone-based digital image colorimeter for detecting tetracycline in milk, Food Chem. 184 (2015) 23–29.

[35] R.E. Masithoh, B. Achmad, L. Zharif, Development of \textquotedblleftSmart Eye\textquotedblright – Smartphone Application – To Determine Image Color and Texture of Tomatoes, Proceeding of the 2nd International Conference on Tropical Agriculture, Springer International Publishing, 2018, pp. 53–60.

[36] N. Novas, J.A. Alvarez-Bermejo, J.L. Valenzuela, J.A. Gázquez, F. Manzano-Agugliaro, Development of a smartphone application for assessment of chilling injuries in zucchini, Biosystems Eng. 181 (2019) 114–127.

[37] J.L. Pérez-Bernal, M. Villar-Navarro, M.L. Morales, C. Ubeda, R.M. Callejón, The smartphone as an economical and reliable tool for monitoring the browning process in sparkling wine, Comput. Electron. Agric. 141 (2017) 248–254.

[38] B. Puschner, R.H. Poppenga, L.J. Lowenstine, M.S. Filigenzi, P.A. Pesavento, Assessment of melamine and cyanuric acid toxicity in cats, J. Vet. Diagn. Invest. 19 (6) (2007) 616–624.

[39] G. Rateni, P. Dario, F. Cavallo, Smartphone-based food diagnostic technologies: a review, Sensors 17 (6) (2017) 1453.

[40] G.M.S. Ross, M G E G Bremer, M.W.F Nielen, Consumer-friendly food allergen detection: moving towards smartphone-based immunoassays, Anal. Bioanal. Chem. 410 (22) (2018) 5353–5371.

[41] T.A. Sergeyeva, D.S. Chelyadina, L.A. Gorbach, O.O. Brovko, E.V. Piletska, S.A. Piletsky, L.M. Sergeeva, A.V. El'skaya, Colorimetric biomimetic sensor systems based on molecularly imprinted polymer membranes for highly-selective detection of phenol in environmental samples, Biopolymers and Cell 30 (3) (2014) 209–215.

[42] T. Sergeyeva, D. Yarynka, E. Piletska, R. Linnik, O. Zaporozhets, O. Brovko, S. Piletsky, A El\textquotesingleskaya, Development of a smartphone-based biomimetic sensor for aflatoxin B1 detection using molecularly imprinted polymer membranes, Talanta 201 (2019) 204–210.

[43] A. Shahvar, M. Saraji, H. Gordan, D. Shamsaei, Combination of paper-based thin film microextraction with smartphone-based sensing for sulfite assay in food samples, Talanta 197 (2019) 578–583.

[44] A.F.S. Silva, F.R.P Rocha, A novel approach to detect milk adulteration based on the determination of protein content by smartphone-based digital image colorimetry, Food Control 115 (2020) 107299.

[45] W. Song, Z. Song, J. Vincent, H. Wang, Z. Wang, Quantification of extra virgin olive oil adulteration using smartphone videos, Talanta 216 (2020) 120920.

[46] K. Su, Y. Pan, Z. Wan, L. Zhong, J. Fang, Q. Zou, H. Li, P. Wang, Smartphone-based portable biosensing system using cell viability biosensor for okadaic acid detection, Sens. Actuators B 251 (2017) 134–143.

[47] J. Tan, J. Xu, Applications of electronic nose (e-nose) and electronic tongue (e-tongue) in food quality-related properties determination: a review, Artif. Intell. Agric. 4 (2020) 104–115.

[48] A.S. Tsagkaris, J.L.D. Nelis, G.M.S. Ross, S. Jafari, J. Guercetti, K. Kopper, Y. Zhao, K. Rafferty, J.P. Salvador, D. Migliorelli, G.I.J. Salentijn, K. Campbell, M.P. Marco,

C.T. Elliot, M.W.F. Nielen, J. Pulkrabova, J Hajslova, Critical assessment of recent trends related to screening and confirmatory analytical methods for selected food contaminants and allergens, TrAC, Trends Anal. Chem. 121 (2019) 115688.

[49] D.D. Uyeh, W. Shin, Y. Ha, T. Park, Food safety applications. In: Smartphone Based Medical Diagnostics, Elsevier, 2020, pp. 209–232.

[50] C. Wang, X. Gao, S. Wang, Y. Liu, A smartphone-integrated paper sensing system for fluorescent and colorimetric dual-channel detection of foodborne pathogenic bacteria, Anal. Bioanal. Chem. 412 (3) (2020) 611–620.

[51] M. Zhang, P. Wu, J. Wu, J. Ping, J. Wu, Advanced DNA-based methods for the detection of peanut allergens in processed food, TrAC, Trends Anal. Chem. 114 (2019) 278–292.

Smartphone-based detection devices for the agri-food industry

Aprajeeta Jha[a], J.A Moses[b], C. Anandharamakrishnan[b]
[a]Department of Agricultural and Food Engineering, Indian Institute of Technology Kharagpur, West Bengal, India
[b]Computational Modeling and Nanoscale Processing Unit, Indian Institute of Food Processing Technology (IIFPT), Thanjavur, Tamil Nadu, India

13.1 Introduction

"Information networks have become a great leveler, and we should use them together to help lift people out of poverty and give them freedom from want."
- Hillary Clinton

Technocrats in every field of life are witnessing a humungous evolution in computing devices, be it basic sciences, modern architecture, medical sciences, environmental sciences, food processing industries, etc. For the first time in the history of mankind, high-end computation devices for various technical applications which were rather exclusive only to experts, are now made available to the common man. 21st century has marked the commencement of a new era of 'Smart-phone', which has remodeled the outlook of human minds in every sphere of life. Smart-phones have brought analytical tools as handy and compact applications, thereby, putting analysis into the hands of the consumer. In a recent couple of years 'Smartphone' has become a rapidly growing phenomenon, and by the end of 2020, the number of smartphone users is estimated to cross 4.78 billion across the globe [34]. With the increasing popularity of smartphones, smartphone-based IoT (Internet to things) solutions are also gaining pace among various R&D teams. It is due to the presence of a diverse range of utility sensors embedded within a smartphone for example accelerometers, gyroscopes, global positioning systems (GPS), audio-visual applications, electronic compasses, and propinquity sensor etc. [65]. Mobility, computing power and the extensive outreach of smartphones has motivated, as well as empowered technocrats to combine information scattered in various fields (like computer science, electronics, and telecommunication, life science, chemical engineering, finance, data science, medical science, supply chain management or environmental science) to cater the direct needs of the common citizen. As a simplest of example; measurement of a distance walked by a person for his fitness regime is now just a click away on his smartphone. Cell phone applications provide us with a smart interface that is capable of sensing the distance walked, speed of walking, heart rate during exercise, as well as can provide descriptive directions regarding the rhythm of the walk. Smartphones in true sense have made it possible for every individual to avail tailor-made utility applications for their specific requirements. Portability, mobility, multi-functional capabilities, and ease of handling have made smartphones based devices the obvious choice of the common man to find one point solution to accommodate various utility functions.

Ever since the inception of human civilization, agriculture and food processing have been one of the most integral domains of scientific research as well as an essential requirement in daily lives. It is the matter of fact that no other field of human scientific endeavors impacts human lives as closely, consistently, and at a far-reaching scale as much as the agricultural and food process engineering operations do. Maintaining food safety at every stage of the agri-food processing chain is among the most non-negotiable, cumbersome, and complex goals for the food engineers. To evaluate the suitability of any food product, it requires extensive and aggressive inspection systems at various processing junctures, starting from farm to godowns, to the retail market, to the consumer, and until consumable food [110]. During this course, evaluation of physical, and biochemical changes induced due to environmental conditions (temperature, relative humidity, moisture content, etc.), the ageing process (such as acid and sugar content, anti-nutritional factors, macro, and micronutrient), microbiological infestations (pathogenic invasion, food poisons, food allergens, etc.) and adulterants (organic and in organic adulteration) are mandatory to maintain safety standards. With growing awareness, consumers have now not only become very particular about the safety standards but also, are inclined to monitor food quality and safety on their own. Therefore, one of the most poignant and upsurging advancements in food processing sector in the last few years have been targeted to bring vivid analytical tools on the user-friendly platforms of smart-phone. For example, in recent times smartphones have found an indispensable place in hands of farmers; its portability complements the dynamic nature of agriculture, and various sensors make them a potential gadget to aid diverse farming tasks such as soil and climate assessment. Smartphone-based detection devices have embarked on a new era of rapid monitoring and food safety analysis. Additionally, the evolution of sophisticated sensors such as paper-based sensors, fluorometric sensors, electrochemical sensors, nano-bio sensors [7,78,81], etc. have redefined the way and the pace of biochemical and microbiological detection. Nowadays, mobile food scanners are widely implemented for non-invasive analysis of macro and micro components, microbial food spoilage, food contamination (adulterants, food poisons, food allergens), track color and odor changes, and to monitor temperature as well as moisture levels, etc. [69,85,111,112].

Smartphones without any doubt have provided the benefit of portable food analysis or 'lab-on-smartphone' platforms with momentous advantages like rapid testing, cost-effectiveness, functional-readiness, information exchange, nominal requisite of diagnostic equipment and manpower [38,48,74]. However, it is worth mentioning that these tools are based on amalgamation various sensors and computer coded programs to follow a model pathway for the detection of any food safety issues. Therefore, a basic understanding of various sensors, the architecture of instrumental interventions involved during the development of such gadgets is a prerequisite for holistic appraisal of this technology. Additionally, the alignment of every soft computing model involved, methodological application of the hardware, and connectivity of the device to information networks are the detrimental factors for accuracy, selectivity, and sensitivity of these devices [55]. The reliability and robustness of smartphone-based sensor technology have yet to be established, to provide solutions for a widespread food analysis regime.

In this queue, with a motivation of presenting elaborate discussion on the applicability of smartphone-based detection devices for the food industry, this chapter has been structured into two sections. Firstly, we shall see the fundamentals of various bio-sensing mechanism, and a brief outline of the feasible approaches implemented for making smartphone-based detection devices, to provide the necessary backdrop to readers about terminologies encountered in food analysis. Secondly, this chapter will present an updated state of the art on various applications of smartphone-based detection devices targeted to diverse food components/processes. Forwarding further, we shall very briefly comprehend the challenges arising during the developmental stage, accuracy, and application of these smartphone-based detection devices.

13.2 Biosensors and their amalgamation with smartphones

13.2.1 A basic overview of biosensors

Sensors can be understood as the devices which are capable of recognizing an analyte and produce a measurable signal, otherwise can be understood as the conversion of analog signals into digital signals. The early 1960s marked the kick-start of biosensor technology and pioneering pillars in the field of sensors was laid down by Clark and Lyons. Furthermore, the remarkable milestone in regards to recognition of biological components was achieved by Updike and Hicks in 1967, with the discovery of first-ever enzyme-based biosensor for the detection of oxidoreductases, polyphenol oxidases, peroxidases, and amino oxidases [55]. Updike and Hicks (1967) implemented immobilization methods, i.e. adsorption of enzymes by Van der Waals forces, ionic bonding, or covalent bonding for the development of an enzyme-based sensor. In 1974 first ever heat-sensitive enzyme sensor was put forth by Klaus Mosbach, which was termed '*Thermistor*'. In 1975, with the banner name of the Yellow Spring Instrument Company, the first glucose sensor was launched commercially by Professor Leland C Clarks, profoundly known as "Father of Biosensors". These sensors belong to the first generation sensor family where the bio-receptors and transducers were enacted independently and required mediator components therefore, these sensors were costly. Fig. 13.1 represents a schematic diagram of sensing components of biosensors.

By the end of 1975, Lubers and Optiz came up with a fiber optic sensor for the detection of oxygen and carbon dioxide. This sensor was termed as 'Optiode' and it marked the beginning second generation of biosensor technology. In line with this discovery, a first-ever commercial biosensor for monitoring biological oxygen demand (BOD) for testing water quality was developed by Nisshin Denki Electric Co. Ltd in the early 1980s. The major advancement in second generation sensor was obliviation of mediator for interaction between transducer and bio-receptors which facilitated compact sizing and reduced costing of these sensors. In the late 1970s,

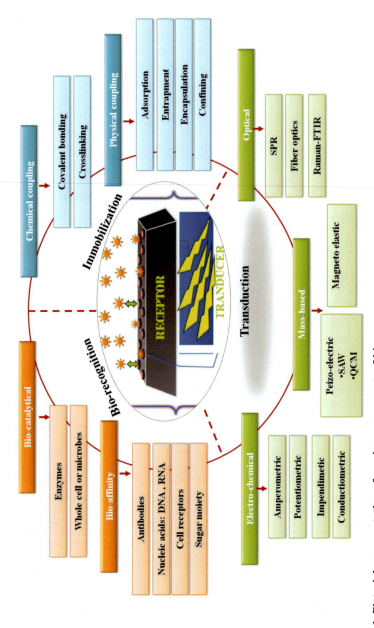

Fig. 13.1 Pictorial representation of sensing components of biosensors.

Clemens came up with a blood glucose sensor. Until the end of the 1990s, the majority of bio-receptor based sensors were used and designed to cater needs of healthcare and medical sciences and were holistically called sensors. The term biosensor was first coined by Cammann in 1991, which refers to the highly specific devices that combine bio-recognition components with physical transducers to capture biological responses into electrical or optical signals. According to Cammann, biosensors can be regarded as a "sub-group of chemical sensors, where biological recognition mechanisms or principles are used as a means of recognizing substances or molecules" [13]. To understand the difference between chemical and biosensors we need to consider the fact that the sensor transduction system generally comprises of i) a molecular recognition system (receptor), ii) transducer, for signal transduction and, iii) detection devices [5]. In cases where the recognition component is a biological element, those devices are termed as a biosensor. However, nowadays biosensors or sensors are a commonly interchangeable term that has found a significant place for detection of a vivid parameter. The bio-receptors are further classified into two subsections based on the nature of receptors a) bio-catalytical (such as enzymes, tissue, and microbial infection), and b) bio-affinity (nucleic acids, cell receptors, antibodies, sugar moiety).

The third generation of biosensors came into the picture with the application of immobilization techniques to prepare cell-based biosensors. Further within due course of time, Liedberg came up with Surface plasmon resonance (SPR) technology for building advanced optical biosensors [113]. An advanced piezoelectric sensor comprising of Quartz crystal microbalance (QCM) and Surface acoustic wave (SAW) transducer; amperometric, conductometric, potentiometric or impedance-based electrochemical sensors integrated to various nucleic acids, cell receptors, antibodies, sugar moiety, enzymes, and microbes are predominantly used in this generation of sensors. fourth generation sensors are can be segregated as an era of multi-connected and portable micro/nano-technology based biosensors. Somewhere close to 1999, the concept of 'Internet of Things (IoT)' or 'Internet of Everything (IoE)' evolved and overpowered the sensor market share [113]. The word *'Internet of Things'* was coined by Kevin Ashton of the Auto-ID Center of MIT [114]. High-end electrical and computational advancements for analog to digital signal conversion devices for Point of Use (POU) and Point of Care (POC) targets are driving the sensors related RandD. BioNMES, Quantum dots, Biochips, microfluidic lab-on-a-chip (LOC) technology also came into play, which largely improved the applicability of sensors in various fields like agriculture and food processing [95]. An outline of the historical chronology of biosensor is briefed in Fig. 13.2. In the upcoming subsections, we shall see a brief outline of the architecture of sensors and smartphone-based sensor technology.

13.2.2 The architecture of smartphone devices

Nowadays, these biosensors are embedded with smartphones, acts as readout, or display devices of the signals with help appropriate hardware and software. The smartphones are enabled with miniature microprocessors and micro-controllers to function

1st generation

- 1962 1st bio-sensor was discovered
- It was an Oxygen concentration detector
- In 1967, a sensor for blood Glucose concentration was discovered
- 1975, 1st ever commercial glucose sensor was re-launched
- 1975 marked initiation of fiber optics based gas detectors c/a 'OPTODES'
- Example; ELISA, Flow cytometry

2nd generation

- Electro-chemical bio sensors based on bio-affinity came into picture
- Independent interaction of bio-recognition sites and transduction system was promoted
- More compact and cost effective sensors came into play
- 1st BOD sensor for water quality test was commercialized
- Predominance of biocatalytic type of bio-sensors

3rd generation

- Cell based bio sensors were predominant
- Application of various physical and chemical immobilization techniques lead in cost effectiveness
- Origin of Surface Plasmon technology
- Progression of mass based transduction systems
- Advanced electrochemical sensor could meet HPLC and MS standards

4th generation

- Biochips came into play
- Mobile sensors with higher specificity, selectivity, accuracy and lower load of detection (LOD) values gained market
- High end microprocessor, microcontroller and miniature soft computing devices was amlgamed to detection components
- Cloud networks for promotion of Point of Care (POC) and IoT is rapidly growing

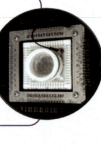

Fig. 13.2 An outline of the historical chronology of biosensor.

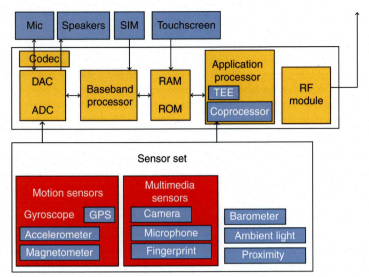

Fig. 13.3 A generalized overview of the internal architecture of smartphone-based detection devices [54].

as detection platforms to amplify and process the electrical signals. A generalized overview of the internal architecture of smartphone-based detection devices is represented in Fig. 13.3. As shown in the figure, smart devices encompass three major components:

- *Processors:* These are very similar to the processing units in a PC and based on domain managed it can be further divided into 3 types a) Application processor b) Base-band processor c) Co-processor [54]. The application processor majorly performs the job of power-saving management, responsible for the administration of embedded sensors, SD cards, and communication modules of the smartphone and responsible for data storage and allocation in a protected area. Functions performed by base-band processor also include keeping track of hardware isolated component which are responsible to maintain WiFi or cellular network connections via SIM card, these connectivity devices ensure 'People to Machine' communication. It remains well-equipped to handle the multi-functionality of various applications simultaneously with the help of real-time operating systems (RTOS). The co-processor is additionally enabled less power consuming electronic module, having its structure and is a competent performing the task on natural language processing (NLP), Contextual computing processing (CCP), and Artificial neural network (ANN) modeling. It is applicable for high-end graphics, broadband signal processing, and encryption/decryption, etc. A few examples of smartphones enabled with coprocessors are Huawei Kirin 970, X8 Huawei Mate 10, Apple M7, and Motorola. A decade before, mobile phone processor capacity was thought to be limited within 1GHz due to its high power requirements [9]. Nevertheless, denying all anticipation Quelcomm had released dual core snapdragon processor (which supports a pair of scorpion core at 1.5 GHz) by the end of 2010 [5]. Today smartphone developers have compiled up numerous smart and powerful processors which can perform complicated computing operations such as iPhone 11 consist of hexa-core Apple A13 Bionic processor with $2 \times 2\ 65$ GHz Lightning high-performance cores and 4×1.8 GHz Thunder efficient CPU cores, Samsung Galaxy octa-core processor that features 4 cores clocked at 1.6 GHz and 4 cores clocked at 1.2 GHz. It comes with 2GB of RAM. The most

updated smartphone processors are comprised of Quad-core technology. The technological surveys indicate that with due course of time there will be significant improvements in the smartphone processor technology.
- ***Sensors:*** We have seen a brief overview of sensors in the previous section, however concerning smartphone sensors (SPS), it can either be physical sensors and virtual sensors [105]. Physical sensors include directly embedded hardware-based sensors into the cell phone set and they extract data in a straight line by evaluating environmental signals. On the other hand, virtual sensors are software-based sensor which import and export data from numerous hardware-based sensor attachments. Based on the sensing mechanism it can again be subdivided into two categories i.e. Proprioceptive (PC) or Exteroceptive (EC). The former sensor is directed into action upon a perception of stimuli that are produced in real-time (for example gyroscope) and later works programmable coding and soft computing such as global positioning system (GPS). Table 13.1 summarizes various types of smartphone embedded sensors used in various applications in food process industries.
- ***Processor and sensor interface modules:*** This section majorly comprises of analog-to-digital converters (ADC), digital-to-analog converters (DAC), memory segment handling instructions given and voice codes. In general, analog types of sensors are embedded in the smartphone devices and therefore, these sensors require an analog-to-digital converter (ADC). To accommodate the entire set-up of sensors and ADC/DAC in a slim smartphone, a system similar to microcontrollers called a system on chip (SoC) technology is implemented. Moreover, the micro-electromechanical system (MEMS) technology and nano-electromechanical system (NEMS) were utilized to design the small sensors.

13.2.3 Data mining strategies for detection devices applicable in food industries

For a clear understanding of the working principle behind smartphone-based sensor systems, let us now dig a little bit into the data mining and computing tools required for designing the architecture of smartphone devices. Artificial intelligence and machine learning for sensor technology have plenty of roles in the agri-food sector such as to monitor every stage of the manufacturing process, estimation of price, inventory management predictions, the supply chain of food products from agricultural farms to the place where consumers receive it. Broadly, machine learning algorithms (MLAs) can be differentiated as supervised algorithms (SA), unsupervised algorithms (UA), and reinforcement algorithms (RA) which further segregate into a variety of branches as shown in Fig. 13.6 [76]. Few of widely usable MLAs with their applications, shortcomings, and advantages are summarized in Table 13.2. For detailed information on machine learning and deep learning, the reader is suggested to go through the works of Berry et al. [10], Han et al. [32], Sun et al. [87].

13.2.4 Classification of smartphone-based detection platforms

In the above sections, we have seen the generalized background of sensors and its wide application, in the present section, we shall see various detection platforms that specifically employ smartphones. State of art indicates that Smartphone-based detection platforms can be classified either as a) information-based tools b) direct application based tools. Information based tools include the opportunities offered

Table 13.1 List of embedded sensors in smartphones and its potential applications in agri-food processing industries

Sensors	Types	Description	Potential applications	References
Accelerometer	Piezo-resistive accelerometers Capacitive accelerometers	Records the three-dimensional acceleration of a body	Hand to mouth gesture recognition to track food intake Food processing and packaging companies improve production line efficiency	[23,29,119]
Time-Temperature sensors	Thermocouple Thermistor Infrared thermometer	Records the ambient temperature w.r.t change in time	Check freshness Monitors air temperatures in storage godowns To ensure the quality and safety of chilled or frozen food	[29,40,107]
Gyroscope	Mechanical gyroscope Optical gyroscopes (including Fiber Optic Gyroscopes (FOGs), Vibration-based gyroscope Ring Laser Gyroscopes (RLG) and Micro-electromechanical system (MEMS) gyroscopes	Estimates the three-dimensional velocity of rotation (rad/s) of a body/device	Detects rotation (spin, turn, etc.) Filling/conveying technology of food products Food Packaging technology	[115,116]
Light sensor	–	Quantifies the degree of illumination	Controls screen brightness To detect color to the product to identify spoilage due to fungal infestation	[78]
Magnetometer/ Inductive sensor	Scalar Magnetometer Vector Magnetometer	It records the directional magnitude geomagnetic field experienced by a device	Handles fluid transfer management in the beverage and dairy industry Magnetic empowered mixing application in food industries	[65,117]

(continued)

Table 13.1 (*Cont'd*)

Sensors	Types	Description	Potential applications	References
Pressure sensors	-	Records the pressure change in mbar or mPa.	Monitors air pressure changes Pressure measurement in various types of equipment like autoclave, pasteurizer, heat exchanger, a high-pressure processing For the survilance of piplines, filters and storage tanks especially employed in dairy, fruit pulp industries, and beverage	[43,65]
Proximity sensors	Optical methods Capacitive methods	It can give an estimation of propinquity (in cm) of an article w.r.t screen position of the device	Detection of position in canning and packaging of food products	[65]
Humidity and moisture sensors	Optical Acoustic Electronic	To record ambient humidity level and moisture content of the product	Greenhouse control Monitors dew point and absolute and relative humidity Food storage monitoring in go-downs Quality assurance of finished food products	[15,118]
Global positioning system	-	It can track the position of objects by contemplating the current latitude and longitude	Spatial data collection Supply-chain management in the food industry Precision farming Information regarding disease outbreaks	[68,97]
Fingerprint identity sensor	Optical capacitive and ultrasonic sound	Fingerprint detection	Identifies a user of product through touching	[30]

Table 13.2 A brief overview of various MLAs and its applications.

MLA platforms	Food component/ product	Advantages	Disadvantages	References
Support vector machine (SVM) algorithm	Sensory quality of apple, beef beer, cocoa, cheese, traces of pesticide in fruit juices and vegetable puree	This can also account for technical difficulty when there is no significant separation between the regions of the different classes of the samples	It requires additional kernel function to solve non-linear models which add up to its difficulty.	[4,6,17,25]
Proximate component analysis (PCA)	Classification of Taquilla, beverages, fruit juices, Moisture analysis in pineapple	Easy analytical behaviors to expose similarity and dissimilarity between the opted variables.	A lot of Informatics resources are necessary. It handles the sample population holistically rather than assigning a class to each sample	[28,53]
Random Forest (RF)	Grape juice, Honey, Green Tea, Rice classification	It can provide resolution to issues related to overfilling and high variance in Decision tree-based algorithms. It provides averaged outputs	To tackle average outputs may result in inaccurate predictions	[24,52,61,67]
k-NN (k-nearest neighbors)	Classification of Turkish olive oils, edible oil, and oil blend detection, cocoa beans, tequila classification	User-friendly method of applying	In the case of the skewed distribution of classes, it may result in an inaccurate classification of the samples	[19,39,62,64,93,94]
ANN (Artificial neural network)	Component analysis of jujube fruit, organic grape juice, the moisture content in potato	It is independent of the input provided to them	It has a fixed number of hidden nodes in the hidden layers and a preset amount of nodes in the first layer	[31,51] (Masoud et al., 2019; Tripathy and Kumar, 2009)
DT/ CART (Decision tree: classification and regression tree)	Grape juice, rice classification	The CART approach can reduce the time and complexity of the analysis.	performs poorly when the training set is small	[11,37,52]

by smartphones to cluster or barter information based on open databases such as the internet or closed databases such as a relational database in the cloud-like PostgreSQL, MySQL, MariaDB, Oracle Database, and SQL Server. these tools require continuous connectivity facilities via the internet, intranet, Bluetooth, wireless personal area networks (WPANs), Zig bee, etc., ([11], Chaudhary et al., 2015; [121]). Broadly all such information tools can be summarized under the Internet of things (IoT) or the Internet for everything (IoE). Secondly, there can be a case where smartphone devices work as sensors via embedded or external sensors attachments (also summarized in Table 13.1). This category of sensors can be called as 'Smartphone Biosensors' (SPB). A classification of smartphone-based detection platforms has been summarized in Fig. 13.4.

13.2.4.1 Internet of things (IoT)

Nowadays, mobility and data repository has become a highly prized quality of any evolving technology. Android platforms have the potential to become extremely useful due to its conjunction with closed or open database management systems to harness information available on the internet. According to Bouge (2014), and unmatched rapid expansion of connected devices was witnessed in the past two decades, the total number of growth of connected devices increased from 500 million to 50 billion from 2003 to 2020 which is more than 6 times of total world population. A huge share of the connected devices was of the smartphone, as of 2020 the number of smartphone devices estimated to cross 4.78 billion worldwide and the number of total connected devices per user has been estimated to be approximately 7 devices (Fig. 13.5). In this light, a recently emerging concept of 'The internet of

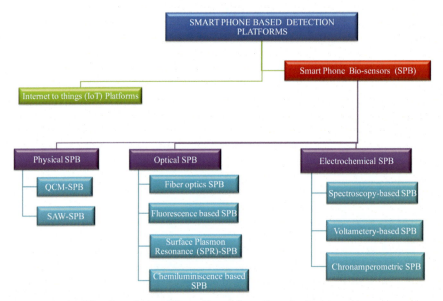

Fig. 13.4 A classification of smartphone-based detection platforms.

Smartphone-based detection devices for the agri-food industry 281

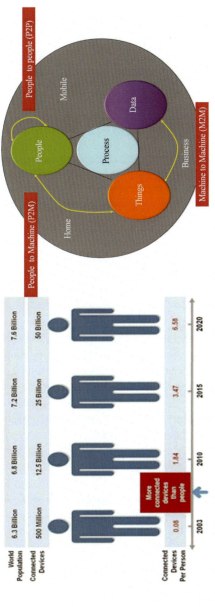

Fig. 13.5 A schematic representation of growing trends of connected devices and IoT working behavior.

things (IoT)" has gained enormous attention from industry and academia especially in the agri-food sector. It can be understood as a concept widespread scheme of a web comprising of a massive amount of objects, which are interlinked and can barter a multitude of information with other objects via universal nodal and internodal interaction schemes through wireless or cabled networks. This extensive platform for the exchange of data can create plenty of network-based services and applications helpful to assist the end-user [121] All the smartphones connected to the Internet can intermingle together for the exchange of data with an added advantage of mobility over fixated devices, which will further fuel such interactions. The Internet of objects comprises 2G/3G/4G wireless sensor networks, blue tooth devices, Zigbee, and RFID etc., [122]. A schematic representation of IoT working behavior is shown in Fig. 13.5. IoT is fundamentally based on free collaboration (between or within) of the following subjects:

- People (involves driving manpower either with professional or personal goals)
- Data (required information satisfying 'People's' goal available on a closed network like intranet and clouds etc., or open-source networks like the internet)
- Things (working devices/platform to facilitate interaction between 'People and Data')

These subjects are generally employed in three workstations, home-based, business/professional workspace-based, and mobile users (lab top, portable devices, or smartphone). The point of communication between aforesaid subject can be through People to Machine (P2M) by a collaboration of graphical interfaces (GUIs), tangible user interfaces (TUIs), touchless gesture user interfaces (TGUIs), voice user interfaces (VUIs), or People to People (P2P) or else via Machine to Machine (M2M). Smartphone-based IoT devices help in coordinating across multiple, disbursed, and often disconnected supply chain actors in the agricultural and food processing sector. The integration of QR code and radio-frequency identification (RFID) tag has been employed to track and map out the instantaneous location and quality of prepackaged food products in the supply chain [43]. They also adopted Extensible Markup Language (XML) for supporting manual or machine-assisted exchange of data within stakeholders. A schematic representation of the structural design of IoT-based tracking and tracing platform is shown in Fig. 13.6.

Mekala et al. [56] adopted IoT technology with cloud computing and Li-Fi to create a smart warehouse. Smartphone devices in collaboration with IoT based ZigBee modules were used to realize the smart monitoring of temperature, humidity, as well as a theft detection feature in the warehouse, creating better surveillance opportunities. Similarly, Mahale et al., [50] utilized smart poultry farm monitoring using IoT and wireless sensor networks. To date, the smartphone-based sensors have been implemented for a variety of smart agriculture regimes such to monitor the color of the soil, to supervise work activity from farm employees. Similarly, a smart IoT based smart irrigation facility for the development of mobile application controlled irrigation capabilities was reported by Vaishali et al. [97]. Inbuilt sensors in the cell phone such as proximity sensors, digital compass, and GPS sensors have been used to track any rollover by farm machinery etc., [110]. For supervision of the dietary content of infant food and automated IoT based sensor was developed by Sundaravadivel et al. [88]. A neural network algorithm (Smart-Log, a novel 5-layer perceptron) along with

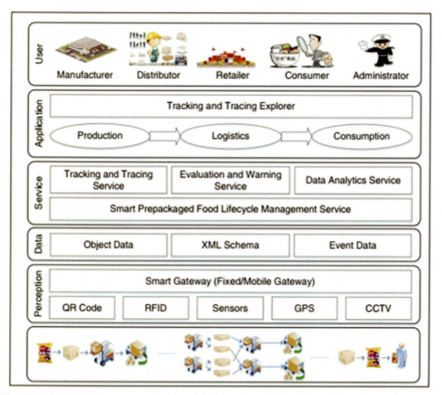

Fig. 13.6 A schematic representation of the structural design of IoT-based tracking and tracing platform employed in agri-food processing (Li et al., 2017).

the Bayesian network approach was designed to monitor 8172 food items of around 1000 meals. Seo et al. [77] fabricated pocket-sized immunosensor for tracing food contamination caused by bacterium *V. parahaemolyticus*. They exercised IoT (internet of things) and lens-free CMOS image sensors attached to a smartphone. An abstract of the work is shown in Fig. 13.7, as it can be seen that smartphone reader was implemented for the testing of food product, further, the test results generated can also be uploaded on to the open-access networks for data sharing.

13.2.4.2 Smartphone biosensors (SPB)

Further down, based on physics involved in the transduction process to produce final digital signals, smartphone-based devices can be segregated into three categories:

I. ***Physical SPB:*** This SPB utilizes a mass-based signal transduction system which may either use piezoelectric sensors or magneto-elastic sensors. Quartz crystal microbalance (QCM), detects changes of the resonant frequency of crystal oscillator electrode due to mass-sensitive variation at even nanogram level. A QCM sensor comprises of disk-shaped AT-cut piezoelectric quartz crystal which is typically encrusted with thin metal film electrodes. In an excited state, crystal oscillates to produce a resonant frequency which decreases proportionally with a change in mass upon the proximity of the analyte [33]. The Bio receptors

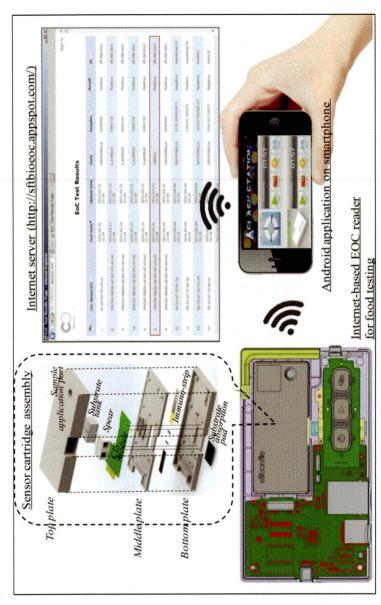

Fig. 13.7 A pocket-sized immunosensor for tracing Food contamination caused by bacterium *V. parahaemolyticus* [77].

attached to QCM can be enzymes, antibodies, nucleic acids, aptamer, and whole cells and recently molecular imprinted polymer (MIP) has also found a place in the detection of food components. QCM- SPB was successfully employed for the detection of pesticides, aflatoxins, water quality, food microbial contamination [26,27,92]. The second category among mass-based SPB is sound acoustic wave (SAW)-SBP. These devices are physically concise, very economical, simple to manufacture, produces a higher degree of effectiveness, and are effortless in use. These sensors record the changes occurring in velocity and/or amplitude of the acoustic waves due to deviation in characteristics propagation path resulted by its contact of any material. Taste-compound sensitive organic coatings have been employed to construct SAW-based electronic tongue for flavor assessment of acetic acid (AA), sucrose (S), and sodium chloride [18]. Yang et al. [103] developed αAstree e-Tongue for detection of umami flavor for the assessment of Monosodium glutamate, disodium inosinate, and guanylate which are generally overexploited as a food additive. Lab-on-Chip (LOC) platform for the detection of salmonella in milk, to contemplate contamination, the amplitude and phase changes were monitored using a surface acoustic wave sensor (Papadakis et al., 2018). Acoustic sensor system based on Hidden Markonikov Model (HMM) developed for chewing and swallowing events during food intake. Food Intake Recognition was performed using an external throat microphone, this sensor was capable to report speed of chewing, water intake, the interval of eating, and snacking times [124]. A Schematic illustration of a) colorimetric SPS b) electrochemical SPS (c) fluorescent SPS devices are shown in Fig. 13.8.

II. *Optical SBP:* This is the most widely applied category of SBP, and the construction of smartphone-integrated microscopes taking advantage of the high-resolution camera equipped on the smartphone was among the initial optical SBP. The major advantage of optical SBP is its versatility of producing direct signals as well as its potential to exploit other techniques to amplify signals facilitate accurate detection of an analyte. The major methodologies for the measurement of optical transduction can be via light scattering, chemilumminance, and colorimetry [27]. Smartphone-attached optical devices for a mobile, compact digital lens-free microscope for microbial imaging in point-of-care diagnostics was reported by Ozcan's group presenting holographic on-chip imaging [75,108]. However, this SBP produced limited spatial resolution only up to the micrometer scale and was not suitable for nanoscale samples. In contrast, fluorescence-based SBP utilizes light-emitting proteins for the evaluation and detection of toxic or pathogenic components in food [48]. Fluorescence-based SBP offers an edge of high sensitivity, temporal and spatial resolution over colorimeter based SBP. Fig. 13.9(a) depicts the basic working theory of the fluorescence-based SBP. To detect nanoscale components, optomechanical attachment for a smartphone was devised to fabricate a mobile fluorescence microscopy facility on a smartphone interface by Wei et al. [102]. The developed fluorescence microscope consisted of a laser diode, interference filter, focus adjustment stage, and external microscope, etc. Added to this queue, [125] exercised a paper-based sensor for blood typing. They implanted hydrophilic bar channels, pre-treated with Anti-A, -B, and -D antibodies on the paper for immuno-sensing of the blood type. Chen et al. [14] developed a smartphone-interfaced lab-on-a-chip device to track residues of 2,2′,4,4′-tetrabromodiphenyl ether (BDE-47) in food, a flame retardant, and environmental contaminant. To regulate sample flow and reaction process the sensor was integrated with micropumps and microfluidic devices, this offered an advantage of the active interface to control test procedures. Su et al. [86] came up with Bioconic e-eye concept based on the iPlate software interface implementing the ELISA technique as shown in Fig. 13.10. Further, Surface Plasmon resonance (SPR) based SBP utilizes the resonant oscillation of conduction electrons of a p-polarized light beam to generate an electron density caused due to the excitement of electron resulting in surface plasmon wave. The phenomenon takes place at the interface between negative and positive

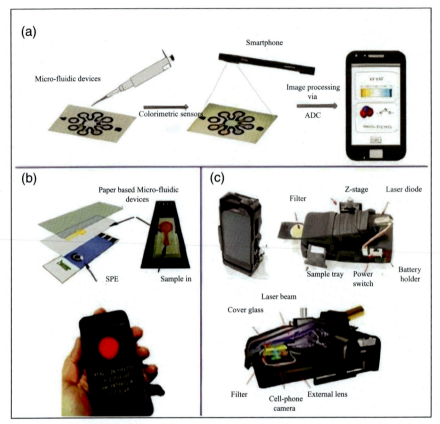

Fig. 13.8 Schematic illustration of (a) colorimetric SPS [47], (b) electrochemical SPS [22] (c) fluorescent SPS [72].

permittivity materials usually made of gold, coated onto a dielectric material. Even the slightest fluctuations occurring in the effective refractive index of an Au metal-dielectric interface can be recorded instantaneously with the help of quartz core [108]. The working of SPR based SBP is represented in Fig. 13.9(b). Concerning food safety, SPR -SBP is widely used for detecting food contamination, allergic components lethal proteins, anti-nutritional factors, and pharmaceutical proteins.

Based on detection modes, SPR sensors can be segregated into four types (i) fiber-optic surface plasmon resonance(FOSPR) (ii) surface plasmon resonance imaging (SPRI) (iii) localized surface plasmon resonance (LSPR) (iv) transmission surface plasmon resonance (TSPR) [109]. According to reports SPR biosensors endow with accurate optical recognition tools for quantifying bio-molecular interactions. Tran et al. [126] implemented Sandwiched type Gold based FOSPR for the detection of food allergen in peanut with a load of detection (LOD) value to be 75 nM. A simple portable imaging smartphone device enabled with surface plasmon resonance (SPR) biochip was proposed by Lee et al. [127] for the detection of imidacloprid pesticides.

Smartphone-based detection devices for the agri-food industry 287

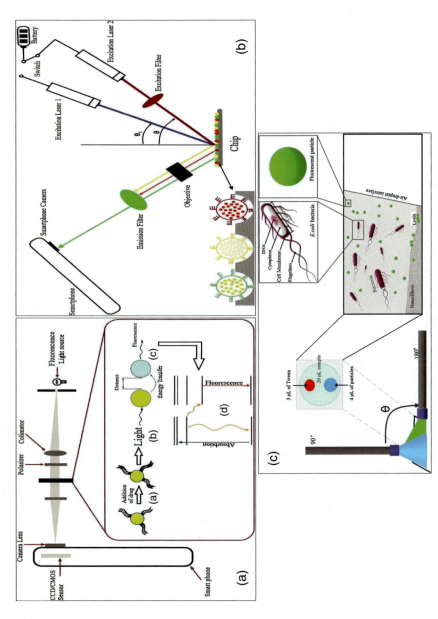

Fig. 13.9 Representation of the working concept of (a) fluorescence-based SBP (b) SPR based SBP (c) chemiluminescence based SBP [78].

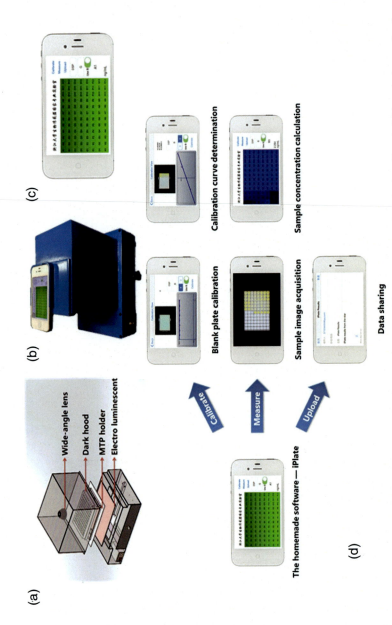

Fig. 13.10 Representation of (a) the experimental arrangement of Bionic e-Eye, (b) a photograph Bionic e-Eye and (c) an in-house developed software iPlate (d) a flow chart to represent functional capabilities of the iPlate [86].

Detection based on chemical/bioluminescence has recently been adopted for food analysis. Chemiluminescence can be understood as the phenomenon of light emission during a chemical/biochemical reaction. These emissive light signals can easily be trapped using a high-end smart-phone photo-camera giving rise to Chemiluminescence based SBP. A pictorial representation of the concept behind Chemiluminescence based SBP is shown in Fig. 13.9(c). A smartphone-based whole-cell biosensor for the detection of ciprofloxacin in milk samples was presented by Lu et al. [128]. Bioluminescent Escherichia coli bio-receptor cells were docked into a Lumi Cell Sense (LCS) interfaced 16 well-biochip, provided with a coating for the migration of oxygen and the whole set-up enacted as Lab-on chip module. Similarly, an iPhone (Apple Inc. Cupertino, CA, USA) which was enabled with a miniaturized bioreactor, a macro lens, a lens barrel, a metal heater tray, and a temperature controller, enclosed in a light-impermeable case was employed for the detection. A load of detection of the aforesaid device was reported to be as low as 8 ng/mL. A the cloth-based electro-chemiluminescence (ECL) signals were read out by using an inexpensive smartphone interfaced cameras comprising of a complementary metal-oxide-semiconductor (CMOS) detectors. ECL biosensor is effectively exploited for the recognition of the presence of lactate traces in human saliva, strongly indicating its prospective in studies related to food metabolism [104].

I. **Electrochemical SBP:** Integration of electrochemical biosensor with a high-end smartphone as the display device has been a promising development to overcome the major challenges of traditional electrochemical biosensors i.e. better sensing performance and compact miniaturized electrode [82]. Fundamentally, electrochemical sensors record minute fluctuation in the current (mA) or potential difference (mV) caused by biochemical reactions taking place within the sensor-sample interface. Based on transduction mechanism electrochemical SBP can be subdivided in Spectroscopy based SPB, Votlametry based SPB and Chronamperometry based SPB. Schematic representation of the main measurement categories of the electrochemical biosensors: spectroscopy-based SPB, Voltammetery based SPB, and chronoamperometry based SPB are given in Fig. 13.11. In the midst of the several categories of electrochemical SBP, the one focused on amperometric measurements of the analyte, has gained significant popularity due to its lower values of LOD and high sensitivity. Additionally, this technique generally exercises the amplification of the enzymatic markers to record antigen or antibody-based interactions via an immuno-sensor device. It is noteworthy that even the first-string biosensor marketed for glucose monitoring by Leyland and Clark in 1975, was based upon amperometric detection principles. These sensors record the current ensuing from the biochemical redox reactions of electro-active components. The rate of reaction is directly proportional to the current signal generated. Rateni et al. [70], has reviewed as a biosensing system for the detection of an adulterant used in livestock feed named clenbuterol (CLB). CLB is a drug that is illegally used to enhance growth rate, reduce fat deposition, and increase protein accumulation in live stocks. For instant detection of the CLB electrochemical device configured into the smartphone device as a biochip was developed. The critical value of this smartphone-based CLB-immunosensor was as low as 0.076 ng mL−1, and the detection time was reported to be 6 min. Similarly, Chip EIA (enzyme immunoassay) involved electro-chemical interaction based on enzyme-linked immunosorbent assay (ELISA), providing more suitable for option point-of-care testing [43]. Presence of Eoli contamination in raw milk, milk products, juices, cheese has been detected by self-assembled monolayer-based bi-enzyme biosensor utilizing amperometric transduction devices [91]. Detection of E. coli has been measured based on the fact that change pH occurs due to ammonia released by urease–E. coli antibody conjugate.

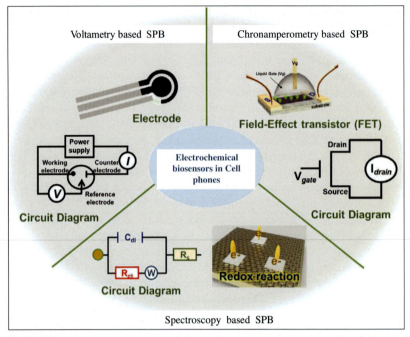

Fig. 13.11 Schematic representation of the main measurement categories of the electrochemical biosensors: impedance-based spectroscopy-based SPB, voltammetry based SPB, and chronoamperometry based SPB. Kwon and Park (2017) [115].

Voltammetric measurements are based on the measurement of potential difference between an indicator and a reference electrode [95]. Pattern recognition device for Brazilian honey samples based on their botanical and geographic origins was performed using cyclic voltammetry with a gold electrode (depicted in Fig. 13.12). This SBP Biosensor platform was connected to potentiostat via USB connection as well as Bluetooth module. The model required for on-site pattern recognition was developed using proximate composition analysis (PCA) on an indigenously developed mobile App [70]. The system was also enabled to distribute information via email, Google Drive, or even social media platforms to either support online processing of electrochemical analysis or for saving data for further usage.

Current technological developments in micro-electro-mechanical systems (MEMS), micro-opto-electromechanical systems (MOEMS), micro-mirror arrays, nano-electro-mechanical systems (NEMS), and magnetic impedance [101], has made a spectroscopic analysis to be available as a portable and miniature detector on the smartphone. Built-in gyro sensor and camera spectroscopy have been used by Liang et al. [44] to perform spectroscopic analysis to study microbial spoilage on meat. Das et al. [20] came up with a smartphone-based wireless spectrometer design comprising of the light source, spectrometer, filters, microcontroller, and wireless circuits in Fig. 13.13. It was employed for a rapid, non-destructive ripeness testing apples on the basis of color difference between yellow and green apple.

Smartphone-based detection devices for the agri-food industry 291

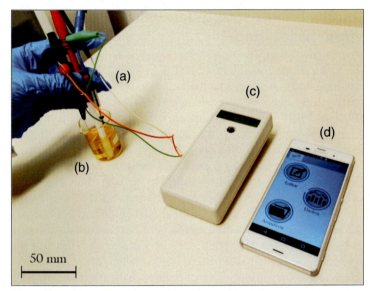

Fig. 13.12 Portable Electrochemical SBP platform deployed for point-of-use analyses of pattern recognition of Brazilian honey. (a) Electrochemical system (b) Sample (c) handheld potentiostat (d) and smartphone [130].

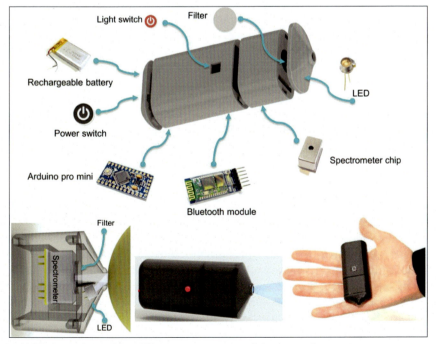

Fig. 13.13 Schematic of the different components of the smartphone spectrometer prototype [20].

13.3 Application of smartphone- based services in agri-food processing

Agriculture is the prime driving force and a major stakeholder in an agri-based economy of developing countries like India, Brazil, Bangladesh, Indonesia, etc., has a critical challenge to interlink information between Further, Surface Plasmon resonance (SPR) based SBP utilizes the resonant oscillation of conduction electrons of a p-polarized light beam to generate an electron density caused due to excitement of electron resulting in surface plasmon wave. The phenomenon takes place at the interface between negative and positive permittivity materials usually made of gold, coated onto a dielectric material such as quartz [108]. The working of SPR based SBP is represented in Fig. 13.9(b). Further, Surface Plasmon resonance (SPR) based SBP utilizes the resonant oscillation of conduction electrons of a p-polarized light beam to generate an electron density caused due to the excitement of electron resulting in surface plasmon wave. The phenomenon takes place at the interface between negative and positive permittivity materials usually made of gold, coated onto a dielectric material such as quartz giving real-time monitoring of small changes in the effective refractive index of an Au metal-dielectric interface [108]. The working of SPR based SBP is represented in Fig. 13.9(b). Concerning food safety SPR -SBP is widely used for detecting food allergens, toxic proteins, marker proteins, antibodies, and pharmaceutical proteins. experts, market components, and farmers. Variety of information be it of farming applications like (Pest and disease control, crop and soil nourishment planning, weather and climate monitoring, market evaluation, expert guidance) or food processing channels (supply chain, storage, quality assurance, adulteration detection) are vital to for both farmer and consumers for ensuring a progressive agri-based economy and total food security. In previous sections we have seen the technology behind smartphone-based detection devices, further in this section, we shall see updated state of art of various applications of smartphone-based services in Agri-food processing. For deep insight about biosensors in food processing and safety, readers may also refer Takhistov, (2005). In this line, a brief pictorial statement of possible applications in the agri-food sector has been summarized in Fig. 13.14

13.3.1 Smartphone applications for irrigation, seed testing, fertilization, soil management, and knowledge sharing

In the recent few decades, the involvement of mobile applications and sensor technology has revolutionized farming practices. Farming involves several predictive operations such as planting, weed removal, manure estimation, fertilization, and taking relevant agricultural assessment which can significantly be improved by the implications of smart farming devices. In this queue, researchers have projected and fabricated cell phone apps for successful supervision and allocation of water resources. A recent expansion of telecommunication tools such as GSM and GPRS have facilitated remote-based supervision of irrigation and water management. Pooja et al. [66] came up with a smart farming strategy to generate records for continuous monitoring of temperature, the moisture of the agricultural produce as well as ambient humidity of

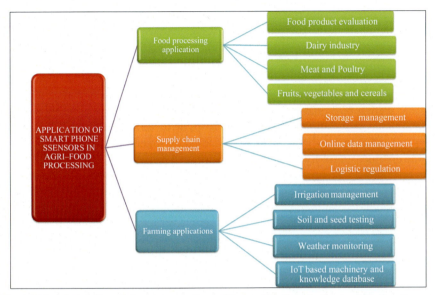

Fig. 13.14 A brief pictorial statement of possible applications in the agri-food sector.

the field, required light intensity and soil testing. An Arduino based automatic plant watering system was designed by Patil et al. [63] which was connected to the internet for accessing the information on the smartphone platform. This system enabled farmers to access information regarding the seed price, required amount of moisture and temperature of crop and soil, nutrient classification of soil, for climatic updates, required pest and weed control measures. Smartphone devices provide a framework for instant data sharing among all stakeholders. An Android-based mobile application called SmartHof employing cloud and fog computing for animal-based indicators (e.g., collar devices in cows, cameras in pigpens) have been put forth by Caprio et al. (2017). This work presented a new advancement in smartphone-based Precision Livestock Farming (PLM) via a low-cost and low-maintenance, Raspberry Pi (R-Pi) computing technology. Sharma et al. [83] employed a machine-learning algorithm to developed a smartphone application based on data collected from SPB, to automatically detect farmworkers' activities (e.g., harvesting, bed making, standstill, and walking) while working in the farm. To detect pest infestation in rice, almonds, tomatoes, apple, watermelon, and cotton crops via image visualization, a smartphone application called ADAMA Bullseye was reviewed by Bruce [12]. Similarly, SIFSS gives users access to the full soil dataset from the Soil Survey of Scotland and SOCiT provides details of organic matter and carbon content present in topsoil by image analysis obtained on color correction card [3]. A paper-based sensor to monitor temperature, humidity, moisture as well as the movement of animals in agricultural fields for crop protection has been linked to smartphones via the Arduino board. It is enabled with an SMS notification facility in case of any animal movement or water requirement detection using Wi-Fi/3G/4G [89]. The presented scheme is capable of creating a two-tier communication facility founded on a two-way associative platform of smartphone devices and open internet connections to promote smooth data transfer. This system provided remarkable advantages of energy autonomy and low cost, signifying its prospective to be a valuable tool for monitoring farming requirements in geologically remote locations. Krishna et al. [41] came up with a Raspberry Pi 2 Model B hardware Mobile Robot by implementing IoT technology and GPS systems. The mobile robot offered a wireless and automated monitoring of temperature, humidity, moisture sensing, scaring birds and animals, spraying pesticides. Galeon et al. [130] summarized the conceptual framework and circuit loops involved in SMS based ICT tools for knowledge sharing networks. They also proposed the urgent requirement of interlinking the enormous quantity of information generated by experts to attend direct needs farm practitioners. It shows a critical need for relevant changes and advancements in a governmental telecommunication policy for the integration of ICT infrastructure with mobile-based knowledge centers. In this background, ICT systems including Knowledge Sharing Systems (KSS) (such as websites, cell phone apps and SMS, digital libraries, social media, and radio broadcast) has been rigorously kicked off by government regulatory bodies, academic and research centers, and local administration. Agriculture and Horticulture Development Board has launched a website, (http://www.ahdb.org.uk/), containing many valuable factsheets. Information about irrigation, pest infections, weeding process, pest management, fertilization and manuring, soil supervision, storage management, and sampling, precision farming can also be accessed via aforesaid channels. Similarly, the International Rice Research Institute (IRRI) has its rice knowledge bank called "Rice doctor" that provides diagnostic tools for the management of rice diseases [74]. Farmers can use audio-visual facilities on mobile via the m-KRISHI app, and m-shayak, based on photos or videos of the plants/crops to articulate their queries to experts [65]. A detailed list of various commercial smartphone apps for different farming applications has been briefed in Table 13.3.

13.3.2 Role of the smartphone in storage management and traceability in the food processing supply chain

One of the most promising roles of Smartphone detection devices and IoT systems lies in the field of product traceability, food supply chain, and storage management. However, the extensive practice of stakeholders to exchange information regarding recorded identification of products manually causes the major obstruction in establishing strong food chain traceability exercises. It obstructs the flow of data on an online platform limiting the scope of quality inspection, product traceability as well as, an open and transparent supply chain management. As an outcome, recently we have witnessed many foodborne diseases like SARS, Swine flu, Corna virus COVID-19. According to the standard definition given by ISO 8402 (1994), traceability can be understood as "the ability to trace the history, application or location of an entity utilizing recorded identification". Therefore, food traceability ought to be thought-out as an imperative element of logistics management in agri-food processing. The collaboration of mobile phone sensors with IoT systems has accelerated improvements in food supply chain management. The due credit of modern consumers being aware and enabled to keep track of food quality and safety data goes to RFID or

Table 13.3 A detailed list of various commercial smartphone apps for different farming applications.

Smartphone App	Country/ Institution	Application	Remarks
1. Crop Protection and Diagnosis			
BioLeaf	Federal University of Mato Grosso do Sul, Brazil	It can detect insects caused lesions on the leaves with the help of image processing of smartphone captured images. It can also quantify defoliation percentage for the total leaf area	Non-destructive identification method
Plantix	Berlin, Germany	Image-based detection of nutritional deficiency and pest infestation	Capable of diagnosing 30 types of crops. It does not require network coverage and may work on offline mode
E-agree	Maharastra, India	To diagnose plant diseases, To find a marketplace and market rates	It can also be used to access soil types of information
CROPROTECT	United Kingdom (UK)	To provide insight about insects, weeding process and crop disease	Additional feature: Two-way communication between farmers and agronomists
ADAMA Bullseye	Israel	To detect pest infestation in rice, almonds, tomatoes, apple, watermelon and cotton crop via image visualization	The life cycle of pests and diseases can also
PMapp	University of Adelaide, Australia	To detect the extent of powdery mildew infection on grape bunches	It offers the facility to store this data for further analysis or email them in XML or CSV format
Plant Disease	Greece	To detect vine diseases	Connected to global positioning service
Agrobase	Luthiana	Used to identify weeds, diseases or pests in various types of crops	It can also help in deciding the dosage of pesticide and herbicide
ImScope		It is as plant health indicators application based on image processing algorithms of smartphone	–

(*continued*)

Table 13.3 (Cont'd)

Smartphone App	Country/ Institution	Application	Remarks
2. Crop Nutrition and Fertilization			
DropLeaf	Brazil	It can estimate the effective coverage area for the application of pesticide	It utilizes the spray cards i.e., images of water-sensitive papers captured by smartphone, to evaluate a spraying technology
Ag PhD Crop Nutrient Deficiencies	Baltic, SD, United States	To identify nutrient deficiencies issues in crops	Can aid in decision making for fertilization tasks
SpraySelect	TeeJet Technologies, Springfield, IL, United States	To decide suitable spray nozzle for pesticide application based on spray characteristics and speed	—
SnapCard	University of Queensland, Australia, and the United States Department of Agriculture (USDA), Texas, United States	It can estimate the required coverage area for pesticide spraying	—
EcoFert	Technical University of Cartagena, Spain	To decide the choice and optimum use of available fertilizers. It can also suggest regarding a suitable mixture of fertilizer as per the requirement of soil and crops	It takes into account the current fertilizer price in the market to decide the type and quantities of fertilizers needed
3. Crop Irrigation			
Grapevine water stress	Adelaide, Australia	Estimates the water condition of the vine and calculates water requirements	This system produces the suggestions based on canopy images of vineyard captured by external thermal camera smartphone

Smartphone-based detection devices for the agri-food industry   297

Smart irrigation Cotton	The University of Georgia, University of Florida and Cotton Incorporated, North Carolina (United States)	Performs soil profiling, crop phenology assessment, and irrigation purpose	It does not give recommendations on irrigation application but information to assist in planning
pCAPS	Elche, Spain	To automate and optimize the calculation of water needs.	
4. Crop Harvest			
vitisFlower	La Rioja, Spain	Estimation of the number of flowers of each bunch of grapes	Utilization of OpenCV library increase the computational efficiency of this tool
vitisBerry	La Rioja, Spain	Early-stage estimation of fruits	It provides phenotypic information of the fruits
fruits	University of Queensland, Australia	To measure fruit size on average, 240 fruits per hour	The application also uses geolocation data
5. Field Mapping and Soil Information			
Agri Precision	Paraná, Brazil	To identify soil nutrients (Mg, Ca, N, P, K)	–
GPS Fields Area Measure	Farmis (Kaunas, Lithuania),	It can calculate the area and perimeter of an agricultural field by analyzing satellite imagery	Can also measure pivot farms
Soil Sampler	Farmis (Kaunas, Lithuania)	To identify soil nutrients (Mg, Ca, N, P, K)	–
Nitrogen Index	USDA, New York, United States	Calculates nitrogen content in the soil	First application developed for Android smartphones
6. Machinery Management[t]			
AgriBus-NAVI	Agri Info Design Hokkaido, Japan	It can provide GPS assistance for driving of farm machinery	–
Farm Navigator	Kaunas, Lithuania	It can provide GPS assistance for driving of farm machinery	Offers specifications like a 3D map-free driving assistant, a night mode driving and activity logging,

(*continued*)

Table 13.3 (Cont'd)

Smartphone App	Country/ Institution	Application	Remarks
FarmManager	Katerini, Greece	It provides field data like field size, geographical location, crop type, etc.	It is a multi-tasking tool
Agroop Cooperation	Lisbon, Portugal	Gives a track of air temperature and humidity; soil temperature and humidity; and solar radiation	–
7. Information System			
AgriApp	Bangalore, India	It offers real-time Agro Advisory to the farmers about 78 different crops	–
Mandi on Mobile	Govt. of UP, India	It facilitates information regarding market rates across every mandi in India	The "Digital Mandi" application provides market estimates for around 3 000 agricultural produce
PlantVillage https://www.plantvillage.org	Penn State, USA	Identification of crop diseases on a cellphone	It is a mobile accessible website
Agriculture and Horticulture Development Board, http://www.ahdb.org.	United Kingdom	Complete information database about for soil analysis, field preparation, pest control, precision farming, crop disease profiling, storage and transportation provisions, etc.,	It is a mobile accessible website

bar codes tagging of products via smartphone. Smartphones enabled with Wireless Sensor Network (WSN) is fit to collect sensing data from physical or environmental conditions (like heavy metal detection) of product storage or transportation facilities (Verma and Kaur, 2016). Smartphone-WSN is capable to employ different strata of interrelated hierarchy featuring the network, to maintain the privacy of communication channel, better interpretation ability, and to establish communication web between sensors and actuator devices [8]. Alfian et al. [110] reported a business model encompassing a global track and trace system that was generated to represent the behavior of the whole supply chain with an added advantage of adaptability and interaction with other supply chains. RFID-based smartphone sensors to enumerate the losses occurring due to quality decay of chilled lamb products. The causative agent of decay was the temperature variations during transportation, and the authors presented an EPCIS-based online system to quantify the deterioration [96]. Verdouw et al. [98] proposed a virtual supply chains flow chart intending to regulate temperature, microbial specifications, and other food quality parameters. Virtual supply chains exercise the concept of food traceability that endow with data repository associated with the position of a food product and its tracking records. This platform acts upon swapping of facts and figures through several internet gateways or cloud proxy machines such as WSN including Bluetooth, Zigbee, Wi-Fi, GPRS, and networked RFID. Even at the end of consumer mobile phone GPRS and WPS system has facilitated the geo-location to identify the inundated grain storage/godown facility in a region. BHUVAN platform has been brought into practice to cater to visualization and collection of spatial data integrated with advanced GIS tools on smartphones. The inter-communication of storage facilities across the country can be established via open-source RDBMS PostgreSQL database for querying and retrieval on the smartphone interface [84]. Directorate of Marketing and Inspection (DMI) is the regulatory agency to monitor cold storages in India by coordinating Research and Development in cold storage, facilitating collection and dissemination of information related to better price realization Recently, DMI is promoting the application of Information and communication technology (ICT) via WiFi, Bluetooth, Websites and mobile applications as a vehicle of extension, to improve efficiency in agricultural marketing [71]. In this context, the mobile electronic auction system has come into play, thanks to extensive outreach and easy handling of smartphones. Mobile Van can serve the whole bidding process of agricultural markets transparently and also provide real-time auctioning information in the market yard with the help of Wi-fi or wireless connection to a smart device [59]. A brief overview of the food supply chain starting from storage nodes until customers including sensor involvement at respective nodal points in the supply chain is depicted in Fig. 13.15.

13.3.3 Smartphone for quality assurance in the food industry

Smartphones are the handiest and accessible tools to ensure food safety with a wide range of food-related biochemical analyte which can be tested via smartphone detection devices. The most vital component of daily food intake is water; however, we mostly tend to ignore quality concerns related to drinking water. In this context,

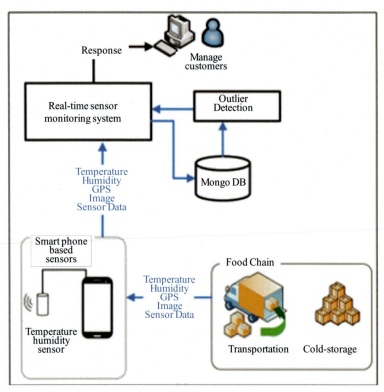

Fig. 13.15 A brief overview of the food supply chain starting from storage nodes until customer including sensor involvement at respective nodal points in the supply chain [110].

for the detection of chromium (Cr), fluoride (F^-), and iron (Fe) in drinking water, a smartphone-based portable instrument using a light-dependent resistor (LDR) sensor was designed by Santra et al. [73]. An Android-application software named Spectruino was used to measure the concentration of the analyte and is priced nominally i.e. INR 1500. Interestingly, natural food pigment i.e., red beetroot was selected as a colorimetric indicator on a portable smartphone sensor (Huawei Mate7 smartphone) for the detection of Cu^{2+} in drinking water. An android application "FSense" which exercises the Mie-scattering principle for the detection and quantification of fluoride concentration in drinking water was introduced by Hussain et al. [36]. In the same line, Hussain et al. [35] employed a portable photometric sensor to quantifying the turbidity of drinking water. An android based application was successfully used to predict the real-time changes in the surface temperature of apple fruit caused due to sunburn [100].

Colorimetric based imaging sensor using a thermal-Red-Green-Blue (RGB) color scheme was applied to customize the app named as "AppSense 1.0" to detect surface color changes. This was a novel advancement in fruit sunburn forecasting in contrast to conventional experimental air temperature data collection by open field weather stations. Near Field Communication (NFC) tag recorded the color scheme of apple samples which was further converted into HSV (hue, saturation, value) color space. This system was employed for the classification of the fruit-based on ripeness grade [42]. Linear discriminant analysis (LDA) and nearest neighbor (kNN) algorithms were reported to be suitable to prepare a grading model to be employed for classification. In the same line, Abasi et al. [1] also constructed a smartphone-based portable instrument to check the degree of ripeness of apple based on moisture content, soluble solids content, pH, and firmness of the sample. For the analysis of apple samples, they applied the Vis/NIR spectral method with the help of optical SPB, on Arduino panel for scheming the procedure of the gadget. Classification of ripe and unripe apples was performed using a decision tree method (DTM) algorithm based on red and green color patterns of the sample. Detection of pesticide residue in food commodities has always been crucial for human health and the assurance of food safety majors. Conventionally, recognition of pesticide residue required well-trained personnel to handle high-end equipment like liquid chromatography-mass spectrometry. Recently a rapid and cost-effective method of pesticide residue diagnosis has been demonstrated by many researchers. A Surface-enhanced Raman scattering (SERS) technology-based SERS chips can be inserted into a smartphone and used for the detection of a variety of pesticide [58]. Molecularly imprinted polymer (MIP) based quartz crystal microbalance (QCM) sensor attached to the cell phone was employed to detect metolcarb a residue of 'Carbamate pesticides'. Metolcarb has been extensively utilized to control rice leafhoppers but if ingested by humans, it can lead to quadriplegic conditions [26]. In contrast to high-performance liquid chromatography, the MIP-QCM sensor produced précised and accurate quantification for metolcarb analysis in apple juice, pear, and cabbage samples. Aflatoxins B1 is considered as a potent poisonous carcinogen produced by certain molds. Sergeyeva et al. [79] also used MIP technology interlinked a biomimetic smartphone-based sensor for the detection of aflatoxin B1. On the other hand, a quartz crystal microbalance (QCM) sensor was utilized for the immunodetection of aflatoxin B1 coupling cargo-encapsulated liposome [92]. Near-field communication (NFC) labeling technology has been recently used to detect food spoilage using a nanostructured conductive polymer wireless sensor [49]. PTS-doped and cross-linked PAni sensors exhibiting high sensitivity toward amines we used, for the detection of Ammonia, putrescine, and cadaverine compounds to indicate the spoilt condition of meat products. In contrary to that, meat freshness detection of pork samples was carried out using a CO_2 sensor and color sensor [21]. Molecularly imprinted polymer membranes in conjugation with smartphones have been exploited for *Fusarium c*ontamination in cereals. Zearalenone-specific MIP membranes were demonstrated to identify estrogen-like toxin zearalenone associated maize, wheat, and rye flour samples. Zeinhom et al. [106] detected the presence of *E. coli* in yogurt and egg portable smart-phone devices. Sandwich ELISA and the specific antibody recognition test were conducted with a fluorescent imager on the smartphone comprising of compact laser-diode-based photo source. The major advantage of this system was the rapid detection of *E. coli* within two hours. Other Egg-based allergens such as Ovalbumin and Lysozyme was also reported to be detected using smartphone sensing devices [46,57,60]. The impacts of smartphone-based sensors are also visible on various dairy products primarily on milk evaluation. For instance, RbST administration to cows increases their milk production, but the administration of that drug is against the law. However, to stop the breaching of the laws regarding usage of RbST in dairy farming, a quantum dot (QD)-coupled detection device comprising of paramagnetic microspheres, has been demonstrated by Ludwig et al. [48]. They developed an Android application, named as "GotMilk," enabled image analysis to be performed on the same cell phone. Similarly, numerous studies have been conducted to establish the leading role of smartphone-based detection devices in the food processing industries.

13.3.4 Major challenges

Smartphone-interfaced detection devices have proved to be a robust and most versatile sensing device, offering widespread application opportunities in different fields. However, this technology is yet in its nascent phase in context with commercial availability of cell phone interfaced sensors and actuator and its practical implementation. Even though R&D activities on mobile sensors have seen a humungous upsurge in the past few decades, there is dire need to develop the whole section of inbuilt sensors on cell-phone interface to address the vast number of issues related to food safety and quality. Not many mobile applications can be easily noticed in the public domain to strongly establish smartphone devices as assertive analytical tools in comparison to conventional diagnostic equipment. The major challenges arising during the developmental stage, accuracy, and application of these smartphone-based detection devices are given as follows:

- Insufficient sensor to deal vast spectrum of analysis
- High power consumption to be met by a smartphone device
- Security and privacy issues about the exchange of data
- Complexity in Signal processing
- Performance limitation due to the complexity of food processing targets and hardware malfunctioning
- The need for external attachments makes it cumbersome and costly
- Insufficient training set; requirement of large scale consumer and farmers training programs.
- Decrease in the area due to the compact design of mobile phones

13.4 Conclusions

The chapter presents the state of art about smartphone-based detection devices for the agri-food processing chain. Undoubtedly, it can be said that smartphones nowadays have become a key player in ensuring food safety at every level agri-food processing network, despite its complex nature. It offers various advantages such as ease of connection, portability, accessibility to all the strata of users be it a consumer, food supply chain administrators, or researchers. Also in recent years, the keen attention of the researcher has been drawn towards the development of machine learning algorithms, big data analysis, and data mining tools to enhance the functionality of smartphones as sensing devices. Smartphone-based sensing devices connected to the Internet can intermingle together for the exchange of data with an added advantage of mobility over fixated devices. 2G/3G/4G wireless sensor networks, Bluetooth devices, Zigbee, and RFID are few in the list which is founding pillars of the vast web of IoT technology. IoT technology with cloud computing has been successfully implemented for a variety of smart agriculture regimes such to monitor soil, to supervise work activity from farm employees, automation of irrigation facilities, and farm machinery. Knowledge-based platforms on smartphone device such as FarmManager, Mandi on mobile, m-KISAN, vitisFlower, Plantix, etc., has empowered farmers to enact precision farming, take an aware decision and avail two-way communication with the experts. The due credit lies in a bag of smartphone devices, to provide economical solution facilities, handy functional facilities, data organization, and strong connectivity to track and trace the quality of the product in a complex web of the food supply chain. Nowadays, mobile food scanners are widely implemented for non-invasive analysis of food products. Freshness indication, detection of food contamination, temperature and moisture scanning of dairy, meat, poultry, fruits, vegetables, cereal grains well as smart sensing of food packaging are among prime applications of smartphone sensors. Nevertheless, smartphone detection devices are still posing the problems of inadequate availability of sensing techniques to deal vast spectrum of food analysis and issues related to power back-up. They also present concerns related to performance in context with the complexity of food processing operations. It is proposed that a large scale consumer and farmers training programs are required to expedite the maximum utility of SPB. In conclusion, state of art suggests that smartphone-based detection devices have emerged as front-line tools in R&D and food industry in context to food safety.

References

[1] S. Abasi, S. Minaei, B. Jamshidi, D Fathi, Development of an optical smart portable instrument for fruit quality detection, IEEE Trans. Instrum. Meas. Acta, Part A: Mol. Biomol. Spectrosc. 153 (2020) 79–86. https://doi.org/10.1016/j.saa.2015.08.006.

[2] C.C. Adley, Past, present and future of sensors in food production, Foods 3 (3) (2014) 491–510.

[3] M. Aitkenhead, D. Donnelly, M. Coull, H. Black, E-smart: environmental sensing for monitoring and advising in real-time. In: International Symposium on Environmental Software Systems, Springer, Berlin, Heidelberg, 2013, pp. 129–142.

[4] N. Alami El Hassani, K. Tahri, E. Llobet, B. Bouchikhi, A. Errachid, N. Zine, N. El Bari, Emerging approach for analytical characterization and geographical classification of Moroccan and French honeys by means of a voltammetric electronic tongue, Food Chem. 243 (2018) 36–42. https://doi.org/10.1016/j.foodchem.2017.09.067.

[5] S. Ali, S. Khusro, A. Rauf, S. Mahfooz, Sensors and mobile phones: evolution and state-of-the-art, Pak. J. Sci. 66 (4) (2014) 385.

[6] J.M. Andrade, D. Ballabio, M.P. Gómez-Carracedo, G. Pérez-Caballero, Nonlinear classification of commercial Mexican tequilas, J. Chemom. 31 (e2939) (2017) 1–14. https://doi.org/10.1002/cem.2939.

[7] P. Arora, A. Sindhu, H. Kaur, N. Dilbaghi, A. Chaudhury, An overview of transducers as platform for the rapid detection of foodborne pathogens, Appl. Microbiol. Biotechnol. 97 (5) (2013) 1829–1840.

[8] M.M. Aung, Y.S Chang, Traceability in a food supply chain: safety and quality perspectives, Food Control 39 (2014) 172–184.

[9] J.J. Barton, Z. Shumin, S.B. Cousins, Mobile phones will become the primary personal computing devices, Proc. 7th IEEE Workshop on Mobile Computing and Systems and Applications (WMCSA), 2006 3–9.

[10] M.J. Berry, G.S. Linoff, Data Mining Techniques: For Marketing, Sales, and Customer Relationship Management, John Wiley and Sons, 2004.

[11] R. Bogue, Towards the trillion sensors market, Sens. Rev. (2014).

[12] T.J. Bruce, The CROPROTECT project and wider opportunities to improve farm productivity through web-based knowledge exchange, Food Energy Secur. 5 (2) (2016) 89–96.

[13] K. Cammann, U. Lemke, A. Rohen, J. Sander, H. Wilken, B. Winter, Chemical sensors and biosensors—Principles and applications, Angewandte Chemie International Edition in English 30 (5) (1991) 516–539.

[14] Y. Chen, G. Fu, Y. Zilberman, W. Ruan, S.K. Ameri, Y.S. Zhang, S.R. Sonkusale, Low cost smartphone diagnostics for food using paper-based colorimetric sensor arrays, Food Control 82 (2017) 227–232.

[15] Z. Chen, C. Lu, Humidity sensors: a review of materials and mechanisms, Sens. Lett. 3 (4) (2005) 274–295.

[16] S. Choudhury, P. Kuchhal, R. Singh, ZigBee and Bluetooth network based sensory data acquisition system, Procedia Comput. Sci. 48 (2015) 367–372.

[17] U. Contreras, O. Barbosa-García, J.L. Pichardo-Molina, G. Ramos-Ortíz, J.L. Maldonado, M.A. Meneses-Nava, P.L. López-de-Alba, Screening method for identification of adulterated and fake tequilas by using UV-vis spesctroscopy and chemometrics, Food Res. Int. 43 (2010) 2356–2362. https://doi.org/10.1016/j.foodres.2010.09.001.

[18] M. Cole, I. Spulber, J.W. Gardner, Surface acoustic wave electronic tongue for robust analysis of sensory components, Sens. Actuators B 207 (2015) 1147–1153.

[19] A. Dankowska, W. Kowalewski, Comparasion of different classifications methods for analyzing fluorescence spectra to characterize type and freshness of olive oils, Eur. Food Res. Technol. (2018). https://doi.org/10.1007/s00217-018-3196-z.

[20] A.J. Das, A. Wahi, I. Kothari, R. Raskar, Ultra-portable, wireless smartphone spectrometer for rapid, non-destructive testing of fruit ripeness, Sci. Rep. 6 (2016) 32504.

[21] I.M.P. de Vargas-Sansalvador, M.M. Erenas, A. Martínez-Olmos, F. Mirza-Montoro, D. Diamond, L.F. Capitan-Vallvey, Smartphone based meat freshness detection, Talanta (2020) 120985.

[22] J.L. Delaney, C.F. Hogan, J. Tian, W. Shen, Electrogenerated chemiluminescence detection in paper-based microfluidic sensors, Anal. Chem. 83 (4) (2011) 1300–1306.

[23] Y. Dong, J. Scisco, M. Wilson, E. Muth, A. Hoover, Detecting periods of eating during free-living by tracking wrist motion, IEEE J Biomed Health Inform 18 (4) (Jul, 2014) 1253–1260 [PubMed: 24058042].

[24] A. Fabris, F. Biasioli, P.M. Granito, E. Aprea, L. Cappelin, E. Schuhfried, I. Endrizzi, PTR-TOF-MS and data mining methods for rapid characterisation of agroindustrial samples: influence of milk storage conditions on the volatile compounds profile of Trentingrana cheese, J. Mass Spectrom. 45 (2010) 1065–1074. https://doi.org/10.1002/jms.1797.

[25] Y. Fan, K. Lai, B.A. Rasco, Y. Huang, Determination of carbaryl pesticide inFuji apples using surface-enhanced Raman spectrocopy coupled with multivariate analysis, LWT - Food Sci. Technol. 60 (2015) 352–357. https://doi.org/10.1016/j.lwt.2014.08.011.

[26] G. Fang, Y. Yang, H. Zhu, Y. Qi, J. Liu, H. Liu, S. Wang, Development and application of molecularly imprinted quartz crystal microbalance sensor for rapid detection of metolcarb in foods, Sens. Actuators B 251 (2017) 720–728.
[27] R.H. Farahi, A. Passian, L. Tetard, T. Thundat, Critical issues in sensor science to aid food and water safety, ACS Nano 6 (6) (2012) 4548–4556.
[28] X. Feng, Q. Zhang, P. Cong, Z. Zhu, Preliminary study on classification of rice and detection of paraffin in the adulterated samples by Raman spectroscopy combined with multivariate analysis, Talanta 115 (2013) 548–555.
[29] G. Fuertes, I. Soto, R. Carrasco, M. Vargas, J. Sabattin, C. Lagos, Intelligent packaging systems: sensors and nanosensors to monitor food quality and safety, J. Sens. 2016 (2016).
[30] X. Gao, N. Wu, Smartphone-based sensors, Electrochem. Soc. Interface 25 (4) (2016) 79.
[31] Y. Guo, Y. Ni, S. Kokot, Evaluation of chemical components and properties of the jujube fruit using near infrared spectroscopy and chemometrics, Spectrochimica Acta. Part A, Mol. Biomol. Spectrosc. 153 (2016) 79–86.
[32] J. Han, J. Pei, M. Kamber, Data Mining: Concepts and Techniques, Elsevier, 2011.
[33] F. Höök, M. Rodahl, P. Brzezinski, B. Kasemo, Energy dissipation kinetics for protein and antibody– antigen adsorption under shear oscillation on a quartz crystal microbalance, Langmuir 14 (4) (1998) 729–734.
[34] H.F. Hsieh, H.T. Hsu, P.C. Lin, Y.J. Yang, Y.T. Huang, C.H. Ko, H.H. Wang, The effect of age, gender, and job on skin conductance response among smartphone users who are prohibited from using their smartphone, Int. J. Environ. Res. Public Health 17 (7) (2020) 2313.
[35] I. Hussain, K.U. Ahamad, P. Nath, Low-cost, robust, and field portable smartphone platform photometric sensor for fluoride level detection in drinking water, Anal. Chem. 89 (1) (2017) 767–775.
[36] I. Hussain, K. Ahamad, P. Nath, Water turbidity sensing using a smartphone, RSC Adv. 6 (27) (2016) 22374–22382.
[37] A.M. Jiménez-Carvelo, A. González-Casado, M.G. Bagur-González, L. Cuadros-Rodríguez, Alternative data mining/machine learning methods for the analytical evaluation of food quality and authenticity–A review, Food Res. Int. 122 (2019) 25–39.
[38] J.C. Jokerst, J.A. Adkins, B. Bisha, M.M. Mentele, L.D. Goodridge, C.S. Henry, Development of a paper-based analytical device for colorimetric detection of select foodborne pathogens, Anal. Chem. 84 (6) (2012) 2900–2907.
[39] A. Kaya, A.S. Keçeli, C. Catal, B. Tekinerdogan, Sensor Failure Tolerable Machine Learning-Based Food Quality Prediction Model, Sensors 20 (11) (2020) 3173.
[40] J.U. Kim, K. Ghafoor, J. Ahn, S. Shin, S.H. Lee, H.M. Shahbaz, J. Park, Kinetic modeling and characterization of a diffusion-based time-temperature indicator (TTI) for monitoring microbial quality of non-pasteurized angelica juice, LWT-Food Sci. Technol. 67 (2016) 143–150.
[41] K.L. Krishna, O. Silver, W.F. Malende, K. Anuradha, Internet of things application for implementation of smart agriculture system, 2017 International Conference on I-SMAC (IoT in Social, Mobile, Analytics and Cloud)(I-SMAC), IEEE, 2017, pp. 54–59.
[42] A. Lazaro, M. Boada, R. Villarino, D Girbau, Color measurement and analysis of fruit with a battery-less NFC sensor, Sens. 19 (7) (2019) 1741.
[43] Z. Li, Z. Li, D. Zhao, F. Wen, J. Jiang, D Xu, Smartphone-based visualized microarray detection for multiplexed harmful substances in milk, Biosens. Bioelectron. 87 (2017) 874–880.

[44] P.-.S. Liang, T.S. Park, J.-.Y. Yoon, Rapid and reagentless detection of microbial contamination within meat utilizing a smartphone-based biosensor, Sci. Rep. 4 (2014) 5953.

[45] H.Y. Lin, C.H. Huang, J. Park, D. Pathania, C.M. Castro, A. Fasano, et al., Integrated magneto-chemical sensor for on-site food allergen detection, ACS Nano 11 (10) (2017) 10062–10069.

[46] L. Liu, H. Bi, Utilising smartphone light sensors to measure egg white ovalbumin concentration in eggs collected from Yinchuan City, China. J. Chem. (2020) 2020.

[47] N. Lopez-Ruiz, V.F. Curto, M.M. Erenas, F. Benito-Lopez, D. Diamond, A.J. Palma, L.F. Capitan-Vallvey, Smartphone-based simultaneous pH and nitrite colorimetric determination for paper microfluidic devices, Anal. Chem. 86 (19) (2014) 9554–9562.

[48] S.K. Ludwig, H. Zhu, S. Phillips, A. Shiledar, S. Feng, D. Tseng, A. Ozcan, Cellphone-based detection platform for rbST biomarker analysis in milk extracts using a microsphere fluorescence immunoassay, Anal. Bioanal. Chem. 406 (27) (2014) 6857–6866.

[49] Z. Ma, P. Chen, W. Cheng, K. Yan, L. Pan, Y. Shi, G. Yu, Highly sensitive, printable nanostructured conductive polymer wireless sensor for food spoilage detection, Nano Lett. 18 (7) (2018) 4570–4575.

[50] R.B. Mahale, S.S. Sonavane, Smart poultry farm monitoring using IOT and wireless sensor networks, Int. J. Adv. Res. Comp. Sci. 7 (3) (2016).

[51] C. Maione, E.S. de Paula, M. Gallimberti, B.L. Batista, A.D. Campiglia, Jr.F. Barbosa, R.M. Barbosa, Comparative study of data mining techniques for the authentication of organic grape juice based on ICP-MS analysis, Expert Syst. Appl. 49 (2016) 60–73. https://doi.org/10.1016/j.eswa.2015.11.024.

[52] C. Maione, B. Lemos Batista, A.D. Campiglia, F. Barbosa Jr., R.M Barbosa, Classification of geographic origin of rice by data mining and inductively coupled plasma mass spectrometry, Comput. Electron. Agric. 121 (2016) 101–107. https://doi.org/10.1016/j.compag.2015.11.009.

[53] S. Martínez-Jarquín, A. Moreno-Pedraza, D. Cázarez-García, R. Winkler, Automated chemical fingerprinting of Mexican spirits derived from Agave (tequila and mezcal) using direct-injection electrospray ionisation (DIESI) and low-temperature plasma (LTP) mass spectrometry, Anal. Methods 9 (2017) 5023–5028. https://doi.org/10.1039/c7ay00793k.

[54] M. Masoud, Y. Jaradat, A. Manasrah, I Jannoud, Sensors of smart devices in the internet of everything (IoE) Era: big opportunities and massive doubts, J. Sens. 2019 (2019).

[55] P. Mehrotra, Biosensors and their applications–A review, J. Oral Biol. Craniofac. Res. 6 (2) (2016) 153–159.

[56] M.S. Mekala, P. Viswanathan, A novel technology for smart agriculture based on IoT with cloud computing, 2017 International Conference on I-SMAC (IoT in Social, Mobile, Analytics and Cloud)(I-SMAC), IEEE, 2017, pp. 75–82.

[57] R.K. Mishra, A. Hayat, G.K. Mishra, G. Catanante, V. Sharma, J.L. Marty, A novel colorimetric competitive aptamer assay for lysozyme detection based on superparamagnetic nanobeads, Talanta 165 (2016) 436–441.

[58] T. Mu, S. Wang, T. Li, B. Wang, X. Ma, B. Huang, J. Guo, Detection of pesticide residues using Nano-SERS chip and a smartphone-based Raman sensor, IEEE J. Sel. Top. Quantum Electron. 25 (2) (2018) 1–6.

[59] J. Nayak, Role of information and communication technology in agricultural marketing in India, Anveshana 6 (1) (2016) 82–96.

[60] S. Neethirajan, X. Weng, A. Tah, J.O. Cordero, K.V. Ragavan, Nano-biosensor platforms for detecting food allergens–New trends, Sens. Biosensing Res. 18 (2018) 13–30.

[61] K. Ni, J. Wang, Q. Zhang, X. Yi, L. Ma, Y. Shi, J. Ruan, Multi-element composition and isotopic signatures for the geographical origin discrimination of green tea in China: a case study of Xihu Longjing, J. Food Compos. Anal. 67 (2018) 104–109. https://doi.org/10.1016/j.jfca.2018.01.005.

[62] E. Ordukaya, B. Karlik, Quality control of olive oils using machine learning and electronic nose, J. Food Qual. (2017) 1–7. ID, 9272404. https://doi.org/10.1155/2017/9272404.

[63] A. Patil, M. Beldar, A. Naik, S. Deshpande, Smart farming using Arduino and data mining, 2016 3rd International Conference on Computing for Sustainable Global Development (INDIACom), IEEE, 2016, pp. 1913–1917.

[64] G. Pérez-Caballero, J.M. Andrade, P. Olmos, Y. Molina, I. Jiménez, J.J. Durán, Miguel-Cruz, Authentication of tequilas using pattern recognition and supervised classification, Trends Anal. Chem. 94 (2017) 117–129. https://doi.org/10.1016/j.trac.2017.07.008.

[65] S. Pongnumkul, P. Chaovalit, N. Surasvadi, Applications of smartphone-based sensors in agriculture: a systematic review of research, J. Sens. 2015 (2015).

[66] S. Pooja, D.V. Uday, U.B. Nagesh, S.G. Talekar, Application of MQTT protocol for real time weather monitoring and precision farming, 2017 International Conference on Electrical, Electronics, Communication, Computer, and Optimization Techniques (ICEECCOT), IEEE, 2017, pp. 1–6.

[67] S. Popek, M. Halagarda, K. Jursa, A new model to identify botanical origin of Polish honeys based on the physicochemical parameters and chemometric analysis, LWT - Food Sci. Technol. 77 (2017) 482–487. https://doi.org/10.1016/j.lwt.2016.12.003.

[68] T. Rafoss, K. Sælid, A. Sletten, L.F. Gyland, L. Engravslia, Open geospatial technology standards and their potential in plant pest risk management—GPS-enabled mobile phones utilising open geospatial technology standards web feature service transactions support the fighting of fire blight in Norway, Comput. Electron. Agric. 74 (2) (2010) 336–340.

[69] R. Rapini, G. Marrazza, Biosensor potential in pesticide monitoring, Comprehensive Analytical Chemistry, 74, Elsevier, 2016, pp. 3–31.

[70] G. Rateni, P. Dario, F. Cavallo, Smartphone-based food diagnostic technologies: a review, Sensors 17 (6) (2017) 1453.

[71] S.U. Rehman, Agricultural Marketing Services in India, APJEM Arth Prabhand: J. Econ. Manage. 1 (3) (2012).

[72] A. Roda, E. Michelini, M. Zangheri, M. Di Fusco, D. Calabria, P. Simoni, Smartphone-based biosensors: a critical review and perspectives, TrAC, Trends Anal. Chem. 79 (2016) 317–325.

[73] D. Santra, S. Mandal, A. Santra, U.K. Ghorai, Cost-effective, wireless, portable device for estimation of hexavalent chromium, fluoride, and iron in drinking water, Anal. Chem. 90 (21) (2018) 12815–12823.

[74] R. Saravanan, S. Bhattacharjee, Mobile Phone Applications for Agricultural Extension in India. Mobile Phones for Agricultural Extension: Worldwide mAgri Innovations and Promise for Future, New India Publishing Agency, New Delhi, 2014, pp. 1–75.

[75] H. Sasikumar, V. Prasad, P. Pal, M.M. Varma, Diffractive interference optical analyzer (DiOPTER), Optical Diagnostics and Sensing XVI: Toward Point-of-Care Diagnostics, 9715, International Society for Optics and Photonics, 2016 971507.

[76] V. Schroeder, E.D. Evans, Y.C.M. Wu, C.C.A. Voll, B.R. McDonald, S. Savagatrup, T.M. Swager, Chemiresistive sensor array and machine learning classification of food, ACS Sensors 4 (8) (2019) 2101–2108.

[77] S.M. Seo, S.W. Kim, J.W. Jeon, J.H. Kim, H.S. Kim, J.H. Cho, S.H. Paek, Food contamination monitoring via internet of things, exemplified by using pocket-sized immunosensor as terminal unit, Sens. Actuators B 233 (2016) 148–156.

[78] S.E. Seo, F. Tabei, S.J. Park, B. Askarian, K.H. Kim, G. Moallem, O.S. Kwon, Smartphone with optical, physical, and electrochemical nanobiosensors, J. Ind. Eng. Chem. 77 (2019) 1–11.

[79] T. Sergeyeva, D. Yarynka, L. Dubey, I. Dubey, E. Piletska, R. Linnik, A. El'skaya, Sensor based on molecularly imprinted polymer membranes and smartphone for detection of fusarium contamination in cereals, Sensors 20 (15) (2020) 4304.

[80] T. Sergeyeva, D. Yarynka, E. Piletska, R. Linnik, O. Zaporozhets, O. Brovko, A. El'skaya, Development of a smartphone-based biomimetic sensor for aflatoxin B1 detection using molecularly imprinted polymer membranes, Talanta 201 (2019) 204–210.

[81] A. Shahvar, M. Saraji, H. Gordan, D Shamsaei, Combination of paper-based thin film microextraction with smartphone-based sensing for sulfite assay in food samples, Talanta 197 (2019) 578–583.

[82] H. Sharma, R. Mutharasan, Review of biosensors for foodborne pathogens and toxins, Sens. Actuators B 183 (2013) 535–549.

[83] S. Sharma, J. Raval, B. Jagyasi, Mobile sensing for agriculture activities detection, 2013 IEEE Global Humanitarian Technology Conference (GHTC), IEEE, 2013, pp. 337–342.

[84] V.K. Sharma, V. Banu, K. Chandrasekar, B.K. Bhattacharya, M.S. Sai, V. Bhanumurthy, Web processing service integrated with mobile application to identify suitable grain storage facility location, Spatial Inf. Res. 25 (1) (2017) 131–140.

[85] S. Srivastava, V. Kumar, M.A. Ali, P.R. Solanki, A. Srivastava, G. Sumana, P.S. Saxena, A.G. Joshi, B.D. Malhotra, Electrophoretically deposited reduced graphene oxide platform for food toxin detection, Nanoscale 5 (7) (2013) 3043–3051.

[86] K. Su, X. Qiu, J. Fang, Q. Zou, P Wang, An improved efficient biochemical detection method to marine toxins with a smartphone-based portable system—Bionic e-Eye, Sens. Actuators B 238 (2017) 1165–1172.

[87] D.W. Sun, Handbook of Frozen Food Processing and Packaging, CRC press, 2016.

[88] P. Sundaravadivel, K. Kesavan, L. Kesavan, S.P. Mohanty, E. Kougianos, Smart-log: a deep-learning based automated nutrition monitoring system in the iot, IEEE Trans. Consum. Electron. 64 (3) (2018) 390–398.

[89] G. Sushanth, S. Sujatha, IOT based smart agriculture system, 2018 International Conference on Wireless Communications, Signal Processing and Networking (WiSPNET), IEEE, 2018, pp. 1–4.

[90] P. Takhistov, Biosensor technology for food processing, safety, and packaging, Handbook of Food Science, Technology, and Engineering, 4 Volume, CRC Press, 2005, pp. 2312–2331 -Set.

[91] H. Tang, A new amperometric method for rapid detection of Escherichia coli density using a self-assembled monolayer-based bienzyme biosensor, Anal. Chim. Acta 562 (2006) 190–196.

[92] Y. Tang, D. Tang, J. Zhang, D Tang, Novel quartz crystal microbalance immunodetection of aflatoxin B1 coupling cargo-encapsulated liposome with indicator-triggered displacement assay, Anal. Chim. Acta 1031 (2018) 161–168.

[93] E. Teye, X. Huang, Novel prediction of total fat content in cocoa beans by FTNIR spectroscopy based on effective spectral selection multivariate regression, Food Anal. Methods 8 (2015) 945–953. https://doi.org/10.1007/s12161-014-9933-4.

[94] E. Teye, X. Huang, F. Han, F. Botchway, Discrimination of cocoa beans according to geographical origin by electronic tongue and multivariate algorithms, Food Anal. Methods 7 (2014) 360–365. https://doi.org/10.1007/s12161-013-9634-4.

[95] M.S. Thakur, K.V. Ragavan, Biosensors in food processing, J. Food Sci. Technol. 50 (4) (2013) 625–641.

[96]   M. Thakur, E. Forås, EPCIS based online temperature monitoring and traceability in a cold meat chain, Comput. Electron. Agric. 117 (2015) 22–30.

[97]   S. Vaishali, S. Suraj, G. Vignesh, S. Dhivya, S. Udhayakumar, Mobile integrated smart irrigation management and monitoring system using IOT, 2017 International Conference on Communication and Signal Processing (ICCSP), IEEE, 2017, pp. 2164–2167.

[98]   C.N. Verdouw, J. Wolfert, A.J.M. Beulens, A Rialland, Virtualization of food supply chains with the internet of things, J. Food Eng. 176 (2016) 128–136.

[99]   N. Verma, G. Kaur, Trends on biosensing systems for heavy metal detection, Comprehensive Analytical Chemistry, 74, Elsevier, 2016, pp. 33–71.

[100] B. Wang, R. Ranjan, L.R. Khot, R.T. Peters, Smartphone application-enabled apple fruit surface temperature monitoring tool for in-field and real-time sunburn susceptibility prediction, Sensors 20 (3) (2020) 608.

[101] T. Wang, Y. Zhou, C. Lei, J. Luo, S. Xie, H. Pu, Magnetic impedance biosensor: a review, Biosens. Bioelectron. 90 (2017) 418–435.

[102] Q. Wei, H. Qi, W. Luo, D. Tseng, S.J. Ki, Z Wan, Fluorescent imaging of single nanoparticles and viruses on a smartphone, ACS Nano 7 (10) (2013) 9147–9155. O.S. Kwon, H.S. Song, T.H. Park, J.S. Jang, Chem. Rev. 119 (2019) 36. http://dx.doi.org/10.1021/acs.chemrev.8b00159.

[103] Y. Yang, Q. Chen, C. Shen, S. Zhang, Z. Gan, R. Hu, Y. Ni, Evaluation of monosodium glutamate, disodium inosinate and guanylate umami taste by an electronic tongue, J. Food Eng. 116 (3) (2013) 627–632.

[104] Y. Yao, H. Li, D. Wang, C. Liu, C. Zhang, An electrochemiluminescence cloth-based biosensor with smartphone-based imaging for detection of lactate in saliva, Analyst 142 (19) (2017) 3715–3724.

[105] S. Zander, B. Schandl, A framework for contextdriven RDF data replication on mobile devices, Proc.6th International Conference on Semantic Systems (I-Semantics), 2010 1–5.

[106] M.M.A. Zeinhom, Y. Wang, Y. Song, M.J. Zhu, Y. Lin, D Du, A portable smart-phone device for rapid and sensitive detection of E. coli O157: H7 in yoghurt and egg, Biosens. Bioelectron. 99 (2018) 479–485.

[107] C. Zhang, A.X. Yin, R. Jiang, J. Rong, L. Dong, T. Zhao, C.H. Yan, Time–Temperature indicator for perishable products based on kinetically programmable Ag overgrowth on Au nanorods, ACS Nano 7 (5) (2013) 4561–4568.

[108] D. Zhang, Q. Liu, Biosensors and bioelectronics on smartphone for portable biochemical detection, Biosens. Bioelectron. 75 (2016) 273–284.

[109] J. Zhou, Q. Qi, C. Wang, Y. Qian, G. Liu, Y. Wang, L. Fu, Surface plasmon resonance (SPR) biosensors for food allergen detection in food matrices, Biosens. Bioelectron. 142 (2019) 111449.

[110] G. Alfian, M. Syafrudin, J. Rhee, Real-time monitoring system using smartphone-based sensors and nosql database for perishable supply chain, Sustainability 9 (11) (2017) 2073.

[111] J. Bunney, S. Williamson, D. Atkin, M. Jeanneret, D. Cozzolino, J. Chapman, The use of electrochemical biosensors in food analysis, Current Research in Nutrition and Food Science Journal 5 (3) (2017) 183–195.

[112] Georgina MS Ross, Monique GEG Bremer, Michel WF Nielen, Consumer-friendly food allergen detection: moving towards smartphone-based immunoassays, Analytical and bioanalytical chemistry 410 (22) (2018) 5353–5371.

[113] C. Dincer, R. Bruch, E. Costa-Rama, M.T. Fernández-Abedul, A. Merkoçi, A. Manz, G.A. Urban, F. Güder, Disposable sensors in diagnostics, food, and environmental monitoring, Advanced Materials 31 (30) (2019) 1806739.

[114] I. Hong, S. Park, B. Lee, J. Lee, D. Jeong, S. Park, IoT-based smart garbage system for efficient food waste management, The Scientific World Journal 2014 (2014).

[115] O. Kwon, T. Park, Applications of smartphone cameras in agriculture, environment, and food: A review, Journal of Biosystems Engineering 42 (4) (2017) 330–338.

[116] J. Collin, P. Davidson, M. Kirkko-Jaakkola, H. Leppäkoski, Inertial sensors and their applications. In: Handbook of Signal Processing Systems, Springer, Cham, 2019, pp. 51–85.

[117] K.N. Choi, Metal detection sensor utilizing magneto-impedance magnetometer, Journal of Sensors 2018 (2018).

[118] S.A. Kolpakov, N.T. Gordon, C. Mou, K. Zhou, Toward a new generation of photonic humidity sensors, Sensors 14 (3) (2014) 3986–4013.

[119] M. Farooq, E. Sazonov, Accelerometer-based detection of food intake in free-living individuals, IEEE sensors journal 18 (9) (2018) 3752–3758.

[120] P.P. Tripathy, S. Kumar, Neural network approach for food temperature prediction during solar drying, International journal of thermal sciences 48 (7) (2009) 1452–1459.

[121] A. Popa, M. Hnatiuc, M. Paun, O. Geman, D. Jude Hemanth, D. Dorcea, Le Hoang Son, S. Ghita, An intelligent IoT-based food quality monitoring approach using low-cost sensors, Symmetry 11 (3) (2019) 374.

[122] G. Alfian, M. Syafrudin, U. Farooq, M.R. Ma'arif, M.A. Syaekhoni, N.L. Fitriyani, J. Lee, J. Rhee, Improving efficiency of RFID-based traceability system for perishable food by utilizing IoT sensors and machine learning model, Food Control 110 (2020) 107016.

[123] Z. Li, G. Liu, L. Liu, X. Lai, G. Xu, IoT-based tracking and tracing platform for prepackaged food supply chain, Industrial Management & Data Systems (2017).

[124] Y. Bi, M. Lv, C. Song, W. Xu, N. Guan, W. Yi, Autodietary: A wearable acoustic sensor system for food intake recognition in daily life, IEEE Sensors Journal 16 (3) (2015) 806–816.

[125] L. Guan, J. Tian, R. Cao, M. Li, Z. Cai, W. Shen, Barcode-like paper sensor for smartphone diagnostics: An application of blood typing, Analytical chemistry 86 (22) (2014) 11362–11367.

[126] D.T. Tran, K. Knez, K.P. Janssen, J. Pollet, D. Spasic, J. Lammertyn, Biosens. Bioelectron. 43 Complete (2013) 245–251.

[127] K.-L. Lee, M.-L. You, C.-H. Tsai, E.-H. Lin, S.-Y. Hsieh, M.-H. Ho, J.-C. Hsu, P.-K. Wei, Nanoplasmonic biochips for rapid label-free detection of imidacloprid pesticides with a smartphone, Biosensors and Bioelectronics 75 (2016) 88–95.

[128] M.-Y. Lu, W.-C. Kao, S. Belkin, J.-Y. Cheng, A smartphone-based whole-cell array sensor for detection of antibiotics in milk, Sensors 19 (18) (2019) 3882.

[129] F. Carpio, A. Jukan, Ana Isabel Martín Sanchez, N. Amla, N. Kemper, Beyond production indicators: A novel smart farming application and system for animal welfare, Proceedings of the Fourth International Conference on Animal-Computer Interaction, 2017, pp. 1–11.

[130] D.H. Galeon, P.G. Garcia Jr, T.D. Palaoag, SMS-Based ICT Tool for Knowledge Sharing in Agriculture, Int J Adv Sci Eng Inf Technol 9 (1) (2019) 342–349.

Point-of-need detection with smartphone

Nuno M. Reis[a,b], Isabel Alves[c], Filipa Pereira[b], Sophie Jegouic[d], Alexander D. Edwards[b,d]
[a]University of Bath, United Kingdom
[b]Capilary Film Technology Ltd, United Kingdom
[c]Loughborough University, United Kingdom
[d]Reading University, United Kingdom

14.1 Introduction

Over the past few decades, we have assisted to remarkable developments in the areas of analytical techniques and molecular diagnostics. This has been mostly driven by modern needs in environmental monitoring, veterinary and human healthcare sectors, with the last being certainly the biggest market driver. Nevertheless, this has happened at almost two different paces in different parts of the globe. In wealthy economies, high precision diagnostics based on highly sophisticated equipment available in centralized pathology laboratories became accessible to the general of population through primary, community and emergency care settings. General Practitioners, hospital consultants or accident, emergency clinicians, to name a few, have now well-established protocols for diagnosis of many acute and chronic diseases based on measurement of one of more analytes/biomarkers in a biological sample. Localized initiatives such as the National Institute for Health and Care Excellence (NICE) in UK have in more recent years established clinical protocols for care interventions, positioning diagnostics testing as an essential step in clinical decisions and management of patients. As a result, personalized medicine has become closer to reality, which also benefited from intense biomarker discovery in the past 2–3 decades. In contrast, molecular testing in developing regions of the globe has remained mostly limited to the colloidal gold rapid lateral flow tests, which comes with its own compromises in terms of e.g. limit of detection, specificity and sensitivity. Whereas in wealthy healthcare systems, for the majority of clinical situations the point-of-need coincides with the point-of-care, mostly in developing economies where a large part of the population has no access to centralized healthcare and/or sophisticated diagnostic capabilities, there is an urgent need in 'democratizing' molecular testing and providing local communities with the diagnostics capabilities of high-performance testing without the investment required by modern centralized pathology laboratory equipment. As a consequence, alternative platforms such as microfluidic devices [1,2] have emerged as tools to decentralize molecular diagnostics and meet WHOs ASSURED criteria [3] (affordable, sensitive, specific, user-friendly, rapid and robust, equipment-free and delivered). It is also widely acknowledged the impact of point-of-need diagnostics in improving life expectancy, where increasingly healthcare decisions are based upon [1,2]. Therefore, one of the biggest current opportunities in global diagnostic testing lies in the use of smartphone interfaces (hardware and software) integrated with modern miniaturized approaches. This chapter describes some of the global unmet medical needs that benefit from point-of-need smartphone testing, including integration of some conventional analytical techniques, some of the key challenges and new portable fluidic capabilities under development. Smartphones are in general not the 'analytical' platform *per si*, requiring a complementary fluidic interface with the (bio)sensing area/volume/device, therefore the reader is also introduced to the latest development miniaturization trends in bioassays. Non-communicable diseases have surpassed communicable (infectious) diseases as the biggest global killers, yet there are clear opportunities for point-of-need smartphone testing on both 'sides of the border', being a good example the current SARS-CoV-2 pandemic. The reader is also directed to relevant references in the emergent field of point-of-need smartphone testing and to two particular case studies that have been significantly progressed by our research team, being smartphone biomarker quantitation [5] and smartphone detection of Urinary Tract Infections (UTIs) [6].

14.2 Modern needs in communicable diseases and bacteria detection

14.2.1 Emerging burden of bacterial infections

Organisms such as bacteria, fungi and viruses are responsible for a wide range of infectious diseases. They can be highly adaptable to extreme conditions, catalyzing their ability to spread and overcome the susceptibility to antimicrobial drugs [7–9]. The treatment of human infectious diseases in England, including costs to the health service, labor market and the individual expenses, are estimated at £30 million per year [7–9]. Moreover, around 25 percent of the population is affected by gastrointestinal infections each year, leading to approximately 1 million GP visits and nearly 29 million days lost from school or work. [7] Although treatable, most of these major infectious diseases represent the second cause of mortality with rates over 50 percent in developing countries [10–12] According to O'Neill [10], drug-resistant strains of tuberculosis (TB), malaria, HIV and certain bacterial infections (the most common caused by *E. coli, Klebsiella* sp. *and Staphyloccocus* sp.) are taking 700 thousand annually lives worldwide, estimated to increase to 10 million per year by 2050.

Bacterial infections are the predominant infectious diseases [8,13] Currently healthcare systems are experiencing a huge demand for effective and portable miniaturized diagnostics tests for bacterial detection and quantification at the point-of-care, in order to help with the fight against antimicrobial resistance (AMR). Yet the diagnostics industry has failed so far to deliver miniaturized biosensors capable of detection and identifying bacteria at the point-of-need. Thus, effective and reliable diagnosis still depends on centralized laboratories fully equipped with sophisticated technologies not accessible or affordable worldwide. In low resource settings, medical or laboratory facilities are still limited and inaccessible to most patients, resulting in high mortality rates caused by communicative and non-communicative diseases. The reality for the majority of population on limited and impoverished income remains very basic health care due to scarce resources such as lack of access to electricity, piped water, no transport network, low economic yield or specialized health care professionals.

It is well acknowledged the remarkable impact of point-of-need diagnostics in improving life expectancy, where increasingly healthcare decisions are based [3,4] This has been the case mostly for the thre 'big killers': HIV, Tubercolosis and Malaria [8,10] . Nevertheless, rural areas still struggle to match the reality of developed countries lacking fully automated and cutting edge facilities to afford sensitive (able to quantify lower concentration of analyte), rapid (in less than 1 h), and reproducible (assay procedure and results do not change with environment) diagnostics.

Effective pathogen detection is essential for the prevention and treatment of human infectious diseases and to tackle the spread of resistant strains worldwide [10,12] The development of suitable diagnostic platforms is crucial for screening of asymptomatic individuals, which needs to follow the ASSURED criteria established by the World Health Organization (WHO): **a**ffordable to everyone, **s**ensitive in any ranges of analyte, **s**pecific for each target to avoid false positives, **u**ser friendly to be used by each patient in a non-invasive and simple way, **r**apid and **r**obust (no fluctuations with environment and very reproducible), **e**quipment-free and **d**eliverable to end-users who need it [14]

Pathogen diagnosis in modern clinical setting is still dependent on culture rich plates which total procedure can take between 48 and 72 h [15,16] Despite its sensitivity, cross contamination is still the major issue of this method. Other techniques rely on polymerase chain reaction (PCR) [17–19], ELISA [17,19–21], aptamers [17,19], antimicrobial peptides (AMPs) [19], peptide nucleic acids (PNA) [12,19], surface plasmon resonance (SPR) [12], impedance [22] or magnetic beads [12,22]. However, the majority of those methods present characteristics which are not suitable for point-of-need applications. Currently, immunological methods are well understood and widely accepted for pathogen detection. Although polyclonal or monoclonal antibodies are easily produced and sold in commercial companies, a major limitation of antibodies includes quality-assured preparation, which poor binding-site and cross reactivity recognition results in decreased sensitivity. [12,19,22]

Microfluidic platforms are emerging as mighty tools to develop decentralized diagnostics that meet the majority of ASSURED criteria. [3] Miniaturized or microfluidic analysis systems, also denominated "micro total analysis systems" (μTAS) or "lab-on-a-chip" (LOC) have grown in popularity due to enhanced analytical performance, less reagents consumption and reducing the time span between sampling and monitorization. [12,23] Microfluidic devices provide a higher surface to volume ratio, a faster rate of mass and heat transfer and the ability to precisely handle very small volumes of bodily fluid, including blood, saliva and urine, ranging from nanoliter to picoliter in microchannel support. [3,12] Moreover, miniaturized versions of high throughput laboratory equipment offer portability, low cost, reliability, power free function (it is not dependent of electricity to be used, for example pregnancy tests present a simple optical results by color change without need of power supply) as well as simple designs for independent patients. [23]

14.2.2 Antimicrobial resistance as a global burden

Empirical antibiotic prescription and overuse of antibiotics are leading to an alarming increase rate of AMR, demanding more rapid and effective point-of-need diagnostics for pathogen detection, avoiding the broad range antibiotic prescription without need. [8,9] *"Getting the right drug, to the right patient at the right time"* is the basis of the problem and not easy to solve. [8] For instance, in the USA, of the forty thousand patients who receive antibiotics to treat respiratory issues annually, twenty seven thousand receive them unnecessarily. [10] Ultimately, technology plays a fundamental role in the development of new rapid diagnostic tools, enabling the decision-making process around antimicrobial drugs to become more accurate. [4,7,11]

Bacterial infectious diseases are the most prevalent and common, with effective treatment based on antibiotics, in contrast to infections caused by viruses or fungi. Nevertheless, the use of antibiotics are promoting animal growth, preventing disease in livestock and other food animals, which contributes to the increase rates of antibiotic consumption daily. [7,11] Additionally they have been used as additives in agriculture and food industrial processes. As a consequence, multidrug resistant bacteria (bacteria with resistant genes) are developing and being transferred to humans through consumption of food or environmental spread (ex. human sewage and runoff water). [9]

Nowadays, diagnosis of a disease and prescription of antimicrobial drugs is still based on expertise of the doctor, or in some cases patients are self-medicating or being recommended antibiotics by pharmacists. [8] In centralized laboratories, bacteria are cultured to confirm the presence of infection and access the correct antibiotic to treat it, a process taking more than 36 h considering the following steps: sample collection, sample transport to centralized labs, culturing samples overnight on agar plates followed by a microscopic examination). In acute illness, patients cannot wait such a long period to receive a treatment, therefore ending up with a broad spectrum antibiotic prescription. [8,11] Regarding low income countries, the scarcity of resources and life conditions are the main causes for the increasing rates of infectious diseases as the majority of population do not have access to treatment. Unfortunately, huge quantities of antimicrobials, in particular antibiotics, are wasted globally on patients who do not need them while others who need them cannot afford them.

The development of rapid and accurate point-of-need diagnostics can create a step change into current situation, enabling any healthcare setting in high or low income areas to use a more informed prescription of antimicrobial drugs, slowing down the development of superbugs (resistant bacterial strains). [2,9] Smartphones are ubiquitous, offering a new route for decentralization of advanced molecular and microbiological testing.

14.2.3 Escherichia coli - the antibiotic resistant superbug

Escherichia coli (E. coli) is one of the gram-negative prokaryotic bacteria that belongs to the Enterobacteriaceae family. [24] Their cell wall consists of a peptidoglycan layer with a cylindrical structure and multiple flagella rearranged around the cell. [24] Extensively studied due to their rapid growth (has a doubling time of approximately 20 min) as facultative anaerobes and they are usually non-pathogenic of the human colonic flora.

E. coli is commonly associated with four clinical syndromes: sepsis, neonatal meningitis, UTIs and enteric intestinal diseases/diarrhea. [24] Furthermore it is used as a biological indicator for water quality and depending on the type of disease, it is divided into different 'pathotypes' responsible for intestinal diseases: enterotoxigenic, enteropathogenic, enteroinvasive, enterohaemorrhagic; enteroaggregative and diffusely adherent. [25] Therefore, monitoring of *E. coli* is required not just in human clinical but also in many water and environmental monitoring situations.

Serological analyses (serotyping) at various levels of complexity are common as a bacterial identification test. The serotyping of *E. coli* is based on their characteristic antigens, which interact with specific antibodies. A bacterial cell surface carries one or more of the following antigens: somatic (O-antigenic polysaccharide or O-antigen) present into lipopolysaccharides (LPS) of the cell outer membrane of Gram-negatives and positives, being generally less defined in the former ones, flagella (H) which are flagellar proteins providing mobility to the bacteria and capsular polysaccharide antigens (K) which are made of carbohydrates of cell capsule. [26] The different antigens or epitopes are polysaccharides (O & K) or proteins (H). Fig. 14.1 presents the cell surface and epitopes, which may be unique for a serogroup or shared, resulting in cross reactions with other serogroups (or serovars) of *E. coli* or even with other Enterobacteriaceae strains.

Species with well-defined antigens are relatively easy to identify based on serology. There are currently known 332 antigens of *E. coli*: 173 O-antigens, 56 H-antigens, and 103 K-antigens. These are valuable in serotyping members of this species. [17,26] Specific diagnosis of *E. coli* serotypes is totally dependent on the clinical sample being cultured in a centralized clinical laboratory or sophisticated lab based technology such as polymerase chain reaction (PCR). [21,29] Furthermore these methods are long (on average more than 36 h), laborious or expensive.

For instance, PCR technique can be used to make copies of a segment of target DNA, generating a large amount of copies of the initial small sample and for example detect *E. coli*. Firstly *E. coli* sample would be lysed and the purified genetic material (eluate) containing the target DNA sequence (ex. a gene only present in K-12) would be placed in a eppendorf containing specific primers to correct pair the bacterial sequence, nucleotides dNTPs and Tac polymerase resistant to high temperatures. The thermal cycles would be carried out in the PCR device, being characterized by 3 phases: Denaturation at 95 °C for 20–30 s to split the double chain of DNA, the annealing between 55 and 65 °C allowing of hybridization the specific primers with complementary extremities in 5'-to-3' direction to each of the single-stranded DNA template and the Extension phase at 72 °C during 10–15 min, where Tac polymerase polymerise the nucleotides in the

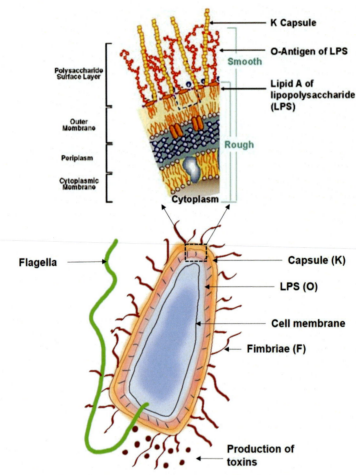

Fig. 14.1 Morphology of *E. coli* and outer surface composition: antigens are represented and demonstrated being respectively: somatic antigenic polysaccharide or O-antigen, present into lipopolysaccharides (LPS) of the cell outer membrane of Gram-negatives and positives, flagella (H) antigen consisting in a protein providing mobility to the bacteria and capsular polysaccharide antigens (K) which are made of carbohydrates of cell capsule (adapted from Alexander et al. [27] with permission of Sage Publishing).

target sequence elongating the sequence and producing the DNA double chain. Those cycles will be repeated during the time wished and the number of DNA target copies (equivalent to 1 *E. coli* cell) formed after a given number of cycles is given by 2^n, where '*n*' is the number of cycles [30,31].

14.2.4 Urinary tract infections (UTIs)

UTIs are a major cause of illness affecting mainly women of any age, children and older men. [32] According with The European Urinalysis Guidelines, the limits for symptomatic UTI caused by *E. coli* is 10^3 CFU/ml. [33–35] There is currently no commercially available test for directly quantify *E. coli* or other UTI-causing bacteria from human urine. [16]

The differentiation of asymptomatic bacteriuria (presence of bacteria in urine) from UTI is subjective and urinary infection symptoms are nonspecific and overlap with numerous symptoms common to non-bacterial infections. [28,33] Generally, the diagnosis of UTIs begins with empirical diagnosis based on typical symptoms (such as pain in back, sometimes blood in urine). Then a quick urine test is performed with colorimetric strip or standard dipstick tests that gives a quick response about the presence of a possible marker of inflammation (presence of nitrite, protein, leukocyte esterase, blood), lacking specificity regarding the correct bacterial strains. [28] This leads to a prescription of a broad range antibiotic. Very often and after a period between 15 days to 6 months, about 50 percent of patients [28,32] will suffer a persistence reoccurrence of the syndromes that can present serious levels as urethritis (urethra infection) and pyelonephritis (severe kidney infection). Very often complicated UTIs are associated to catheter use in hospital, leading to serious pyelonephritis with sepsis, renal damage in young children and further high-level of antibiotic resistance. [28,32,35]

A reduction of unnecessary use of antimicrobials is intrinsically correlated to rapid and accurate diagnostics, which could slow down the pace of multi-drug resistant pathogen. Globally, extended-spectrum beta-lactamase (ESBL) Enterobacteriaceae, fluoroquinolone-resistant *Pseudomonas* and methicillin-resistant *staphylococcus aureus* (MRSA) are among the most challenging public health issues. [13,16]. Fig. 14.2 shows the distribution of ESBL geographically according level of resistance to the last generation of antibiotics (cephalosporins).

14.3 Modern needs in non-communicable diseases

With worldwide deaths from non-communicable diseases surpassing those from infectious diseases, there is an urgent need in 'democratizing' diagnostics for non-communicable diseases by making available high-performance point-of-need testing such as ELISA diagnostics. There is a clear correlation between life expectancy and access to modern diagnostics, this is clear in certain diseases such as cancer where an early diagnosis increases the rates of 5-years survival. Consequently, it is not a surprise that since 2001 cardiovascular diseases and cancer head the top tables in respect to mortality in the developing regions of the globe. [36,37] As reviewed elsewhere, from the 16 million deaths by cardiovascular diseases in 2001, 13 million occurred in low-income and middle income countries, compared with just 3 million in high income countries. [38] This is associated to rapid changes in life style and increasing life expectancy for those living in developing regions, in addition to chronic diseases linked to infectious conditions such as HIV, malaria and TB prevalent in those regions. [39,40]

Diagnosis of many non-communicable diseases currently is currently done based on blood testing and measurement of certain biomarkers through techniques such as

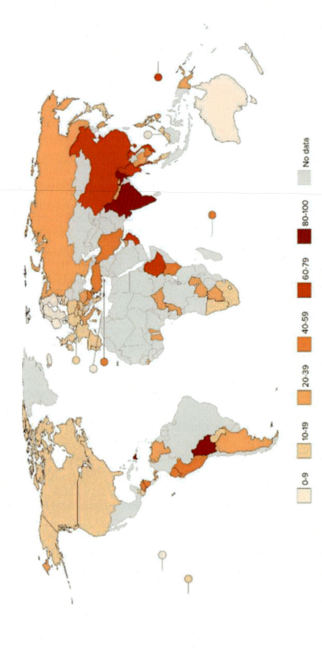

Fig. 14.2 Percentage of extended-spectrum beta-lactamase resistant to the third-generation of cephalosporin, by country from 2011 to 2014, adapted from O'neill [8] (with source permission from CDDEP 2015 and WHO 2014 [13]).

ELISA. Developing portable and affordable point-of-need tests capable of easily and accurately detecting non-communicable health conditions and in particular measuring protein biomarkers in the range of concentrations that is clinically relevant, is now more urged than ever before, and this should be regarded as a global challenge. Regular testing could facilitate health check-ups, or offer a more cost-effective testing alternative to centralized pathology laboratory measurement by facilitating diagnosis at the comfort of home, in community centers, or simply enabling testing in surgeries and hospitals that lack appropriate diagnostic equipment required for effective diagnosis of non-communicable diseases. Early diagnosis enables early treatment decreasing the number of deaths worldwide and the overall cost of patients treatment. [40]

The number of POC tests for non-communicable diseases currently available is very limited, this is certainly linked to the difficulty in developing robust tests capable of measuring very low concentrations of protein biomarkers in biological samples (e.g. whole blood, serum or urine) in a format that is compact, affordable and disposable. On the extreme scenario, POC tests are expected to meet WHOs ASSURED policy regarded as the international standard for developing POC tests. [41] Consequently, measurement of protein biomarkers is currently performed in centralized pathology laboratories using expensive and bulky equipment, in bioassay formats that take several hours to complete and involve very complex fluid handling and pipetting. [42,43] Many studies have reported successful approaches to miniaturize point-of-need ELISA testing of protein biomarkers, some of these are summarized in Section 14.5, yet the reader is also referred to the many high-quality critical review articles especially in specialist journals from analyical chemistry and miniaturization, including Barbosa and Reis [44].

14.4 Point-of-need integration of conventional analytical techniques

14.4.1 Microbiological bacteria detection

Culture and colony counting-based microbiological methods are used as reference, "gold standard" for bacteria identification and quantitation owing to its high sensitivity, ideal specificity, and good reliability. [21] Although powerful, this approach is not suitable for point-of-need testing and rapid screening due to labor-intensive manipulation, time-consuming culture, requiring highly trained people and laboratory facilities. [17,45] Polymerase chain reaction (PCR) is a culture-free technique allowing rapid and multiplexed assay of pathogenic bacteria. [21] However, it is only applicable to DNA and RNA and usually requires cell disruption and nucleic acid extracting, thus limiting its usage in POC testing.

In the last few years, molecular recognition, such as antibody, aptamers, polypeptide, bacteriophage and proteins, has become more suitable for direct assaying whole cells of pathogenic bacteria. [21] Some of those methods present requirements as high assay speed, simple manipulation and easy development of portable point-of-need devices, which become an attractive potential to be applied in microfluidic platforms. [17,46–48]

Regarding *E. coli* infections, namely UTIs induce the symptomatic disease and cause serious health effects to the patient. Lately, lateral flow assays, colorimetric strips to detect presence of nitrite and μPADs (microfluidic paper analytical devices) are the approach applied in the available commercial kits to detect an UTI in urine sample, in parallel to the laborious task of urine culture. Urine is a complicated biofluid that usually requires sample pretreatment, such as purification and/or enrichment steps prior to analytical steps to determine specific components in the urine. [16,49,50] Nevertheless benchtop-based protocols increase the prevalence of cross contamination cultures, demanding in average large sample volume. Thus, there is a need for automated, miniaturized, inexpensive and easy-to use microdevices for urinalysis. An Israel based company, healthy.io has managed to bring the urinalysis 'dip stick' to the next level by integrating image interrogation with a smartphone camera (Fig. 14.3). Though it still presenting the same limitations of a conventional urinalysis test strip (such as reduced specificity for UTIs), they claimed the performance is comparable to a lab test. According to information available on their website (healthy.io) they have developed products for UTI and Chronic Kidney Disease (CKD) and so far they have secured CE-marked for sale in the EU and 2 FDA 510(k) clearances.

Sample preparation is a major stubling block in point-of-need testing. Bodily fluids as blood, urine or saliva present complex matrix components interfering in diagnostic performance. For instance, it is estimated the yearly occurrence of 150 million of UTIs worldwide, where so far there is no commercial available device able to quantify *E. coli* directly from human urine. [16,35] In turn, diagnosis of UTIs is still very challenging, relying on multiple biomarkers detection which do not gather consensus by the medical community. Symptomatic signs are very often unclear and demanding

Fig. 14.3 Overview of healty.io smartphone urinalysis test for UTIs (shared under consent of Healthy.io).

further clinical tests. Furthermore, some symptoms can be masked by other clinical conditions which for example an immunosuppressed patients (eg. patients with HIV and cancer) may not develop fever being the source of infection sometimes impossible to identify. In UTI, the culture plate remain the gold standard of clinical assessment however it is not rapid for enabling early treatment. [4,23,28,32].

14.4.2 Immunoassays

Immunoassays are the 'gold standard' in detection and quantitation of clinically relevant protein biomarkers. Recently, new biosensors for pathogen detection were also developed based on the principles of immunoassays or antibody recognition. [51] Immunoassays offering a high degree of selectivity in many applications, demand signal amplification mainly in lower concentration of analyte/bacterial sample, achieved by use of enzymes. [52] In addition, multi-step assays are preferable due high sensitivity and specificity offered and suitable for miniaturization format compatible with POC. [1] This section provides an overview of fundamentals of immunoassays and shift towards the point-of-need.

14.4.3 Fundamentals of immunoassays

Immunoassays are a powerful bioanalytical tool developed to measure the presence of an analyte through antigen-antibody interaction. The sensitivity and specificity of the immunoassays is highly dependent on the choice of antibodies and their affinity to the target molecule. Enzyme-linked immunosorbent assay (ELISA) and enzyme immunoassay (EIA) are immunological techniques to describe the same technology. ELISA is a plate-based assay technique where the target antigen must be immobilized on a solid surface and then complexed with an antibody that is linked to an enzyme. The signal detected from this techniques can be radioactive, colorimetric or fluorescent. [17,53]

Antibody (Ab), also known as immunoglobulin or glycoprotein, is produced mainly by plasma cells (also called B cells or B-lymphocytes) that are used by the immune systems to identify and neutralize pathogens such as bacteria and viruses. Furthermore, it has high ability for biorecognition and to bind molecules and complexes, denominated antigens. Therefore are indispensable molecules for broad application including diagnosis and disease prevention. [17,54]

Animals are routinely used as host organisms to produce polyclonal and monoclonal antibodies. An Ab molecule presents a "Y" shape molecule, as shown in Fig. 14.4., consisting of two pairs of identical polypeptide chains, named light and heavy chains linked by disulphide (-SS-) bonds, responsible for stabilizing the molecular structure. The two variable domains on the light (VL) and heavy (VH) chains make up the antigen (Ag) recognition and binding site.

The amino acid sequence of antigen binding site is highly variable and this contributes to the broad recognition of the antibody to a wide range of target molecules. The antigen fragment (Fab) contains the variable domains of light and heavy chains, plus the first constant domains, while the constant fragment (Fc) is composed by constant

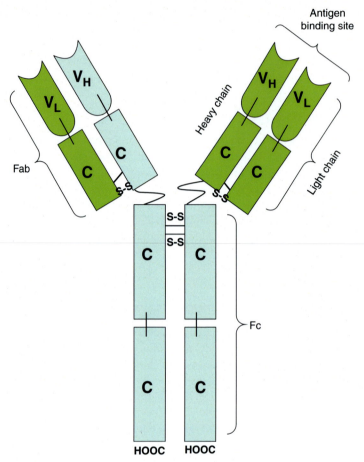

Fig. 14.4 Schematic representative of Immunoglobulin (IgG) structure molecule. On the whole antibody (IgG) molecule, the heavy chain is blue and the light chain is green. Antigen binding receptor is located in Fab domain while Fc is composed by constant domains C and disuphide bonds (-SS) which confers stability of IgG and bonds light to the heavy chains. Adapted from Zourob [17].

domains (C) and disulphide bonds (-SS-), which binds to the Fc receptors located on many mammalian cells.

Immunoglobulin exist as 5 different classes: IgA, IgG, IgM, IgE, and IgD (rare) and subclasses which slightly vary between humans and other species. [17,53] For immunoassay applications, IgG class is highly desirable due to their stability, binding affinity, high retention time on cell surface receptors and low cross reactivity. [55]

The antibody's ability to recognize and bind with high affinity to specific antigenic sites (denominated epitopes or complementarity determining regions

(CDRs)) is exploited for qualitative and quantitative measurement of the antigens, even in a complex mixture. To obtain a successful immunoassay, the production and selection of a suitable antibody is imperative, which depends on the assay parameters. [17,56] Conventional immunoassay methods such as lateral flow and ELISA are the most applied in commercially available immunoassays and therefore widely used. [57]

A polyclonal antibody (PAb) is a heterogeneous mixture of antibody molecules arising from a variety of constantly evolving B-lymphocytes, so that even successive bleeds from one animal are unique. PAbs recognize multiple antigens or multiple epitopes located on the same antigen, in contrast, a monoclonal antibody recognizes only a specific epitope on an antigen. [17,57] For quantitation of protein biomarkers, monoclonal antibodies tend to be the preferred choice due to the high avidity and (in the case of a sandwich immunoassay) the need to avoid competition of the antibody pair to the same epitote (i.e. area of binding of antibody on the antigen protein). For bacterial detection, PAbs more commonly used by early immunologists and microbiologists for their ability to react with a variety of epitopes of bacteria. PAbs are also more stable over a broad pH and salt concentration, whereas MAbs can be highly susceptible to small changes in both. [57]

Commercial assays use polyclonal antibodies due to the high cost involved in the production of monoclonal antibodies and because they also found to be superior in capturing and concentrating target molecules. [55] The type of antibodies (PAb vs. MAb) to be used depends on the specific application. Whenever possible, monoclonal antibodies are preferred due their high specificity, however the production of a target-specific and high performance antibody depends on the proper strategy in selecting and delivering the antigenic molecules. [17,54,57]

14.4.4 Antibody-antigen interaction

The strength of interaction between a single epitope and antigen binding site is called its affinity and it is determined by the sum of multiple non-covalent bonds. Each antibody-antigen interaction has a distinct affinity, which affects the reaction kinetics and therefore the speed and sensitivity of the immunoassay. Whereas the affinity of an antibody reflects its binding energy to a single epitope, avidity reflects the overall binding intensity between antibodies and a multivalent antigen presenting multiple epitopes. Avidity is determined by the affinity of the antibody for the epitope, the number of antibody binding sites, and the geometry of the resulting antibody-antigen complexes. [44,55,57,58]

Equilibrium binding equation (Eq. (14.1)) is represented for the law of mass reaction, where [Ag] is the concentration of antigen, [Ab] the concentration of antibody, [Ag-Ab] the concentration of antigen-antibody complex and K_{eq} is the equilibrium or affinity constant, which represents the ratio of bound and unbound analyte – antibody: [44,55]

$$[Ag]+[Ab]\overset{K_{eq}}{\leftrightarrow}[Ag-Ab] \quad (14.1)$$

The equilibrium binding is a fundamental key to understand and maximize antibody performance in immunoassays. According to the Scatchard's model [55], the Ab-Ag equilibrium is given by Eq. (14.2):

$$\frac{B}{F} = K_{eq}(N - B) \qquad (14.2)$$

where B and F represent the concentration of bound and free antibody, K_{eq} is the affinity constant, and $(N - B)$ is the concentration of unoccupied sites. If binding is weak the equilibrium will be shifted to the left. If binding is strong, equilibrium will be shifted to the right. [55]

14.4.5 Immunoassay configurations

Homogeneous methods have been developed for large and small analytes using both competitive and non-competitive protocols. They are characterized by adding all sample and reagents in a single step where rate of binding reaction is not limited by slow diffusion, decreasing incubation time and reducing automation requirements. [52] For instance, direct ELISA is implemented in one step when the analyte contacts the sensor's receptor layer. Direct ELISA is typically used in analysis of an antigen immune response, in contrast to heterogeneous assays where immunoassay are developed in multi steps, allowing to constituents be removed easily by a wash step, minimizing variations in signal caused by nonspecific effects of the sample matrix. [52] This section is mostly based on colorimetric or fluorescence immunoassays, which are easier to integrate with a smartphone though optical interrogation with the smartphone's camera.

Immunological methods combine a specific antibody-antigen interaction and an efficient catalysis between enzyme-substrate resulting a chromogenic, radioactive or fluorescent signal. [56] The specific association of antigens and antibodies is dependent on the lock and key complex (for ex. enzyme - substrate, antigen - antibody) and affinity between the two molecules on the solid surface. There are three different heterogeneous ELISA assays formats [56]: competitive, indirect and sandwich ELISA (described by Fig. 14.5).

In competitive ELISA (Fig. 14.5A), the primary antibody is mixed in a separate tube with various dilutions of bacteria (or antigen) and added to the wells containing immobilized antigen. Only the free, unbound antibody will bind the immobilized antigen. A secondary antibody-enzyme conjugate and substrate system is added for colorimetric detection. The signal reaction measured is inversely proportional to the concentration of bacteria. Despite cross reactivity be a challenge typical in ELISA, this format is more susceptible to signal interference, where non target analytes in a sample compete for binding sites on antibody, resulting in inaccurate analyte concentration being determined. It is commonly used for detecting small antigens that cannot be bound by two different antibodies or when only one antibody is available for the antigen of interest. [17,55,56].

In indirect ELISA (Fig. 14.5B), the antigen from the sample is previously immobilized in the wells of a microtiter plate and then added to the antibody solution. A secondary antibody conjugated with substrate modifying enzyme, e.g., horseradish

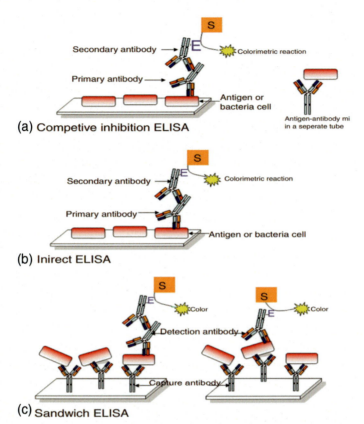

Fig. 14.5 **Heterogeneous ELISA formats user for bacteria or analyte (protein) colorimetric detection (E = enzyme; S = substrate).** (a) Competitive ELISA. (b) Indirect ELISA. (c) Sandwich ELISA, adapted from Zourob [17] (with permission of Springer).

peroxidase (HRP) or alkaline phosphatase (AP), is added to bind to the primary antibody, and the reaction is developed with a suitable substrate to produce a color reaction. If a fluorescent molecule (e.g., FITC, Atthophos®) is used, the reaction is quantified by the amount of fluorescence emitted proportional to concentration of antigen [17,56]. Indirect ELISA is suitable for quantification of total antibody concentration present in samples.

Sandwich assays (Fig. 14.5C) tend to be more sensitive and robust and therefore tend to be the most commonly used. It works similarly to indirect ELISA, except that capture antibody (capAb) is firstly immobilized in microtiter plates, followed by antigen incubation. The reaction is developed adding a detection antibody (detAb) forming a complex sandwich with capAb-antigen-detAb. Afterwards enzyme is added and respective enzymatic subtract (FITC, Atthophos®) for fluorescence signal quantitation. This format is preferable when it is aimed to quantify lower concentrations of analyte and reduce the non-specific binding. This method is highly suitable for complex matrix samples analysis and normally does not need pre-treatment sample prior to measurement. [55,56]

Table 14.1 Summary of advantages and disadvantages of possible immunoassay configurations [59–63].

Types of ELISA	Advantages	Disadvantages
Direct	Faster Less prone to error due require no steps	Immunoreactivity with enzymes High background No flexibility in pair of antibodies. Minimal signal amplification-reducing signal sensitivity
Indirect	Wide commercial availability of detAb Versatile High sensitivity Allows signal amplification	Cross-reactivity from detAb Background Extra incubation step is required in the procedure
Competitive	Less variability More consistent assay No sample pre treatment	Complex procedure steps Limited to small antigens that cannot bind to capAb and detAb as in sandwich assay
Sandwich	High sensitivity High specificity High throughput in complex matrix samples	Cross-reactivity might occur

For better understanding each ELISA format, Table 14.1 summarizes the main advantages and disadvantages.

14.4.6 Immunoassay performance

The success of assay development is intrinsically related to key parameters that clinically characterized the performance. Specificity, sensitivity, precision or reproducibility, calibration model and limits of detection and quantification are some core characteristics to be accessed early on development. [59,62,63]

Precision, also known as reproducibility or variability of the assay, gives the level of confidence of the assay performance. It expresses the coefficient of variation (CV) for each specific analyte concentration point. When precision is evaluated in the same assay experiment is named intra-assay, if it is accessed in different assay runs is denominated inter-assay variability. The limit of acceptance recommended for intra-assay precision is ≤10 percent and inter-assay precision between 20–25%percent. [18,52,54]

Assay specificity is highly dependent of the specificity of antibody directed against the bacteria/analyte. This parameter evaluates the capability of assay to detect or quantify unequivocally the target, even in presence of other components in the sample analyzed. Biological samples present complex matrices that can highly increase non specificity. For instance urine is composed by around 95 percent water and 5 percent of solids that are generally urea, uric acid, chloride, sodium potassium, creatinine and other dissolved ions, inorganic and organic compounds (proteins, hormones and metabolites). Therefore, sample preparation is often an extra step considered in attempt to reduce any interference by pH, viscosity sample, and presence of urine constituents or even the presence of other bacterial strains common in UTI as *Pseudomonas sp, Klebsiella sp.* and *Enterobacteriaceae* sp. [18,53,64]

Calibration models enables determine the response-error relationship denominated "curve fitting" or just "response curve" between experimental data and parameters estimated. Four parameter logistic model (4PL) is the reference model for many immunoassays providing an accurate representation of the fitting between experimental measured response and theoretical response expected by the following equation [18]:

$$y = d + \frac{a-d}{1+\left(\frac{x}{c}\right)^b} \tag{14.3}$$

where x is the concentration of analyte known, y is the cromogenic or fluorescence signal quantified, a the minimum value that can be obtained by the assay, d is the maximum colorimetric or fluorescent value obtained from the assay, c the point of inflection of sigmoidal curve, (halfway between a and d), b the slope of the curve related to the steepness at inflection point. [66]

The limit of quantitation is provided by the highest and lowest concentration of analyte in a sample denominated validated range or dynamic range. They are determined from the lower limit of detection (LLoD) to the upper limit of detection (ULoD) which requirements for each dynamic range vary according assay and analyte target. Accuracy and precision of detection dynamic range is determined from the lower limit of detection quantified, from Eq. (14.4), which means the lowest concentration of an analyte detected in a sample, but not necessarily quantified. [18,56]

$$LLoD = Blank + (3*\sigma) \tag{14.4}$$

Where '*Blank*' represent the signal quantified by the negative control or diluent sample without antigen, σ is the standard deviation of measurement and lower limit of quantitation is given by Eq. (14.5) and represent the lowest concentration of analyte determined in a sample with acceptable precision and accuracy according protocol conditions. [66]

$$LLoQ = Blank + (10*\sigma) \tag{14.5}$$

Providing those values gives insight into positional biases and the precision profile. Nevertheless sensitivity evaluates the capacity of the assay to detect the lowest analyte concentration which is different of value of cut off or blank (sample without analyte). This parameter is assessed by comparing several replicates being the limit of sensitivity determined by lower limit of detection according Eq. (14.4).

14.4.7 Immunoassays standard platform and considerations for translating to point-of-need

Microtiter plate (MTP) remains the 'gold' standard platform for ELISA worldwide used (see Fig. 14.6A). It is fabricated from polystyrene and the dominant format is the 96-well, arranged in 8 rows by 12 columns. It presents a maximum volume capacity of 300 µl with 80 percent of volume surface effectively usable (Fig. 14.6B). The biggest limitation of the MTP is that it depends on a non-portable and expensive microtiter reader, also it does not in general provide the possibly of fluid/reagent flow, making it very laborious in terms of fluid handling and also time consuming (each immunoassay step needs long incubations, normally more than 1 h due transport limitations) and demands high reagent volumes. [52,59]

Enzymes are considered biocatalysts accelerating bioconversions and transformations of subtracts in specific products being the most widely used in antibody conjugation. In immunoassays, the most versatile and common enzymes are horseradish peroxidase (HRP) and alkaline phosphatase (AP). Those enzymes are bound to detection antibody via biotin. The substrate conversion is respectively H_2O_2 and AttoPhos® to produce a colorimetric and fluorescent signal. Like the antibodies, enzymes are susceptible to pH, and temperature variation which can affect immunoassay performance. [52,68] Therefore, understanding optimal reaction conditions regarding enzyme concentration, buffer diluent and time incubation are fundamental to achieve maximum rate and perform immunoassay in less time with high throughput and sensitivity. The kinetics of enzymatic reaction is described by a linear rate with substrate concentration conversion until reaches saturation at maximum. This means all active sites of enzyme are occupied which intrinsic enzymatic rate per second is described by Kinetics of Michaelis–Menten model [68]:

$$v_0 = \frac{v_{max} \cdot [S]}{K_M + [S]} \quad (14.6)$$

$$v_{max} = K_s \cdot C_E \quad (14.7)$$

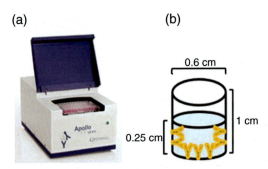

Fig. 14.6 Current gold-standard in heterogeneous immunoassays. (a) Microplate reader and microtiter plate (MTP) adopted from Berthold Technologies [67]. (b) Structure and dimensions of a single well of a 96-well plate filled with 100 µl of reagent.

where v_0 the initial reaction velocity, [S] the substrate concentration, C_E the enzyme concentration, v_{max} the maximum reaction velocity, K_M correspond to the substrate concentration where $v = \frac{1}{2}v_{max}$, and K_s the enzyme turnover number or dissociation enzyme-substrate constant.

The biggest advantage of using an enzymatically amplified immunoassay is the fact that enzymes can effectively amplify the (optical or electrochemical) signal. In case of colorimetric or fluorescence detection, miniaturization of an immunoassay leads to significant reduction in light path distance compared to a MTP well and enzymes offer a strategy to overcome those optical limitation [69] when it comes to integration with e.g. a smartphone.

14.4.8 Antibody immobilization and relevance of surface area

Surface chemistry is paramount for antibody immobilization and the key step in high performance of heterogeneous immunoassays. [51] [44] Surface modifications include at first capture antibody immobilization and blocking (surface passivation) of the remaining available surface sites. Those steps influence remarkably the equilibrium antibody-antigen and prevent nonspecific binding. [70] Therefore immobilization technique can interfere in the density and distribution of capture antibody (capAb), that must provide proper antibody orientation, allowing CDRs availability in order to maximize affinity for bacteria target. [51] [71]

In heterogeneous immunoassays, antibodies are usually immobilized either onto the surface of the channel walls or onto microbeads. Immobilization onto surfaces requires additional micro-fabrication processing steps and might be affected by low reproducibility and reliability. Immobilization onto microbeads is very often preferred due offer larger surface-area-to-volume ratio and therefore more sensitive immunoassays. [72]

The use of coatings is very common during miniaturization of an immunoassay, yet the specifications depend on characteristics of microfluidic surface and the type of interaction antibody-antigen such as passive adsorption surfaces, covalent binding or surface attaching by 3D groups, (shown in Figs. 14.7A, B and C respectively). [44]

The binding between capture antibody and antigen must be considered, in order to be strong enough in comparison to surface-antigen affinity or surface-antibody affinity. Passive adorption is the simplest method where antibody interacts directly with surface without any external crosslink. Despite some reports suggesting weak binding and random antibody orientation [73], passive antibody adsorption has been successfully applied in FEP-Teflon® microarrays [74–78], glass surfaces hydrophobised with Teflon® [79] and polysterene channels including MTP. [80]

Covalent immobilization is very often applied into microfluidic surfaces, offering more stability, higher surface coverage and promote better antibody orientation, crucial for sentive immunoassays. Consequently there is an increase of complexity in manufacturing process by chemistries involved in surface modification. Majority of platforms (glass, silicone, polymeric) involve organofunctional alkoxysilane molecules as APTES, functional groups from proteins as epoxy groups from serine or threonine amino acids (-OH), NH_2 and COOH (amine) and glutaraldehyde (GA). [72,73]

Fig. 14.7 Antibody immobilization techniques. (a) Passive adsorption by intermolecular forces (b) Covalent bond (c) Surface-attaching head group by e.g. PEG (spacer it is often used for improving protein activity), adapted from Kim D. et al. [72] (with permission of PubMed Central).

In recent studies our research team has fully characterized adsorption of antibodies on Teflon®FEP using both passive adsorption [81] and covalent immobilization [82]. We have also developed a range of fluoropolymer microfludiic immunoassays, covering detection/quantitation of both protein biomarkers and bacteria cell, that are easily, interrogated optically, including a smartphone imaging [6,75]. Fluoropolymers present some challenges regarding surface modification and covalent immobilization of antibodies in comparison to silicon, glass or PDMS. Covalent immobilization in microcapillary films (MCF) strips exploiting crosslinkers such as high-molecular weight polyvinyl alcohol (PVOH), NHS-ester groups, APTES, glutaraldehyde, and maleimide was successfully reported firstly by Reis et al. [83] and Pivetal et al. [82] When in presence of limited surface, antibody immobilization occurs with 3D groups attached on planar surfaces. Among these 3D structures, hydrogels such as polyacrylamide gel and polyethylene glycol (PEG) gel provide hydrophilic environments conducive to good protein stability and retained protein activity. [72,73]

14.5 Point-of-need trends in immunoassays miniaturization

Miniaturization of sandwich immunoassays demands simple fluid handling systems or automated steps in order to achieve accurate response. It remains challenging the delivery of cost effective point-of need tests whilst maintaining complex interactions for achieving sensitive tests and integration of cutting edge technology for high throughput. [84]

The starting point for manufacturing point-of-need microfluidic platforms is the capacity of those platforms being mass produced by cheap materials without compromising performance. [44] Nevertheless it demands non-opaque microfluidic materials compatible for low cost optical detection, ideal for imaging with a smartphone camera. Ultimately, design and geometry are key aspects for successful adoption and commercialization of microfluidic point-of-need tests. Scalable manufacturing process at industrial replication requires adjusting all of the important geometrical parameters of a microfluidic structure for a successful functioning in the intended application. [85]

Microfluidic production with transparent substrantes mostly relies on gold standard techniques such as injection molding whereas it allow to replicate by casting of soft elastomers such as poly-(methyl methacrylate) and polycarbonate, poly(dimethyl siloxane). Most microfluidic biomarker and pathogen sensing systems are based on those polymeric materials reviewed in Section 14.5 due their low cost and fully optimized protocols for development of bioassays in parallel with glass or paper based devices. [17]

In addition, multistep assays demand multiple interactions achieved by external accessories as micro-pumps and micromechanical valves for increase automation which might increase costs drastically. Indeed, most research in this field is being directed towards eliminating or minimizing the need for external accessories and power and enhance performance of liquid driving systems. Nevertheless automation would offer a reduction of external accessories and costs of manufacturing and therefore minimizing human error, dead volumes and increasing reproducibility. [79,84,86] To implement that, fluidic operations in miniaturized devices can be operated by pressure driven, centrifugal forces, eletrokinetic, passive flow systems. [22,44]

Centrifugal based platforms are typically produced with a footprint disc shape containing channels and microchambers, relying on spinning frequency to drive fluid movements. Regarding sample analysis, some units are added to these lab-on-disks microfluidic platforms such as sample up-take, reagent supply, mixing and incubation sample. These compact disk (CD) devices are often applied in biomedical applications as biomarkers or infectious diseases detection (see Fig. 14.8A). [87]

Lee et al. [88] reported an innovative approach to detect antibody and antigen of hepatitis B virus. A whole blood sample is injected into device fully automated with a detector, motor and laser diode for valve control integrated. [88] Similarly, Olanrewaju and team [15] have developed microfluidic capillaric circuit containing capillary pumps and micro valves for rapid *E. coli* detection. However CD based devices present shortcomings regarding environment conditions might affect negatively performance [89]. Temperature and humidity surroundings influence the migration speed of reagents and sample and recognition of antigen as well. Moreover, low concentration samples might be difficult due poor hydrophobic barriers, which might contribute to dead volumes and sample interference. [89]

Pressure driven are the most common and versatile method used in fluid control. Generally, it requires the external devices as flow or syringe pumps to deliver or stop flow according steps of immunoassays. Since those devices are easily connected to the microcapillaries, the flow is typically laminar due to small dimensions of platform handled. However, the use of pumps increases costs and demands power supply, compromising portability and suitability for low income settings.

Barbosa et al. [74] have reported a cost effective microfluidic technology interface named Multiple Syringe Aspirator (MSA) capable of loading simultaneous 80 microarrays through 1 ml syringes using a simple rotation of a central knob (see Fig. 14.8B). In addition Reis and co-authors [83] have launched a gravity driven dipstick for one step assay, consisting in MCF strips coated with PVOH. Pressure driven flow occurring due to the gravitational potential of the fluid's height is also relevant for development of point-of-need devices requiring non-skilled workflorce. This could be achieved using e.g. a funnel where reagents are deposited sequentially, however the biggest limitation in respect to enzymatic immunoassays are the washings. With the high turnover of some of the enzymes used in ELISA such as HRP, a single enzyme molecule left beyond unbound is sufficient for giving high background and reduce the performance (especially LoD) of the test.

Magnetic forces exploit a magnetic field, where fluid actuation is performed driving multi step reagents into microfluidic devices. [79] For example Lab on a chip (LOC) devices using magnetic beads conjugated with antibodies for immunomagnetic separation (IMS) and impedance spectrometry (IS) techniques for capture, separation or detection of analytes.

Yang and team [91] have reported a LOC for separation of *E. coli* K-12 from synthetic urine in a chamber containing micro magnetic beads conjugated with

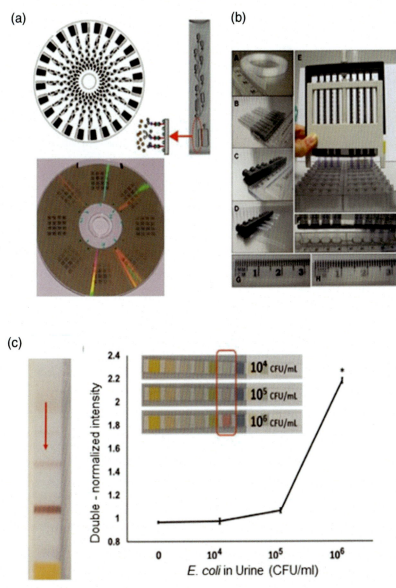

Fig. 14.8 Fluid handling designs commonly used in optical microfluidic platforms.
(a) Heterogeneous multiplexed immunoassay on a digital microfluidics platform, with 24 experiments on a centrifugal microfluidic platform. Adopted from Gorkin et al. [87] and Lai et al. [90]. (b) Multiple Syringe Aspirator (MSA) fully assembled, with a plastic frame, a syringe plug holder with a central knob allowing fluid aspiration by rotation through capillaries action adapted from Barbosa et al. [74] (c) Commercial urinalysis test strips were used with urine samples with $10^4, 10^5$ and 10^6 CFU/mL *E. coli* adapted from Cho et al. [16] (with permission of Royal Society of Chemistry and Elsevier).

anti-*E. coli* antibody, with an limit of detection of 3.4×10^4 CFU/mL. The LOC consists in a concentration and sensing chambers disposed in series and integrating an impedance detector. Clogging of channels is a limitation specially observed in bead-based microfluidic assays. [79,91]

Passive flow systems are independent of any external device or magnetic force, being the fluid transport driven by surface properties of microfluidic platform. Chemical gradients on surface, osmotic pressure, gravity and capillary action are some forces promoting the fluid handling. The biggest challenges of this method are related to control of flow rate, volume and incubation time. For instance capillary action has been used in dipstick assays for over 30 years in lateral flow pregnancy tests and colorimetric strip tests for urinalisys (Fig. 14.8C). [44,71]

The first paper-based sandwich ELISA was developed to test for human chorionic gonadotropin (hCG) in a human pregnancy assay. [92] The commercial urinalysis test strips are a common example used to detect the presence of an UTI in urine samples. [16,49] Despite very simple to use and capable to identify an UTI infection very quickly, dipsticks present limitations in specificity of bacterial strain, are not quantitative and completely accurate giving false positives and being deeply dependent on the correct sample collection.

14.6 Modern point-of-need fluidic capabilities

14.6.1 *Microfluidic platforms for bacterial detection and quantitation at the point-of-need*

Sensitive, specific and rapid antigen and pathogen detection methods are crucial in human diagnostics [22,51,77], bioterrorism defense [17], food safety [21] and drug production. [45,46] Microfluidic devices that follow WHOs ASSURED policy struggle to achieve lower LoD for bacteria with high sensitivity and specificity unable to early detection of such infections. The clinical threshold for symptomatic UTI caused by *E. coli* is 10^3 CFU/ml [28,32,33], which is intrinsically difficult to achieve in a conventional 'dip stick' test or a microfluidic test, the last related to the very small sample volumes used. Currently, diagnosis industry has failed to find any microfluidic platform that suits those ASSURED requirements in a format compatible to point-of-need. Therefore healthcare systems face a technological gap by the absence of any commercially available test capable to detect directly from human urine *E. coli* or other UTI-causing bacteria. [15,16]

Novel sensing and bioassay approaches has been recently proposed for rapid pathogen, including the capture of whole pathogen cells or molecular fragments for further amplification [93] and identification, with detection methods using a variety of transducing technologies (optical, electrochemical, surface plasmon resonance and piezoelectric) [2,20,88,90,92,94–100] Sensitive detection methods such as immunoassays are broadly applied for biomolecules. [96]

Microfluidic platforms are suitable for this purpose, as they allow fluid volumes manipulation in pico- and nanoliter range with easy accurate transport and cells positioning. [96] Pathogen sensing demands disposable systems to eliminate the risk of cross-contamination, so there is a need to minimize the cost of materials involved in microfabrication of technological devices.

Controversially, many drugs companies producing affordable generic antibiotics present challenges to improving diagnostics, having no commercial interest in the advent of rapid microfluidic diagnostics. [8] Improving diagnosis has the impact of improving life quality and average life expectancy but would act to limit the antibiotic prescription.

Mass device fabrication in a cost-effective way and technological challenges as sample pre-treatment, compatibility with fluorescence detection and chemical resistance explain why use of microfluidic diagnostics in clinicians' offices and patient's homes is not yet widespread. [8] Prize initiatives in the UK, USA and EU have been important catalysts in raising attention for the need for rapid POC diagnostics. Longitude prize is one example to promote an accurate, rapid, affordable and easy to use anywhere in the world. [101] But to sustain innovation in the medium and long term, and to encourage uptake of the resultant technology, further and more sustained interventions are needed. [8,11].

Therefore, finding new, cost effective, and simple approaches for simple signal detection and readout systems, fluid actuation and storing reagents on chip are essential for broad point-of-need diagnostics commercialization. [102] Miniaturization of immunoassays increases the challenges regarding surface tension, reduced volume of sample in lower concentration samples and influence the antibody-antigen kinetics.

Biological fluids (blood, urine, saliva) present increased levels of viscosity that bring additional surface tension into microfluidic platforms, affecting sample distribution and accuracy of results. In addition, handling lower sample volumes means less bacterial cells per sample which is challenging for achieving lower LoD and high sensitivity, although analyte concentration is the same. Shear stress from reagents and sample flow might represent a drawback in quantitation systems using antibodies immobilized in surface wall, being fundamental understanding the binding between bacteria-capture antibody. [28,87,100]

On the other hand, miniaturization of immunoassays exploits the potential of reduced diffusion distance between capture antibody and antigen, enabling scientists to reduce time of immunoassay and therefore reduce equilibrium time. Therefore increasing the reactivity of the system with incremented surface-area-to-volume ratio, speeding up and improving LoD of the immunoassays. Development of quantitative heterogeneous immunoassays at microscale has promoted the reduction of environmental space and need for sophistated facilities, making them affordable and practical anywhere. [4,100]

The current microfluidic devices design according to ASSURED criteria struggle to quantify low LoD with high sensitivity and specificity in a short time window to enable early detection of such infections without extra steps. [2,14,103] Despite the limitations highlighted above, several microfluidic devices were developed with the capability of performing sensitive *E. coli* detection and quantitation from biological or synthetic samples. [12,16,22] Majority of those devices use immunoassays or other detection techniques described in Section 14.2.1.

Table 14.2 summarizes some of the most relevant microfluidic assays reported in the literature for pathogen detection. The review focused in target bacteria, method of detection in sensing platform, limit of detection (LoD), pretreatment steps, assay time and the type of sample. The most commonly targeted pathogens in those biosensing platforms are: *E. coli, Salmonella sp., Staphylococcus aureus, Pseudomonas sp.* and *Klebsiella pneumonia*. All these bacteria are commonly recognized as superbugs of antibiotics resistance and responsible for a wide range of symptomatic infections as UTIs and sepsis. [104] For instance, Chang et al. [50] have presented a microfluidic chip embedded with antimicrobial peptides (AMPs) able to detect 10 cells/ml in 20 min. Wang et al. [105] have shown a microfluidics fluorescence assay able to quantify 50 CFU/ml of *E. coli* in 30 min testing blood and buffer but with extra filtering step. Golberg et al. [95] and Yoo et al. [19] have reported microfluidic fluorescence assay approaches with LoD below 10^3 CFU/ml, however those approaches require sample pretretament or long assay times. Cho et al. [16] and Olanrewaju et al. [15] shown fastest microfluidic platforms to quantify *E. coli* with LoD matching the clinical threshold for an UTI in less than 10 min of total assay. It is important to highlight that *E. coli* O157:H7 is often presented as bacteria case study, reported in those microfluidic platforms due the huge commercial availability of antibodies, however they are not cause–related to symptomatic UTIs. In our recent study, Alves and Reis [6] we have reported smartphone detection of *E. coli* K12 in synthetic urine in less than 21 min without any sample treatment. This is one of the two case studies further detailed in Section 14.9.Table 14.2 Summary of microfluidic assay previously reported for pathogen detection in literature. Reported techniques are compared regarding target bacteria, method of detection, sensing platform, limit of detection (LoD), pretreatment, assay time and the type of sample.

14.6.2 Materials for microfluidic platforms

The very first diagnostic microfluidic devices were fabricated by silicon and glass, although lately microfluidic pathogen sensing devices are based on polymeric materials, such as poly-(methylmethacrylate) (PMMA), polycarbonate (PC) and poly(dimethylsiloxane) (PDMS). Polymers like PDMS are known as rubber –like elastomers offering optical transparency, potential for down-scaling, disposability, they are a cheaper material compared to silicon or glass. [51,71,114] Besides polymers offer the potential of easy manufacturing, reconfiguration, microfabrication and injection molding. PDMS was firstly introduced by George Whiteside's group [102] in optical microfluidic devices fabrication, being widely applied in several microfluidic fields. [115] On the other hand PDMS properties present some limitations regarding not compatibility with high temperatures (PDMS degradation above 200 °C) [116], challenging integrations with eletrodes on its surface, possibility to react and absorb hydrophobic molecules and tendency to adsorb proteins on their surface. [117]

Fluoropolymers are a good alternative to the conventional PDMS, glass or silicon devices. It present unique optical and dielectrical properties, flexibility and chemical inertness. [82,118] Initially, fluoropolymer films were exploited for the simple

Table 14.2 Summary of microfluidic assay previously reported for pathogen detection in literature. Reported techniques are compared regarding target bacteria, method of detection, sensing platform, limit of detection (LoD), pretreatment, assay time and the type of sample.

Reference	Pathogen	Detection method	LoD - E. coli	Pre-treatment	Assay time	Sample
Liao et al. 2006 [106]	E. coli, P. mirabilis, K.pneumoniae, E.aerogenes, Pseudomonas sp., Enterococcus sp.	Micro-fabricated electrode array	2.6×10^8 CFU/ml	Lysis	45 min	Urine
Boehm et al. 2007 [107]	E. coli (BL21(DE3))	Impedance-based microfluidic biosensor	10^4 CFU/ml	–	–	synthetic
Lackza et al. 2011 [108]	E. coli and Salmonella sp.	EIS microelectrode Capacitive Immunosensor	10^4–10^5 cells/ml	–	1 h	PBS
Bercovici et al. 2011 [109]	E. coli	Microfluidic isotachophoresis with FRET	10^6 CFU/ml	Centrifugation, lysis and dilution	15 min	Urine
Yang et al. 2011 [106]	E. coli	Microfluidic cell impedance assay	3.4×10^4 CFU/ml	–	100 min	synthetic urine
Sanvicens N. 2011 [110]	E. coli O157:H7	fluorescent quantum dot-based antibody array	10 CFU/ml	–	2 h	PBS
Safavieh et al. 2012 [20]	E. coli	Microfluidic LAMP with electrochemical detection	48 CFU/ml	Filtration	60 min	filtered urine
Zhu et al. 2012 [48]	E. coli O157:H7	Quantum dot immunoassay	5–10 CFU/ml	–	> 1.5 h	2 percent gelatine–PBS, food matrix
Wang et al. 2012 [105]	E. coli	Microfluidic fluorescence assay	50 CFU/ml	Filtering	30 min	PBS, blood and food

(continued)

Table 14.2 (Cont'd)

Reference	Pathogen	Detection method	LoD - E. coli	Pre-treatment	Assay time	Sample
David et al. 2013 [45]	E. coli O157:H7	Ac electrical impedance lab on a chip	10^2 cells/ml	–	–	HBS buffer
Yoo et al. 2014 [19]	E. coli	Microfluidic fluorescence assay	10^3 CFU/ml	–	30 min + 45 min staining	PBS
Golberg et al. 2014 [95]	E. coli	Microfluidic fluorescence assay	150 CFU/ml	Filtering, incubation on-chip enrichment	8 h	water with feces
Stratz et al. 2014 [96]	E. coli O157:H7	Immunoassay- based analysis on-Chip Enzyme Quantification	Single E. coli / β-glucosidase	–	3 h	culture medium
Safavieh M. 2014 [94]	E. coli	Ribbon cassette LAMP with colorimetry	30 CFU/ml	–	<1 h	water
Rajendran V. 2014 [111]	Salmonella spp. and E. coli O157	Smartphone based bacterial detection using biofunctionalized fluorescent nanoparticles	< 10^5 CFU/mL	–	> 10 min	PBS
Cho et al. 2015 [16]	E. coli and N. gonorrhoeae	Smartphone using microfluidic paper analytical device (µPAD) was	10 CFU/ml	1 percent Tween 80	> 5 min	urine
Shih et al. 2015 [49]	E. coli DH-5α	Paper based ELISA	10^4–10^5 cells/ml	–	5 h	culture medium
Chang et al. 2015 [50]	E. coli O157:H7	Microfluidic chip embedded with AMPs	10 cells/ml	–	20 min	PBS

Reference	Target	Method	Detection limit	Sample preparation	Time	Sample matrix
Angus et al. 2015 [112]	E. coli K12	surface-heated droplet PCR	10^3 genome copies per sample	—	19 min	buffer
Chang et al. 2015 [50]	E. coli, S. aureus, P. syringae, Enterococcus sp., Staphylococci sp.	Colorimetric/ PCR on a chip	10^2 CFU/ml	Automated washing	30 min + 40 min PCR	Human fluid
Kokkinis et al. 2016 [11]	E. coli K12	Bitinylated antibodies functionalized with magnetic microparticles	Positive/ negative signal	Centrifugal separation	≥ 90 min	PBS-Tween 20 (0.01 percent v/v)
Olanrewaju A. et al. 2017 [15]	E. coli O157:H7	microfluidic capillaric circuit (CC) with antibody-functionalized microbeads	1.2×10^2 CFU/ml	—	<7 min	synthetic urine
Alves and Reis, 2019 [6]	E. coli K12	Microfluidic ELISA detection with optical smartphone interrogation	240 CFU/ml	None	<21 min	synthetic urine
Sultan Ilayda et al. 2020 [113]	E. coli ATCC 25,922	Label-free smartphone quantitation of bacteria by darkfield imaging of light scattering in fluoropolymer micro capillary film allows portable detection of bacteriophage lysis	10^4 CFU per microdevice	Pre-enrichment	> 2.5 h	LB medium

production of valves and pumps in glass microfluidic devices, similarly to PDMS counterparts. [119] The microengineered MCF material (detailed in Section 14.6.3) is a novel microfluidic platform which exploit the low cost manufacturing process of fluoropolymer FEP-Teflon® and benefit of potential properties such as optical transparency, resistance to weathering, higher electric conductivity, low friction and non-stick characterizes, and withstand at extreme temperatures (FEP degradation temperature above 260–300 °C). [120]

14.6.3 Fluorinated microcapillary film (MCF)

Immunoassays can be conducted using a MCF which is a novel and cheap microengineered material made from fluorinated ethylene propylene (FEP Teflon®) by melt extrusion process. [121] The refractive index of fluoropolymers (similar to that of water) makes it an ideal microfluidic substrate for optical colorimetric or fluorescence smartphone detection, for favoring excellent signal-to-noise ratio [69,74,121]. Fluorinated ethylene propylene is a copolymer of hexafluoropropylene and tetrafluoroethylene which contains strong carbon-fluorinated bonds. [118] The MCF was first presented as a cost-effective microfluidic immunoassay platform by Edwards et al. [121] It consists in a flat plastic ribbon containing 10 embedded capillaries (Fig. 14.9) with mean internal diameter 206 ± 12.2 μm. [121] Fluoropolymer MCF is exclusively manufactured by Lamina Dielectrics Ltd (Billingshurst, West Sussex, UK). The external dimensions of the ribbon are 4.5 ± 0.10 mm wide and 0.6 ± 0.05 mm thick.

The flat film geometry allows simple yet effective immobilization of antibodies for immunoassays by passive adsorption on the plastic surface of the microcapillaries due their hydrophobic surface. [121] The flat geometry is also responsible for the exceptional optical transparency providing a short path length through the wall with no curvature to refract the light. Similarly, the MCF has a refractive index of 1.338, very close to water (1.333). [121] This produces minimal optical refraction at the water-substrate interface allowing simple optical detection of colorimetric substrates, resulting in high signal-to-noise ratio (SNR) which is fundamental for sensitive signal quantitation. The MCF platform has demonstrated to be capable of performing up to ten different parallel microfluidic assays when dipped into a single sample whilst providing relevant optical information. It is possible to perform multiplex immunoassays and collect different reagents in each capillary, individual capillaries can be fitted with a fine needle and syringe. [76] It can provide flow, presents flexible signal detection able to be detected by cost effective equipment as smartphone, flatbed scanner or camera; is portable and therefore suitable for point-of-care diagnostics.

Moreover, its surface-area-to-volume ratio allows the limit of detection to increase in comparison with MTP. [74,75,122] A clear example is the lab-in-briefcase concept for prostate cancer biomarker detection reported by Barbosa et al. [74] showing a major capability from this material demonstrating 80 measurements at same time in less than 15 min due to the significantly reduced incubation times.

The MCF provides affordability and portability which would be ideally suited to low resource health settings as measurements can take place outside the laboratory. On the other side, the material enables to work with high concentrations of antibodies

Point-of-need detection with smartphone 341

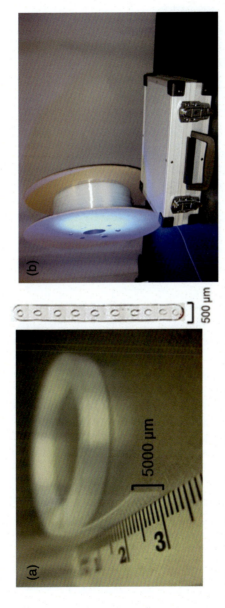

Fig. 14.9 Overview of the fluoropolymer Microcapillary Film. (a) Microphotograph of 5 m long MCF reel and cross section showing the ten ~200 μm channels embedded in the FEP polymer. (b) Overview of a 1 km long reel next to the "Lab in a brifcase" presented by Barbosa and co-authors [74] (Acknowledgements to Dr Ana I Barbosa for providing this picture with her permission).

due to high surface area to volume ratio (approximately 200 cm^{-1}) in comparison to approximately 15 cm^{-1} a 96 well ELISA MTP well, given by Eq. (14.8):

$$\frac{SA}{V} = \frac{4}{D_h} \qquad (14.8)$$

where D_h is the equivalent hydraulic diameter of each microcapillary, assuming each strip is 30 mm long and MTP well with dimensions described in Fig. 14.6.

The flow in microfluidic devices is generally laminar [123] and therefore characterized by low Reynolds number, Re described by Eq. (14.9), meaning the viscous forces are prevalent compared to inertial forces.

$$Re = \frac{\rho.v.D_h}{\mu} \qquad (14.9)$$

where ρ (kg/m^3) is the fluid density, v (m/s) is the velocity of the fluid, μ (Ns/m^2) is the fluid viscosity, and D_h(m) is the hydraulic diameter of the channel. As reference, flow in a pipe is laminar for $Re \langle 2100$, transient flow when $2100 < Re < 4000$ and turbulent when $Re \rangle 4000$. [124,125]The higher hydraulic diameter, which in a circular capillary is equal to its diameter, the lower is the Reynolds number at the same flow rate. The volumetric flow rate will also affect the flow regime in a pipe, as according to Eq. (14.10) affecting superficial flow velocity, where Q (m^3/s) is the volumetric flow rate, u (m/s) is the flow velocity, and A (m^2) is the cross sectional area.

$$Q = u.A \qquad (14.10)$$

Laminar flow presents a typical velocity profile in tubes, showing maximum velocity in the center and zero velocities at the walls of the tube (Fig. 14.10). In laminar flow, molecules move parallel to each other (no-slip wall boundary) and no mixing occurs. Another consideration to take into account in microfluidic immunoassays is the molecular orientation of antibodies and their concentration on the surface area. Buijs et al. [126] suggested a relationship between the adsorbed amount and the molecular orientation on the surface, based on the dimensions of antibody molecules.

Thus 200 ng/cm^2 would represent a monolayer with antibodies in a "flat-on" orientation, 260 ng/cm^2 in an "end on" orientation with Fab fragments in line, and 550 ng/cm^2 also in an "end-on" orientation with Fab fragments close together and parallel.

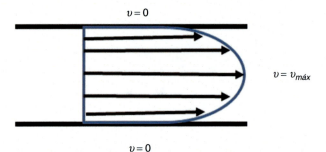

Fig. 14.10 Velocity profile of a fluid in laminar flow regime with no-slip wall boundary.

According to Barbosa et al. [127], the antibody adsorption to FEP-Teflon was found to be similar to protein adsorption onto hydrophobic surfaces and other fluorinated surfaces.

A study characterizing antibody adsorption onto FEP-Teflon microcapillaries showed a maximum surface density of ~400 ng/cm^2 at 40 µg/ml of IgG solution, demonstrating maximum binding capacity happened above half monolayer. [127] Therefore MCFs adsorption studies agreed with previous studies performed on other surfaces by Buijs et al. [126] demonstrating reliability for immunoassay development in point-of-care diagnostic. [127]

An E. coli cell has a diameter of about ≈ 1 µm, a length of ≈ 2 µm, and a volume of ≈1.3 µm^3 $\left(A = \pi r^2 \cdot \left(L - \frac{2}{3} \cdot r^2\right)\right)$, considering a capsular shape [128]. Taking into account the internal diameter is around 200 µm, in a 30 mm strip length, each capillary has a volume of ≈94,200 µm^3 ($V = L*\pi*r^2$). This means a MCF strip can theoretically capture 72,461 bacteria per strip however a fundamental study by Alves and Reis [129] showed immunocapture of bacteria cells is not 100 percent efficient as expected, with flow and bacteria cells settling playing a major role.

High analytical efficiency is intrinsically related with the size of microfluidics platform. The Scatchard's model [76] (shown in Eq. (14.2)) shows that higher concentrations of antibody favor the formation of antibody-antigen complex in an antibody-antigen binding reaction, so higher numbers of immobilized antibodies will capture more analytes (antigens) in the microfluidic system. Although SVR is important for diagnostic tests, the total antibody immobilization surface area is relevant for sensitivity, cost and portability of point-of-need bioassay. [53,80] Furthermore, microfluidic devices can be affected for different conditions such as the surface tension of the samples or biological fluids as blood or urine, the reduced volume on low concentration samples and the antibody-antigen kinetics.

14.7 Positioning of smartphone technology in point-of-need testing

In miniaturized systems, the detection and quantitation of low amount of analyte is fundamental for achieving sensitive (bio)assays. The detection mode of choice can condition the sensitivity but also the portability and cost of the point-of-need test. The choice of a smartphone for optical or electrochemical interrogation of the test helps solving many of the issues that has limited moe significant developments in point-of-need testing. For example, the current big demand in miniaturized immunoassays aims to implement off-the-shelf technology, reducing usage of bulky equipment such as MTP reader which is, totally dependent on centralized labs. Moreover low resource settings need urgently sensitive, power free and inexpensive microfluidic POC test for healthcare. [23,130,131]

In heterogeneous immunoassays there is a trend for use optical detection systems to detect colorimetric, fluorescence or chemiluminescent signals. Fluorescence still the most widely applied since ensures more sensitive and specific assays [132]. Some examples listed in Table 14.2 are the works by Kokkinis et al. [11], Zhu et al. [48], Golberg et al. [95], Chang et al. [50], Wang et al. [105] and Yoo et al. [19].

Optical techniques present beneficial characteristics over electrochemical detection by their high sensitivity, high quantitative throughput, adaptability with benchtop protocols and multiplexed detection of several targets detection in a single sample. [133,134] Cost effective readout systems such as smartphones, scanners and cameras present potential to deliver the requirements of an ideal point-of-need test, addressing portability and being power free enabling fluorescence readings by using additional materials as torch or LED to excite or detect fluorescence (only requires battery power but are not plugged in to electricity). [134]

A smartphone can be considered is a miniaturized device with internal memory, high quality camera lens, that can be easily connected to wifi and well spread worldwide offering similar and cheaper potential than analytical devices such as microscope or spectrophotometer. [135,136] Moreover, a smartphone is widspread among all ages, with more than 40 000 mobile health applications nowadays, representing an opportunity to increase acessibility to mobile healthcare diagnosis. [133,137,138] Indeed, the treatment of cancers, infectious or chronic diseases could be improved tremendously by the use of routine tests in impoverished areas. [139] In this way, smartphone ($\approx$£800/unit) or tablets ($\approx$£100/unit) would serve as a point-of-need diagnostic tool for the rapid detection of protein biomarkers, pathogens or environmental analytes, reducing total costs of healthcare systems, providing portability and affordability. [134,137,138] Note, for example smartphone would be a portable device for analysis of several point-of-care tests ($\approx$1000), being easily transported between rural areas. An even more affordable option for fluorescent detection would be the implementation of a raspeberry Pi ($\approx$£20/unit).

Prior to achieving sensitivity and throughput optical conditions, limited optical aberrations and incongruences from differential ambient light and transmitted light capture are still key points to evaluate. [3] Due to increasing needs in point-of-need testing, several prognostic approaches have been developed using smartphone technology such as blood test detecting glucose, haemoglobin levels, protein, virus, bacteria and drugs. [139–141] Zhu et al. [48] developed a smartphone-based detection for visualizing bacteria (Fig. 14.11A). The systems consisted in a cost-effective bacterial detection platform with anti-*E. coli* O157:H7 antibodies immobilized on the interior surface of a capillary tube and attached to a cell phone. Proof-of-concept have demonstrated a LoD of 5–10 CFU/mL, in less than 2 h

Cho and team [16] have reported a smartphone based μPAD conjugated with fluorescent antibody-conjugated particles to quantify *E. coli* from urine. They present a very rapid system with a LoD of 10 CFU/ml in urine of *E. coli* however human urine samples spiked with *E. coli* or *N. gonorrhoeae* were incubated for 5 min with 1 percent Tween 80. Park et al. [131] have developed a smartphone-based detection of *Salmonella* on paper microfluidics pre-loaded with antibody anti-*Salmonella Typhimurium* conjugated latex microparticles and dried out subsequently (Fig. 14.11B). Rajendran V. and colleagues [111] have developed a smartphone based bacterial detection using biofunctionalized fluorescent nanoparticles to detect

Point-of-need detection with smartphone 345

Salmonella sp. and *E. coli* O157:H7. Despite being multiplexed, this system presents a LoD of 10^5 CFU/ml of *E. coli*. One of the earliest works reporting microscopic smartphone fluorescence capabilities for imaging individual cells like bacteria or microparticles was the work of Wei et al. [143] (Fig. 14.11C). Shen et al. [136] have developed a point-of-care colorimetric detection approach with a smartphone to measure pH variation in urine. Coskun and team [142] have reported an albumin tester, running on a smart-phone that image and automatically analyses fluorescent assays confined within disposable test tubes for sensitive and specific detection of albumin in urine. Although very impressive LoD, all these assays still present big assay time (>2 h), aim to detect other *E. coli* strains not responsible for UTIs or demand additional steps of sample preparation to be conducted in synthetic or real samples. In an attempt to explore cost effective readout systems aiming portability for low resource settings, a rapid and sensitive bacterial detection assay was integrated by our research team, which is further detailed as case study 2 in section 9.

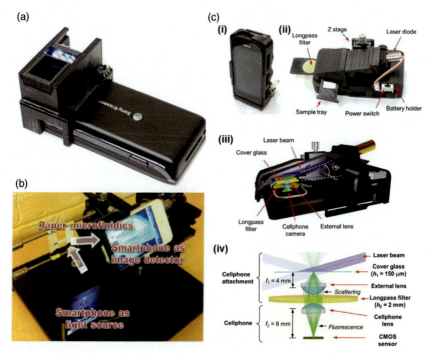

Fig. 14.11 Microfluidic platforms for bacterial detection using smartphone as readout system. (a) Quantum dot enabled detection of *E. coli* using a cell-phone adopted from Zhu et al. [48] (b) Smartphone quantifies *Salmonella* from paper microfluidics adopted from Park et al. [131] (c) Cell-phone-based fluorescence microscope by Wei et al. [143], one of the earliest works with capabilities of imaging individual bacteria cells, herein demonstrated with fluorescence imaging of 1 μm diameter green fluorescent beads. Reprints (adapted with permission from Royal Society of Chemistry and American Chemical Society).

14.8 Camera requirements for smartphone diagnostics and digital imaging of microfluidic bioassays

This section summarizes some of the learning's in respect to the importance and choice of optics and camera settings for the performance of a point-of-need microfluidic ELISA, mostly involving clinically relevant cardiac and cancer protein biomarkers.

Unsurprisingly for those familiar with bioassay development and optical imaging, the type of camera and hardware performance have a significant influence over bioassay quantitation. Technical development driven by mass consumer markets for digital cameras has led to a wide range of imaging sensors being widely available, and a corresponding wide range of digital cameras both standalone and embedded within mobile phones. There is an inevitable trade-off between cost, performance and ease-of-use. Some examples of different types of digital cameras that are widely available are summarized in Table 14.3, ranging from smartphone cameras- perhaps the most widely available – to the stripped down optoelectronic components found within smartphones that are now available either as an industrial component, or through suppliers making these components available to engineers and hobby electronics markets.

The minimum requirement for bioassay quantitation using a digital camera – including smartphone imaging – is that the device can be clearly imaged in focus, and with an appropriate range of intensities that reflect the assay signal. This is especially important when imaging an array of parallel microcapillaries, as any greyscale pixel blurring will directly reduce the signal-to-noise ratio for each individual microcapillary. For colorimetric images, a loss in intensity corresponding to the dye absorbance must be quantified, and for fluorescence, an increase in intensity proportional to the concentration of target (Fig. 14.12A). Surprisingly simple cameras- including the cheapest consumer devices such as toy cameras intended for children- can take images of microfluidic devices with signal proportional to assay concentration, however if the individual microchannels cannot be resolved, it is not possible to quantify individual data points (Fig. 14.12B). The addition of a close focus lens allows a far smaller field of view, and correspondingly higher resolution image of the device, permitting clear and quantifiable signal measurement. The cost of this additional close-focus lens is a much smaller field of view. A higher performance camera- for example a mid-range smartphone costing ~10x more – is capable of high resolution images of much larger fields of view, so that many more microdevices can be recorded simultaneously (Fig. 14.12C). When a range of cameras from digital SLR with the highest resolution sensor and optics, through different smartphones, to the cheapest toy camera are compared, the reduction in image quality is obvious, affecting signal quantitation in individual microdevices. It is important therefore to ensure that the digital camera selected has adequate field of view and resolution to quantify the target effectively.

On separate experiments we have quantified the role of exposure time and numerical aperture on the analytical sensitive and signal-to-noise for imaging a converted fluorescence substrate. Several MCF strips were loaded with fluorescein at varying concentrations and imaged at the combination of settings shown in Fig. 14.12 and Table 14.4. Though 'intuition' would perhaps recommend increasing exposure time

Table 14.3 Comparison of digital imaging detection hardware that can be used for colorimetric or fluorescence detection of optical microfluidic biosensing strips.

Product type	Consumer: mobile phone camera			Consumer: digital camera		Industrial camera/ machine vision	Singleboard camera component
	Smartphone Flagship	Smartphone Mid-range	Smartphone Budget	Feature phone	Compact	SLR/ mirrorless		
Class								
Price range	£500->1000	£150-300	£50-150	£25-50	£20-200	£250->>2000	>£150	£22
Example	iPhone 10 Samsung Galaxy S	Moto G5 iPhone 6S					IDS microeye range	Raspberry Pi camera v2
Typical sensor size (width)	>5 m	3-5 mm	3-5 mm	>3 mm	5-10 mm	>10 mm	>3 mm	3.7 mm
Typical sensor resolution (MPixels)	>14	6-13	<=5	1.4-5	5-16	>14	> = 0.8	8
Typical focal length (35mm equivalent)	<25 - >40	25-40	~30		20-150 mm	Wide range	Wide range	~30 mm
Typical aperture	<F1.5	~F2.5	>F2.8		>F2.0	<F1.2	Depends on lens	F2.0

348 Smartphone-Based Detection Devices

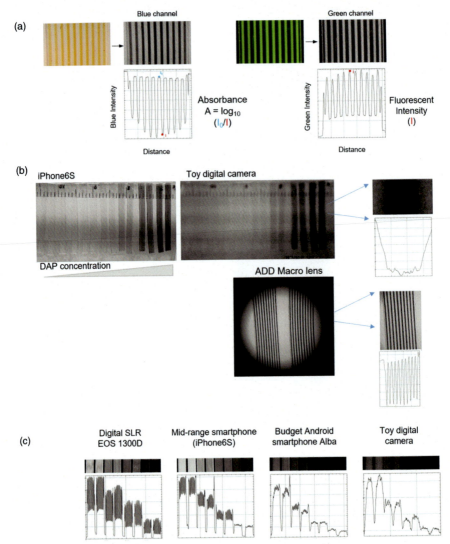

Fig. 14.12 Camera resolution and field of view affect the number of microfluidic devices that can be recorded. (a) Absorbance imaging of colorimetric test strips (left hand side, with converted OPD) and fluorescence strips (right hand side, with converted AP substrate). (b) Dilution series of DAP (i.e. final product of OPD enzymatic conversion) imaged with iPhone6s vs a toy digital camera, using a macrolens. (c) Greyscale images and profile plot for a dilution series of DAP with strips imaged with a range of cameras. Note the direct link between quality of the sensor and the ability to resolve individual microcapillaries.

Table 14.4 Influence of camera settings on analytical sensitivity and dynamic range of fluorescence measurement in microfluidic devices.

Aperture:	F4.0	F5.6	F8.0	F4.0	F5.6	F8.0
Exposure Time		LoD (nM)			Optimal Measurement Range	
30″	Saturated	10	1	—	1–500 nM	1–500 nM
15″	0.1	0.01	10	1–500 nM	1–500 nM	1 nM–2 uM
8″	0.001	1	1	1–500 nM	1 nM–2 uM	1 nM–2 uM
4″	0.5	1	20	1 nM–2 uM	1 nM–2 uM	1 nM–12 uM
2″	0.1	15	5	1 nM–2 uM	1 nM–12 uM	1 nM–12 uM
1″	10	1	50	1 nM–12 uM	1 nM–12 uM	20 nM–60 uM
1/2″	2	20	250	1 nM–12 uM	20 nM–60 uM	0.1–60 uM
1/4″	50	250	600	20 nM–60 uM	20 nM–60 uM	0.5–300 uM
1/8″	300	900	2000	0.1–300 uM	0.1–300 uM	0.5 uM–1.5 mM
1/15″	1000	2000	2000	0.5 uM–1.5 mM	0.5 uM–1.5 mM	0.5 uM–1.5 mM
1/30″	1000	1000	5000	0.5 uM–1.5 mM	0.5 uM–1.5 mM	2 uM–1.5 mM
1/60″	900	6000	15,000	0.5 uM–1.5 mM	2 uM–1.5 mM	12 uM–1.5 mM
1/125″	3000	15,000	25,000	2 uM–1.5 mM	12 uM–1.5 mM	12 uM–1.5 mM
1/250″	7000	20,000	80,000	12 uM–1.5 mM	12 uM–1.5 mM	60 uM–1.5 mM

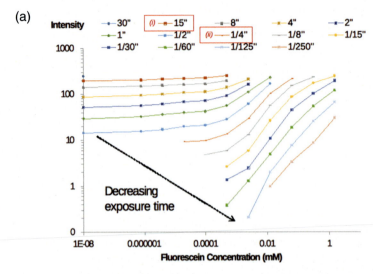

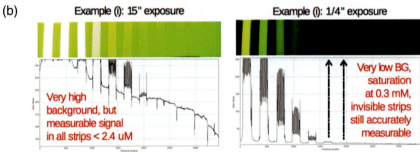

Fig. 14.13 Camera settings affect sensitivity and dynamic range of bioassay measurements. A Effect of exposure time on fluorescence intensity signal for a range of fluorescein concentrations. Note only a narrow range offers a good dynamic range and good signal-to-noise ratio. (b) Two contrasting scenarios as flagged in (a) showing a long (15s) exposure vs short (¼ s) exposure. Whereas longer exposure enables higher signal intensity at lower concentrations, the background signal is also amplified, yielding a poor signal-to-noise ratio.

so that smaller concentrations can be detected as it happens with chemiluminescent substrate imaged with a deep-cooled CCD camera, smaller exposure times in the order of less than a second actually yielded maximum working range and best signal-to-noise ratio (Fig. 14.13A). This is further supported by the two contrasting scenarios shown in Fig. 14.13B. Naturally, the limit of detection (determined as the signal intensity of the blank plus 3 standard deviations) and the measurement range both were highly dependent on both exposure time and numerical aperture of the camera as summarized in Table 14.4. A point-of-need test needs to suit a range of concentrations of analyte ideally without requiring dilution of the sample, at the same time it needs to provide a limit of detection and quantitation within the clinically range. In that respect, the results herein summarized suggest the clinical performance of a point-of-care test is highly dependent not only on the molecular and biosensing aspects of the test but also on the optical imaging setup.

14.9 Example of point-of-need smartphone tests developed by our research team

This section provides an overview of smartphone detection approaches recently developed within our research team, with focus on UTIs and protein biomarker quantitation.

14.9.1 Case study 1: smartphone detection of PSA

Back in 2016, our research team has reported rapid smartphone ELISA quantitation of PSA (an established biomarker for prostate cancer) using fluoropolymer MCF strips, the main results were summarized in Fig. 14.14. We compared colorimetric against fluorescence detection using chromogenic and fluorescence substrate and enzymes, respectively, and in both cases the limit of detection determined from the full response curve was below the clinical threshold used for biopsy (typically 1–4 ng/ml). One of the key steps that enabled reduced variability was the ability of normalizing the Absorbance or fluorescence signal by imaging a reference strip and the test strip, with the macrolens, in the same image. The reference strip contained the product of the enzymatic conversion (see capillaries on the right hand side of RGB and greyscale images). As camera settings such as exposure time are usually automatically set for most smartphones, variances in signal intensity can be eliminated by plotting the response curve in terms of absorbance or fluorescence signal ratios of the test strip to the reference strip.

Though the biggest contribution in terms of speed and performance of the rapid microfluidic ELISA relied on the nature of the Microcapillary Film strips (optical transparency, large surface area to volume and short diffusion distance), smartphone imaging enabled this high-performance ELISA testing to be done without the need of any power source. Someone with some level of training in a small diagnostics/pathology lab or on the field would be able to carry out this high-performance ELISA test without access to centralized and sophisticated lab equipment. This realized our vision of a portable, power-free ELISA for global point-of-need use without the compromises of existing lateral flow testing.

14.9.2 Case study 2: smartphone detection of UTIs

One of the last development within our research team was a miniaturized smartphone test for timely and sensitive bacterial identification and quantitation. This was mostly driven by the urgency to overcome the rapid diagnostic demands to tackle *E.coli* infections and spread of resistance strains. Microfluidic test strips were mass manufactured from a plastic MCF fabricated from Teflon® FEP, containing 10 embedded capillaries each having 200 μm i.d. Our research group has previously demonstrated the ability of running parallel bioassays (protein biomarkers) simultaneous (see case study 1 on Section 14.9.1), or using capillary action for one-step bacteria testing [78,82].

The excellent optical transparency of fluoropolymer MCF material is ideal for naked eye detection or measurement with portable, inexpensive optoelectronic equipment including a smartphone camera. We used 40 mm long, MCF strips coated with

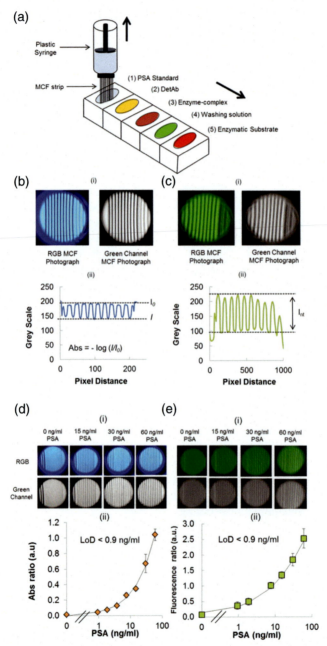

Fig. 14.14 Overview of colorimetric and fluorescence smartphone quantitation of PSA cancer biomarker using MCF strips. (**a**) Illustration of multi-step ELISA used for quantitation of PSA with MCF strips. (**b**) and (**c**) Imaging and image analysis of colorimetric and fluorescence test strips. (**d**) and (**e**) Full response curves for colorimetric and fluorescence ELISA smartphone quantitation of PSA. Adapted from Barbosa et al. [75] (under permission of Elsevier).

a polyclonal antibody against *E.coli*. MCF strips were optically interrogated using a smartphone as shown in Fig. 14.15. Sample and ELISA reagents (biotinylated antibody, Strep-AP and AttoPhos) where aspirated with 1ml syringes.

Our study reported for the first time a smartphone-based optical fluorescence microfluidic immunoassay capable of detecting and quantifying *E.coli* in <25 min in the clinical range relevant for UTIs (10^3–10^5 CFUs/ml), as shown by the full response curve in synthetic urine spiked with *E.coli* from $10°$ to 10^8 CFU/ml (Fig. 14.15D). This was accomplished without any sample preparation step. The assay showed good reproducibility with a recovery in the range of 80–120 percent, and intra-and inter-assay precisions below 20 percent, therefore commensurate with a high-performance sandwich ELISA. The response curves showed a limit of detection of <240 CFU/ml for the 10-bore microfluidic strips, equivalent to just 24 CFUs per capillary. Further details about assay development and optimization can be found at Alves and Reis, [6] The smartphone immunoassay detection of E.coli demonstrated similar performance in terms of recovery, signal-to-noise and inter- and intra-assay CV values to a high-performance, lab-based biomarker ELISA, which opens up the opportunity to develop rapid biosensing strategies for bacteria identification and quantitation without relying on bacteria growth, in contrast to microbiological techniques.

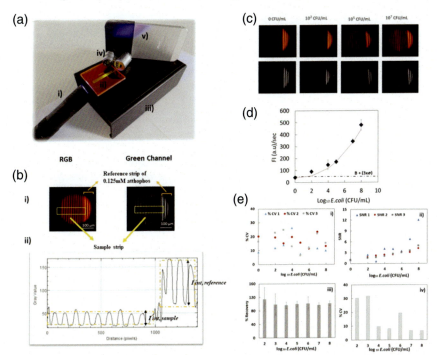

Fig. 14.15 (**a**) Smartphone components used for *E. coli* quantitative fluorescent immunoassay and (**b**) signal image analysis. (**c**) Smartphone fluorescence images of *E. coli* in synthetic urine with (**d**) a full response curve and respective (**e**) Inter-variability and intra-assay variability showing performance compatible with a high-precision MTP ELISA. Adapted from Alves and Reis [6] (under permission of Elsevier).

14.10 Conclusions

Smartphones are incredible digital vehicles for delivering high precision analytical detection at the point-of-need, enabling cheap and portable advanced hardware capabilities (such as optical interrogation with the camera) but also remote software interface, including remote server connection and embedded image analysis and signal processing. Although current smartphones are not designed to handle fluids or biological samples, the development of smartphones coincides with the timing of booming development of disruptive analytical microfluidic technologies. The many successful stories reported in literature and some of those overviewed in this chapter suggest smartphone-microfluidics interface is capable of transforming point-of-need testing in years to come.

References

[1] C.D. Chin, V. Linder, S.K. Sia, Commercialization of microfluidic point-of-care diagnostic devices, Lab Chip 12 (2012) 2118–2134, doi:10.1039/c2lc21204h.
[2] C.D. Chin, V. Linder, S.K. Sia, Lab-on-a-chip devices for global health: past studies and future opportunities, Lab Chip 7 (2007) 41–57, doi:10.1039/b611455e.
[3] S. Nayak, N.R. Blumenfeld, T. Laksanasopin, S.K. Sia, Point-of-care diagnostics: recent developments in a connected age, Anal. Chem. 89 (2017) 102–123, doi:10.1021/acs.analchem.6b04630.
[4] C.P. Price, Regular review: point of care testing, BMJ 322 (2001) 1285–1288, doi:10.1136/bmj.322.7297.1285.
[5] A.I. Barbosa, A.P. Castanheira, N.M. Reis, Sensitive optical detection of clinically relevant biomarkers in affordable microfluidic devices: overcoming substrate diffusion limitations, Sens. Actuators, B Chem 258 (2018) 313–320, doi:10.1016/j.snb.2017.11.086.
[6] I.P. Alves, N.M. Reis, Microfluidic smartphone quantitation of Escherichia coli in synthetic urine, Biosens. Bioelectron. 145 (2019) 111624, doi:10.1016/j.bios.2019.111624.
[7] European Comission, One health action plan against antimicrobial resistance, (2017).
[8] J. O'Neill, Review on antimicrobial resistance. tackling a global health crisis: rapid diagnostics : stopping unnecessary use of antibiotics, Indep. Rev. AMR. (2015) 1–36.
[9] J.A. Ayukekbong, M. Ntemgwa, A.N. Atabe, The threat of antimicrobial resistance in developing countries: causes and control strategies, Antimicrob. Resist. Infect. Control. 6 (2017) 1–8, doi:10.1186/s13756-017-0208-x.
[10] J. O'Neill, Tackling drug-resistant infections globally: final report and recommendations, (2016).
[11] G. Kokkinis, B. Plochberger, S. Cardoso, F. Keplinger, I. Giouroudi, A microfluidic, dual-purpose sensor for in vitro detection of Enterobacteriaceae and biotinylated antibodies, Lab Chip 16 (2016) 1261–1271, doi:10.1039/c6lc00008h.
[12] A.M. Foudeh, T. Fatanat Didar, T. Veres, M. Tabrizian, Microfluidic designs and techniques using lab-on-a-chip devices for pathogen detection for point-of-care diagnostics, Lab Chip 12 (2012) 3249–3266, doi:10.1039/c2lc40630f.
[13] E., P., Center for disease Dynamics, the State of the World 's Antibiotics, State World 's Antibiot, 2015.

[14] D. Mabey, R.W. Peeling, A. Ustianowski, M.D. Perkins, Diagnostics for the developing world, Nat. Rev. Microbiol. 2 (2004) 231.

[15] A.O. Olanrewaju, A. Ng, P. Decorwin-Martin, A. Robillard, D. Juncker, Microfluidic capillaric circuit for rapid and facile bacteria detection, Anal. Chem. 89 (2017) 6846–6853.

[16] S. Cho, T.S. Park, T.G. Nahapetian, J.Y. Yoon, Smartphone-based, sensitive µPAD detection of urinary tract infection and gonorrhea, Biosens. Bioelectron. 74 (2015) 601–611, doi:10.1016/j.bios.2015.07.014.

[17] M. Zourob, S. Elwary, A. Turner, Principles of bacterial detection, 2008. doi:10.1093/pcp/pcw027.

[18] WCS JWA Findlay, J.W .L M.N.K., G.D. Nordblom, I. Das, B.S. DeSilva, R.R. Bowsher, Validation of immunoassays for bioanalysis: a pharmaceutical industry perspective, J. Pharm. Biomed. Anal. 21 (2000) 1249–1273, doi:10.1016/S0731-7085(99)00244-7.

[19] J.H. Yoo, D.H. Woo, M.S. Chun, M.S. Chang, Microfluidic based biosensing for Escherichia coli detection by embedding antimicrobial peptide-labeled beads, Sensors Actuators, B Chem 191 (2014) 211–218, doi:10.1016/j.snb.2013.09.105.

[20] M. Safavieh, M.U. Ahmed, M. Tolba, M. Zourob, Microfluidic electrochemical assay for rapid detection and quantification of Escherichia coli, Biosens. Bioelectron. 31 (2012) 523–528, doi:10.1016/j.bios.2011.11.032.

[21] J.W.F. Law, N.S.A. Mutalib, K.G. Chan, L.H. Lee, Rapid metho ds for the detection of foodborne bacterial pathogens: principles, applications, advantages and limitations, Front. Microbiol. 5 (2014) 1–19, doi:10.3389/fmicb.2014.00770.

[22] W. Su, X. Gao, L. Jiang, J. Qin, Microfluidic platform towards point-of-care diagnostics in infectious diseases, J. Chromatogr. A. 1377 (2015) 13–26, doi:10.1016/j.chroma.2014.12.041.

[23] V. Gubala, L.F. Harris, A.J. Ricco, M.X. Tan, D.E. Williams, Point of care diagnostics: status and future, Anal. Chem. 84 (2012) 487–515, doi:10.1021/ac2030199.

[24] J.T. Poolman, Escherichia coli, Int. Encycl. Public Heal. (2017) 585–593, doi:10.1016/B978-0-12-803678-5.00504-X.

[25] W.H. Ewing, Isolation and identification of escherichia coli serotypes associated with diarrheal diseases, 1963.

[26] R. Stenutz, A. Weintraub, G. Widmalm, The structures of Escherichia coli O-polysaccharide antigens, FEMS Microbiol. Rev. 30 (2006) 382–403, doi:10.1111/j.1574-6976.2006.00016.x.

[27] C. Alexander, E.T. Rietschel, Bacterial lipopolysaccharides and innate immunity, J. Endotoxin Res. 7 (2001) 167–202.

[28] E.H.-.P. Guido Schmiemann, E. Kniehl, K. Gebhardt, M.M. Matejczyk, Diagnosis of urinary tract infections, Dtsch Arztebl Int 34 (2010) 361–367, doi:10.3238/arztebl.2010.0361.

[29] M.J. Espy, J.R. Uhl, L.M. Sloan, S.P. Buckwalter, M.F. Jones, E.A. Vetter, J.D.C. Yao, N.L. Wengenack, J.E. Rosenblatt, F.R.C. Iii, T.F. Smith, Real-time PCR in clinical microbiology : applications for routine laboratory testing, 19 (2006) 165–256. doi:10.1128/CMR.19.1.165.

[30] Z. Fu, S. Rogelj, T.L. Kieft, Rapid detection of Escherichia coli O157:H7 by immunomagnetic separation and real-time PCR, Int. J. Food Microbiol. 99 (2005) 47–57, doi:10.1016/j.ijfoodmicro.2004.07.013.

[31] Ana, Flores-Mireles, J.N. Walker, M. Caparon, Urinary tract infections: epidemiology, mechanisms of infection and treatment options, Nat. Rev. Microbiol. 13 (2015) 269–284, doi:10.1038/nrmicro3432.

[32] M. Grabe, R. Bartoletti, T.E.B. Johansen, T.C.G. Associate, M. Çek, B.K.G. Associate, K.G. Naber, Guidelines on urological infections, (2015).

[33] O. Aspevall, H. Hallander, V. Gant, T. Kouri, European guidelines for urinalysis: a collaborative document produced by European clinical microbiologists and clinical chemists under ECLM in collaboration with ESCMID, Clin. Microbiol. Infect. 7 (2001) 173–178, doi:10.1046/j.1198-743X.2001.00237.x.

[34] C.A. Hammett-stabler, L.R. Webster, A Clinical Guide to urine drug testing, North Carolina, 2008.

[35] W.E. Stamm, S.R. Norrby, Urinary Tract Infections: disease Panorama and Challenges, J. Infect. Dis. 183 (2001) S1–S4, doi:10.1086/318850.

[36] D. Yach, C. Hawkes, C.L. Gould, K.J. Hofman, The Global burden of chronic diseases, JAMA 291 (2004) 2616, doi:10.1001/jama.291.21.2616.

[37] R. Beaglehole, D. Yach, Globalisation and the prevention and control of non-communicable disease: the neglected chronic diseases of adults, Lancet 362 (2003) 903–908, doi:10.1016/S0140-6736(03)14335-8.

[38] A.D. Lopez, C.D. Mathers, M. Ezzati, D.T. Jamison, C.J. Murray, Global Burden of Disease and Risk Factors, World Bank, 2006.

[39] P. Kanavos, The rising burden of cancer in the developing world, Ann. Oncol. 17 (Suppl 8) (2006) viii15–viii23, doi:10.1093/annonc/mdl983.

[40] D.S. Celermajer, C.K. Chow, E. Marijon, N.M. Anstey, K.S. Woo, Cardiovascular disease in the developing world: prevalences, patterns, and the potential of early disease detection, J. Am. Coll. Cardiol. 60 (2012) 1207–1216, doi:10.1016/j.jacc.2012.03.074.

[41] TDR, Mapping the Landscape of Diagnostics for Sexually Transmitted Infections, WHO, 2015.

[42] D.A. Giljohann, C.A. Mirkin, Drivers of biodiagnostic development, Nature 462 (2009) 461–464, doi:10.1038/nature08605.

[43] P. Yager, T. Edwards, E. Fu, K. Helton, K. Nelson, M.R. Tam, B.H. Weigl, Microfluidic diagnostic technologies for global public health, Nature 442 (2006) 412–418, doi:10.1038/nature05064.

[44] A.I. Barbosa, N.M. Reis, A critical insight into the development pipeline of microfluidic immunoassay devices for the sensitive quantitation of protein biomarkers at the point of care, Analyst 142 (2017) 858–882, doi:10.1039/c6an02445a.

[45] S. David, C. Polonschii, M. Gheorghiu, D. Bratu, A. Dobre, E. Gheorghiu, Assessment of pathogenic bacteria using periodic actuation, Lab Chip 13 (2013) 3192–3198, doi:10.1039/c3lc50411e.

[46] W. Kong, J. Xiong, H. Yue, Z. Fu, Sandwich fluorimetric method for specific detection of staphylococcus aureus based on antibiotic-affinity strategy, Anal. Chem. 87 (2015) 9864–9868, doi:10.1021/acs.analchem.5b02301.

[47] M.S. Chang, J.H. Yoo, D.H. Woo, M.S. Chun, Efficient detection of Escherichia coli O157:H7 using a reusable microfluidic chip embedded with antimicrobial peptide-labeled beads, Analyst 140 (2015) 7997–8006, doi:10.1039/c5an01307k.

[48] H. Zhu, U. Sikora, A. Ozcan, Quantum dot enabled detection of Escherichia coli using a cell-phone, Analyst 137 (2012) 2541–2544, doi:10.1039/c2an35071h.

[49] C.M. Shih, C.L. Chang, M.Y. Hsu, J.Y. Lin, C.M. Kuan, H.K. Wang, C. Te Huang, M.C. Chung, K.C. Huang, C.E. Hsu, C.Y. Wang, Y.C. Shen, C.M. Cheng, Paper-based ELISA to rapidly detect Escherichia coli, Talanta 145 (2015) 2–5, doi:10.1016/j.talanta.2015.07.051.

[50] W.H. Chang, C.H. Wang, C.L. Lin, J.J. Wu, M.S. Lee, G. Bin Lee, Rapid detection and typing of live bacteria from human joint fluid samples by utilizing an integrated microfluidic system, Biosens. Bioelectron. 66 (2015) 148–154, doi:10.1016/j.bios.2014.11.006.

[51] J. Mairhofer, K. Roppert, P. Ertl, Microfluidic systems for pathogen sensing: a review, Sensors (Switzerland) 9 (2009) 4804–4823, doi:10.3390/s90604804.

[52] D. Wild, The Immunoassay Handbook: Theory and Applications of Ligand Binding, ELISA and Related Techniques, 4th ed., Elsevier, Oxford, 2013, doi:10.1016/B978-0-08-097037-0.00001-4 D.B.T.-T.I.H. (Fourth E. Wild (Ed.).

[53] K.L. Cox, V. Devanrayan, A. Kriauciunas, J. Manetta, C. Mantrose, S. Sittampalam, Immunoassay Methods, Assay Guid. Man. 43 (2012) NBK92434 [bookaccession].

[54] E.A.P. David R. Davies, S. Sheriff, Antibody -antigen complexes, Annu. Rev. Biochem. 59 (1990) 439–473.

[55] Edited by Norman H.L., Chiu and T.K. Christopoulos, Advances in immunoassay technology, 2012. doi:10.2307/488965.

[56] S.S. Deshpande, Enzyme immunoassaysm from concept to product development exquisite specificity–The monoclonal antibody revolution introductory immunobiology stress-inducible cellular responses, 25 (1997) 1997.

[57] N.S. Lipman, L.R. Jackson, L.J. Trudel, F. Weis-Garcia, Monoclonal versus polyclonal antibodies: distinguishing characteristics, applications, and information resources, ILAR J. 46 (2005) 258–268, doi:10.1093/ilar.46.3.258.

[58] T. Klonisch, G. Panayotou, P. Edwards, A.M. Jackson, P. Berger, P.J. Delves, T. Lund, I.M. Roitt, Enhancement in antigen binding by a combination of synergy and antibody capture, Immunology 89 (1996) 165–171, doi:10.1046/j.1365-2567.1996.d01-722.x.

[59] CRC Press, in: Brian law (Ed.), Immunoassay: A Practical Guide, CRC PRESS, London, 1996.

[60] S. Hosseini, P. Vazquez, M. Rito-Palomares, S.O. Martinez-Chapa, Advantages, disadvantages and modifications of conventional ELISA, Springer Briefs Appl. Sci. Technol. (2018) 67–115, doi:10.1007/978-981-10-6766-2_5.

[61] J.R. Crowther, The ELISA guidebook, 2009. doi:10.1134/S000629790909017X.

[62] J. Tate, G. Ward, Interferences in immunoassay - Google Scholar, Clin. Biochem. Rev. 25 (2004) 105–120, doi:10.1016/B978-0-08-097037-0.00027-0.

[63] Bio-Rad Laboratories, ELISA basics guide, 2017.

[64] D.S. Hage, Immunoassays, Anal. Chem. 67 (1995) 455–462, doi:10.1021/ac00108a030.

[65] 4PL model. Available at: https://www.myassays.com/four-parameter-logistic-regression.html.

[66] A. Shrivastava, V. Gupta, Methods for the determination of limit of detection and limit of quantitation of the analytical methods, Chronicles Young Sci 2 (2011) 21–25, doi:10.4103/2229-5186.79345.

[67] Berthold Technology. Available at: https://kem-en-tec-nordic.com/berthold-technologies. (Accessed: 20th August 2018).

[68] M.G. Joaquim, M.S. Cabral, Maria Raquel Aires-Barros, Engenharia Enzimática, Lidel, Lisboa-porto, 2003.

[69] A.P. Castanheira, A.I. Barbosa, A.D. Edwards, N.M. Reis, Multiplexed femtomolar quantitation of human cytokines in a fluoropolymer microcapillary film, Analyst 140 (2015) 5609–5618, doi:10.1039/c5an00238a.

[70] J.R. Crowther, The ELISA guidebook., 2000.

[71] D. Mark, S. Haeberle, G. Roth, F. Von Stetten, R. Zengerle, Microfluidic lab-on-a-chip platforms: requirements, characteristics and applications, Chem. Soc. Rev. (2010) 1153–1182, doi:10.1039/b820557b.

[72] D. Kim, A.E. Herr, Protein immobilization techniques for microfluidic assays, Biomicrofluidics 7 (2013), doi:10.1063/1.4816934.

[73] F. Rusmini, Z. Zhong, J. Feijen, Protein immobilization strategies for protein biochips, Biomacromolecules 8 (2007) 1775–1789, doi:10.1021/bm061197b.

[74] A.I. Barbosa, A.P. Castanheira, A.D. Edwards, N.M. Reis, A lab-in-a-briefcase for rapid prostate specific antigen (PSA) screening from whole blood, Lab Chip 14 (2014) 2918–2928, doi:10.1039/c4lc00464g.

[75] A.I. Barbosa, P. Gehlot, K. Sidapra, A.D. Edwards, N.M. Reis, Portable smartphone quantitation of prostate specific antigen (PSA) in a fluoropolymer microfluidic device, Biosens. Bioelectron. 70 (2015) 5–14, doi:10.1016/j.bios.2015.03.006.

[76] A.P. Castanheira, A.I. Barbosa, A.D. Edwards, N.M. Reis, Multiplexed femtomolar quantitation of human cytokines in a fluoropolymer microcapillary film, Analyst 149 (2015) 5609–5618, doi:10.1039/c5an00238a.

[77] V.K. Rajendran et al., Gravity-driven microfluidic particle sorting device with hydrodynamic separation amplification, Lab Chip. 79 (2017) 1369–1376. doi:10.1021/ac061542n.

[78] A.I. Barbosa, J.H. Wichers, A. Van Amerongen, N.M. Reis, Towards one-step quantitation of prostate-specific antigen (PSA) in microfluidic devices : feasibility of optical detection with nanoparticle labels, (2017) 718–726. doi:10.1007/s12668-016-0390-y.

[79] R.S. Sista, A.E. Eckhardt, V. Srinivasan, M.G. Pollack, S. Palanki, V.K. Pamula, Heterogeneous immunoassays using magnetic beads on a digital microfluidic platform, Lab Chip. 8 (2008) 2188–2196, doi:10.1039/b807855f.

[80] J. Kai, A. Puntambekar, N. Santiago, S.H. Lee, D.W. Sehy, V. Moore, J. Han, C.H. Ahn, A novel microfluidic microplate as the next generation assay platform for enzyme linked immunoassays (ELISA), Lab Chip 12 (2012) 4257–4262, doi:10.1039/c2lc40585g.

[81] A.I. Barbosa, A.S. Barreto, N.M. Reis, Transparent, hydrophobic fluorinated ethylene propylene offers rapid, robust, and irreversible passive adsorption of diagnostic antibodies for sensitive optical biosensing, ACS Appl. Bio Mater. 2 (2019) 2780–2790, doi:10.1021/acsabm.9b00214.

[82] J. Pivetal, F.M. Pereira, A.I. Barbosa, A.P. Castanheira, N.M. Reis, A.D. Edwards, Covalent immobilisation of antibodies in Teflon-FEP microfluidic devices for the sensitive quantification of clinically relevant protein biomarkers, Analyst 142 (2017) 959–968, doi:10.1039/C6AN02622B.

[83] N.M. Reis, J. Pivetal, A.L. Loo-Zazueta, J. Barros, A.D. Edwards, Lab on a stick: multi-analyte cellular assays in a microfluidic dipstick, Lab Chip 16 (2016) 2891–2899, doi:10.1039/C6LC00332J.

[84] L.A. Sasso, T. Road, I.H. Johnston, J.D. Zahn, Automated microfluidic processing platform for multiplexed magnetic bead immunoassays Lawrence, Microfluid Nanofluidics 13 (2012) 603–612, doi:10.1007/s10404-012-0980-0.Automated.

[85] A. Waldbaur, H. Rapp, K. Länge, B.E. Rapp, Let there be chip - Towards rapid prototyping of microfluidic devices: one-step manufacturing processes, Anal. Methods. 3 (2011) 2681–2716, doi:10.1039/c1ay05253e.

[86] M. Zimmermann, H. Schmid, P. Hunziker, E. Delamarche, Capillary pumps for autonomous capillary systems, Lab Chip 7 (2007) 119–125, doi:10.1039/b609813d.

[87] R. Gorkin, J. Park, J. Siegrist, M. Amasia, B.S. Lee, J.M. Park, J. Kim, H. Kim, M. Madou, Y.K. Cho, Centrifugal microfluidics for biomedical applications, Lab Chip 10 (2010) 1758–1773, doi:10.1039/b924109d.

[88] Y. K.C., C.K. Beom Seok Lee, J.-.N. Lee, J.-.M. Park, J.-.G. Lee, S. Kim, A fully automated immunoassay from whole blood on a disc, Lab Chip 9 (2009) 1548–1555.

[89] J. Gilmore, M. Islam, R. Martinez-Duarte, Challenges in the use of compact disc-based centrifugal microfluidics for healthcare diagnostics at the extreme point of care, Micromachines. 7 (2016). doi:10.3390/mi7040052.

[90] S. Lai, S. Wang, J. Luo, L.J. Lee, S.T. Yang, M.J. Madou, Design of a compact disk-like microfluidic platform for enzyme-linked immunosorbent assay, Anal. Chem. 76 (2004) 1832–1837, doi:10.1021/ac0348322.

[91] Y. Yang, S. Kim, J. Chae, Separating and detecting escherichia coli in a microfluidic channel for urinary tract infection applications, J. Microelectromechanical Syst. 20 (2011) 819–827, doi:10.1109/JMEMS.2011.2159095.

[92] J.L. Vaitukaitis, Development of the home pregnancy test, Annu. New York Acad. Sci. 1038 (2004) 220–222, doi:10.1196/annals.1315.030.

[93] I.P. Alves, N.M. Reis, Immunocapture of Escherichia coli in a fluoropolymer microcapillary array, J. Chromatogr. A. 1585 (2019) 46–55, doi:10.1016/j.chroma.2018.11.067.

[94] M. Safavieh, M.U. Ahmed, E. Sokullu, A. Ng, M. Zourob, A simple cassette as point-of-care diagnostic device for naked-eye colorimetric bacteria detection, Analyst 139 (2014) 482–487, doi:10.1039/c3an01859h.

[95] A. Golberg, G. Linshiz, I. Kravets, N. Stawski, N.J. Hillson, M.L. Yarmush, R.S. Marks, T. Konry, Cloud-enabled microscopy and droplet microfluidic platform for specific detection of Escherichia coli in water, PLoS One 9 (2014), doi:10.1371/journal.pone.0086341.

[96] S. Stratz, K. Eyer, F. Kurth, P.S. Dittrich, On-chip enzyme quantification of single Escherichia coli bacteria by immunoassay-based analysis, Anal. Chem. 86 (2014) 12375–12381, doi:10.1021/ac503766d.

[97] M. Ikami, A. Kawakami, M. Kakuta, Y. Okamoto, N. Kaji, M. Tokeshi, Y. Baba, Immuno-pillar chip: a new platform for rapid and easy-to-use immunoassay, Lab Chip 10 (2010) 3335–3340, doi:10.1039/c0lc00241k.

[98] E.D. Goluch, J.M. Nam, D.G. Georganopoulou, T.N. Chiesl, K.A. Shaikh, K.S. Ryu, A.E. Barron, C.A. Mirkin, C. Liu, A bio-barcode assay for on-chip attomolar-sensitivity protein detection, Lab Chip 6 (2006) 1293–1299, doi:10.1039/b606294f.

[99] A.K. Yetisen, M.S. Akram, C.R. Lowe, Lab Chip (2013) 2210–2251, doi:10.1039/c3lc50169h.

[100] L. Lafleur, D. Stevens, K. McKenzie, S. Ramachandran, P. Spicar-Mihalic, M. Singhal, A. Arjyal, J. Osborn, P. Kauffman, P. Yager, B. Lutz, Progress toward multiplexed sample-to-result detection in low resource settings using microfluidic immunoassay cards, Lab Chip 12 (2012) 1119–1127, doi:10.1039/c2lc20751f.

[101] Longitude prize. Available at: http://longitudeprize.org. (Accessed: November 2015).

[102] G.M. Whitesides, The origins and the future of microfluidics, Nature 442 (2006) 368–373 http://dx.doi.org/10.1038/nature05058.

[103] R.W. Peeling, Diagnostics in a digital age: an opportunity to strengthen health systems and improve health outcomes, Int. Health. 7 (2015) 384–389, doi:10.1093/inthealth/ihv062.

[104] U.C. Ogbu, O.A. Arah, World Health Organization, Int. Encycl. Public Heal. (2016) 461–467, doi:10.1016/B978-0-12-803678-5.00499-9.

[105] S. Wang, F. Inci, T.L. Chaunzwa, A. Ramanujam, A. Vasudevan, S. Subramanian, A.C. Fai Ip, B. Sridharan, U.A. Gurkan, U. Demirci, Portable microfluidic chip for detection of Escherichia coli in produce and blood, Int. J. Nanomedicine. 7 (2012) 2591–2600, doi:10.2147/IJN.S29629.

[106] J.C. Liao, M. Mastali, V. Gau, a Marc, A.K. Møller, D. a Bruckner, T. Jane, Y. Li, J. Gornbein, E.M. Landaw, E.R.B. Mccabe, B.M. Churchill, a David, M. a Suchard, J.T. Babbitt, D. a Haake, Use of electrochemical DNA biosensors for rapid molecular identification of uropathogens in clinical urine specimens use of electrochemical DNA biosensors for rapid molecular identification of uropathogens in clinical urine specimens, J. Clin. Microbiol. 44 (2006) 561–570, doi:10.1128/JCM.44.2.561.

[107] D.A. Boehm, P.A. Gottlieb, S.Z. Hua, On-chip microfluidic biosensor for bacterial detection and identification, Sens. Actuators, B Chem 126 (2007) 508–514, doi:10.1016/j.snb.2007.03.043.

[108] O. Laczka, J.M. Maesa, N. Godino, J. del Campo, M. Fougt-Hansen, J.P. Kutter, D. Snakenborg, F.X. Muñoz-Pascual, E. Baldrich, Improved bacteria detection by coupling magneto-immunocapture and amperometry at flow-channel microband electrodes, Biosens. Bioelectron. 26 (2011) 3633–3640, doi:10.1016/j.bios.2011.02.019.

[109] M. Bercovici, G.V. Kaigala, K.E. MacH, C.M. Han, J.C. Liao, J.G. Santiago, Rapid detection of urinary tract infections using isotachophoresis and molecular beacons, Anal. Chem. 83 (2011) 4110–4117, doi:10.1021/ac200253x.

[110] N. Sanvicens, N. Pascual, M.T. Fernández-Argüelles, J. Adrián, J.M. Costa-Fernández, F. Sánchez-Baeza, A. Sanz-Medel, M.P. Marco, Quantum dot-based array for sensitive detection of Escherichia coli, Anal. Bioanal. Chem. 399 (2011) 2755–2762, doi:10.1007/s00216-010-4624-5.

[111] V.K. Rajendran, P. Bakthavathsalam, B.M. Jaffar Ali, Smartphone based bacterial detection using biofunctionalized fluorescent nanoparticles, Microchim. Acta. (2014), doi:10.1007/s00604-014-1242-5.

[112] S.V. Angus, S. Cho, D.K. Harshman, J.Y. Song, J.Y. Yoon, A portable, shock-proof, surface-heated droplet PCR system for Escherichia coli detection, Biosens. Bioelectron. 74 (2015) 360–368, doi:10.1016/j.bios.2015.06.026.

[113] S.İ. Dönmez, S.H. Needs, H.M.I. Osborn, A.D. Edwards, Label-free smartphone quantitation of bacteria by darkfield imaging of light scattering in fluoropolymer micro capillary film allows portable detection of bacteriophage lysis, Sensors Actuators, B Chem 323 (2020), doi:10.1016/j.snb.2020.128645.

[114] S. haeberle, R. Zengerle, Microfluidic platforms for lab-on-a-chip applications, Lab Chip 7 (2007) 1094–1110, doi:10.1007/978-3-642-18293-8_22.

[115] D. Mark, S. Haeberle, G. Roth, F. Von Stetten, R. Zengerle, Microfluidic lab-on-a-chip platforms: requirements, characteristics and applications, NATO Sci. Peace Secur. Ser. A Chem. Biol. (2010) 305–376, doi:10.1007/978-90-481-9029-4-17.

[116] M. Liu, J. Sun, Q. Chen, Influences of heating temperature on mechanical properties of polydimethylsiloxane, Sensors Actuators, A Phys 151 (2009) 42–45, doi:10.1016/j.sna.2009.02.016.

[117] A.J.F. Alvaro Mata, S. Roy, Characterization of polydimethylsiloxane (PDMS) properties for biomedical micro/nanosystems, Biomed. Microdevices. 7 (2005) 281–293.

[118] S. Ebnesajjad, 1 - Introduction to Fluoropolymers, William Andrew Publishing, Oxford, S.B.T.-F. (Second E. Ebnesajjad (Ed.) 2015, pp. 1–6, doi:10.1016/B978-1-4557-3197-8.00001-8.

[119] W.H. Grover, M.G. Von Muhlen, S.R. Manalis, Teflon films for chemically-inert microfluidic valves and pumps, Lab Chip 8 (2008) 913–918, doi:10.1039/b800600h.

[120] J.W. Gooch, Teflon FEP, Encycl. Dict. Polym. (2011) 731, doi:10.1007/978-1-4419-6247-8_11595.

[121] A.D. Edwards, N.M. Reis, N.K.H. Slater, M.R. Mackley, A simple device for multiplex ELISA made from melt-extruded plastic microcapillary film, Lab Chip 11 (2011) 4267–4273, doi:10.1039/C0LC00357C.

[122] A.P. Castanheira, A.I. Barbosa, A.D. Edwards, N.M. Reis, Multiplexed femtomolar quantitation of human cytokines in a fluoropolymer microcapillary film, Analyst 140 (2015) 5609–5618, doi:10.1039/c5an00238a.

[123] P.V. Danckverts, Continuous flow systems. Distribution of residence times, Chem. Eng. Sci. (1952) 3857–3866.

[124] C.J. Geankoplis, Transport Process and Unit Operations, 3rd ed., Prentice-Hall International, Inc., 1983

[125] O. Reynolds, On the dynamical theory of incompressible viscous fluids and the determination of the criterion, Philos. Trans. R. Soc. A Math. Phys. Eng. Sci. 186 (1895) 123–164, doi:10.1098/rsta.1895.0004.

[126] J. Buijs, W. Norde, J.W.T. Lichtenbelt, Changes in the secondary structure of adsorbed IgG and F(ab')$_2$ studied by FTIR spectroscopy, Langmuir 12 (1996) 1605–1613, doi:10.1021/la950665s.

[127] A.I. Barbosa, The development and optimisation of a novel microfluidic immunoassay platform for point of care diagnostics, (2015).

[128] E. coli size. http://book.bionumbers.org/how-big-is-an-e-coli-cell-and-what-is-its-mass/. [Online] [Cited: June 22, 2016.], (2016) 2016.

[129] I.P. Alves, N.M. Reis, Immunocapture of Escherichia coli in a fluoropolymer microcapillary array, J. Chromatogr. A. (2019), doi:10.1016/j.chroma.2018.11.067.

[130] S.C.B. Gopinath, T.H. Tang, Y. Chen, M. Citartan, T. Lakshmipriya, Bacterial detection: from microscope to smartphone, Biosens. Bioelectron. 60 (2014) 332–342, doi:10.1016/j.bios.2014.04.014.

[131] T.S. Park, W. Li, K.E. McCracken, J.Y. Yoon, Smartphone quantifies Salmonella from paper microfluidics, Lab Chip 13 (2013) 4832–4840, doi:10.1039/c3lc50976a.

[132] B. Kuswandi, J.H. Nuriman, W. Verboom, Optical sensing systems for microfluidic devices: a review, Anal. Chim. Acta. 601 (2007) 141–155, doi:10.1016/j.aca.2007.08.046.

[133] M.C. Pierce, S.E. Weigum, J.M. Jaslove, R. Richards-Kortum, T.S. Tkaczyk, Optical systems for point-of-care diagnostic instrumentation: analysis of imaging performance and cost, Ann. Biomed. Eng. 42 (2014) 231–240, doi:10.1007/s10439-013-0918-z.

[134] F.B. Myers, L.P. Lee, Innovations in optical microfluidic technologies for point-of-care diagnostics, Lab Chip 8 (2008) 2015–2031, doi:10.1039/b812343h.

[135] D. Erickson, D. O'Dell, L. Jiang, V. Oncescu, A. Gumus, S. Lee, M. Mancuso, S. Mehta, Smartphone technology can be transformative to the deployment of lab-on-chip diagnostics, Lab Chip 14 (2014) 3159–3164, doi:10.1039/c4lc00142g.

[136] L. Shen, J. a Hagen, I. Papautsky, Point-of-care colorimetric detection with a smartphone, Lab Chip 12 (2012) 4240–4243, doi:10.1039/c2lc40741h.

[137] V. Oncescu, D. O'Dell, D. Erickson, Smartphone based health accessory for colorimetric detection of biomarkers in sweat and saliva, Lab Chip 13 (2013) 3232–3238, doi:10.1039/c3lc50431j.

[138] H. Zhu, S.O. Isikman, O. Mudanyali, A. Greenbaum, A. Ozcan, Optical imaging techniques for point-of-care diagnostics, Lab Chip 13 (2013) 51–67, doi:10.1039/c2lc40864c.

[139] A. Roda, E. Michelini, M. Zangheri, M. Di Fusco, D. Calabria, P. Simoni, Smartphone-based biosensors: a critical review and perspectives, TrAC - Trends Anal. Chem. 79 (2016) 317–325, doi:10.1016/j.trac.2015.10.019.

[140] M. Bates, A. Zumla, Rapid infectious diseases diagnostics using Smartphones, Ann. Transl. Med. 3 (2015) 1–5, doi:10.3978/j.issn.2305-5839.2015.07.07.

[141] J. Il Hong, B.Y. Chang, Development of the smartphone-based colorimetry for multi-analyte sensing arrays, Lab Chip 14 (2014) 1725–1732, doi:10.1039/c3lc51451j.

[142] A.F. Coskun, R. Nagi, K. Sadeghi, S. Phillips, A. Ozcan, Albumin testing in urine using a smart-phone, Lab Chip 13 (2013) 4231–4238, doi:10.1039/c3lc50785h.

[143] Wei, et al., Fluorescent imaging of single nanoparticles and viruses on a smart phone, ACS Nano (2013).

Point-of-care diagnostics with smartphone

Haleh Ayatollahi[a,b]
[a]Health Management and Economics Research Center, Health Management Research Institute, Iran University of Medical Sciences, Tehran, Iran
[b]Department of Health Information Management, School of Health Management and Information Sciences, Iran University of Medical Sciences, Tehran, Iran

15.1 Introduction

Recently, continuous developments in the field of physics, chemistry, and biotechnology have resulted in significant evolution in point-of-care diagnostics (POCD) [1]. There are many definitions for point-of-care diagnostics; however, it is mainly defined as quick acquisition of test results out of the laboratory, or in patient site so that the patient can be treated more effectively at the earliest time [2,3] World Health Organization (WHO) recommended developing and using point-of-care diagnostic devices, especially for developing countries in which medical infrastructure is under constraint [4]. A point-of-care diagnostic device is a portable version of an analyzer that can be easily used by healthcare providers, patients, and their careers or families to analyze a small biological or chemical sample [1,5]. It is rapid and can provide the results in a short time. A major advantage of POC diagnostic devices is their low cost compared to the cost of conventional diagnostic tests [2]. Moreover, they can be used for managing and monitoring patient health conditions in underserved regions [6]. As one of the challenges in public health is to improve quality of care in underserved areas, providing people in these regions with point-of-care testing (POCT) can help to control many diseases [7]. POCT is an important tool in reducing burden of diseases and mortalities by providing accurate, convenient and affordable diagnostic tests [8], in particular, where resources are inadequate and medical care is not easily accessible [9].

The development of POCD devices has been affected by applying new ideas since the early 1990's. These ideas were mainly related to miniaturization, integration, improving analytical capabilities, and global health applications [10]. For example, early POC technologies required additional devices for diagnostic and post-analytical purposes. As a result, developing easy to use and cost-effective POC technologies was regarded as a priority [11]. Currently, POCD has moved towards using smart and mobile devices, such as smartphones, which have revolutionized providing healthcare services [1,12]. The smartphones are valuable tools to support healthcare and many heath applications have been created to be installed on mobile phones [9]. These devices help to increase the speed of diagnosis in a more cost-effective manner [4].

Smartphone-Based Detection Devices. DOI: https://doi.org/10.1016/B978-0-12-823696-3.00017-9
Copyright © 2021 Elsevier Inc. All rights reserved.

Todays, mobile applications are able to carry out different types of laboratory tests without using complicated test procedures and attending in a large laboratory environment [13]. This technology is capable of providing personalized medicine to anyone regardless of location and time constraints and its wireless network can help to access healthcare services at the point of need [9]. Other features of smartphones, such as sensors, cameras, processors, the high level of seamless connectivity, robust functionalities and portability also make this technology appropriate to be used as an analytical diagnostic device [6,12]. The use of smartphone as a point-of-care diagnostic device can also be considered an example of mobile health (mHealth) and telemedicine technologies to offer more healthcare services in remote areas that are far from professional test centers [12].

15.2 Point-of-care testing (POCT)

Generally, laboratory testing is the most favorable diagnostic method in medicine. However, setting up a laboratory test center is often costly, mainly in terms of the equipment and professional workforce expenditure. Accordingly, waiting time for receiving test results may take several days, which may cause disease outbreaks or disease progression. To overcome these challenges, point-of-care testing (POCT) emerged as a modern clinical testing method [8,9]. It refers to gathering medical data and test results where the patient is examined [2]. The POCT was first introduced in 1962 for rapid analysis of blood glucose and the advancement was continued by developing rapid pregnancy test in 1977 [2].

Some point-of-care tests are paper-based. However, paper-based results are difficult to distribute, translate, navigate and update [7]. POCT should be completed within 10–30 min and should ensure the patient that the follow-up care will be provided after the test. Therefore, it seems that smartphone-based technology can be an alternative choice to paper-based results and can support the patient by sending and receiving data at the point of need [14,15].

An ideal POC test should be cost-effective, sensitive and specific, simple, and standalone (independent of external power or equipment) [3]. As suggested by World Health Organization (WHO), 'ASSURED' criteria should be considered for POC tests in low-resource countries. It means that the POC test should be affordable, user-friendly (easy to use with a minimum need for training), specific (few false positive), sensitive (few false negatives), equipment free, fast and robust, and delivered [2]. WHOs REASSURED framework is another guideline which shows more practical and technical characteristics of a perfect smartphone-based point-of-care diagnostic device. The main differences between this framework and ASSURED criteria are related to synchronized connectivity and ease of sample gathering. It is notable that these guidelines are "ideal" features and they can be tailored depending on the technology and the field of study [4].

POC tests can also be used by patients at their homes, as a part of their self-care or self-management programs. Compared to laboratory-based testing, which requires expensive analytical procedures and laboratory facilities; POC testing offers a faster and cheaper diagnostic test result. Currently, different types of POC testing tools, such as POC tools for testing pregnancy, blood gas and electrolytes analysis, blood glucose, bacteria screening, urine testing, and forensic medicine are available in the market to support continuous patient monitoring and timely therapeutic interventions [9].

15.3 Detection methods

Clinical diagnostic tests are categorized into chemistry/biochemistry, microscopy and molecular-based tests [12]. A smartphone-based POCT performs these tests through applying two detection methods. The first method performs in vivo tests, in which built-in or external sensor of a smartphone is used to detect biological signals. The second method is used to perform in vitro tests which are followed by biochemical reactions [11].

In vivo tests collect health data without using a body specimen. This method can be divided into smartphone-based POCD with built-in sensors and smartphone-based POCD with external sensors. Some sounds, like heart rate and biological signals like two-dimensional (2D) color images can be collected by smartphone internal devices and the data collected by smartphone's built-in sensors are usually limited to images and sounds. However, a number of new external devices have been created to be integrated into the smartphones to extract more clinical data, such as body temperature and functional images of different parts of a body [11]. By contrast, in vitro tests are biochemical tests that identify organisms (e.g., cells and microbes) and biological components (e.g., proteins) from sweat, blood, saliva, water, urine, or food. Smartphone-based microscopy, smartphone-based strip, tube, specimen and microfluidic tests are used for in vitro tests to detect or measure biological components [11]. Smartphones are also used for bright field testing, fluorescent detection, colorimetric sensing, electrochemical sensing, phase imaging, light scattering and spectroscopy [8,5,6,12,15].

15.4 Applications of smartphone-based POCD/POCT

A number of smartphone-based POCD applications have been developed to date [1]. Surgical treatment and diagnostic applications, ophthalmic applications, biochemical and electrochemical applications, environmental monitoring applications, biomedical applications, colorimetric and spectrometry applications are some of the examples of using smartphone-based POCD in medical sciences [13]. Medical imaging is another application of smartphone-based POCD which helps to create images of the human body for clinical purposes. Recently, significant advances in smartphone medical imaging have been realized that made the clinical diagnosis easier [11,12]. These images are similar to those taken by conventional medical instruments [3]. Smartphone-based POCD/POCT has also been used in diagnosing and monitoring different diseases and health conditions. Automated smartphone-based semen analyzer for male infertility screening [16], infectious disease diagnostics services [17], sickle cell disease diagnosis [18], POC diagnostic services during coronavirus 2019 (COVID-19) pandemic [19], diabetes monitoring [20], automated ovulation testing [21], POC diagnostic systems for cystic fibrosis (CF) [22], POC diagnostics devices for retinal vasculature analysis [23], POC diagnosis of influenza A (H1N1) virus [24], cardiovascular diseases (CVDs) health status monitoring [25], and hormonal diagnostics [26] are some of the applications of this technology.

15.5 Benefits of smartphone-based POCD/POCT

Generally speaking, effective and rapid POC diagnostics play a key role in public health [9,11]. Before the emergence of smartphones, diagnostic technology depended on battery powered handheld devices that might be heavy and should be connected to other laboratory equipment [15]. However, smartphone-based point-of-care (POC) applications have overcome previous challenges and introduced a number of benefits, such as patient health status monitoring, quick testing, reducing the cost of medical and laboratory instruments [13], decreasing the time duration between sample acquisition and test results reporting [2], eliminating mislabeling and mishandling the test samples and their results [6], timely diagnosis, preventing the diseases spread as well as continues monitoring of chronic diseases, such as diabetes [11]. Point-of-care diagnostics could also control infectious diseases in low and middle income countries by giving adequate data to the healthcare providers for faster and better decision making and timely treatments [14]. Their impact on epidemiology and public health should not be underestimated, as accurate geographic and demographic data can be collected along with the diagnostic test results [12].

Other advantages of smartphone-based point-of-care applications are related to low limit of detection (LOD), the possibility of using built-in sensors (e.g., microphone and camera), low power consumption, reliable sensitivity, high specificity, cost-effective designs, fast testing, and high selectivity [13]. As smartphones are equipped with Wi-Fi, Bluetooth connections and the Global System for Mobile Communication (GSM), data can be easily transferred to other computers or remote servers for further usage in various telemedicine applications [3] and a remote test site can easily seek for professional guidance from a centralized laboratory [11]. The real-time connectivity of smartphones can also improve the process of patient care by sending the results to the healthcare professionals and receiving their advice via smartphones [4].

The use of smartphones as an imaging platform in the clinic or at home can help on-site data acquisition and management. These images can be sent from the patient site to the healthcare professionals at a convenience time [12]. Moreover, it is also possible to diagnose diseases by taking samples from the patient body. Smartphone-based point-of-care (POC) applications have made sample collection easier for the users [8], and overcome the constraints of conventional laboratory testing efficiently [1]. Overall, healthcare providers can take benefit from smartphone-based point-of-care

(POC) applications to diagnosis a disease faster than before, and patients can take advantage of these applications by getting involved in managing and controlling their health condition and reducing the risk of disease outbreaks [6,10].

Similarly, POC tests are appropriate for constant monitoring of a condition or performing follow-up tests to identify the progress of a disease or a treatment. Effective implementation of POC testing creates a more streamlined and faster workflow [10], reduces turn-around time and completes the diagnosis in a single visit. The test results can be stored in central databases and shared with healthcare providers via new technologies such as a cloud server [2]. Point-of-care testing (POCT) helps clinicians to make medical decisions as early as possible and set up their care plans to improve quality of care [1]. Other advantages of POCT are simple operation without the need for professional training, simple and cost-effective production of needed tools, ease of use especially in low-resource regions, and low energy consumption [8].

15.6 Development considerations

Generally, the development of mobile-based point-of-care diagnostics includes four steps: needs and value assessment, technology development, preclinical verification, and clinical validation and field trials [2]. Before, designing and implementing smartphone-based POCD applications, the type of the disease that can be diagnosed by collecting specimens and a suitable application for this purpose should be defined [8]. Sometimes, technical parameters are overemphasized in designing a POC device which does not necessarily increase the clinical effectiveness. Therefore, assessing the real diagnostic needs and the potential value of the device is of paramount importance. Moreover, the results of preclinical verification are based on the laboratory resources and a decreased performance might be experienced during field evaluation compared to the initial laboratory verification [2,15]. Therefore, the field condition should be simulated during laboratory verification to achieve successful results [15]. A preclinical validation and field trails are also necessary to prove that a device is ready to be implemented at the point-of-care [2]. The preclinical validation tests should be performed in a similar condition that the device intends to be used [4].

To test the samples, sometimes an "adapter" or "cradle" is used which is an external hardware attachment and may have different shapes. Usually, two approaches are used for designing an adapter. The first approach focuses on examining a specific sample and its characteristics. The second one uses the optical design criteria and different types of samples can be tested by using this method. The first approach is useful for the research purposes, and the second approach is more suitable for commercial and industrial purposes [13]. Smartphone adapters are essential components of point-of-care diagnostics and should be able to complete fast and accurate tests even for the low concentration specimens. Therefore, before designing an adapter, adequate knowledge in designing optical systems and related applications are required. The cost of the adapter, the integration of optical and mechanical parts and other accessories should also be taken into account [13].

Currently, the smartphone applications that can be used by users should be compatible with the operating system of their own smartphone [13]. However, the development of equipment-free devices that are compatible to any smartphones is preferable [11]. Overall, several parameters, such as safety, compactness, illumination sources, material, cost, imaging optics, interchangeability, and functional robustness should be considered during system design [13].

15.7 mHealth and smartphone-based POCD/POCT

eHealth is defined as the use of information and communication technologies (ICT) in the healthcare settings [27]. According to World Health Organization, eHealth has four components that include mobile health (mHealth), health information systems (HIS), telemedicine, and distance learning (eLearning). Therefore, mHealth is a subset of eHealth that delivers health care services via mobile and wireless technologies [28].

To date, mHealth interventions have been used in many areas, such as health monitoring, disease surveillance and health education. These interventions reduce staff workload and medical errors and can assist clinicians in timely decision making and diagnosis. However, mHealth intervention is not just a single intervention, but it is a complex set of connected ecosystem which must be considered before applying the new technologies. It means that the data collected from individual mHealth devices should be securely stored in the related databases or easily shared with other mHealth applications. Moreover, several evaluation studies need to be conducted to demonstrate its benefits and potential risks to patients, healthcare providers and healthcare systems in terms of economic, social and clinical outcomes [17]. Usability of mhealth applications and devices is another challenging issue which has been regarded as an essential component in technology design. It refers to the use of a product by specific users to achieve predefined goals while they feel effectiveness, efficiency and satisfaction [29]. In fact, a system that is difficult to operate may fail, and this point should be taken into account when developing mHealth systems [30].

According to the literature, the success of mhealth projects, such as smartphone-based POCT depends on the system integration into the health care processes. This technology should be compatible with the current activities, and care models should be taken into account to increase the acceptability of the applications by the end-users [29,31,32]. Obviously, such a successful integration requires overcoming a number of operational and administrative challenges which are common to many eHealth applications and services.

15.8 Challenges of smartphone-based POCD/POCT

To date, several smartphone-based POCD devices have been developed and implemented for monitoring different physiological parameters, such as pulse rate, blood pressure, blood glucose, electrocardiogram, weight, and physical activities [4]. Although smartphone-based POCD/POCT has many advantages, several technical and non-technical challenges may hinder introducing and using new POC devices [10]. For example, the confidentiality and privacy of patient data are the main concerns [1,4] and data confidentiality principles should be respected [11]. In addition, setting international cloud-computing standards and guidelines for managing 'Big Data' are necessary [1]. Cyber-attacks are other serious issues which must be taken into account by improving cybersecurity, network control systems, and internet-based infrastructures [6].

Other concerns are related to the accuracy and reliability of POC testing procedures and the sensitivity of biosensors. In fact, the test results should be explicit and accurate [8,11]. To overcome these challenges, setting regulatory policies have been suggested. These policies should be able to support accountability and security of data, and ensure the users that the accuracy and reliability of POC devices as well as smartphone applications are guaranteed [15].

While smartphone-based POCD devices have the potential to facilitate diagnostic procedures, they need to show a streamlined workflow for sample processing which can be easily used even in underserved areas. Another issue is related to the processing capacity. In fact, screening multiple pathogens from a single sample, finding the causative organism which might be a bacterium or a virus, and detecting multiple analytes simultaneously are important, as similar clinical symptoms might be produced by different pathological agents [11,15]. Moreover, multiple tests can avoid misdiagnosis [11].

The cost of smartphone-based POCT is another major issue. As the use of patient testing is increasing in developing countries mainly due to insufficient healthcare facilities in low-resource regions, smartphone-based POCT should be available at a low cost to be used by people and healthcare professionals [15]. Other costs of POCT include cost of materials, substance consumption, sample processing, transportation and refrigeration. As a result, it is necessary to develop devices that can use specimens without complex processing procedures [2]. In addition, further reduction in the cost of advanced molecular testing that needs expensive substances and samples that need to be stored and transferred in specific conditions require paying more attention. Finally, commercialization approaches need to be applied to gain cost-effective and robust platforms [12].

The quality of the App can also influence user's perception of the proposed system. Therefore, the App should be easy to use and compatible with the technological and commercial standards for future upgrades and functional improvements [15]. Testing techniques and the capacity of smartphone-based POCT should also vary to be able to test different biological makers such as blood, saliva, urine, sweat and tears [13].

According to the literature, only a few devices have passed a proof-of-concept phase and are currently used at the patient site. Therefore, it is emphasized that before developing a new technology, assessing the real requirements, costs, and geographical area should be taken into account. The preclinical and clinical evaluations are also extremely important to ensure that the device performance is comparable with clinical reference standards [4].

The power supply is another challenge that should be addressed properly, since POC testing performance depends considerably upon power availability. In some developing countries, the electricity distribution network is not appropriate and the use of batteries is not cost-effective. Therefore, power supply is an important issue for stand-alone POC devices [9]. In low-resource regions, Internet connectivity is another serious issue which mainly occurs due to the lack of affordable communication networks and platforms [6]. Other challenges are related to developing a quality assessment guideline, health system integration, data management solutions [9], funding, user acceptance, and market adoption [6]. It is notable that in low-resource settings, some of these challenges might be regarded as serious obstacles which may hinder designing and implementation of POC devices [6].

In terms of the evaluation, it is often difficult to evaluate the quality of smartphone-based POCD/PCDT from the trained or untrained users' perspectives. As a result, some solutions, such as appointing point-of-care coordinators to ensure the consistency of procedures, correct documentation, regulatory compliance, and end-user assistance are suggested. [10]

15.9 Future development of smartphone-based POCD/POCT

Smartphone-based POC diagnostics can be developed in different aspects. In future, this technology can empower patients to monitor and manage their health more easily [1]. Different types of portable biosensors need to be produced to be connected to the smartphones. The development of new image processing algorithms [11], more complicated and accurate biochemical test techniques, and increasing the number of analytes that can be identified by portable devices are other opportunities for future research [13].

The app users usually look for those applications that are reliable and meet their needs. Therefore, creating a centralized database of approved applications is beneficial to the users. These applications should be matched with different operating systems of smartphones. In fact, developing new algorithms for being used in the smartphone apps should be reported by the researchers to the decision makers to be approved and used in practice [15]. The issue of power management needs to be further investigated. As power supply is a major challenge, especially in developing countries, adopting new power supply systems including portable biofuel cells, wireless energy transfer, body energy harvesters, and energy-harvesting modules, and other electric materials are suggested to overcome energy and environmental crisis. Meanwhile, more research about self-powered POCT can be beneficial for future development [3,9].

The advancement in developing sensor and biosensors is also necessary for improving POC platforms [6]. Moreover, miniaturization and system integration should be among the research priorities to accelerate the practical applications of POC diagnostic devices [4,8].

To develop better smartphone-based POC devices, home testing tools that are simple and easy to use as well as new technologies that operate without external sensors can provide the researchers with new opportunities for future research. Obviously, new applications should be compatible with different operating systems of mobile phones [4].

The developers of POC devices should pay more attention to designing systems that are able to be connected to ehealth and electronic health record systems to share data [10]. The collected data can be analyzed by using machine learning techniques to predict disease patterns which are useful for public health and epidemiological interventions [4]. This integration will also revolutionize personalized medicine and has a powerful impact on care quality [13]. Other ideas for future developments are applying deep learning algorithms, big data, cloud computing, Internet of Things (IoTs), 5G networks, machine learning, imaging algorithms, open source software platforms, the near field communication (NFC) technology, and wearable sensors [6,8]. The application of POCT also needs to be further expanded to be used in biochemical technology, food safety, environmental safety, and other aspects [8].

In the future, the adoption of 3D printing technology and its integration with smartphone sensing technology can be useful for rapid production of low-cost and sensitive POC applications [6,15]. Nevertheless, the academic studies, such as cost-benefit analysis and evaluation studies are always necessary to show the capabilities and performances of devices [4]. It is also expected that smartphone-based POCD/POCT becomes a main component of the mHealth ecosystem and can contribute to evolve evidence-based medicine [12].

15.10 Conclusion

Accurate and timely diagnosis is critical to provide patients with effective care plans and prevent disease progress or outbreaks. Smartphone-based POCD/POCT technology plays a significant role in disease detection, especially in remote areas. Various types of detection methods, including brightfield testing, fluorescent detection, colorimetric sensing, electrochemical sensing and microscopy have transformed the detection of diseases by smartphone-based POCD/POCT. While there are several technical and non-technical challenges ahead of developing new technologies, it seems that more smartphone-based POCD/POCTs are needed to support personalized medicine. This technology along with eHealth and mHealth technologies will provide new possibilities to diagnose and control diseases and can improve quality of care, efficiency, and effectiveness of the healthcare system.

References

[1] S. Vashist, Point-of-care diagnostics: recent advances and trends, Biosensors (Basel) 7 (4) (2017) 62. https://doi.org/10.3390/bios7040062.

[2] A.N. Konwar, V. Borse, Current status of point-of-care diagnostic devices in the Indian healthcare system with an update on COVID-19 pandemic, Sen. Int. 1 (2020). https://doi.org/10.1016/j.sintl.2020.100015.

[3] S. Zhang, Z. Li, Q. Wei, Smartphone-based cytometric biosensors for point-of-care cellular diagnostics, Nanotech. Prec. Eng. 3 (1) (2020) 32–42. https://doi.org/10.1016/j.npe.2019.12.004.

[4] A. Malekjahani, S. Sindhwani, A.M. Syed, W.C.W Chan, Engineering steps for mobile point-of-care diagnostic devices, Acc. Chem. Res. 52 (9) (2019) 2406–2414. https://doi.org/10.1021/acs.accounts.9b00200.

[5] B. Purohit, A. Kumar, K. Mahato, P. Chandra, Smartphone-assisted personalized diagnostic devices and wearable sensors, Curr. Opin. Biomed. Eng. 13 (2020) 42–50. https://doi.org/10.1016/j.cobme.2019.08.015.

[6] M. Zarei, Portable biosensing devices for point-of-care diagnostics: recent developments and applications, TrAC, Trends Anal. Chem. 91 (2017) 26–41. https://doi.org/10.1016/j.trac.2017.04.001.

[7] N. Dell, G. Borriello, Mobile tools for point-of-care diagnostics in the developing world, ACM DEV '13: Proceedings of the 3rd ACM Symposium on Computing for Development, Bangalore, India, 2013. https://doi.org/10.1145/2442882.2442894.

[8] J. Liu, Z. Geng, Z. Fan, J. Liu, H. Chen, Point-of-care testing based on smartphone: the current state-of-the-art (2017–2018), Biosens. Bioelectron. 132 (2019) 17–37. https://doi.org/10.1016/j.bios.2019.01.068.

[9] S. Choi, Powering point-of-care diagnostic devices, Biotechnol. Adv. 34 (3) (2016) 321–330. https://doi.org/10.1016/j.biotechadv.2015.11.004.

[10] S. Nayak, N.R. Blumenfeld, T. Laksanasopin, S.K. Sia, Point-of-care diagnostics: recent developments in a connected age, Anal. Chem. 89 (1) (2017) 102–123. https://doi.org/10.1021/acs.analchem.6b04630.

[11] X. Xu, A. Akay, H. Wei, S. Wang, B. Pingguan-Murphy, B. Erlandsson, X. Li, W. Lee, J. Hu, L. Wang, Advances in smartphone-based point-of-care diagnostics, Proc. IEEE 103 (2) (2015) 236–247. https://doi.org/10.1109/JPROC.2014.2378776.

[12] I. Hernández-Neuta, F. Neumann, J. Brightmeyer, T. Ba Tis, N. Madaboosi, Q. Wei, A. Ozcan, M. Nilsson, Smartphone-based clinical diagnostics: towards democratization of evidence-based health care, J. Intern. Med. 285 (1) (2019) 19–39. https://doi.org/10.1111/joim.12820.

[13] Alawsi, Al-Bawi, A review of smartphone point-of-care adapter design, Eng. Rep. 1 (2019).

[14] J. Long, H.E. Parker, K. Ehrlich, M.G. Tanner, K. Dhaliwal, B. Mills, Frugal filtering optical lenses for point-of-care diagnostics, Biomed. Opt. Express 11 (4) (2020) 1864–1875. https://doi.org/10.1364/BOE.381014.

[15] V.K. Rajendran, P. Bakthavathsalam, P.L. Bergquist, A. Sunna, Smartphone technology facilitates point-of-care nucleic acid diagnosis: a beginner's guide, Crit. Rev. Clin. Lab. Sci. (2020). https://doi.org/10.1080/10408363.2020.1781779.

[16] M.K. Kanakasabapathy, M. Sadasivam, A. Singh, C. Preston, P. Thirumalaraju, M. Venkataraman, C.L. Bormann, M.S. Draz, J.C. Petrozza, H. Shafiee, An automated smartphone-based diagnostic assay for point-of-care semen analysis, Sci. Transl. Med. 9 (382) (2017). https://doi.org/10.1126/scitranslmed.aai7863.

[17] C.S. Wood, M.R. Thomas, J. Budd, T.P. Mashamba-Thompson, K. Herbst, D. Pillay, R.W. Peeling, A.M. Johnson, R.A. McKendry, M.M. Stevens, Taking connected mobile-health diagnostics of infectious diseases to the field, Nature 566 (7745) (2019) 467–474. https://doi.org/10.1038/s41586-019-0956-2.

[18] P.T. McGann, C. Hoppe, The pressing need for point-of-care diagnostics for sickle cell disease: a review of current and future technologies, Blood Cells Mol. Dis. 67 (2017) 104–113. https://doi.org/10.1016/j.bcmd.2017.08.010.

[19] N. Farshidfar, S. Hamedani, The potential role of smartphone-based microfluidic systems for rapid detection of COVID-19 using saliva specimen, Mol. Diagn. Ther. 24 (4) (2020) 371–373.

[20] Wang, C. Lio, H Huang, A feasible image-based colorimetric assay using a smartphone RGB camera for point-of-care monitoring of diabetes, Talanta 206 (2020).

[21] V. Potluri, P. Kathiresan, H. Kandula, P. Thirumalaraju, M. Kanakasabapathy, S. Pavan, D. Yarravarapu, A. Soundararajan, K. Baskar, R. Gupta, N. Gudipati, J. Petrozza, H. Shafiee, An inexpensive smartphone-based device for point-of-care ovulation testing, Lab Chip 1 (2019). https://doi.org/10.1039/C8LC00792F.

[22] C. Zhang, J. Kim, M. Creer, J. Yang, Z. Liu, A smartphone-based chloridometer for point-of-care diagnostics of cystic fibrosis, Biosens. Bioelectron. 97 (2017) 164–168. https://doi.org/10.1016/j.bios.2017.05.048.

[23] X. Xu, W. Ding, X. Wang, R. Cao, M. Zhang, P. Lv, F. Xu, Smartphone-based accurate analysis of retinal vasculature towards point-of-care diagnostics, Sci. Rep. 6 (2016). https://doi.org/10.1038/srep34603.

[24] X. Qiu, S. Ge, P. Gao, K. Li, S. Yang, S. Zhang, X. Ye, N. Xia, S. Qian, A smartphone-based point-of-care diagnosis of H1N1 with microfluidic convection PCR, Microsyst. Technol. 23 (7) (2017) 2951–2956. https://doi.org/10.1007/s00542-016-2979-z.

[25] J. Hu, X. Cui, Y. Gong, X. Xu, B. Gao, T. Wen, T.J. Lu, F. Xu, Portable microfluidic and smartphone-based devices for monitoring of cardiovascular diseases at the point of care, Biotechnol. Adv. 34 (3) (2016) 305–320. https://doi.org/10.1016/j.biotechadv.2016.02.008.

[26] P. Matías-García, J. Martinez-Hurtado, A. Beckley, M. Schmidmayr, V. Seifert-Klauss, Hormonal smartphone diagnostics. In: investigations of early nutrition effects on long-term health, in: P. Guest (Ed.), Methods in Molecular Biology, Vol. 1735, Humana Press, 2018. https://doi.org/10.1007/978-1-4939-7614-0_38.

[27] McDonald, C.L. Backman, A. Beckley, M. Schmidmayr, V. Seifert-Klauss, G. Macdonald, A. Townsend, P. Adam, S. Kerr, eHealth technologies, multimorbidity, and the office visit: qualitative interview study on the perspectives of physicians and nurses, J. Med. Internet Res. 1735 (1) (2018).

[28] J. Lewis, P. Ray, S.T. Liaw, Recent worldwide developments in eHealth and mHealth to more effectively manage cancer and other chronic diseases - A systematic review, Yearb. Med. Inform. 1 (2016) 93–108. https://doi.org/10.15265/iy-2016-020.

[29] D.C. Christodouleas, B. Kaur, P. Chorti, From point-of-care testing to eHealth diagnostic devices (eDiagnostics), ACS Cent. Sci. 4 (12) (2018) 1600–1616. https://doi.org/10.1021/acscentsci.8b00625.

[30] L. Wallis, P. Blessing, M. Dalwai, S.D. Shin, Integrating mHealth at point of care in low- and middle-income settings: the system perspective, Glob. Health Action 10 (2017). https://doi.org/10.1080/16549716.2017.1327686.

[31] Y. Licher, J. Visser, G. Van, Diehl JC- Formulating design recommendations for the acceptance of the use and results of point-of-care testing in low-and middle-income countries: a Literature review, Proceedings of the 22nd International Conference on Engineering Design (ICED19), 2019 5–8.

[32] K. Verhees, L. Simonse, Care model design for e-Health: integration of point-of-care testing at Dutch general practices, Int. J. Environ. Res. Public Health 15 (1) (2017).

16 Smartphone-based sensors in health and wellness monitoring – Perspectives and assessment of the emerging future

Himadri Sikhar Pramanik[a], Arpan Pal[b], Manish Kirtania[a], Tapas Chakravarty[c], Avik Ghose[b]

[a]Marketing Transformation - Research, Tata Consultancy Services, Salt Lake City, Kolkata, India
[b]Tata Consultancy Services Research & Innovation, Kolkata, West Bengal, India
[c]TCS Research and Innovation, Tata Consultancy Services, Bengaluru, India

In the context of prevailing health and wellness scenario it is important to understand the present state and maturity of smartphone-based sensing as a viable emergent healthcare solution. This chapter is based on review of prevailing literature, study of commercial adoptions, application of mobile and emergent technology in healthcare bounded by subject matter expertise. This will help identify present capabilities, conceptualize enhanced applications of new technologies in navigating the pandemic and other global health issues. In embracing emergent technology particularly for healthcare, it is imperative to minimize associated risks. Multidisciplinary perspectives, evaluation of solution efficiency, considerations for privacy, security, associated legislations, capability building are all related considerations. Prevailing theories of technology adoption under circumstance of health vulnerability and other risks may need reconsideration from newer theory building perspectives.

The chapter provides an overview of smartphone-based capabilities in sensing human health indicators and how this is further enhanced in integration with wearables and implantables. An indicative classification is attempted of the key prevailing applications, while elaborating some key examples that we have keenly observed. We conclude the chapter with perspectives on the future – how we view newer technologies and capabilities to emerge and evolve health sensing.

16.1 Introduction to smartphone-based sensing of health and wellness

Over the last decade smartphones have evolved to enable multiple attractive and comfortable healthcare, wellness and diagnostic solutions. Technology researchers are increasingly integrating functional knowledge from varied areas like – telemedicine [1], biotechnology, sensor and physical systems, chemical and environmental sciences to develop solutions that can be supported on smartphones and allied wearables. Capabilities to directly measure a wide range of physical quantities, wireless connectivity with intelligent eco-systems, ease of operation, scale of smartphone diffusion at relative economy have all contributed towards m-health proliferation.

Smartphone-based solutions leverage data from multiple and vital bio-signals. These include electrocardiogram, heart rate, blood pressure, cardiac electrophysiology for cardiac functioning. Measuring signals from skin and blood are administered in many cases – include respiration rate, blood oxygen saturation, blood glucose, capnography, body temperature. Other observables like motion, location, ambient parameters, that enable these solutions to work with a fair degree of accuracy and effectiveness. In many cases this is achieved by smartphones working in collaboration with intelligent wearable and implantable devices.

Prevailing smartphone-based sensing applications may be broadly categorized as participatory and opportunistic sensing at an individual and collective level [2]. With increasing emergence of artificial intelligence and machine learning, sensing is likely to extend present paradigms towards pervasive and contextual personalized health [3]. In developing and designing these solutions some of the key considerations are power and infrastructure dependency, feasibility, reliability and security, comfort and ergonomics [4, 5]. Moreover, there are considerations of social acceptability, ethics, compliance to regulations, market dynamics, security, technological advances through innovations, among others.

Smartphones and allied wearables today are high capability computing devices with memory, high quality camera, microphone and operating systems. This enables advanced analytical sensing capabilities and their applications are becoming increasingly specific and refined - as reported in the literature [6, 7]. Diagnostics apparatus like microscopic, spectrophotometer, accelerometer, gyroscope, GPS (Global Positioning System) can be or are already built into smartphone through sensors and other low-cost devices with increasing technology miniaturization. In addition to in-built sensing, smartphones are inter-operating across other healthcare devices to augment sensing and other relevant capabilities. These capabilities provide attractive connected, portable solutions globally including in developing nations. These innovative solutions provide for an easy and available alternative to the mainstream medical traditions [8].

Easy access to monitoring health and well-being is likely to become a prime focus of individuals, organizations and governments as we navigate and recover from a global pandemic crisis. Beyond individual health and wellness, the collective power of sensor devices will serve to identify location of patients, containment, quarantine and others from a community health perspective. Smartphones along with other allied wearables and accessories can serve as point-of-care platforms as well. The wearable health devices help to monitor health status both at an activity/fitness level for self-health tracking and at a medical science level. At a medical science level, the data available through smartphones and allied wearables can make connected intelligent ecosystems more robust, targeting effective real time remedies. Prevailing literature indicates application of smartphone-based health solutions for monitoring cardiovascular activity, eye health, respiratory and lung health, skin health, sleep, ear and

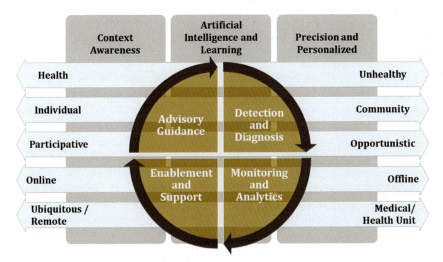

Fig. 16.1 Landscape of smartphone and wearable based sensing for health and well-being.

cognitive function and mental health, among others [9]. General scenario of such applications can be for a patients or even a healthy person in home/remote or within a clinic /hospitals for assessing activity, anomaly and pattern detection, prediction, support diagnosis and care including raising of any alarm in critical ailment conditions [10].

Fig. 16.1 provides a clear representation of the landscape of smartphone and wearable based health and sensing solutions. Technology capabilities make these devices more context aware, intelligent autonomous with the objectives of enhancing precise and personalized interventions. Fig. 16.1 is developed through a qualitative review of existing solutions which have gained some scale. These include solutions which are either being taken to markets or being discussed in public data sources for academic, scientific practices. The key smartphone and wearable based solutions post 2012 have been considered for developing this landscape view.

The range of solutions can primarily classify into four purposes they serve. These solution purposes as evident from our study include: **(1)** Health Advisory and Guidance, **(2)** Health Condition Detection and Diagnosis, **(3)** Continuous Monitoring and Analytics followed with necessary communications and **(4)** Necessary Enablement and Support towards prognosis and therapy. These form the broad classifications of purposes being served by these devices presently. The context of application may be varied leading to diversity of sensing and data analysis mechanisms being deployed. The context of application may be for a healthy person or someone who is suffering from an ailment, it may be individual or even at a community collective level. The sensing may be participative or opportunistic, over network connectivity or otherwise and may be ubiquitous or in predefined medical zones with required infrastructure. The landscape view indicated here helps to classify existing solutions and to understand how newer solutions may emerge. Some existing solutions interestingly operate to achieve more than one purpose and the context of application is also multiple.

16.2 Key observables – sensing body health and wellness signals and beyond

Human bodies generate multiple different physiological signs that can be measured. These can be electrical, biochemical, activity-based, among others. The bio-signals are extracted to better understand health status, wellness. These signals hold a lot of information for health diagnostics and subject reaction to external factors [11]. The prime instances of bio-signals being observed and measured through smartphones and allied technology sensors are heart rate, blood pressure, respiratory rate, blood oxygen saturation and body temperature. Some others include capnography, stroke volume, pain, level of consciousness and urine output. Besides the core medical parameters there are other important parameters monitored, such as skin perspiration and actigraphy. These are deployed to evaluate neurological function, rehabilitation procedures, posture, motion control and sports/activity performance.

Heartrate (HR) is a routine measurement. HR monitoring provides information on physiologic status indicating changes in heart cycle. This is a standard vital sign measured in both healthcare and fitness/sport activities. Blood pressure is considered an important cardiopulmonary parameter; indicates the pressure exerted by blood on arterial walls. Yu-Pin Hsu and Young [12] developed a novel technique for BP measurement. This is based on pulse wave velocity. It uses two micro electro-mechanical sensors installed in two adjacent body parts (may be wrist and neck). Woo et al. [13] proposed an experimental watch-type prototype. The prototype uses a pressure sensor near the radial artery. This technique results in fairly accurate blood pressure measurement on personal smartphones. The sensing capability can be deployed to ensure real-time continuous monitoring of BP through wearable devices.

Respiratory rate ambulatory monitoring is an important technique. This has been effective in detection of symptoms of respiratory diseases. This method is deployed for maladies like sleep apnea syndrome, chronic obstructive pulmonary disease and asthma. Such interventions have improved the administration of treatments for mentioned ailments. There are three primary methods to obtain understanding of respiratory functions [14]. These are primarily elastomeric plethysmography (EP), impedance plethysmography (IP) and respiratory inductive plethysmography (RIP). Besides these three methods, other technologies are also deployed to get respiratory waveform. For example, there are deployment instances of accelerometers; extractions from the ECG signal; derivatives from pulse oximetry polymer-based transducers sensors; and use of optical fibers, among others. Al-Khalidi [15] conducted a comprehensive review on methods used for respiration rate sensing using smartphone and related allied wearables. Monitoring signals relating to human circulatory systems especially blood tend to be minimally invasive. Continuous Glucose Monitoring (CGM) [16] device capable to measure body glucose levels use an adhesive patch with a needle. The CGMs send data wirelessly into a wearable insulin pump to release insulin into the human body. Interconnectedness among these devices also sends data wirelessly to mobile devices for continuous monitoring. However, there are several non-invasive techniques as well. For example, to measure blood oxygen saturation, wearable solutions are deployed. Photoplethysmography (PPG) has gained significant popularity in medical environment.

Skin perspiration is not a conventional clinical parameter. But it is a physiological sign, that can be effectively used to analyze human reaction to several situations. Life situations cause neurological reactions from autonomic nervous system (ANS). This in turn triggers and may increase skin sweating. The change in moisture on the human body surface changes the electrical conductance of the skin. This can be used to measure the quantity of sweat produced by galvanic skin response (GSR) techniques. The nervous system is responsible in control of other physiological parameters like heart rate, respiration and blood pressure. Deploying GSR alongside the acquisitions of some other vital body signals are being leveraged as health and wellness intervention. For example, skin perspiration and heart rate variability can be used in conjunction. Collective signals and the variations can be used to classify mental states, helping in the distinction, also in detection of mental stress. In sports, skin perspiration continuous monitoring is considered an important physiological sign. There are multiple applications in this area and relating to human behavior as well. For example, capnography is a non-invasive and cost-effective method. It is used to evaluate human ventilation, indicating the carbon monoxide levels present in the respiration cycle [17]. Skin temperature varies within a range of temperatures. Human thermoregulation mechanisms regulate core temperature of the body. Skin temperature is affected by multiple factors including blood circulation, heart rate and metabolic rate. Sensing these helps understand deviations from a health perspective. Different wearable systems have been developed to measure body temperature. These wearables form skin-like arrays with precision temperature sensors or come in the form of wearable adhesive devices to enable continuous body temperature sensing. These applications are particularly relevant today in containing prevailing health crisis, large population based thermal screening, to understand and isolate patients in communicable diseases.

Body movements have multiple applications in medical rehabilitation, posture evaluation and sport performance. Motion analysis is widely used in actigraphy [18]. In medical rehabilitation, it is important to monitor mobility. Smartphones are being exploited routinely in specific therapeutic exercises in order to evaluate movements and help in exercise techniques and motion rehabilitation improvements. This has multiple beneficial impacts including, maximizing patient recovery. Monitoring methods to evaluate human rest and activity cycles provide insight of daily activities and routine. These passive sensing of activity is deployed in elderly care, mental health management and other health and well-being programs. Ambient parameters are the environmental parameters in each subjects' surroundings and have a high relevance in several human body monitoring areas. Primary sensors in analysis of ambient systems measure temperature, light, humidity and sound level. The continuous monitoring of air pollutants is gaining ground in urban deployments. Purification in cases of airborne disease is a critical step for control. Measuring air quality is also important due to its association with many cardiac and pulmonary diseases. Outdoor daily activities can be continuously monitored with ambience sensors. It helps analyze surrounding characteristics that human body is being subjected to during sports or therapeutic rehabilitation exercises. Temperature and humidity are important to evaluate dehydration of patients. Effective ambience sensors can make possible estimates into the occupancy and crowded nature of spaces, activity of subjects, and even estimate metabolic rates. These work more effectively mainly in indoors environments as contribution of distractors from external factors can be limited. This will become a key deployment in multiple close spaces including offices, factories, shopping malls as the world recovers while establishing stringent norms of social distancing. According to a study made by Jin et al. [19] it is possible to distinguish several daily indoor activities through light, temperature and humidity measurement patterns. The parameters such as ambient light, sound and temperature are important in study and evaluation of sleep quality and quantity.

Beyond, individual health and wellness monitoring at a physiological level, community health is likely to become increasingly focused. This will be especially relevant to industry and government as global economy gets into revival post the pandemic crisis. This is particularly more relevant for communicable diseases. Individuals carrying smartphones and allied wearables in public spaces will be observed for a comprehensive societal community health check perspective. Such collective community level health and well-being tracking may be considered as broader "human-sensing" the process of extracting information regarding people in some environment and context.

16.3 Sensing capability and mechanism of smartphones, allied wearables and implantables

Smartphones today are fitted with several sensors. Observables measured by smartphone-sensors coupled with device usage patterns provide precise indications on physical and mental health. Understanding of the observables improves even more with sensing and analysis over a long period of time. This is achieved by application of analytical models to understand patterns and through machine learning and intelligence. Disease management and prevention are feasible through collection and analysis of health-related data to the extent where patient-reported observables may be viewed as an enabler for precision medicine [20]. Smartphone and allied wearable-based sensing applications are broadly categorized as participatory and opportunistic sensing. Application of prevailing solutions are evidenced at an individual and collective level based on active or passive involvement. While physical sensors are capability deterministic this draws heavily from advances in ubiquitous and context-aware computing as well. Beyond the smartphone and wearable based sensors device activity-based analytics also plays significantly in determining outcomes. Collection of bio-signals either in a participative or opportunistic manner may happen at an individual/collective level within such landscape. This data can be aggregated further for community health interventions. Our observations indicate sensing capabilities of smartphones are growing. This may be viewed as a continuum as indicated in the Fig. 16.2. The capabilities of solutions are rendered based on the nature of sensing.

At a limited level there is only data based on device usage, the inbuilt sensing capabilities of smart phone can leverage device analytics and more. Smartphone base

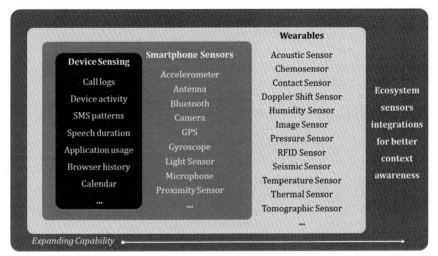

Fig. 16.2 Expanding solution capability. Beyond basic to smart devices further enhanced by wearables and ecosystem sensing.

sensors function as a detector of necessary inputs to feed into healthcare solutions. For example, an accelerometer or gyroscope help determine the speed of movement in space as well as speed of rotation, respectively. The smartphone-based antenna helps in detecting nearby cellular towers and relaying the signal to the broadband processor for voice/SMS/data communication. Bluetooth detecting and communicating helps in transmission of data. Global Positioning Systems (GPS) are key for community health sensing, containment and quarantine applications. Light sensor, microphone, proximity sensors help in greater context awareness to deliver more personalized precise solutions.

The capabilities of a smartphone are further elevated by integrating smart interoperable wearables alongside. These wearables and some of them implantable render smartphone as an instrumental interface [21]. Evidences indicate exploitation of smartphones as an interface device via Wi-Fi, Bluetooth and micro-USB with wearable based sensing and analytical instruments [22]. Devices may be controlled by an App and the results are automatically stored in a smartphone. Smartphones deploy advanced capabilities in analytics as well. Beyond smartphone and allied devices, the ecosystem and context sensing may happen by leveraging existing sensor infrastructures. In such scenario machine to machine communications may generate macro perspectives and more contextual solutions. This will be increasingly leveraged for community health management.

Wearable devices like the smartwatch are increasingly becoming part of mainstream mobile healthcare. Smart watch accelerometer is used to detect various activities using machine learning based sequence classification techniques [23]. Sequence classification varies from usual classification technique in the sense that there is a correlation between instances of data that needs to be captured by the classifier. There are examples that illustrate how a large-scale trial for Atrial Fibrillation can be undertaken using lead ECG from a smart-watch device. Wearables like an in-sole [24] is used for home-based gait monitoring for patients recovering from trauma, surgery of stroke, or patients with chronic motor conditions like Parkinson's Disease or Arthritis. As a next wave, hearing aids and earphones with sensors are also making their way, which have the potential to tap into brain signals like electroencephalogram (EEG) [25] for predicting episodes of epilepsy [26]. There is significant innovation in customizing wearables, making them comfortable and aligned to ergonomic guidelines. The wearables range from macro to micro. Examples of macro-wearables include jackets, harness, digital skin, bands, among others. Examples of micro-wearables include patches, earplugs, glasses, and small implantables.

Beyond wearables, implantables and ingestibles technologies are transformational in mobile health application outcomes, providing real-time guidance to improve health management and tracking [27]. Smart pills monitor the time at which medicine is taken. Powered by chemical reaction within the stomach it sends signals to body patch to monitor heart rate, blood pressure, pH, and temperature. Pacemakers used to treat irregular heartbeats, provides low-energy electrical pulses to restore the normal rhythm. If the heartbeat cannot be restored to a normal rhythm corrective actions are being taken. Pulse generator, which is used for deep brain stimulation is implanted in the brain provides electrical signals to control movement, particularly as an intervention for Parkinson's disease.

Also, particularly noteworthy is the role of emerging Artificial Intelligence (AI). AI when applied to assess observables, discovers patterns – renders many new capabilities. Interestingly, Poduval [28] provides a good overview of how AI is being used in healthcare in general and orthopedics domain. Observations indicate proliferations of similar trends in mobile and wearable health domain. If we look at Artificial Intelligence, it can be divided into learning systems, pattern recognition systems and reasoning systems. Elaborating some examples to illustrate AI role in mobile/wearable based healthcare systems. Patterns of EEG signal captured from geriatric subjects can show early signs of mild cognitive impairment [29]. Deep learning on medical images have shown effectiveness in detection of multiple conditions from cancer to glaucoma [30]. Knowledge based expert systems have proven effective in detecting conditions that have a very subjective differential diagnosis like mental health issues [31]. It is likely, AI will continue to play an important part in mobile based healthcare, in both providing actionable insights to the physicians as well as digital therapeutics [32]. Essentially the capability of physical device based sensing are augmenting with capabilities in analytics including machine intelligence and autonomous practices. However, incorporation of AI in healthcare needs to address issues and concerns around privacy, trust, liability and ethics before they are deployed at scale. As such AI also needs to become 'explainable' from causality perspective in order to become a true 'helping-hand' tool for doctors and healthcare providers.

Depending on the sensor capabilities and application the actual sensing mechanism varies considerably. Three key sensor networks have been applied across solutions. These are Body Area Network (BAN), Body Sensor Network (BSN) and Wireless Sensor Network (WSN). While there is varied differences among these three, there are multiple instances where the distinction is diffused. These three are used in different types of smartphone and wearable devices according to its architecture. The inter connection of sensors creates a network of sensors (BAN), which are sent to smartphone processing which in many cases is also portable. Such data may be shared with other intelligent network through wireless connectivity for storage and analytical processing. WSNs usually involve large numbers of low-cost, low-power and tiny sensor nodes, with each node having a predefined set of components (sensors, microcontroller, memory among others) granting that each node has sensing, computing, storage and communication capabilities. Multiple benefits can be realized by connecting prime sensors establishing a network. It enables centralization of data in a single unit gathering information from different sensors. The data can be analyzed to learn patterns and derive insights. This can enable sending data to external networks for remote processing and necessary medical interventions. Patients can be monitored remotely. Establishing networks enhance greater control, synchronization, scheduling and programming of whole healthcare systems. These are significant for infrastructure and necessary inter-operability across devices. Necessary system adaptations can be coordinated aligned to prevailing patient conditions and external environment. These advantages culminate in optimal resources usage.

Wireless communication is a key asset to enable systems go mobile and ubiquitous. A Wireless Body Sensor Network (WBSN) defines an autonomous system which is used to monitor the daily life activities including health and well-being. This consists of intelligent sensor nodes which do not hamper the daily life activities. It is useful in detecting chronic diseases like heart attack, asthma, diabetes etc. Early variations enable these sensors to warn patients to potentially avoid emergency conditions. The sensor nodes in WBSN consume less power and serve multiple applications in healthcare. With the help of WBSNs monitoring activities, movements and vital human body signals from a remote location by the means of internet is possible. Privacy secured aggregation and analytics of individual data can be used to understand health variables at a community and location basis. The smartphones and allied devices have enhanced capabilities in detection and to serve as an instrumental interface in some cases.

16.4 Classification of smartphone and wearable applications for healthcare

Based on observations across key solutions exploiting smartphone and allied wearables we include a classification in Table 16.1. We have documented the broad health condition being addressed by these solutions as the basis of classification. We observe primary applications of smartphone and wearables for **(A)** Mental Health and Well-Being, **(B)** General Health and Activity, **(C)** Key Organ Health and Well-being. Based on our study we capture the prevailing solution capabilities and the relative scale of adoption. Scale of adoption is commented based on availability of commercially viable products in the market and by evaluating credible advocacies and opportunities for technology adoption. Also indicated the mechanism of vital sign sensing through sensors and other observables. It is evident from this collation that most solutions

Table 16.1 Summary of key applications of smartphones and wearables in healthcare.

(A) Mental Health and Well-Being

Health Condition	Application of Smartphone and Wearable Solutions	Measured Vital Sign	Sensors Used	Other Observables Context Awareness	Relative Scale of Adoption
Bipolar Disorder [33]	Correlate physical activity (state and state change) with symptoms of bipolar disorder. Social Rhythm Metric (SRM) scale for personalized models.	Activity, SRM	Accelerometer, Microphone, Light sensor, GPS	Call logs, SMS Patterns	Low
Depression [34], [35]	Improving depressive symptoms, Predict from location and phone usage data. Correlate location with symptoms. Context sensitive cognitive behavioral therapy-based micro-interventions	Activity	Accelerometer, GPS	Call logs, SMS Patterns, Calendar	High
Diseases like Schizophrenia, Autism [36]	Feasibility of passive sensing, detect out-of-home activities to understand social functioning	Activity	Accelerometer, Bluetooth, GPS, Light sensor, Microphone	Call logs, SMS Patterns, Calendar	Low
General Mental Health [37]	Predict mood, Correlate smartphone data with depressive symptoms. Measure daily stress levels, mental health status from smartphone data like sleep, mobility speech duration, geospatial activity, kinesthetic activity and social anxiety	Daily Activity, Speech duration	Accelerometer, Bluetooth, GPS, Light sensor, Microphone	Device activities like Call logs, SMS Patterns, Calendar	High
Addiction [38], [39]	Assess smartphone addition, Location data to send timely messages to smokers	Phone activity	GPS	Smartphone daily use time, speech, Application usage, Diet	Low
Stress [40]	Assess stress levels from smartphone data	Phone activity, Speech Duration	Accelerometer, Light sensor, GPS	Device activities like Call logs, SMS Patterns, Calendar, Diet	Medium

(B) General Health and Activity				
Sleep [41], [42]	Sleep quality prediction from smartphone data, Sleep monitoring device, predict sleep time, duration, and deprivation, iphone based results compared with polysomnography, Movement during sleep	Daily Activity including sleep, Movement, Patterns, Anomalies	Accelerometer, Light sensor, Microphone, Proximity sensor	Ambient variables, Application usage, Device activity, Sleep duration	Medium
General Health and Well Being [43]	Recommendations for calorie loss from accelerometer and location data, Assess sociability, Overall health status	Daily Activity, Diet	Bluetooth, Accelerometer, GPS	Application usage, call logs, Pressure, Temperature, Humidity, Light	High
Activity [44]	Human activity and gait recognition, Measure of multiple daily activities like sitting, walking, jogging. Also measure of ascending and descending stairs at different paces. Fall detection and notification system, Dynamic knee extension, Knee range of motion	Step length, Step width, acceleration, speed, vector, Diet	Accelerometer, gyroscope, magnetometer, light sensor, GPS, Sensors to measure Temperature, Pressure, Humidity, Light	Call log, SMS & Application Usage, Device activity, humidity, temperature, and barometric pressure, sensing other environmental parameters	High
Geriatrics [45], [46]	Physical activity from accelerometer to motivate older adults for fitness, mitigate loneliness by messaging, Fall detection and notification system	Daily Activity	Accelerometer, Proximity Sensors, GPS	Application usage, call logs, SMS	Medium
(C) Organ Health and Well-being				
Heart/ Cardiovascular Diseases [47], [48], [49]	Sense physical activity for those with chronic heart failure-constant monitoring, Smartphone camera sensors are deployed to measure heart rate and variability from PPG signals derived from videography of bare skin	Daily Activity, Heart Rate (HR), Variability in Heart Rate (HRV) and Pulse rates	Accelerometer, Proximity Sensors, GPS, Light Sensors, Camera	Call Logs, Video	Medium

(continued)

Table 16.1 (Cont'd)

Pulmonary Health [50], [51]	Distinct breathing pattern, Spectrogram of the cough sound isolated, Realize a low-cost spirometer in a mobile platform. Assessing the flow and volume of air inhaled and exhaled by lungs for breathing, Parameters such as forced vital capacity, forced expiratory volume, peak expiratory flow, and key ratios. Personalized tool for lung cancer risk assessment	Inhalation and Exhalation Patterns, Coughs, Volume-Time, Flow volume	Microphone, Bluetooth, Associated Wearable Chest Belts, GPS	Spirometric flow curves, Tobacco use, occupational environment and family history	Low
Ophthalmic Health [52]	Smartphone camera for retinal fundus imaging, Portable ophthalmic imaging system. Used to identify visual impairment in children	Eye images	Light sensors, Camera	Application usage, call logs, SMS	Low
Skin Health [53]	Smartphone-based Teledermoscopy, Smartphones deployed to test both in vitro and in vivo for the Hemoglobin index and Melanin index. Captured cellular details of human skin with a smartphone, low-cost confocal microscopes are deployed for investigations	Images	Light sensors, Camera	Application usage, call logs, SMS	Low
Diabetes [54]	Diabetes apps for Android smartphones suggesting medications or calculate prandial insulin dosages	Self-monitoring of blood glucose	Data inputs-app based	Device interactions	Medium
Hearing impairment [55]	Applied for hearing loss screening, Smartphone-based audiometer, Smartphone based hearing aids allow users to control volume and frequency-gain response. These can be modulated and adjusted for patient comfort.	Speech, Ambient Noise	Microphone, Bluetooth, Associated wearable	Device Activity, Speech duration and intelligibility	Low
Skeletal System [56], [57]	Interventions to prevent osteoporosis by increasing calcium intake or physical activity	Daily activity, Diet	Accelerometer, Proximity Sensors, GPS	Application, call logs, SMS	Low

include multiple sensing approaches to generate results. The summarized view in Table 16.1 help readers gain a perspective on the range of prevailing applications and understand new frontiers of extension and explorations.

16.5 Highlights of key case-studies in smartphone-based health & wellness sensing

In this section, we elaborate on some case studies where mobile phones and connected wearable devices have been used in studies related to healthcare and wellness monitoring and assessment. In most such cases, as we shall observe the mobile phone has been used more as an access device and edge computing device as outlined by Ghose [58]. While this is a common practice, the mobile itself can also be used as a sensing device as indicated by Chandel [59].

The cases presented here are aligned to our earlier classification of applications in categories of (**A**) *mental,* (**B**) *general and* (**C**) *organ health and well-being* as indicated in Table 16.1.

(A) **Mental Health and Well-Being**

Monitoring mental and associated physical well-being Management of mental health is a natural premise for mobile based health monitoring and assessment because people interact with their mobile phones on a regular basis during their daily life and simple inferences like social isolation, emotional disposition or activity levels can be derived from long term mobile usage patterns. Mobile apps also become a source of ubiquitous mental health assessment via a variety of Human Computer Interaction (HCI) modalities. A well-versed illustration is provided by Petrucka [60]. Moreover, youth today have substantially longer and deeper interactions with their mobile devices than adults. They are psychologically a vulnerable and less understood group. Hence, using mobile based mental health management for adolescents and young adults is extremely relevant. This paradigm is explored by Kenny [61]. Grist [62] provides a good systematic review of all such applications presented in research. Most of them are instrument based that capture patient reported outcomes in various gamified manner. For example, 'Journey' [63] is a game designed to induce a positive 'affect', which means induce a positive emotional valence in the user via meaningful interactions. However, there are major challenges in creating mobile applications for mental health including regulatory compliance, adoption, observation bias amongst others. The challenges are well articulated by Olff [64].

In this section we will look at a few case studies on monitoring physical and mental well-being parameters of employees working in potentially hostile environments from the physical and/or mental well-being perspective. We provide two case studies of a lone worker scenario and workplace stress monitoring scenario.

Monitoring the physical well-being of a lone worker in hazardous conditions: The work by Mukhopadhyay [65], discusses the need for ambulatory heart rate measurement using the Photoplethysmography (PPG) signal from a wrist worn device for a lone worker operating in hazardous conditions. Since, the worker is performing daily duties, it becomes important we can measure heart rate in the presence of arm/hand motion. The solution provides an approach of using the 3-axis accelerometer on the device as a motion signal generator and then use simple signal processing methods and decision tree-based classifiers to detect the heart rate on the wearable device itself. As data can be streamed to a mobile device, a more sophisticated approach illustrated by Ahmed [66] is used to measure heart rate more accurately. The updated method uses a technique called singular spectrum analysis (SSA). In this technique, a signal is converted to its Henkel matrix and a Singular Value Decomposition (SVD) is performed on the same. Each of the singular vectors now represent one independent component in the signal. Now, the hypothesis being made here is that noise that is out of band in frequency domain can be removed by removing similar singular vectors from both domains. The in-band noise can then be removed using a Wiener filter, which is the variant of an adaptive filter.

Monitoring stress-levels of employees in a workplace: Here, we consider a group of employees performing desk jobs in a typical office environment. However, work related stress is a major factor effecting their mental well-being. The work by Das [67] goes on to show that the Galvanic Skin Resistance (GSR) sensor on a wrist wearable can be coupled with analysis on a mobile phone to provide a measure of mental workload for a given task. A further elaborate research on the topic yields that a multi-modal approach may be needed to study employee behavior and mine the factors that lead to building up of stress at workplaces. Multiple sensors on wearable devices are used to get user location, activity and physiology along with mental effort. This helps to build a complete model for employee stress analysis which is an important parameter in well-being.

(B) **General Health and Activity**

General health and activity sensing may range around varied applications. Here we elaborate on the geriatrics and communicable diseases. Both are increasingly relevant, with aging economies and limited care givers and in the wake of global health crisis with an evolving pandemic.

Geriatric care – Ambient Assisted Living: Caring for elderly populations who are ageing can be driven by technology as shown in Ghose [68]. A multitude of sensors and technology have been proposed to enable ubiquitous monitoring of elderly subjects. A system is created which enables sensing and analytics for elderly population. In this scenario the wearable device connected to the subject's personal mobile device acts as an important instrument in understanding the activities of daily living as shown in Jaiswal [69]. Arguably, continuous detection of activity of an elderly subject would require that the systems be able to operate at power-levels that allow the wearable and other personal devices like the mobile last for an entire day. In a geriatric facility, algorithms like fall detection, immobility detection and activity levels as described by Chandel [70], needs to be combined with learning models for detecting upper body gestures and the whole system needs to be optimized for sensor consumption, network usage and computation with respect to power. While fall and immobility are emergency conditions and needs immediate attention, some inferences like throughout the day activity levels, detection of episodes of eating, smoking etc. can be intervened from immediate computation and hence can be offloaded to the mobile device for computing.

Probably, the next important point to ponder is what kind of actionable insights can be provided based on this data, apart from emergency response like fall detection. The work done by Choudhury [71] looks at using the data of daily behavior to identify early cases of Mild Cognitive Impairment (MCI) in the elderly. The work is based on the principal that human behavior is repetitive and every individual, even in the condition of not being supervised has an inherent "routine" which is followed. However, when memory starts becoming weaker, the authors make a hypothesis that as memory degrades, the routine disintegrates, and the behavior becomes less orderly. They show the efficacy of the solution against the administration of the Abbreviated Mental Test (AMT) instrument. Similarly, the work by Chandel [72] shows how co-analysis of activity and physiology can lead to the analysis of cardiac fatigue which is an early indication of heart-failure. The heart rate, breathing rate, breathing power is collected before and after an activity. The Metabolic Equivalent of Task (MET) for the activity is computed and along with the mentioned parameters the recovery time to basal heart rate is used to define the computational model of fatigue.

Beyond life-style wearable devices, some specialized wearable devices which can connect to a mobile phone like an insole as described by Chandel [73] can be used to monitor the gait parameters in home setting for an elderly patient. Such technology can help patients having muscular, skeletal, spinal and neuro-motor disorders like Parkinson's Disease (PD), along with patients recovering from stroke and rehabilitating from knee replacement or trauma – as may be highly prevalent in elderly population.

Communicable Disease Management: Mass digitization of health using mHealth technologies will lead to geo-tagging of symptoms that can lead to early identification of epidemics and pandemics. In the wake of the prevailing pandemic and other likely health crisis, one major application of mHealth is contact tracing whereby we know the physical proximity history of a patient for mining suspected infections for proactive quarantine. However, such proximity analysis immediately raises privacy concerns, Abeler [74] provides an analysis of how such technologies can be made privacy friendly and compliant to such regulations. Ghose [75] provides a method for determining proximity between mobile devices using Bluetooth. Mobile based infrastructure is aptly utilized to understand wellness in various human contexts. In post COVID-19 scenarios, we see how mobile devices are being used for proximity tracking thereby enabling contact tracing. Bluetooth based proximity tracking as illustrated by Ghose indicates how distance estimates can be improved using commercial Bluetooth devices by defining a mathematical model for spatial frequencies in

Bluetooth channel. This idea is extended in Chandel [76] which uses BLE (Bluetooth Low Energy). Such contact tracing allows us to monitor people who have met potential patients during the estimated incubation period.

(C) **Organ Health and Well-being**

Monitoring Personal Well-Being at Home: In a beyond COVID-19 world, due to social distancing norms and high probability of infections in hospital OPD (Outpatient Department), home based monitoring will take a pivotal role for chronic patients and other ailments with issues like hypertension and diabetes etc. Hence, technology-enabled need adoption and implementation at scale for personal well-being. Parameters like Blood Pressure is possible to measure using wearable devices as shown by Ahmed [77], Banerjee [78].

Key attributes of remote monitoring systems to be effective are user-friendliness, reliability, privacy-protection and cost-effectiveness. While wearable based sensing together with mobile gateway is an attractive solution, significant research is continuing to make the measurements accurate and consistent. Recent work by Mukherjee et al. [79] is a step in that direction. Here the authors focus on three components related to cardiac health namely Heart Rate, Heart Rate Variability and Blood Pressure. A wrist wearable reflective type PPG sensor along with smartphone has been proposed. A virtual network computing (VNC) server is used to enable remote access. In order to accurately decipher BP from PPG signal, authors have established the correlation between PPG maxima (or minima) with Systolic Blood Pressure (or Diastolic Blood Pressure) using the variations of skin tone of the subjects under test. The training dataset consisted of 70 volunteers and the trial set consisted of 20 new volunteers. The accuracy obtained is shown to be at 98.5 percent so far.

Remote home monitoring will also be successful if we are able to monitor personal well-being in a privacy preserving setting while letting the person carry on normal daily routine. Electromagnetic (EM) imaging is a promising technology which enables to detect human activities (and vital signs) in their natural settings. A step forward in this direction leads to the design of small, reliable, energy efficient and customizable millimeter wave (MMW) radar units forming a network through mobile gateway.

16.6 Focus on emergent technologies

We believe there are multiple technology elements that will determine the growth and proliferation of smartphone-based health solutions. These may involve artificial intelligence, analytics, edge computing, connectivity and sensing capabilities. We identify three key such future and emergent technologies for the next frontier in smartphone-based sensing. Included here is a brief commentary on the emergent technologies that will serve as key enabler going forward.

(i) **Radar based vitals, movement and in-body imaging:** Radar based human sensing offers three distinct advantages like ability to sense at distance, privacy preserving and not affected by ambience. In addition, this is a non-ionizing sensor and can be used for frequent measurements. This unobtrusive sensing system can detect human presence; decipher activity/gesture as explained by Giggie [80, 81] as well as continuously monitor cardio-pulmonary signature [82, 83].

For activity detection, the proposed system consists of a panel of networked radars which are spatially separated radars. These radars jointly help in musculoskeletal and cardiopulmonary activity detection. Continuous Wave (CW) radars have been designed to measure both breathing rate as well as heart rate. Going forward, these radars units will be configured to get connected to smartphones while ensuring data security and privacy compliance. Radar based sensing have also created new possibilities for in-body imaging (in the domain of medical imaging). Ultra-wide band (UWB) radars are known to penetrate human body due to its high bandwidth. It is possible to utilize this feature to build internal images like X-ray and MRI. However, unlike X-ray and MRI, radar-based imaging is likely to offer coarse spatial resolution and high temporal resolution.

A future is envisaged where miniature, low power ultra-wide band (UWB) devices are built as external attachment to smartphones which will in-turn reconstruct the image. Such a device will push advanced medical imaging techniques to home space thereby greatly enhancing a citizen's quality of life. Radars are becoming parts of mobile devices as Google Pixel brings in the project soli radar in emerging versions. Similarly, Apple iPhone 11 is armed with an UWB radar. These sensors are primarily mounted for detecting gestures but can be possibly used to monitor vitals and activity in a non-contact fashion, which makes them ideal for monitoring in scenarios of infectious diseases as well.

(ii) **Analysis of Volatile Organic Compounds:** While a radar system can detect physical changes in the environment, a Volatile Organic Compound (VOC) sensor can detect chemical changes in the ambience which leaves a chemical trail (chem-trail) in the human volatiles. Various diseases have their own volatile markers as shown by Anton [84]. The study indicates a wide range of bio-chemical base of breath molecules. It is envisaged that just like today's mobile devices come with eyes (camera) and ears (microphone), the mobiles of tomorrow will be equipped with noses (VOC sensors).

(iii) **Hyperspectral Sensing:** Another modality of well-being sensing is the application of hyperspectral sensing. From food quality estimation as depicted in Mithun [85], to brain mapping using NIR as provided by Masako [86], to detecting arterial blockages as shown by Imani [87]. Currently a few prototype NIR devices are available that connect to mobile device for data aggregation and analytics.

16.7 Conclusion

The prevailing pandemic has changed entire behavioral pattern of humans, economies, societies and governments globally. The crisis has disrupted surrounding healthcare and other eco-systems. Physical, in-contact practices are limited and will be increasingly contained as much as possible. What this means for digital health is that tele-health and remote monitoring are going to become more normative. Trials are going to be largely decentralized and digital therapeutics is likely to be the new primary care for chronic disease management. Continuous monitoring of symptoms in physical distancing is going to be crucial for human safety and survival. There is likely to be a more diffused approach towards individual, collective and well-being of communities – with significant influence and overlap among these. Individual health data will need to be aggregated while securing privacy to enable macro-level interventions for health and well-being. In all these trends, the mobile devices are going to play a crucial role, playing the gateway to wearable, implantable and ingestible sensors and continue to act like a sensing device itself.

In this chapter we attempted introduction to smartphone-based sensing of health and wellness. The chapter provides a classificatory perspective into the available solutions by indicating a landscape of applications and intent. Included key observables in humans based on available sensing capabilities of smartphones and allied devices. While in many cases there is focus on a single vital sign like body temperature or heart rate, there are also instances on multi-sign sensing and monitoring for greater accuracy. We discuss how sensing capabilities are enhanced by deploying smartphones along with smart wearables, implantables and ingestables and how solutions are increasingly adopting advanced analytics, artificial intelligence-based learning. The chapter includes summary view across many smartphone and wearable applications for healthcare in the domain of mental, general and organ health and well-being. We envisage the mobile device to be equipped with further sensors like radars that pick up human movement and vitals, to hyperspectral cameras and perhaps electronic noses that can sense the ambience. As mobile computing power continues to increase it will also become a critical health analytics platform for the future.

Beyond technology of sensing and capabilities associated with it, there are multiple other allied disciplines. These include considerations of social acceptability, ethics, privacy, compliance to regulations, market dynamics, security, trust, technological advances through innovations, among others. The role of multiple agents like individuals, doctors, healthcare providers, innovators, government and other advocacies need greater explorations. It would be interesting to understand how theoretical constructs of technology adoption alongside relevant agents play moderating and mediating influence on the proliferation of new-age mobile health and wellness sensing and management.

References

[1] B.M.C. Silva, J J P C Rodrigues, I. de la Torre Díez, M. López-Coronado, K Saleem, Mobile-health: a review of current state in 2015, J. Biomed. Inform. 56 (2015) 265–272. https://doi.org/10.1016/j.jbi.2015.06.003.

[2] Teixeira, T., Dublon, G., & Savvides, A. (2010). A survey of human-sensing:methods for detecting presence, count, location, track, and identity (Vol. 1).

[3] N. Oliver, O. Mayora, M. Marschollek, Machine learning and data analytics in pervasive health, Methods Inf. Med. 57 (4) (2018) 194–196. https://doi.org/10.1055/s-0038-1673243.
[4] F. Seoane, I. Mohino-Herranz, J. Ferreira, L. Alvarez, R. Buendia, D. Ayllón, C. Llerena, R. Gil-Pita, Wearable biomedical measurement systems for assessment of mental stress of combatants in real time, Sens. (Switzerland) 14 (4) (2014) 7120–7141. https://doi.org/10.3390/s140407120.
[5] T. Yilmaz, R. Foster, Y. Hao, Detecting vital signs with wearablewireless sensors, Sens. 10 (12) (2010) 10837–10862. https://doi.org/10.3390/s101210837.
[6] K.E. McCracken, J.Y. Yoon, Recent approaches for optical smartphone sensing in resource-limited settings: a brief review, Anal. Methods 8 (36) (2016) 6591–6601. https://doi.org/10.1039/c6ay01575a.
[7] S. Kanchi, M.I. Sabela, P.S. Mdluli, Inamuddin, K Bisetty, Smartphone based bioanalytical and diagnosis applications: a review, Biosens. Bioelectron. 102 (2018) 136–149. https://doi.org/10.1016/j.bios.2017.11.021.
[8] F. Xu, A. Akay, H. Wei, S. Wang, B. Pingguan-Murphy, B.E. Erlandsson, X. Li, W. Lee, J. Hu, L. Wang, Advances in smartphone-based point-of-care diagnostics, Proc. IEEE 103 (2) (2015) 236–247. https://doi.org/10.1109/JPROC.2014.2378776.
[9] S. Majumder, M.J. Deen, Smartphone sensors for health monitoring and diagnosis, Sens. (Switzerland) 19 (9) (2019). https://doi.org/10.3390/s19092164.
[10] H. Banaee, M.U. Ahmed, A. Loutfi, Data mining for wearable sensors in health monitoring systems: a review of recent trends and challenges, Sens. (Switzerland) 13 (12) (2013) 17472–17500. https://doi.org/10.3390/s131217472.
[11] J.C.D Duarte, Wearable health devices—vital sign monitoring, systems and technologies, Sen. 18 (2018).
[12] Y.P. Hsu, D.J. Young, Skin-coupled personal wearable ambulatory pulse wave velocity monitoring system using microelectromechanical sensors, IEEE Sens. J. 14 (10) (2014) 3490–3497. https://doi.org/10.1109/JSEN.2014.2345779.
[13] S.H. Woo, Y.Y. Choi, D.J. Kim, F. Bien, J.J. Kim, Tissue-informative mechanism for wearable non-invasive continuous blood pressure monitoring, Sci. Rep. 4 (2014). https://doi.org/10.1038/srep06618.
[14] l Guo, L. Berglin, U. Wiklund, H. Mattila, Design of a garment-based sensing system for breathing monitoring, Text. Res. J. 83 (5) (2013) 499–509. https://doi.org/10.1177/0040517512444336.
[15] F.Q. Al-Khalidi, R. Saatchi, D. Burke, H. Elphick, S. Tan, Respiration rate monitoring methods: a review, Pediatr. Pulmonol. 46 (6) (2011) 523–529. https://doi.org/10.1002/ppul.21416.
[16] N. Babu et al., Continuous Glucose Monitoring. J. Global Trends. Pharm. Sci. 11 (2) (2020) 7562–7568. https://www.researchgate.net/publication/340982796_CONTINUOUS_GLUCOSE_MONITORING_DEVICES_A_SYSTEMATIC_REVIEW.
[17] K. Wac, C. Tsiourti, Ambulatory assessment of affect: survey of sensor systems for monitoring of autonomic nervous systems activation in emotion, IEEE Trans Affect Comput 5 (3) (2014) 251–272. https://doi.org/10.1109/TAFFC.2014.2332157.
[18] S. Mohammed, I. Tashev, Unsupervised deep representation learning to remove motion artifacts in free-mode body sensor networks, 2017 IEEE 14th International Conference on Wearable and Implantable Body Sensor Networks, BSN 2017, Institute of Electrical and Electronics Engineers Inc, 2017, pp. 183–188. https://doi.org/10.1109/BSN.2017.7936037.
[19] M. Jin, H. Zou, K. Weekly, R. Jia, A.M. Bayen, C.J. Spanos, Environmental sensing by wearable device for indoor activity and location estimation, IECON Proceedings

[19] (Industrial Electronics Conference), Institute of Electrical and Electronics Engineers Inc, 2014, pp. 5369–5375. https://doi.org/10.1109/IECON.2014.7049320.

[20] V.P. Cornet, R.J. Holden, Systematic review of smartphone-based passive sensing for health and wellbeing, J. Biomed. Inform. 77 (2018) 120–132. https://doi.org/10.1016/j.jbi.2017.12.008.

[21] S. Kanchi, M.I. Sabela, P.S. Mdluli, Inamuddin, K Bisetty, Smartphone based bioanalytical and diagnosis applications: a review, Biosens. Bioelectron. 102 (2018) 136–149. https://doi.org/10.1016/j.bios.2017.11.021.

[22] G. Andreoni, C.E. Standoli, Perego, Defining requirements and related methods for designing sensorized garments, Sensors 16 (2016).

[23] G.M. Weiss, J.L. Timko, C.M. Gallagher, K. Yoneda, A.J. Schreiber, Smartwatch-based activity recognition: a machine learning approach, 3rd IEEE EMBS International Conference on Biomedical and Health Informatics, BHI 2016, Institute of Electrical and Electronics Engineers Inc, 2016, pp. 426–429. https://doi.org/10.1109/BHI.2016.7455925.

[24] I. González, J. Fontecha, R. Hervás, J. Bravo, An ambulatory system for gait monitoring based on wireless sensorized insoles, Sens. (Switzerland) 15 (7) (2015) 16589–16613. https://doi.org/10.3390/s150716589.

[25] A.J. Casson, Wearable EEG and beyond, Biomed Eng Lett 9 (1) (2019) 53–71. https://doi.org/10.1007/s13534-018-00093-6.

[26] Y. Salant, I. Gath, O. Henriksen, Prediction of epileptic seizures from two-channel EEG, Med. Biol. Eng. Comput. 36 (5) (1998) 549–556. https://doi.org/10.1007/BF02524422.

[27] H.C. Koydemir, A. Ozcan, Wearable and implantable sensors for biomedical applications, Annu. Rev. Anal. Chem. 11 (2018) 127–146. https://doi.org/10.1146/annurev-anchem-061417-125956.

[28] M. Poduval, A. Ghose, S. Manchanda, V. Bagaria, A. Sinha, Artificial intelligence and machine learning: a new disruptive force in orthopaedics, Indian J Orthop 54 (2) (2020) 109–122. https://doi.org/10.1007/s43465-019-00023-3.

[29] J. Snaedal, G.H. Johannesson, T.E. Gudmundsson, N.P. Blin, A.L. Emilsdottir, B. Einarsson, K. Johnsen, Diagnostic accuracy of statistical pattern recognition of electroencephalogram registration in evaluation of cognitive impairment and dementia, Dement. Geriatr. Cogn. Disord. 34 (1) (2012) 51–60. https://doi.org/10.1159/000339996.

[30] J.G. Lee, S. Jun, Y.W. Cho, H. Lee, G.B. Kim, J.B. Seo, N. Kim, Deep learning in medical imaging: general overview, Korean J. Radiol. 18 (4) (2017) 570–584. https://doi.org/10.3348/kjr.2017.18.4.570.

[31] W.M. Wang, C.F. Cheung, W.B. Lee, S.K. Kwok, Knowledge-based treatment planning for adolescent early intervention of mental healthcare: a hybrid case-based reasoning approach, Expert Syst. 24 (4) (2007) 232–251. https://doi.org/10.1111/j.1468-0394.2007.00431.x.

[32] S. Makin, The emerging world of digital therapeutics, Nature 573 (2019).

[33] A. Grünerbl, G. Bahle, J. Weppner, P. Oleksy, C. Haring, P. Lukowicz, Towards Smart Phone Based Monitoring of Bipolar Disorder, Proceedings of the Second ACM Workshop on Mobile Systems, Applications, and Services for HealthCare, 2012.

[34] G. Eysenbach, J. Proudfoot, B. Dear, M.N. Burns, M. Begale, J. Duffecy, D. Gergle, C.J Karr, E. Giangrande, D.C. Mohr, Harnessing Context Sensing to Develop a Mobile Intervention for Depression, J. Med. Internet Res. 13 (3) (2011) 55.

[35] F. Wahle, T. Kowatsch, E. Fleisch, M. Rufer, S. Weidt, Mobile Sensing and Support for People with Depression: A Pilot Trial in the Wild, JMIR Mhealth Uhealth 4 (3) (2014) 111.
[36] D. Ben-Zeev, R. Wang, S. Abdullah, R. Brianv, E.A. Scherer, L.A. Mistler, M. Hauser, J.M. Kane, A. Campbell, T. Choudhury, Mobile Behavioral Sensing for Outpatients and Inpatients with Schizophrenia, Psychiatr. Serv. 67 (5) (2016) 558–591.
[37] J. Asselbergs, J. Ruwaard, M. Ejdys, N. Schrader, M. Sijbrandij, H. Riper, Mobile phone-based unobtrusive ecological momentary assessment of day-to-day mood: an explorative study, J. Med. Internet Res. 18 (3) (2016) 72.
[38] H.J. Kwon, Development of smartphone-based ECL sensor for dopamine detection: Practical approaches, Results in Chemistry 2 (2020) 100051.
[39] F. Naughton, S. Hopewell, N. Lathia, R. Schalbroeck, C. Brown, C. Mascolo, et al., A context-sensing mobile phone app (Q Sense) for smoking cessation: a mixedmethods study, JMIR mHealth uHealth 4 (2016) 106.
[40] H. Lu, D. Frauendorfer, M. Rabbi, M.S. Mast, G.T. Chittaranjan, A.T. Campbell, D. Gatica-Perez, T. Choudhury, StressSense: Detecting stress in unconstrained acoustic environments using smartphones, In Proceedings of the 2012 ACM Conference on Ubiquitous Computing—UbiComp, 2012, p. 351.
[41] R. Nandakumar, S Gollakota, N. Watson, Contactless Sleep Apnea Detection on Smartphones, GetMobile: Mob. Comput. Commun. 19 (2015) 22–24.
[42] V. Natale, M. Drejak, A. Erbacci, L. Tonetti, M. Fabbri, M. Martoni, Monitoring sleep with a smartphone accelerometer, Sleep Biol. Rhythms 10 (2012) 287–292.
[43] M. Rabbi, A. Pfammatter, M. Zhang, B. Spring, T. Choudhury, Automated personalized feedback for physical activity and dietary behavior change with mobile phones: a randomized controlled trial on adults, JMIR Mhealth Uhealth 14 (3) (2015).
[44] M. Seera, C.K. Loo, C.P. Lim, A hybrid FMM-CART model for human activity recognition, In Proceedings of the IEEE International Conference on Systems, Man and Cybernetics—SMC, San Diego, CA, USA, 2014, pp. 182–187.
[45] Y. Lee, H. Yeh, K. Kim, O.A. Choi, Real-time Fall Detection System Based on the Acceleration Sensor of Smartphone, Int. J. Eng. Bus. Manag. 10 (2) (2018) 315–326.
[46] S. Abu-Ghanem, O. Handzel, L. Ness, M. Ben-Artzi-Blima, K. Fait-Ghelbendorf, M. Himmelfarb, Smartphone-based audiometric test for screening hearing loss in the elderly, Eur. Arch. Otorhinolaryngol. 273 (2) (2015) 333–339.
[47] A. Bánhalmi, J. Borbás, M. Fidrich, V. Bilicki, Z. Gingl, L. Rudas, Analysis of a Pulse Rate Variability Measurement Using a Smartphone Camera, J. Healthc. Eng. (2018) 15. eHealth Solutions for the Integrated Healthcare.
[48] S.A. Siddiqui, Y. Zhang, Z. Feng, A. Kos, A Pulse Rate Estimation Algorithm Using PPG and Smartphone Camera, J. Med. Syst. 40 (5) (2016) 126.
[49] N. Koenig, A. Seeck, J. Eckstein, A. Mainka, T. Huebner, A. Voss, S. Weber, Validation of a New Heart Rate Measurement Algorithm for Fingertip Recording of Video Signals with Smartphones, Telemedicine and e-Health 22 (8) (2016).
[50] E.C. Larson, M. Goel, M. Redfield, G. Boriello, M. Rosenfeld, S.N. Patel, Tracking lung function on any phone, Proceedings of the 3rd ACM Symposium on Computing for Development, 2013, p. 29.
[51] Z. Szanto, I. Benko, L. Jakab, G. Szalai, A. Vereczkei, The use of a smartphone application for fast lung cancer risk assessment, Eur. J. Cardiothorac. Surg. 51 (6) (2017) 1171–1176.

[52] B.C. Toy, D.J Myung, L. He, C.K. Pan, R.T. Chang, A. Polkinhorne, D. Merrell, D. Foster, M.S. Blumenkranz, Smartphone-Based Dilated Fundus Photography and Near Visual Acuity Testing as Inexpensive Screening Tools to Detect Referral Warranted Diabetic Eye Disease, Retina 36 (5) (2016) 1000–1008.

[53] S. Kim, D. Cho, J. Kim, M. Kim, S. Youn, J.E. Jang, M. Je, D.H. Lee, B. Lee, D.L. Farkas, J.Y. Hwang, Smartphone-based multispectral imaging: system development and potential for mobile skin diagnosis, Biomed. Opt. Express 28 (7) (2016) 5294–5307.

[54] A.P. Demidowich, K. Lu, R. Tamler, Z. Bloomgarden, An evaluation of diabetes self-management applications for Android smartphones, J. Telemed. Telecare. (2012) 235–238.

[55] F. Chen, S. Wang, J. Li, H. Tan, W. Jia, Z. Wang, Smartphone-Based Hearing Self-Assessment System Using Hearing Aids with Fast Audiometry Method, IEEE Trans. Biomed. Circuits Syst. 13 (1) (2019) 170–179.

[56] A.K. Subasinghe, S.M. Garland, A. Gorelik, J.D. Wark, Using Mobile Technology to Improve Bone-Related Lifestyle Risk Factors in Young Women With Low Bone Mineral Density: Feasibility Randomized Controlled Trial, JMIR Form. Res. 3 (1) (2019).

[57] H. Slater, B. F.Dear, M.A. Merolli, L.C. Li, Use of eHealth technologies to enable the implementation of musculoskeletal Models of Care: Evidence and practice, Best Pract. Res.: Clin. Rheumatol. 30 (2016) 483–502.

[58] A. Ghose, C. Bhaumik, D. Das, A.K. Agrawal, Mobile healthcare infrastructure for home and small clinic, Proceedings of the International Symposium on Mobile Ad Hoc Networking and Computing (MobiHoc), 2012 15–20. https://doi.org/10.1145/2248341.2248347.

[59] V. Chandel, A.D. Choudhury, A. Ghose, C. Bhaumik, AcTrak-unobtrusive activity detection and step counting using smartphones, International Conference on Mobile and Ubiquitous Systems: Computing, Networking, and Services, 2013 447–459.

[60] Petrucka, S. Bassendowski, H. Roberts, T James, mHealth: a vital link for ubiquitous health, Online J. of Nurs. Inf. (OJNI) 17 (2013) 2675–2681.

[61] R. Kenny, B. Dooley, A. Fitzgerald, Developing mental health mobile apps: exploring adolescents' perspectives, Health Inf. J 22 (2) (2014) 265–275. https://doi.org/10.1177/1460458214555041.

[62] R. Grist, J. Porter, P. Stallard, Mental health mobile apps for preadolescents and adolescents: a systematic review, J. Med. Internet Res. 19 (5) (2017) e176. https://doi.org/10.2196/jmir.7332.

[63] V. Agrawal, M. Duggirala, S. Chanda, Journey: a game on positive affect, CHI PLAY 2018 - Proceedings of the 2018 Annual Symposium on Computer-Human Interaction in Play Companion Extended Abstracts, Association for Computing Machinery, Inc, 2018, pp. 373–379. https://doi.org/10.1145/3270316.3271532.

[64] M. Olff, Mobile mental health: a challenging research agenda, Eur J Psychotraumatol 6 (2015). https://doi.org/10.3402/ejpt.v6.27882.

[65] S. Mukhopadhyay, N. Ahmed, D. Jaiswal, A. Sinharay, A. Ghose, T. Chakravarty, A photoplethysmograph based practical heart rate estimation algorithm for wearable platforms, WearSys 2017 - Proceedings of the 2017 Workshop on Wearable Systems and Applications, co-located with MobiSys 2017, Association for Computing Machinery, Inc, 2017, pp. 23–28. https://doi.org/10.1145/3089351.3089354.

[66] N. Ahmed, V. Sharma, A. Chowdhury, S. Mukhopadhyay, A. Ghose, A weiner filter based robust algorithm for estimation of heart rate from wrist based photoplethysmogram, UbiComp/ISWC 2019- - Adjunct Proceedings of the 2019 ACM International Joint Conference on Pervasive and Ubiquitous Computing and Proceedings of the 2019

[67] P. Das, D. Chatterjee, A. Ghose, A. Sinha, A system for remote monitoring of mental effort, IEEE International Conference on Consumer Electronics - Berlin, ICCE-Berlin, Vols. 2016-, IEEE Computer Society, 2016, pp. 222–226. https://doi.org/10.1109/ICCE-Berlin.2016.7684760.

[68] A. Ghose, P. Sinha, C. Bhaumik, A. Sinha, A. Agrawal, A.D. Choudhury, UbiHeld - Ubiquitous healthcare monitoring system for elderly and chronic patients, UbiComp 2013 Adjunct - Adjunct Publication of the 2013 ACM Conference on Ubiquitous Computing, 2013 1255–1264. https://doi.org/10.1145/2494091.2497331.

[69] D. Jaiswal, K. Muralidharan, A. Ghose, Continuous activity recognition using smart watches, UbiComp/ISWC 2018 - Adjunct Proceedings of the 2018 ACM International Joint Conference on Pervasive and Ubiquitous Computing and Proceedings of the 2018 ACM International Symposium on Wearable Computers, Association for Computing Machinery, Inc, 2018, pp. 662–665. https://doi.org/10.1145/3267305.3267681.

[70] V. Chandel, A. Sinharay, N. Ahmed, A. Ghose, Exploiting IMU sensors for IoT enabled health monitoring, IoTofHealth 2016 - Proceedings of the 1st Workshop on IoT-Enabled Healthcare and Wellness Technologies and Systems, co-located with MobiSys 2016, Association for Computing Machinery, Inc, 2016, pp. 21–22. https://doi.org/10.1145/2933566.2933569.

[71] A. Chowdhury, S. Bhattacharya, A. Ghose, B. Krishnan, Early detection of mild cognitive impairment using pervasive sensing, Proceedings of the Annual International Conference of the IEEE Engineering in Medicine and Biology Society, EMBS, Institute of Electrical and Electronics Engineers Inc, 2019, pp. 5456–5459. https://doi.org/10.1109/EMBC.2019.8856435.

[72] V. Chandel, D.S. Jani, S. Mukhopadhyay, S. Khandelwal, D. Jaiswal, A. Pal, C2P: an unobtrusive smartwatch-based platform for automatic background monitoring of fatigue, HumanSys 2017 - Proceedings of the 1st International Workshop on Human-Centered Sensing, Networking, and Systems, Part of SenSys 2017, Association for Computing Machinery, Inc, 2017, pp. 19–24. https://doi.org/10.1145/3144730.3144732.

[73] V. Chandel, S. Singhal, V. Sharma, N. Ahmed, A. Ghose, PI-Sole: a low-cost solution for gait monitoring using off-the-shelf piezoelectric sensors and IMU, Proceedings of the Annual International Conference of the IEEE Engineering in Medicine and Biology Society, EMBS, Institute of Electrical and Electronics Engineers Inc, 2019, pp. 3290–3296. https://doi.org/10.1109/EMBC.2019.8857877.

[74] J. Abeler, M. Bäcker, U. Buermeyer, H. Zillessen, Covid-19 contact tracing and data protection can go together, JMIR Mhealth Uhealth 8 (4) (2020). https://doi.org/10.2196/19359.

[75] A. Ghose, C. Bhaumik, T. Chakravarty, BlueEye - A system for proximity detection using bluetooth on mobile phones, UbiComp 2013 Adjunct - Adjunct Publication of the 2013 ACM Conference on Ubiquitous Computing, 2013 1135–1142. https://doi.org/10.1145/2494091.2499771.

[76] V. Chandel, N. Ahmed, S. Arora, A. Ghose, Inloc: an end-to-end robust indoor localization and routing solution using mobile phones and ble beacons, 2016 International Conference on Indoor Positioning and Indoor Navigation, IPIN 2016, Institute of Electrical and Electronics Engineers Inc, 2016. https://doi.org/10.1109/IPIN.2016.7743592.

[77] N. Ahmed, R. Banerjee, A. Ghose, A. Sinharay, Feasibility analysis for estimation of blood pressure and heart rate using a smart eye wear, WearSys 2015 - Proceedings of

the 2015 Workshop on Wearable Systems and Applications, Association for Computing Machinery, Inc, 2015, pp. 9–14. https://doi.org/10.1145/2753509.2753511.

[78] R. Banerjee, A. Ghose, A. Dutta Choudhury, A. Sinha, A. Pal, Noise cleaning and Gaussian modeling of smart phone photoplethysmogram to improve blood pressure estimation, ICASSP, IEEE International Conference on Acoustics, Speech and Signal Processing - Proceedings, Vols. 2015-, Institute of Electrical and Electronics Engineers Inc, 2015, pp. 967–971. https://doi.org/10.1109/ICASSP.2015.7178113.

[79] R. Mukherjee, S.K. Ghorai, B. Gupta, T. Chakravarty, Development of a wearable remote cardiac health monitoring with alerting system, Instrum. and Exp. Tech. 63 (2) (2020) 273–283. https://doi.org/10.1134/S002044122002013X.

[80] A. Gigie, S. Rani, A. Chowdhury, T. Chakravarty, A. Pal, An agile approach for human gesture detection using synthetic radar data, UbiComp/ISWC 2019- - Adjunct Proceedings of the 2019 ACM International Joint Conference on Pervasive and Ubiquitous Computing and Proceedings of the 2019 ACM International Symposium on Wearable Computers, Association for Computing Machinery, Inc, 2019, pp. 558–564. https://doi.org/10.1145/3341162.3349332.

[81] A. Gigie, S. Rani, A. Sinharay, T. Chakravarty, Novel approach for vibration detection using indented radar, Progress In Electromagn. Res. C 87 (2018) 147–162. https://doi.org/10.2528/pierc18071702.

[82] A. Ray, Rani Khasnobish, Chakravarty, Live demonstration: unobtrusive and continuous monitoring of respiration employing a dual CW radar assembly, Accepted in IEEE ISCAS (2020).

[83] S. Rani, A. Khasnobish, R. Rakshit, A. Gigie, T. Chakravarty, Optimum channel selection of dual radar for respiration detection - A time domain approach, Proceedings of IEEE Sensors, Vols. 2019-, Institute of Electrical and Electronics Engineers Inc, 2019. https://doi.org/10.1109/SENSORS43011.2019.8956905.

[84] Anton, A., & Smith, D. (2013). *Volatile biomarkers: non-invasive diagnosis in physiology and medicine*.

[85] B.S. Mithun, S. Shinde, K. Bhavsar, A. Chowdhury, S. Mukhopadhyay, K. Gupta, B. Bhowmick, S. Kimbahune, Non-destructive method to detect artificially ripened banana using hyperspectral sensing and RGB imaging, Proceedings of SPIE - The International Society for Optical Engineering, Vol. 10665, SPIE, 2018. https://doi.org/10.1117/12.2306367.

[86] O. Masako, D. Ippeita, Automated cortical projection of head-surface locations for transcranial functional brain mapping, Neuroimage (2005) 18–28. https://doi.org/10.1016/j.neuroimage.2005.01.018.

[87] S.M. Imani, A.M. Goudarzi, P. Valipour, M. Barzegar, J. Mahdinejad, S.E. Ghasemi, Application of finite element method to comparing the NIR stent with the multi-link stent for narrowings in coronary arteries, Acta Mech. Solida Sin. 28 (5) (2015) 605–612. https://doi.org/10.1016/S0894-9166(15)30053-7.

Smartphone-based detection of explosives

17

Arpana Agrawal[a], Chaudhery Mustansar Hussain[b]
[a]Department of Physics, Shri Neelkantheshwar Government Post-Graduate College, Khandwa, India
[b]Department of Chemistry and Environmental Science, New Jersey Institute of Technology, Newark, N J, USA

17.1 Introduction

Explosives and related materials are extensively employed in military for homeland security from the foreign elements [1]. However, in recent years, there has been a rapid and global increase in the number of explosive disasters that has threatened the mankind and hence it is the need of the hour to develop effective and dedicated detection systems to monitor them continuously before their occurrence [2]. For such purposes, effective sensors are required to be installed in several public places to keep an eye on such events and alarm the authority about them. Till now, the existing systems are still inadequate to monitor this task and hence the advanced systems are required. Traditionally, specially trained dogs were considered as the only effective and mobile way to trace the threat or explosive materials using their very sensitive smelling power. However, this method requires a lot of time to train the dogs and there can a performance variation with the passage of time. Later on, various analysis methods including spectroscopic techniques such as Raman spectroscopy [3], surface-enhanced Raman spectroscopy [4], ion-mobility spectrometry (IMS) [5], Infrared (IR) spectroscopy [6], laser-induced thermal emissions [7], optical emission spectroscopy [8], etc. were employed for such purposes. Such systems are highly expensive, sophisticated and their inherent bulky nature and poor portability restricts their appropriateness for on-site applicability.

In contrast to this, the detection of explosives using smartphone technology is rapidly increasing because of being the most acceptable electronic device and readily available with the large global population at low cost and adopting such technology will serve as a sensor carrier. Several features of smartphone technology that has made this device as the most acceptable platform includes high storage capacity, bluetooth functioning, multiple communications mediums, low power consumption, global positioning systems, audio-video recording features, accelerometers, high video and picture resolutions, in-built camera of high quality, excellent multiprocessing speeds and their integration along with several developing software or allied applications. All such fascinating features contribute to formulate smartphone technology as an excellent raised area to develop miniaturized mobile sensors with high signal processing capabilities in real-time. It should be noted here that the smartphone technology has become an important part of our day-to-day life and has a wide range

Smartphone-Based Detection Devices. DOI: https://doi.org/10.1016/B978-0-12-823696-3.00013-1
Copyright © 2021 Elsevier Inc. All rights reserved.

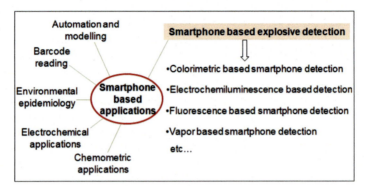

Fig. 17.1 Smartphone based applications with particular focus on explosive detection systems based on different working principles.

of utilities in the field of food and beverage producers [9,10], education [11], health care: detection and clinical diagnostic systems [12–14], microscopy/imaging [15], barcode reading [16,17], etc. Fig. 17.1 shows the various smartphone based applications with particular focus on smartphone based explosive detection systems coupled with different principles.

Apart from this, operating systems (for example, Android or iPhone etc.) of the smartphone are other important features that are required to be checked before their use for any application. Additionally, the availability of open source Android mobile operating system facilitates further research on smartphone based sensors. In comparison to this Android phone based operating systems, the iPhones are much more protected platform and does not readily allows any kind of modifications in the source code. The basic principle adopted for utilizing smartphones as explosive detectors involves the collection of data of the unknown material and analyzing then by comparing with that of the preexisted database. Upon finding any matching between the data obtained and the database, electrical signals will be generated that will intimate the coprocessor regarding the existence of explosives in the surrounding. Accordingly, an alarm either in terms of message, call or buzzer will be generated using the smartphone depending upon the programming. It is worthy to mention here that the mobile based explosive detection systems generally integrate the potency of the mobile phones and gas sensors owing to the fact that several explosive have their own unique smell or are based on colorimetric detection using smartphones. Zhang and Liu [18] have presented a detailed review on the integration of various sensors with that of the smartphone for mobilized biochemical detections. They have summarized smartphone based optical biosensors including fluorescence biosensors, colorimetric biosensors, surface plasmon resonance biosensors and smartphone-based electrochemical biosensors including amperometric biosensors, potentiometric biosensors, or impedimetric biosensors. Alcaidinho et al. [19]., have also discussed the collaboration of mobile with human and canine police explosive detection teams for the detection of explosives and there are many more reports that have discussed the adaptability of smartphone based explosive detection system over the other means for the detection of explosives.

Accordingly, the present chapter summarizes the various available smartphone based systems for the detection of explosives along with few recent results. Additionally, the various types of explosive materials, their conventional detection techniques with particular focus on the spectroscopic methods along with their advantages and limitations have also been discussed. Superiority of smartphone based explosive detection systems over the conventional methods for the detection of explosives has also been reported.

17.2 Types of explosives and explosive detection

There are several commonly used explosives which can be classified and identified on the basis of material density, vapor detection and elemental composition [20]. Table 17.1 lists the various classes of commonly used explosives including nitroaromatics, nitroamines, nitroesters, acid-salts, ammonium picrate/picric acid and organic perocides. Each of this class has been further consists of several explosives as summarized in Table 17.1. Structures of few commonly used explosives are also shown in Fig. 17.2.

Table 17.1 Different classes of explosive materials with their types.

	Types of explosives
Class	**Types**
Nitroaromatics	2,4,6-Trinitrotoluene (TNT)
	1,3,5-Trinitrobenzene (TNB)
	1,3-Dinitrobenzene (DNB)
	2,4-Dinitrotoluene (2,4-DNT)
	2,6-Dinitrotoluene (2,6-DNT)
	2,4,6-trinitro-phenylmethylnitramine (Tetryl)
	2-amino-4,6-dinitrotoluene (2AmDNT)
	4-amino-2,6-dinitrotoluene (4AmDNT)
	Nitrotoluene (3 isomers) (NT)
	Nitrobenzene (NB)
	Ethylene glycol dinitrate (EGDN)
Nitroamines	1,3,5-trinitro-1,3,5-triazine (RDX)
	Octahydro-1,3,5,7-tetranitro-1,3,5,7-tetrazocine (HMX)
	Nitro guanidine (NQ)
Nitroesters	Nitroglycerin (glycerol trinitrate) (NG)
	Pentaerythritol tetra nitrate (PETN)
Acid salt	Ammonium nitrate
	Urea nitrate
Ammonium Picrate/ Picric Acid	Ammonium 2,4,6-trinitrophenoxide/2,4,6-trinitrophenol (AP/PA)
Organic peroxides	Triacetate triperoxide (TATP)
	Hexamethylene triperoxide diamine (HMTD)

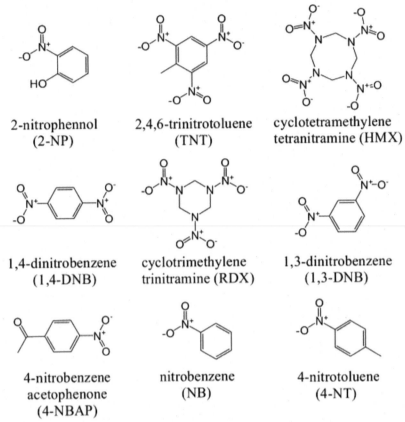

Fig. 17.2 Structure of few commonly used explosive materials including TNT, DNT, HMX, Tetryl, DDNP (diazodinitrophenol), RDX, PA, DMNB (2,3-dimethyl-2,3-dinitrobutane), TATP [21].

In general, two different kinds of approaches are employed for the detection of explosives i.e. trace detection and bulk detection approach [22]. Bulk detection of explosives is based on either determining the properties of nuclear explosives or imaging the explosives and hence requires an observable amount of the explosive materials. On the other hand, trace detection approach requires minute quantity of explosive particles and explosive vapors and the corresponding detection technique can be termed as particulate detection and vapor detection, respectively. Particulate detection involves the direct handling (using hands and gloves) of the explosive by pasting them over the surface of the sensors while in case of vapor detection, detection takes place from the vapors that are released from the explosives and hence direct touching is not required, making this approach much more superior that of particulate detection or bulk detection approaches. The several methods that are employed for the detection of explosive materials include vapor based organic polymer sensors, X-ray detection systems, well trained dogs and trained bees, metal detection systems, neutron activation systems and various spectroscopic techniques. Each explosive material has its own characteristic vapor elements which can be detected using vapor-based sensors which are generally made up of organic polymers and the basic working principle of such sensors relies upon an increase in the conductivity of the sensor on increasing the gas concentrations which will be then converted into the output signals. In contrast to this, X-ray detection systems allow the detection of explosives by examining the density of the explosive materials. Trained dogs and bees are also one of the traditional approached to identify explosive materials, however, provide training is a time consuming task and requires a lots of skills which can vary with the passage of time. Metal detection systems are based on the detection of explosives hidden or packed within the metallic container. However, the use of this approach is also very limited because various modern explosive materials are now-a-days loaded into plastic casings. Apart from this, spectroscopic techniques are a quite easily adoptable approach for the identification of explosives. However, these spectroscopic techniques do not allow the portable, instantaneous, on-site and quick detection of explosive substances and hence are not in general adopted for all the situations.

17.3 Spectroscopic techniques to detect explosives

This section will discuss the various conventional spectroscopic approaches engaged for the detection and discrimination of explosives. Such spectroscopic techniques includes IMS [23,24], Raman spectroscopy [25–27], Terahertz (THz) Spectroscopy [28–31], THz time domain spectroscopy (THz-TDS) [32,33], laser induced breakdown spectroscopy (LIBS) [34,35], mass spectroscopy [36], Fourier Transform Infrared (FTIR) spectroscopy [37], IR spectroscopy [38], nuclear magnetic resonance (NMR) spectroscopy [39], etc. Such spectroscopic techniques employ specific range of electromagnetic (EM) radiations, for example, Raman spectroscopy utilizes visible part of EM radiation, THz spectroscopy and IR spectroscopy works under terahertz, and IR regime of EM radiations, respectively. IMS extensively used at the airports for the rapid identification of explosives and relies upon the time required by the ionized molecules to be collected after drifting through the electric field. In this technique, the molecule is irradiated by the radiation source and gets ionized due to the corona discharge. These generated ions will then drifted in the field-free region as a consequence of electric field gradient and finally get collected. The time required to drift through the electric field will then be utilized to identify the material. Lee et al. [40]., have demonstrated the detection of five commonly used explosives namely RDX, HMX, TNT, DNT and PETN using coupled IMS–Mass spectroscopy and have reported the average drift time of the order of few milliseconds. Identification of explosive using IMS has also been reported by Tabrizchi and ILbeigi [41]. It is worthy to mention here that each individual explosive material has some unique absorption feature in the THz frequency regime which can be employed as a finger print to detect the explosives. EM radiations with frequency ranging from 0.1–10 THz are employed in THz spectroscopy or THz-TDS. In this technique, ultrashort laser pulses are made to incident on the explosive material and the transmitted light is detected and examined which will thus help to identify the explosive [42]. Sleiman et al. [43]., measured the absorption spectra of cyclotrimethylenetrinitramine, PETN, and their mixtures by THz-TDS. Various explosives such as HMX, RDX, NTO, TNT, PETN etc. were also identified via recording the emission spectra of specific elements present in the explosives using LIBS and is reported by Myakalwar et al. [44]. LIBS is an emission spectroscopy where an intense laser pulse of high energy is irradiated onto the explosive material which leads to plasma ignition as a result of the interaction of the laser pulse with that of the explosives. As a consequence of the plasma plume generation, thermal vaporization of the material gets initiated and hence plasma gets expanded leading to the occurrence of Bremsstrahlung process and finally the emission of the specific element of the explosive material gets started which are then recorded to identify the explosive material. Fig. 17.3 shows the normalized LIBS spectra of various explosives with certain offset for proper visualization. Raman spectroscopy is also another important technique to endow with a structural fingerprint for explosive detections and depends upon the inelastic scattering of photons. Since the energy has to be conserved, there occurs a shift at either lower or higher frequency regime giving rise to stokes and antistokes Raman lines, respectively. Raman spectra of several explosives has been recorded and demonstrated by Hwang et al. [45]., and Jin et al. [46].

All these spectroscopic techniques require several sophisticated and bulky instruments with well trained technical support and chemicals/solvents, quite time

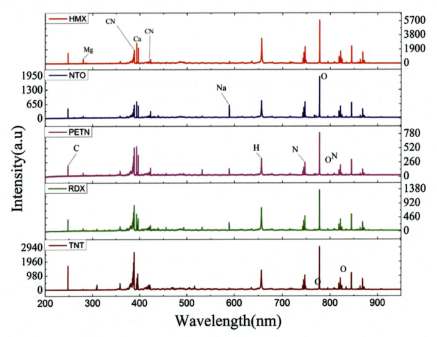

Fig. 17.3 Normalized LIBS spectra of various explosives including HMX, NTO, PETN, RDX and TNT.

consuming, presumptive, and most importantly, does not allows rapid and real time monitoring of the explosives because of the immobility of such bulky instruments and hence much more advanced explosive detection systems are being currently utilized which are based upon smartphone based technology.

17.4 Smartphone-based detection of explosives

Smartphone based explosive detection systems generally comprise of various sensors/detectors interfaced with that of the smartphone. The most commonly used sensors for the purpose of explosive detection includes colorimetric arrays based sensors which are coupled with portable and easy to handle smartphones to provide efficient and highly sensitive way to identify and discriminate several explosives. This colorimetric approach for explosive detection has several advantages over other existing techniques such as less time consuming, simplicity of operation, does not require sophisticated instrumentation and hence much more technical know-how is not necessitate etc. Several researchers have adopted this technique to fabricate colorimetric array based smartphone detectors for explosives. Salles et al. [47]., have also demonstrated a cost effective and disposable colorimetric based paper sensor for the identification and discrimination of explosives using a smartphone. They have examined five different type of explosives, namely, TATP, HMTD, 4-amino-2-nitrophenol (4A2NP), NB, and PA and have used an Apple iPhone 4s which serve as the colorimetric detector and is based on red, green, blue (RGB) analysis. The detector array was scanned and analyzed using smartphone's built-in camera and Apple Xcode 4.3.2 application. Specific Application Program Interface then measures the standard deviation for each RGB component of each spot and then converts the image pixel into 24-bit data ans such a sensor can trace the explosive upto micro-range. Finch et al. [48]., have also demonstrated a methodology for the development of chemical, biological and explosive (CBE) sensing device using android based smartphone. These explosives were detected using colorimetric test kit and using a novel smartphone adapter along with graphical user interfacing, it has become possible to identify CBE materials.

A sensitive, low cost and portable smartphone based colorimetric sensing device for the selective detection of TNT using Amine-trapped polydimethylsiloxane (PDMS) membranes has also been illustrated by Tang et al. [49]. PDMS membranes were prepared by loading a mixture of PDMS based and curing agent onto a silicon substrate containing an anti-adhesion layer of 1H,1H,2H,2H-Perfluordecyltrichlorosilane (PFDTS). This layer was then peeled off from the silicon substrate to make a PDMS membrane of thickness of 0.01 mm, which was then dipped into the solution of amino-silanes and ethanol. This amine-trapped PDMS was then interacted with TNT and was examined using ultra-violet (UV–Vis) spectroscopy where the absorption peak is sensitive to the concentration of the TNT. The detection of TNT was also demonstrated by Zhang et al. [50]., where a new wood based chemical sensor for their colorimetric visualization via the camera of smartphone has been illustrated. This mesoporous wood based sensor was fabricated via the reaction of wood soaked in 3-aminopropyltriethoxysilane (APTES) and was based on the formation of Meisenheimer complex between TNT- and APTES-modified wood, leading to a prominent color changeover from light yellow to brown and was then employed for the detection of TNT explosive in water, soil and air [50]. For water containing TNT, aqueous solution of TNT with varying concentrations is prepared and casted onto the sensor surface. As a result, a change in color is observed for different concentrations and the image is captured using a smartphone. These images were then analyzed using Color Helper software and the intensity of the RGB component of the TNT contained target wood and reference wood. This method allows rapid and selective determination of TNT in water [50].

Very recently, Liu et al. [51]., have also reported a colorimetric reagent design strategy for the detection of improvised explosives including TNT, 2,4-DNT, elemental sulfur (S) and potassium permanganate (KMnO4) and can trace with a mass as low as 1.45 nanogram level. This colorimetric reagent design is based on the specific nucleophilic substitution reaction and the base-catalyzed oxidation induced electron-transfer which guarantee the distinctive characteristic color change with various target explosives. Colorimetric assay visualization of nitramine explosives including RDX and HMX using smartphones has also been reported by Xie et al. [52]., which is based on compartmentalizing incompatible tandem reactions in Pickering emulsions.

Other commonly used smartphone based explosive detection strategy includes explosive vapor detection systems as demonstrated by Verma and Goudar [22]. They have proposed a mobile based explosive vapor detection system to save the humankind. For such detection systems, they have employed explosive vapor detection where a chemical vapor sensor is loaded in the smartphone having a transmitter and a receiver to transmit the explosive vapors in the surrounding atmosphere and to collect the specific sensed vapor of several explosive materials, respectively. This detector is based on the fact that different explosive materials liberate unique chemical vapor in the surrounding and have unmatched chemical signature. Advanced feature of radar system of the mobile phone and the signature of stored explosive helps to examine the presence of traces of vapors of explosive chemicals in the surroundings. This sensor is activated using a smartphone based application and whenever there is a matching between the stored explosive signature and the explosive vapor in the surroundings, the radar section will then gets activated to keep trace the explosive material and alert the user about its presence. Fig. 17.4 shows the overall working of the chemical vapor sensor where once the detector gets activated, the data of the presence of certain explosive will be then sent to the processing unit to process the data and then forwarded to memory unit where the explosive signatures are stored in the database. This memory unit will compare both the data received with the database and if matching is found between the two, the radar section alerts the user provided the explosive is in the range of radar section. Hence, provides an early warning to the user.

Electrochemiluminescence and fluorescent based smartphone detectors for explosive identification has also been proposed [53,54]. Li et al. [53]., have developed a silica nanopores derived smartphone based electro-chemiluminescence system via the integration with smartphone having universal serial bus-on the go (USB-OTG) and built-in camera for capturing the luminescence for the detection of nitroaromatic explosives. Portable paper based and fluorescent Chitosan film based sensors coupled

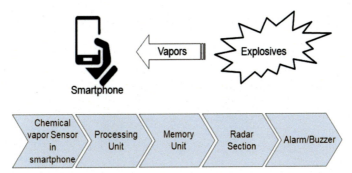

Fig. 17.4 Illustration of the working process of chemical vapor based smartphone detection system for explosives.

with smartphone has also been developed by Tawfik et al. [54]., for the detection of 2,4,6-trinitrophenol (TNP) explosive at picolevels using a multiple emitting Amphiphilic conjugated polythiophene coated CdTe quantum dots (QDs). It should be noted here that TNP if exposed for a shorter time may cause eye or skin irritation and is highly detrimental for respiratory system and kidneys upon long time exposure. For the determination of TNP, Tawfik et al. [54]., have developed four biocompatible paper sensors based on nanohybrids, namely, cationic polythiophene nanohybrids (CPTQDs), nonionic polythiophene nanohybrids (NPTQDs), anionic polythiophene nanohybrids (APTQDs), and thiophene copolymer nanohybrids (TCPQDs). The reaction scheme for the growth of various amphiphilic thiophene monomers and insitu-polymerization scheme for the growth of multicolor emissive amphiphilic conjugated polythiophene-coated CdTe QDs are shown in Figs. 17.5(A) and 17.5(B), respectively.

It has been reported that the molecular interactions including hydrogen bonding, electrostatic and π-π interaction driven inner filter effect is responsible for the detection of TNP. Fig. 17.6(A) schematically illustrates the detection mechanism of TNP using TCPQDs sensor. This synthesized coated CdTe QDs sensors are multicolored owing to the coating of QDs with differently charged polythiophene ligands which gives rise to different fluorescence emission and different quantum yields along with high quantum efficiency, high sensitivity, excellent photostability and low cytotoxicity. Fig. 17.6(B) shows the overlapped fluorescence emission spectra of synthesized CPTQDs, NPTQDs, APTQDs, and TCPQDs and the UV–Vis absorbance spectra of various nitroaromatic explosive materials examined under a wavelength range of 200 nm-800 nm (plotted in double y-axis). The morphology of the synthesized TCPQD nanohybrids having an average particle size of 49.5 nm has also been depicted by performing transmission electron microscopy (TEM) and is shown in Fig. 17.6(C).

The efficiency of TCPQDs sensor has also been investigated using fluorescence emission spectroscopy (Fig. 17.6(D)) showing a decrease in the fluorescence intensity upon increasing the concentration of TNP with a slight red shift. This decrease in fluorescence intensity is attributed to electron transfer between the explosive and

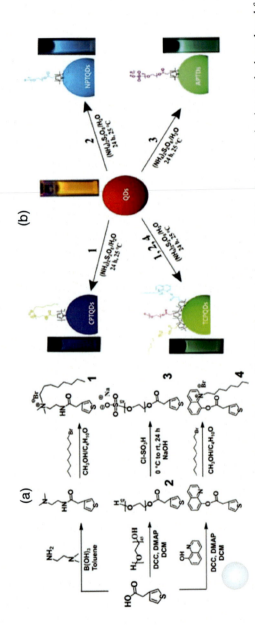

Fig. 17.5 (a) Reaction scheme for the synthesis of various amphiphilic thiophene monomers; (b) In-situ polymerization method employed for the synthesis of multicolor emissive amphiphilic conjugated polythiophene-coated CdTe QDs. (Reproduce from Tawfik et al. [54]).

Smartphone-based detection of explosives 409

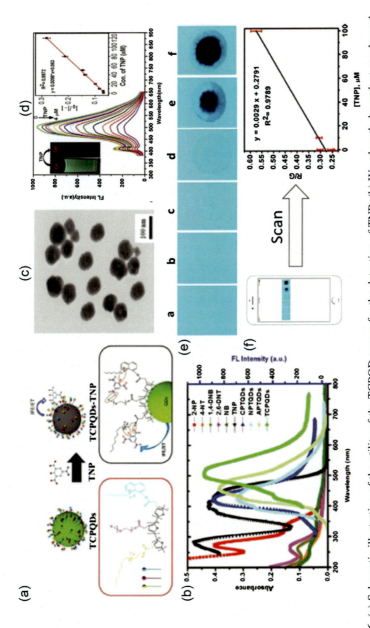

Fig. 17.6 (a) Schematic illustration of the utility of the TCPQD sensor for the detection of TNP; (b) Wavelength dependent overlapped fluorescence emission of synthesized CPTQDs, NPTQDs, APTQDs, and TCPQDs and absorbance of various nitroaromatic explosive; (c) TEM profile of TCPQDs; (d) Fluorescence emission spectra of TCPQDs in PBS buffer solution for various TNP concentrations. Right inset shows the plot of $(I_o/I - 1)$ against the TNP concentration and the left inset shows the color change of the sensor solutions in PBS before and after adding TNP upon illumination of UV lamp; (e) Color of fluorescent paper test strips under UV light with varying TNP concentrations; (f) Calibration plot using our visual sensor based on paper strips in combination with a smartphone application for TNP detection; (Reproduce from Tawfik et al. [54].)

the sensor. The inset (right) of Fig. 17.6(D) shows the plot of $(I_0/I - 1)$ as a function of TNP concentration ($\sim 10^{-9}$ M) while the left inset shows the change in the color of sensor solution in phosphate buffer saline (PBS) upon irradiation of UV light (365 nm) before and after adding TNP. For the visualization of TNP, TCPQD nanohybrid was casted on the paper surface and the color of the fluorescence paper test strip was also examined under same illumination source with varying TNP concentrations as shown in Fig. 17.6(E), and such a sensor can detect TNP traces upto 1×10^{-9} M In order to develop a portable paper-based sensor, smartphone was used to capture the image which was then scanned using "PAD analysis" application and rely on the difference in the intensities of Red (R) and Green (G) color of the images, depicting the various TNP concentration. Fig. 17.6(F) illustrates the plot of intensity ratio of red and green color as a function of TNP concentration which clearly shows an increase in the R/G color intensity ratio upon increasing the TNP concentration. Lopez-Ruiz et al. [55]., have also demonstrated a computer vision-based portable measurement system which can obtain and process the imaging of the sensor array and able to send the obtained information to a smartphone via WiFi link (as shown in Fig. 17.7) for the identification of nitroaromatic compounds including PA, 2,4-DNT, DNB, TNB, TNT, and tetryl, 3,5-dinitrobenzonitrile, and 2-chloro-3,5-dinitrobenzotrifluoride.

Mobile based explosive detection system using gED mobile application consisting of multiple transducers using DASH7 technology was proposed by Muthukumaresan et al. [56]. These transducers serve as nodes to position the hotspots and whenever the mobile is within the range of any explosive sample, this gED mobile application will activate the ultrasonic transducers. This transducer will then receive the signals from the explosive samples and then the application compares this obtained signal with that of the database. Upon identification of any kind of threat, a warning message alerts the users provided the user is within the sampled area. Temporal and frequency responses of smartphone accelerometer has been demonstrated by Thandu et al. [57]., for the detection of explosives. Liu et al. [58]., have also demonstrated the fabrication of smartphone based sensing system for the detection of volatile organic compounds

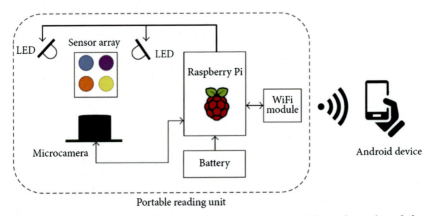

Fig. 17.7 Block diagram of the instrument including the portable reader unit and the programmed Android mobile device. (Reproduce from Lopez-Ruiz et al. [55]).

using zinc oxide, graphene and nitrocellulose modified interdigital electrodes to obtain the impedance responses of the volatile organic compounds. Zhang et al. [59]., have also proposed a detection system for TNT using impedance monitoring. They have employed TNT-specific peptides modified printed electrodes to obtain impedance response of TNT and using this system, TNT can be detected at much lower concentrations $\sim 10^{-6}$ M very recently, a practical approach for the development of smartphone based electrochemiluminescence sensor for the detection of dopamine has also been illustrated by Kwon et al. [60].

Overall, smartphone based detection systems are portable making them highly appropriate for on-site applicability and serves as the most fascinating approach for the real time, rapid, sensitive and selective detection of various explosives.

17.5 Conclusions

In conclusion, the present chapter provides an overview of the utility of smartphones for the purpose of detection of explosives. Few recent studies for smartphone based detection of explosives including electrochemiluminescence-based smartphone detectors, fluorescence-based smartphone detectors, smartphone based explosive vapor detection system, smartphone based colorimetric sensing devices for sensitive and selective detection of explosives has been comprehensively discussed. Several types of explosive materials have been summarized along with the various conventional spectroscopic techniques such as ion mobility spectroscopy, Raman spectroscopy, terahertz Spectroscopy, terahertz time domain spectroscopy, laser induced breakdown spectroscopy, mass spectroscopy, Fourier Transform Infrared spectroscopy, infrared spectroscopy, nuclear magnetic resonance spectroscopy etc., that are employed for their detection and discrimination. However, it is the need of the hour to develop portable and much more sensitive detectors for the real time monitoring of the explosives and hence it is highly recommended to customize the smartphone based explosive detection system to help the government for speedy detection and discrimination of explosives in order to save the mankind.

Acknowledgement

I heartily thank Mr Jatin Gandhi for his constant support and encouragement.

References

[1] https://en.wikipedia.org/wiki/Explosion.
[2] K.C. To, S. Ben-Jaber, I.P. Parkin, Recent developments in the field of explosive trace detection, ACS Nano 14 (9) (2020) 10804–10833.
[3] I. Malka, S. Rosenwaks, I., Bar, Photo-guided sampling for rapid detection and imaging of traces of explosives by a compact Raman spectrometer, Appl. Phys. Lett. 104 (22) (2014) 221103.

[4] B.D. Piorek, S.J. Lee, M. Moskovits, C.D. Meinhart, Free-surface microfluidics/surface-enhanced Raman spectroscopy for real-time trace vapor detection of explosives, Anal. Chem. 84 (22) (2012) 9700–9705.

[5] U. Gaik, M. Sillanpää, Z. Witkiewicz, J. Puton, Nitrogen oxides as dopants for the detection of aromatic compounds with ion mobility spectrometry, Anal. Bioanal. Chem. 409 (12) (2017) 3223–3231.

[6] J.R. Castro-Suarez, L.C. Pacheco-Londoño, J. Aparicio-Bolaño, S.P. Hernández-Rivera, Active mode remote infrared spectroscopy detection of TNT and PETN on aluminum substrates, J. Spectrosc. 2017 (2017) 1–12.

[7] N.J. Galán-Freyle, L.C. Pacheco-Londono, A.M. Figueroa-Navedo, S.P. Hernandez-Rivera, Standoff detection of highly energetic materials using laser-induced thermal excitation of infrared emission, Appl. Spectrosc. 69 (5) (2015) 535–544.

[8] R.L. Vander Wal, C.K. Gaddam, M.J. Kulis, An investigation of micro-hollow cathode glow discharge generated optical emission spectroscopy for hydrocarbon detection and differentiation, Appl. Spectrosc. 68 (6) (2014) 649–656.

[9] Y. Chen, G. Fu, Y. Zilberman, W. Ruan, S.K. Ameri, Y.S. Zhang, et al., Low cost smart phone diagnostics for food using paper-based colorimetric sensor arrays, Food Control 82 (2017) 227–232.

[10] J.L.D. Nelis, A.S. Tsagkaris, M.J. Dillon, J. Hajslova, C.T. Elliott, Smartphone-based optical assays in the food safety field, TrAC, Trends Anal. Chem. 129 (2020) 115934.

[11] A. Singhal, R.S. Pavithr, Degree certificate authentication using QR Code and Smartphone, Int. J. Comput. Appl. 120 (16) (2015).

[12] Y.R. Park, Y. Lee, G. Lee, J.H. Lee, S.Y. Shin, Smartphone applications with sensors used in a tertiary hospital—Current status and future challenges, Sensors 15 (5) (2015) 9854–9869.

[13] I. Hernandez-Neuta, F. Neumann, J. Brightmeyer, T. BaTis, N. Madaboosi, Q. Wei, et al., Smartphone-based clinical diagnostics: towards democratization of evidence-based health care, J. Intern. Med. 285 (1) (2019) 19–39.

[14] W. Zhao, S. Tian, L. Huang, J. Guo, K. Liu, L. Dong, Smartphone-based Biomedical sensory system, Analyst 145 (8) (2020) 2873–2891.

[15] Q. Wei, H. Qi, W. Luo, D. Tseng, S.J. Ki, Z. Wan, et al., Fluorescent imaging of single nanoparticles and viruses on a smart phone, ACS Nano 7 (2013) 9147–9155.

[16] M.R. Hyder, T. Khan, Automatic expiry date notification system interfaced with smart speaker, Int. J. Eng. Sci. Inven. 9 (7) (2020) 14–20.

[17] M. Maringer, N. Wisse-Voorwinden, P. van't Veer, A. Geelen, Food identification by barcode scanning in the Netherlands: a quality assessment of labeled food product databases underlying popular nutrition applications, Public Health Nutr. 22 (7) (2019) 1215–1222.

[18] D. Zhang, Q. Liu, Biosensors and bioelectronics on smartphone for portable biochemical detection, Biosens. Bioelectron. 75 (2016) 273–284.

[19] J. Alcaidinho, L. Freil, T. Kelly, K. Marland, C. Wu, B. Wittenbrook, et al., Mobile Collaboration for Human and Canine Police Explosive Detection Teams, Proceedings of the 2017 ACM Conference on Computer Supported Cooperative Work and Social Computing, 2017 925–933.

[20] X. Zhang, R. Zhang, J. Wang, L. Wang, An adaptive particle swarm optimization algorithm based on aggregation degree, Recent Adv. Electr. Electron. Eng. 11 (4) (2018) 443–448.

[21] M.J. Kangas, R.M. Burks, J. Atwater, R.M. Lukowicz, P. Williams, A.E. Holmes, Colorimetric sensor arrays for the detection and identification of chemical weapons and explosives, Crit. Rev. Anal. Chem. 47 (2) (2017) 138–153.

[22] P. Verma, R.H. Goudar, Mobile phone based explosive vapor detection system (MEDS): a methodology to save humankind, Int. J. Syst. Assur. Eng. Manage. 8 (1) (2017) 151–158.

[23] C.K. Hilton, C.A. Krueger, A.J. Midey, M. Osgood, J. Wu, et al., Improved analysis of explosives samples with electrospray ionization-high resolution ion mobility spectrometry (ESI-HRIMS), Int. J. Mass Spectrom. 298 (1–3) (2010) 64–71.

[24] N. Sivakumar, M. Joseph, P. Manoravi, P.R. Vasudeva Rao, B. Raj, Development of an ion mobility spectrometer for detection of explosives, Instrum. Sci. Technol. 41 (1) (2013) 96–108.

[25] F. Zapata, C. García-Ruiz, Determination of nanogram microparticles from explosives after real open-air explosions by confocal Raman microscopy, Anal. Chem. 88 (13) (2016) 6726–6733.

[26] M. Ghosh, L. Wang, S.A. Asher, Deep-ultraviolet resonance Raman excitation profiles of NH_4NO_3, PETN, TNT, HMX, and RDX, Appl. Spectrosc. 66 (9) (2012) 1013–1021.

[27] N. Nuntawong, P. Eiamchai, S. Limwichean, B. Wong-Ek, M. Horprathum, V. Patthanasettakul, et al., Trace detection of perchlorate in industrial-grade emulsion explosive with portable surface-enhanced Raman spectroscopy, Forensic Sci. Int. 233 (1–3) (2013) 174–178.

[28] L.A. Skvortsov, Standoff detection of hidden explosives and cold and fire arms by terahertz time-domain spectroscopy and active spectral imaging, J. Appl. Spectrosc. 81 (5) (2014) 725–749.

[29] S. Han, K. Bertling, P. Dean, J. Keeley, A.D. Burnett, Y.L. Lim, et al., Laser feedback interferometry as a tool for analysis of granular materials at terahertz frequencies: towards imaging and identification of plastic explosives, Sensors 16 (3) (2016) 352.

[30] L.M. Lepodise, J. Horvat, R.A. Lewis, Terahertz spectroscopy of 2, 4-dinitrotoluene over a wide temperature range (7–245 K), J. Phys. Chem. A 119 (2) (2015) 263–270.

[31] N. Palka, M. Szala, E. Czerwinska, Characterization of prospective explosive materials using terahertz time-domain spectroscopy, Appl. Opt. 55 (17) (2016) 4575–4583.

[32] H. Ping, Q. Rui, T. Yuxiang, Study on terahertz spectroscopic of energetic ion salt and oxidizer, J. Microw. Power Electromagn. Energy 49 (1) (2015) 21–28.

[33] K. Choi, T. Hong, K. Ik Sim, T. Ha, B. Cheol Park, J. Hyuk Chung, et al., Reflection terahertz time-domain spectroscopy of RDX and HMX explosives, J. Appl. Phys. 115 (2) (2014) 023105.

[34] A.H. Rezaei, M.H. Keshavarz, M.K. Tehrani, S.M.R. Darbani, A.H. Farhadian, S.J. Mousavi, et al., Approach for determination of detonation performance and aluminum percentage of aluminized-based explosives by laser-induced breakdown spectroscopy, Appl. Opt. 55 (12) (2016) 3233–3240.

[35] E.N. Rao, S. Sunku, S.V. Rao, Femtosecond laser-induced breakdown spectroscopy studies of nitropyrazoles: the effect of varying nitro groups, Appl. Spectrosc. 69 (11) (2015) 1342–1354.

[36] T.H. Ong, T. Mendum, G. Geurtsen, J. Kelley, A. Ostrinskaya, R. Kunz, Use of mass spectrometric vapor analysis to improve canine explosive detection efficiency, Anal. Chem. 89 (12) (2017) 6482–6490.

[37] A. Cuisset, S. Gruet, O. Pirali, T. Chamaillé, G. Mouret, Synchrotron FT-FIR spectroscopy of nitro-derivatives vapors: new spectroscopic signatures of explosive taggants and degradation products, Spectrochim. Acta Part A 132 (2014) 838–845.

[38] A.V. Ewing, S.G. Kazarian, Infrared spectroscopy and spectroscopic imaging in forensic science, Analyst 142 (2) (2017) 257–272.

[39] E. Balcı, B. Rameev, H. Acar, G.V. Mozzhukhin, B. Aktaş, B. Çolak, et al., Development of Earth's field nuclear magnetic resonance (EFNMR) technique for applications in security scanning devices, Appl. Magn. Reson. 47 (1) (2016) 87–99.

[40] J. Lee, S. Park, S.G. Cho, E.M. Goh, S. Lee, S.S. Koh, et al., Analysis of explosives using corona discharge ionization combined with ion mobility spectrometry–mass spectrometry, Talanta 120 (2014) 64–70.

[41] M. Tabrizchi, V. ILbeigi, Detection of explosives by positive corona discharge ion mobility spectrometry, J. Hazard. Mater. 176 (1–3) (2010) 692–696.

[42] M. Markiewicz-Keszycka, X. Cama-Moncunill, M.P. Casado-Gavalda, Y. Dixit, R. Cama-Moncunill, P.J. Cullen, C. Sullivan, Laser-induced breakdown spectroscopy (LIBS) for food analysis: a review, Trends Food Sci. Technol. 65 (2017) 80–93.

[43] J.B. Sleiman, B. Bousquet, N. Palka, P. Mounaix, Quantitative Analysis of Hexahydro-1, 3, 5-trinitro-1, 3, 5, Triazine/Pentaerythritol Tetranitrate (RDX–PETN) Mixtures by Terahertz Time Domain Spectroscopy, Appl. Spectrosc. 69 (12) (2015) 1464–1471.

[44] A.K. Myakalwar, N. Spegazzini, C. Zhang, S.K. Anubham, R.R. Dasari, I. Barman, et al., Less is more: avoiding the LIBS dimensionality curse through judicious feature selection for explosive detection, Sci. Rep. 5 (2015) 13169.

[45] J. Hwang, N. Choi, A. Park, J.Q. Park, J.H. Chung, S. Baek, et al., Fast and sensitive recognition of various explosive compounds using Raman spectroscopy and principal component analysis, J. Mol. Struct. 1039 (2013) 130–136.

[46] J.H. Chung, S.G. Cho, Nanosecond gated raman spectroscopy for standoff detection of hazardous materials, Bull. Korean Chem. Soc. 35 (12) (2014) 3547.

[47] M.O. Salles, G.N. Meloni, W.R. De Araujo, T.R.L.C. Paixão, Explosive colorimetric discrimination using a smartphone, paper device and chemometrical approach, Anal. Methods 6 (7) (2014) 2047–2052.

[48] A.S. Finch, M. Coppock, J.R. Bickford, M.A. Conn, T.J. Proctor, D.N. Stratis-Cullum, Smart phones: platform enabling modular, chemical, biological, and explosives sensing, Chemical, Biological, Radiological, Nuclear, and Explosives (CBRNE) Sensing XIV, 8710, International Society for Optics and Photonics, 2013, p. 87100D.

[49] N. Tang, L. Mu, H. Qu, Y. Wang, X. Duan, M.A. Reed, Smartphone-enabled colorimetric trinitrotoluene detection using amine-trapped polydimethylsiloxane membranes, ACS Appl. Mater. Interfaces 9 (16) (2017) 14445–14452.

[50] Y. Zhang, Y. Cai, F. Dong, L. Bian, H. Li, J. Wang, et al., Chemically modified mesoporous wood: a versatile sensor for visual colorimetric detection of trinitrotoluene in water, air, and soil by smartphone camera, Anal. Bioanal. Chem. 411 (30) (2019) 8063–8071.

[51] Y. Liu, J. Li, G. Wang, B. Zu, X. Dou, One-step instantaneous detection of multiple military and improvised explosives facilitated by colorimetric reagent design, Anal. Chem. 92 (20) (2020) 13980–13988.

[52] Z. Xie, H. Ge, J. Du, T. Duan, G. Yang, Y. He, Compartmentalizing incompatible tandem reactions in pickering emulsions to enable visual colorimetric detection of nitramine explosives using a smartphone, Anal. Chem. 90 (19) (2018) 11665–11670.

[53] S. Li, D. Zhang, J. Liu, C. Cheng, L. Zhu, C. Li, et al., Electrochemiluminescence on smartphone with silica nanopores membrane modified electrodes for nitroaromatic explosives detection, Biosens. Bioelectron. 129 (2019) 284–291.

[54] S.M. Tawfik, M. Sharipov, S. Kakhkhorov, M.R. Elmasry, Y.I. Lee, Multiple emitting amphiphilic conjugated polythiophenes-coated CdTe QDs for picogram detection of trinitrophenol explosive and application using chitosan film and paper-based sensor coupled with smartphone, Adv. Sci. 6 (2) (2019) 1801467.

[55] N. López-Ruiz, M.M. Erenas, I. de Orbe-Payá, L.F. Capitán-Vallvey, A.J. Palma, et al., Computer vision-based portable system for nitroaromatics discrimination, J. Sensors 2016 (2016).

[56] M.B. Muthukumaresan, S. Sakthivel, S.J. Kumar, gED–mobile based explosive detection, Int. J. Appl. Eng. Res. 10 (37) (2015).

[57] S.C. Thandu, On Temporal and Frequency Responses of Smartphone Accelerometers for Explosives Detection, 2014.

[58] L. Liu, D. Zhang, Q. Zhang, X. Chen, G. Xu, Y. Lu, et al., Smartphone-based sensing system using ZnO and graphene modified electrodes for VOCs Detection, Biosens. Bioelectron. 93 (2017) 94–101.

[59] D. Zhang, J. Jiang, J. Chen, Q. Zhang, Y. Lu, Y. Yao, et al., Smartphone-based portable biosensing system using impedance measurement with printed electrodes for 2, 4, 6-trinitrotoluene (TNT) detection, Biosens. Bioelectron. 70 (2015) 81–88.

[60] H.J. Kwon, E.Ccopa Rivera, M.R.C. Neto, D. Marsh, J.J. Swerdlow, R.L. Summerscales, et al., Development of smartphone-based ECL sensor for dopamine detection: practical approaches, Results Chem. 2 (2020) 100029.

Future of smartphone-based analysis

18

Rüstem Keçili[a], Fatemeh Ghorbani-Bidkorbeh[b], Ayhan Altıntaş[a,c], Chaudhery Mustansar Hussain[d]

[a]Anadolu University, Yunus Emre Vocational School of Health Services, Eskişehir, Turkey
[b]Shahid Beheshti University of Medical Sciences, School of Pharmacy, Department of Pharmaceutics, Tehran, Iran
[c]Anadolu University, Faculty of Pharmacy, Department of Pharmacognosy, Eskişehir, Turkey
[d]New Jersey Institute of Technology, Department of Chemistry and Environmental Science, Newark, NJ, United States

18.1 Introduction

There are an estimated 3.5 billion smartphone users all over the world [1] and it is gradually increasing every year. Owing to its diverse capability such as advanced computing power, fast data processing ability, higher resolution of the camera, the smartphone is a versatile handheld computer beyond a simple phone device.

On the other hand, powerful miniaturized devices exhibit a number of advantages including facile operation, portability and rapid analysis time have the great potential to ensure the execution of complicated and time-consumed analytical processes in the field without the need for use of expensive analytical instruments or the researcher who has high levels of expertise [2–7]. Nanomaterials [8–17] display unique properties and play very important role in the design and development of these miniaturized systems such as lab-on-a-chip [18–20].

Rather than simply using the advanced capabilities for voice communications and web browsing, smartphones technology has found a broad range of successful applications in different areas [21–34].

This chapter provides an overview of the recent developments on the design, construction and smartphone-based devices in healthcare, environmental and food applications. Since smartphones display a number of advantages such as portability, operability and integration into the various analytical devices, they can be successfully employed as powerful tools in the design and development of new sensor platforms. Integration of smartphones into the different types of selective and sensitive sensor platforms enables develop user-friendly, portable and sensitive devices. The chapter includes interesting examples reported on health monitoring, diagnostics, environmental monitoring and food applications of smartphone technology as well as major challenges and future perspectives of the smartphone-based devices.

18.1.1 Health monitoring and diagnostic applications based on smartphone technology

In the past years, a number of smartphone-based healthcare applications has exploded as evidenced by over 50 k mobile health applications submitted to FDA [35]. Scientists around the world reviewed these advances in details. For example, two research groups, Patrick and colleagues and Wang and co-workers offered respective review opinions reported in 2008 and 2009, respectively [36,37]. Early applications effectively applied the smartphone cameras to carry out the biomedical imaging as demonstrated in the reported review by Xie et al. [38]. Recently, the smartphone technology fastly evolved. Modern smartphones now feature high-resolution camera imagers with adaptive lenses, multiple cameras to assist with depth analysis, optical and electronic image stabilization and a number of software algorithms working to bring out a true image. These progresses are bound to improve and extend biomedical imaging applications of smartphone-based devices. For example, the research group Ozcan designed and fabricated smartphone-based devices for the powerful fluorescence imaging and microscopy which can be successfully employed for the sensitive detection of a single virus as well as carry out the automated cytometric analyses [39].

One of the most exciting proof of smartphone capabilities for health diagnostics was developed by the Laksanasopin and colleagues where they used a smartphone with a custom dongle to for the sensitive detection of HIV virus [40]. This research was extensively featured in the media along with science editorials affirming that the area of smartphone technology-based diagnostic tools was of immediate relevance and crucial effect.

The application of smartphone technology-based diagnostic devices can be divided into two categories which are in vivo testing and in vitro testing [41]. In vivo testing, employs the built-in sensors is employed to perform health monitoring.

On the other hand, in-vitro testing based on smartphone technology investigates the application of smartphones beyond images and sounds. These applications are based on custom or off-the-shelf hardware and dongles paired up with a smartphone application. The hardware and dongles are combined with specialized sensors which enables previously unattainable information including temperature, ultrasound imaging etc. Some other examples include applying smartphones in conjunction with simpler blood pressure monitors, glucose monitors and coagulation monitors [42].

In-vitro applications of smartphones also play an important role as a user-input and computational device for the analysis, storage and transmission of the achieved data from the sensor hardware. Beyond simple health analysis and monitoring, smartphones are now being utilized to assist or perform complex biochemical assays for the diagnosis of infectious diseases and other healthcare applications.

A number of successful examples of smartphone technology to assist with POC diagnostic testing was reported in the literature. For example, İlyas et al. reported the development of a facile, fast, cheap and effective smartphone-based system for the detection of Sickle cell disease [43]. This fabricated smartphone-based system utilizes

Fig. 18.1 Smartphone-based platform for sickle cell blood imaging. (Republished with permission from [43]).

an external lens which can be easily integrated into the smartphone camera to record the obtained images from various blood samples inside a microchip. Fig. 18.1 shows the developed smartphone-based platform for sickle cell blood imaging.

In this study, the obtained images were fastly processed by employing a MATLAB program and the total number of sickled cells was automatically determined. To investigate the efficiency of the developed setup, normal blood sample was used as well as the Sickle cell disease patients' samples. The developed algorithm for image processing was successfully employed for the accurate quantification of the sickled cells in deoxygenated blood samples. The achieved results obtained by using the developed smartphone-based platform were compared and validated with the achieved data by using optical microscope.

In another interesting research carried out by Barnes and co-workers [44], a smartphone-based smartphone-based real-time loop-mediated isothermal amplification (smaRT-LAMP) platform was designed and fabricated for the sensitive determination of pathogen ID in urinary sepsis patients. The free and custom-built mobile phone application enabled the smartphone to serve as a stand-alone system for quantitatively diagnosis, providing the real time determination of genome copy-number of bacterial pathogens (Fig. 18.2). A comparative analysis of bacteria urine samples from sepsis patients revealed that the efficiency of the developed smaRT-LAMP matched

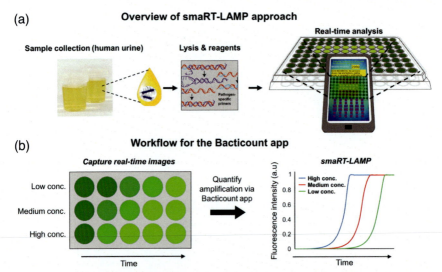

Fig. 18.2 The schematic depiction of the (a) overview of SmaRT-LAMP approach for the direct testing of specimens in urine samples from sepsis patients. The applied assay includes sample collection (human urine), the addition of bacterial cell lysis/reagents and real-time analysis b applying the smartphone. (b) workflow for the Bacticount app that enables the effective analyses of the fluorescence data continuously achieved from multiple samples by using camera of the smartphone (left panel) and further applies these obtained results to real time determination of the genome copy-number of bacterial pathogens (right panel) (Republished with permission from [44]).

that of clinical diagnostics at the admitting hospital in a fraction of the time (~1 h vs. 18–28 h). Among patients with bacteremic complications of their urinary sepsis, pathogen ID from the urine matched that from the blood that potentially enables the early diagnosis of pathogen after hospital admission. In addition, the achieved data confirmed that developed smaRT-LAMP does not display false positives in sepsis patients with clinically negative urine cultures.

In a crucial work reported by Chandra [45], a promising personalized smartphone-based electrochemical sensor system for the sensitive and efficient detection of COVID-19 was proposed. According to the authors it can possible to design and fabricate a powerful biosensor towards the target virus COVID-19 by combination of a complete disposable sensing module with a smartphone-based application platform for personalized diagnosis (Fig. 18.3). Sensing surface can be efficiently optimized according to the marker molecules, then, it can be improvised for point of care (POC) diagnosis in real clinical samples. This type of miniaturized system can provide a rapid and affordable sensing process not only to sentively detect but also for efficiently monitoring of the COVID-19 in large scale. A smartphone-based "cloud" directory can also enable real-time surveillance through geo-tagging which constitutes the process of defining, creating and provisioning a set of geolocation information to a smartphone.

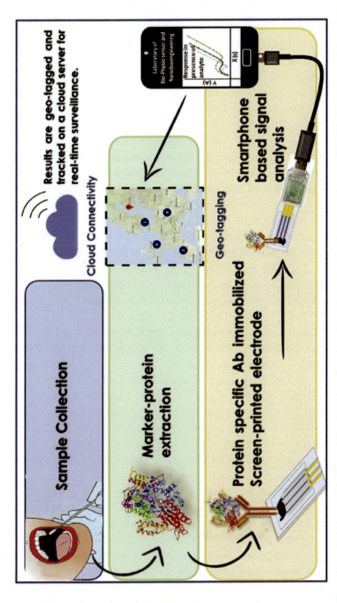

Fig. 18.3 The schematic representation of the suggested smartphone-based sensitive detection of COVID-19 and cloud-based real-time surveillance. (Republished with permission from [45]).

18.1.2 Environmental monitoring based on smartphone technology

Smartphone tehnology was also successfully applied for the sensitive detection of various pollutants such as bacteria and heavy metals etc. in environmental samples.

In a study performed by Park and Yoon [46], smartphone-based effcient detection of Escherichia coli in field water samples was successfully carried out by applying paper microfluidics. In their study, a three-channel paper chip was prepared with a negative control channel preloaded with bovine serum albumin (BSA)-conjugated beads and 2 detection channels towards E. coli preloaded with anti-E. coli antibody-conjugated beads for the sensitive detection at low and high concentration levels. In this study, the field water samples were introduced to the paper chip by dipping or pipetting and the antigens from E. coli moved through the fibers of paper via capillary force. Anti-E. coli antibody-conjugated beads confined within the paper fibers, immunoagglutinate in the existence of E. coli antigens while BSA-conjugated beads did not. The quantification of the extent of immunoagglutination was performed by evaluating Mie scatter intensity from the images obtained at an optimized angle and distance using a smartphone. The achieved data exhibited that there is a great agreement with the data from MacConkey plate i.e., the count of viable E. coli. The obtained results also confirmed that the limit of detection is single-cell-level and the total analysis time was achieved as 90 s.

In another important research reported by Liu and colleagues [47], the sensitive and onsite detection of toxic Hg^{2+} ions was successfully conducted by using the developed smartphone- based optical fiber fluorescence sensor. In their research, the smartphone-based sensor was composed of a semiconductor laser for fluorescence signal excitation, a fiber probe modified with CdSe/ZnS quantum dots (QDs) towards the target Hg^{2+} ions, a laser and smartphone. The obtained data exhibited that the developed smartphone- based optical fiber fluorescence sensor showed excellent sensitivity and recognition behavior for Hg^{2+} ions with a quite wide range of detection between 1 nM and 1000 nM. The detection limit was obtained as 1 nM.

Dong et al., an ultra sensitive smartphone-based sensor platform was designed and developed for the efficient recognition of pyrophosphate in environmental and food samples [48]. For this purpose, a colorimetric sensor system based on smartphone through anti-aggregation of silver nanoparticles modified with polyvinylpyrrolidone caused by Pb^{2+} was designed and fabricated. Pb^{2+} has ability to effectively induce the aggregation of AgNPs/ polyvinylpyrrolidone because of its cross-link impact that lead to color change of AgNPs/ polyvinylpyrrolidone dispersion. However, once Pb^{2+} ions are added into the AgNPs/ polyvinylpyrrolidone dispersion containing pyrophosphate, the AgNPs/polyvinylpyrrolidone can keep well dispersed without color change. It is because that pyrophosphate is more prone to chelate with Pb^{2+} via coordination bond. Therefore, it leads to a formation of a stable complex. This sensing mechanism was verified by UV–vis, transmission electron microscope (TEM) and dynamic light scattering (DLS) techniques. The achieved results exhibit that the developed smartphone-based sensor leads to a change of color from blue to yellow for the sensitive detection of pyrophosphate in canned meat and water samples. The obtained data displayed that the developed smartphone-based colorimetric sensor exhibits great sensitivity towards the target compound pyrophosphate in environmental and food samples. The limit of detection was achieved as 1.0 μM and 0.2 μM analyzed by naked-eyes and smartphone-based sensor, respectively.

18.1.3 Smartphone-based food applications

Smartphone-based systems have also potential to revolutionize food safety control. Various interesting applications are summarized in the following:

Chaisiwamongkhol et al. reported the design and development of a smartphone-based sensor for the colorimetric and sensitive detection of sibutramine, an appetite suppressant used for many years but it was banned by The United States Food and Drug Administration (FDA) and European Medicines Agency (EMA) due to its cardiovascular risks, in suspected food supplement products using Au nanoparticles [49]. In this study, the aggregation of Au nanoparticles stabilized with citrate in the existence of the target compound sibutramine caused to a color change from wine red to blue that was successfully verified by using transmission electron microscopy (TEM) and UV–Vis spectrophotometry techniques. The change in the color of the solution was visible by the naked eye and could be effectively monitored using a smartphone. The achieved results indicated that the ratio of green and red colors in a photo using a smartphone is linearly related to the concentration of the target sibutramine in the range between 5 μM and 15 μM with the $R^2 = 0.979$. The quantification limit and detection limit values were achieved as 3.47 μM and 1.15 μM, respectively,

In another interesting research [50], Jin and colleagues reported the design and fabrication of a smartphone-based sensor for the on-site and sensitive detection of the target organophosphorus pesticide paraoxon in apple-banana juice samples. In their research, the researchers proposed a concept of multi-enzyme cascade system-based hydrogel kit integrated with a smartphone detector for on-site detection of the target pesticide (Fig. 18.4). For this purpose, the prepared a target-responsive hydrogel (TRhg)-based kit by embedding MnO_2 nanoflakes (NFs) into the sodium alginate hydrogel. Based on the considerable activity of mimic oxidase, MnO_2 nanozyme induced color reaction by applying the sensing probe 3,3′,5,5′-tetramethylbenzidine (TMB). Choline oxidase (ChO) and acetylcholinesterase (AChE) played a crucial role in the efficient catalysis of the acetylcholine (ACh) hydrolysis to generate H_2O_2 that triggered decomposition of MnO_2 NFs, further blocked the TMB oxidation. the target organophosphorus pesticide paraoxon as an inhibitor of AChE suppressed the production of H_2O_2, that significantly decreased the decomposition of MnO_2 NFs which results in the color response of the kit. The obtained images by using the smartphone can be further analyzed through an application program. The achieved results confirmed that the developed smartphone-based sensor can be successfully employed for the on-site and sensitive detection of the target organophosphorus pesticide paraoxon. The limit of detection was found as 0.5 ng mL^{-1}.

In another crucial work carried out by Coşkun and co-workers [51], a smartphone-based personalized food allergen testing system which employed the colorimetric assays of test samples was designed and developed. Using the android software

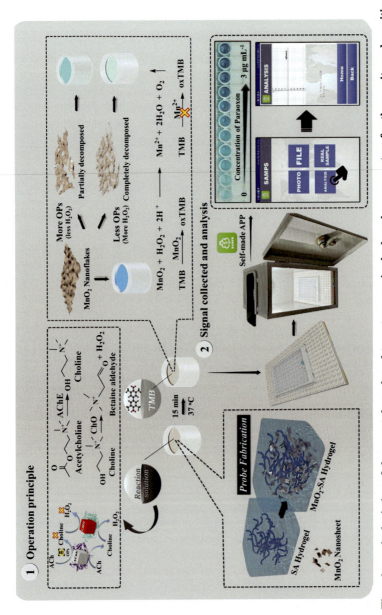

Fig. 18.4 The schematic depiction of the smartphone-based target-responsive hydrogel sensor system for the on-site and sensitive detection of target organophosphorus pesticide. (Republished with permission from [50]).

exists on the Samsung Galaxy S II model smartphone, a 3D-printed attachment was designed for operating at the field conditions and LED (Light Emitting Diodes, a peak wavelength: 650 nm, the bandwidth: 15 nm) was utilized to illuminate both a test tube and a control tube. Arbitrary rectangular type of the frame (i.e., 300 × 300 pixels) of acquired images of the tube were processed to determine the value of relative allergen concentration which is in the sample. For the depiction of the ability of this form, a standard food allergy test kit was applied. For the system calibration, different amount of food samples should be digitally quantified before using the device. Through the real food test (i.e., peanut included in the food), the quantity of the allergen in the food samples was measured and an allergy test kit exhibited a similar result of that of smartphone. These food allergen testing results can be uploaded to servers for sharing the information and it may be very useful for allergic individuals globally

18.2 Conclusions and future perspectives

The smartphone technology is undoubtedly a revolutionary invention. The rapid advancements in electronics technology not only lead to reduce of the dimensions but also significantly increase in the functions of smartphones which enables them to be highly portable, facile operation and ability to their integration into the sensor platforms. These unique properties received the great interest from the researchers who work in different fields.

In this chapter, we provide the various successful applications (i.e. health monitoring and diagnostics, environmental monitoring and food applications), future opportunities and challenges of the smartphones. The smartphone technology came a long route in complementing and competing with conventional instruments in laboratories which are expensive, bulky, cumbersome and relatively more complicated to operate. Although excellent progresses were achieved in this technology, there are still some challenges needs to be overcome in the design and fabrication of smartphone-based sensor systems. For example, the obtained signals in smartphone-based optical sensors were transmitted through digital camera and various optical attachments in the form of images. In the next step, the analysis of the obtained image information can be conducted through the smartphone. Therefore, much attention and efforts should be put on the minimization of these optical attachments on smartphones via simplification of the optical components and routes to get lighter and smaller devices. For example, some microscopic attachments can be integrated into the smartphone. For this purpose, innovative imaging approaches including near-field imaging or shadow imaging approaches can be employed in order to decrease weight and bulk of the device.

On the other hand, most of the smartphone-based sensor platforms require optical signal processing through custom applications (apps). Setting up a viable app is a main challenge for smartphone-based sensor systems. These apps should be facile for untrained people to understand and operate.

Other main issues need to be considered are the minimization of inter-phone variation, optimizing protocols of sample extraction process and integration of the smartphone software into the operational apps while limiting the use of auxiliary parts. Although a number of approaches to resolve these issues exhibit promising results, the solutions need to be efficiently combined.

It is strongly believed that after efficiently overcome the current challenges and successfully addressing the issues need to be considered, the smartphone-based sensor platforms will play an excellent role in the design and fabrication on novel sensor systems for POC analysis and have an even more positive effect on the future of health monitoring and novel diagnostic systems, environmental monitoring and food technology.

References

[1] S. O''Dea, Number of Smartphone Users Worldwide; Dec 10, 2020. https://www.statista.com/statistics/330695/number-of-smartphone-users-worldwide/(accessed on 05 February 2021).

[2] K. Khachornsakkul, W. Dungchai, Development of an ultrasound-enhanced smartphone colorimetric biosensor for ultrasensitive hydrogen peroxide detection and its applications, RSC Adv. 10 (2020) 24463–24471.

[3] Editor(s) S. Büyüktiryaki, Y. Sümbelli, R. Keçili, C.M. Hussain, Lab-on-chip platforms for environmental analysis, in: Paul Worsfold, Colin Poole, Alan Townshend, Manuel Miró (Eds.), Encyclopedia of Analytical Science3rd Edition, Academic Press, 2019, pp. 267–273.

[4] Editor(s) R. Keçili, F. Ghorbani-Bidkorbeh, İ. Dolak, C.M. Hussain, Chapter 1 - Era of nano-lab-on-a-chip (LOC) technology, in: Chaudhery Mustansar Hussain (Ed.), Handbook on Miniaturization in Analytical Chemistry, Elsevier, 2020, pp. 1–17.

[5] Editor(s) Ö. Biçen Ünlüer, F. Ghorbani-Bidkorbeh, R. Keçili, C.M. Hussain, Chapter 12 - Future of the modern age of analytical chemistry: nanominiaturization, in: Chaudhery Mustansar Hussain (Ed.), Handbook on Miniaturization in Analytical Chemistry, Elsevier, 2020, pp. 277–296.

[6] J. Sengupta, C.M. Hussain, Graphene and its derivatives for analytical lab on chip platforms, TrAC Trend Anal. Chem. 114 (2019) 326–337.

[7] Editor(s) R. Keçili, S. Büyüktiryaki, C.M. Hussain, Chapter 8 - Micro total analysis systems with nanomaterials, in: Chaudhery Mustansar Hussain (Ed.), Handbook of Nanomaterials in Analytical Chemistry, Elsevier, 2020, pp. 185–198.

[8] C.M. Hussain, Magnetic nanomaterials for environmental analysis, in: C.M. Hussain, B. Kharisov (Eds.), Advanced Environmental Analysis-Application of Nanomaterials, The Royal Soc. of Chem., 2017

[9] R. Keçili, C.M. Hussain, Recent progress of imprinted nanomaterials in analytical chemistry, Int J Anal Chem (2018) 8503853.

[10] R. Keçili, S. Büyüktiryaki, C.M. Hussain, Advancement in bioanalytical science through nanotechnology: past, present and future, TrAC Trend Anal. Chem. 110 (2019) 259–276.

[11] S. Büyüktiryaki, R. Keçili, C.M. Hussain, Functionalized nanomaterials in dispersive solid phase extraction: advances & prospects, TrAC Trend Anal. Chem. 127 (2020) 115893.

[12] D. Sharma, C.M. Hussain, Smart nanomaterials in pharmaceutical analysis, Arab. J. Chem. 13 (2020) 3319.

[13] C.M. Hussain, Nanomaterials in Chromatography: Current Trends in Chromatographic Research Technology and Techniques, Elsevier, 2018.

[14] C.M. Hussain, R. Keçili, Modern Environmental Analysis Techniques for Pollutants, 1st Edition, Elsevier, 2019.

[15] D. Rawtani, P.K. Rao, C.M. Hussain, Recent advances in analytical, bioanalytical and miscellaneous applications of green nanomaterial, TrAC, Trends Anal. Chem. 133 (2020) 116109.

[16] R. Keçili, G. Arli, C.M. Hussain, Chapter 14 - Future of analytical chemistry with graphene, in: Chaudhery Mustansar Hussain (Eds.), Comprehensive Analytical Chemistry, 91, Elsevier, 2020, pp. 355–389.

[17] S. Büyüktiryaki, R. Keçili, C.M. Hussain, Chapter 2- Modern age of analytical chemistry: nanomaterials, in: Chaudhery Mustansar Hussain (Eds.), Handbook of Nanomaterials in Analytical Chemistry, Elsevier, 2020, pp. 29–40.

[18] C. Kiang Chua, A. Ambrosi, M. Pumera, Graphene based nanomaterials as electrochemical detectors in Lab-on-a-chip devices, Electrochem. Commun. 13 (5) (2011) 517–519.

[19] A. Kobuszewska, D. Kolodziejek, M. Wojasinski, E. Jastrzebska, T. Ciach, Z. Brzozka, Lab-on-a-chip system integrated with nanofiber mats used as a potential tool to study cardiovascular diseases (CVDs), Sens. Actuators B 330 (2021) 129291.

[20] I.L. Celine Justino, A.P Teresa R.-S., Armando C. Duarte, Chapter 14 - Nanomaterials in lab-on-chip chromatography, in: Chaudhery Mustansar Hussain (Eds.), Nanomaterials in Chromatography, Elsevier, 2018, pp. 387–400.

[21] KartickChandra Majhi, Paramita Karfa, Rashmi Madhuri, Chapter 6 - Smartphone-based nanodevices for in-field diagnosis, in: Suvardhan Kanchi, Deepali Sharma (Eds.), Nanomaterials in Diagnostic Tools and Devices, Elsevier, 2020, pp. 159–187.

[22] Akbar Hasanzadeh, Chapter 12 - Smartphone-based microfluidic devices, in: Michael R. Hamblin, Mahdi Karimi (Eds.), Biomedical Applications of Microfluidic Devices, Academic Press, 2021, pp. 275–288.

[23] Zheng Li, Shengwei Zhang, Qingshan Wei, Chapter 5 - Smartphone-based flow cytometry, in: Jeong-Yeol Yoon (Ed.), Smartphone Based Medical Diagnostics, Academic Press, 2020, pp. 67–88.

[24] S.K. Vashist, J.H.T. Luong, Chapter 16 - Smartphone-based immunoassays, in: Sandeep K. Vashist, John H.T. Luong (Eds.), Handbook of Immunoassay Technologies, Academic Press, 2018, pp. 433–453.

[25] Anna Pyayt, Chapter 6 - Smartphones for rapid kits, in: Jeong-Yeol Yoon (Eds.), Smartphone Based Medical Diagnostics, Academic Press, 2020, pp. 89–102.

[26] D.S.Y. Ong, M. Poljak, Smartphones as mobile microbiological laboratories, Clin. Microbiol. Infect. 26 (4) (2020) 421–424.

[27] Mei-.Ting Liu, Jing Zhao, Shao-.Ping Li, Application of smartphone in detection of thin-layer chromatography: case of salvia miltiorrhiza, J. Chromatogr. A 1637 (2021) 461826.

[28] J.T. Moehling, D.H. Lee, M.E. Henderson, M.K. McDonald, P.H. Tsang, S. Kaakeh, E.S. Kim, S.T. Wereley, T.L. Kinzer-Ursem, K.N. Clayton, J.C. Linnes., A smartphone-based particle diffusometry platform for sub-attomolar detection of vibrio cholerae in environmental water, Biosens. Bioelectron. 167 (2020) 112497.

[29] Ze Wu, J. Lu, Q. Fu, L. Sheng, B. Liu, C. Wang, C. Li, T. Li, A smartphone-based enzyme-linked immunochromatographic sensor for rapid quantitative detection of carcinoembryonic antigen, Sens. Actuators B 329 (2021) 129163.

[30] W. Luo, J. Deng, J. He, Z. Han, C. Huang, Y. Li, Q. Fu, H. Chen, A smartphone-based multi-wavelength photometer for on-site detection of the liquid colorimetric assays for clinical biochemical analyses, Sens. Actuators B 329 (2021) 129266.

[31] R. Bandi, M. Alle, C.-W. Park, S.-Yi. Han, Gu-J. Kwon, N.-H. Kim, J.-C. Kim, S.-H. Lee, Cellulose nanofibrils/carbon dots composite nanopapers for the smartphone-based colorimetric detection of hydrogen peroxide and glucose, Sens. Actuators B 330 (2021) 129330.

[32] X. Kou, L. Tong, Y. Shen, W. Zhu, Li Yin, S. Huang, F. Zhu, G. Chen, G. Ouyang, Smartphone-assisted robust enzymes@MOFs-based paper biosensor for point-of-care detection, Biosens. Bioelectron. 156 (2020) 112095.

[33] P. Teengam, W. Siangproh, S. Tontisirin, A. Jiraseree-amornkun, N. Chuaypen, P. Tangkijvanich, C.S. Henry, N. Ngamrojanavanich, O. Chailapakul, NFC-enabling smartphone-based portable amperometric immunosensor for hepatitis B virus detection, Sens. Actuators B 326 (2021) 128825.

[34] Z. Liu, Q. Hua, J. Wang, Z. Liang, J. Li, J. Wu, X. Shen, H. Lei, X. Li, A smartphone-based dual detection mode device integrated with two lateral flow immunoassays for multiplex mycotoxins in cereals, Biosens. Bioelectron. 158 (2020) 112178.

[35] S. Mcinerney, Can you diagnose me now? a proposal to modify the fda's regulation of smartphone mobile health applications with a pre-market notification and application database program recommended citation, Mich. J. L. Reform 48 (2015) 1073.

[36] K. Patrick, W.G. Griswold, F. Raab, S.S. Intille, Health and the mobile phone, Am. J. Prev. Med. 35 (2) (2008) 177–181.

[37] H. Wang, J. Liu, Mobile phone based health care technology, Recent Pat, Biomed. Eng. 2 (1) (2009) 15–21.

[38] Q.-M. Xie, J. Liu, Mobile phone based biomedical imaging technology: a newly emerging area, Recent Patents Biomed. Eng. 3 (1) (2010) 41–53.

[39] A. Ozcan, Mobile phones democratize and cultivate next-generation imaging, diagnostics and measurement tools, Lab Chip 14 (17) (2014) 3187–3194.

[40] T. Laksanasopin, et al., A smartphone dongle for diagnosis of infectious diseases at the point of care, Sci. Transl. Med. 7 (273) (2015) 4 273re1.

[41] X. Xu, et al., Advances in smartphone-based point-of-care diagnostics, Proc. IEEE 103 (2) (2015) 236–247.

[42] S.K. Vashist, P.B. Luppa, L.Y. Yeo, A. Ozcan, J.H.T. Luong, Emerging technologies for next-generation point-of-care testing, Trends Biotechnol. 33 (11) (2015) 692–705.

[43] S. Ilyas, Mazhar Sher, E. Du, Waseem Asghar, Smartphone-based sickle cell disease detection and monitoring for point-of-care settings, Biosens. Bioelectron. 165 (2020) 112417.

[44] L. Barnes, DouglasM. Heithoff, ScottP. Mahan, GaryN. Fox, Andrea Zambrano, Jane Choe, Lynn N. Fitzgibbons, Jamey D. Marth, Jeffrey C. Fried, H.Tom Soh, Michael J. Mahan, Smartphone-based pathogen diagnosis in urinary sepsis patients, EBioMedicine 36 (2018) 73–82.

[45] P. Chandra, Miniaturized label-free smartphone assisted electrochemical sensing approach for personalized COVID-19 diagnosis, Sensors Int. 1 (2020) 100019.

[46] T.-S. Park, J.-Y. Yoon, Smartphone detection of escherichia coli from field water samples on paper microfluidics, IEEE Sens. J. 15 (3) (2015) 1902–1907.

[47] T. Liu, Wenqi Wang, Dan Jian, Jiahao Li, He Ding, Dingrong Yi, Fei Liu, Shouyu Wang, Quantitative remote and on-site Hg^{2+} detection using the handheld smartphone based optical fiber fluorescence sensor (SOFFS), Sens. Actuators B 301 (2019) 127168.

[48] C. Dong, Xuehua Ma, Nianxiang Qiu, Yujie Zhang, Aiguo Wu, An ultra-sensitive colorimetric sensor based on smartphone for pyrophosphate determination, Sens. Actuators B 329 (2021) 129066.

[49] K. Chaisiwamongkhol, Shakiroh Labaidae, Sunisa Pon-in, Sakchaibordee Pinsrithong, Thanthapatra Bunchuay, Apichai Phonchai, Smartphone-based colorimetric detection using gold nanoparticles of sibutramine in suspected food supplement products, Microchem. J. 158 (2020) 105273.

[50] R. Jin, Fanyu Wang, Qingyun Li, Xu Yan, Mengqi Liu, Yue Chen, Weirong Zhou, Hao Gao, Peng Sun, Geyu Lu, Construction of multienzyme-hydrogel sensor with smartphone detector for on-site monitoring of organophosphorus pesticide, Sens. Actuators B 327 (2021) 128922.

[51] A.F. Coskun, J. Wong, D. Khodadadi, R. Nagi, A. Tey, A. Ozcan, Cell-Phone Based Food Allergen Testing, Optical Soc. of America, 2013 CTu2M. 7.

Index

Page numbers followed by "*f*" and "*t*" indicate, figures and tables respectively.

A

Accelerometer, 201
Active contents in drugs, investigation of, 131
Aflatoxins, 37–40, 40*f*, 258–264
    smartphone-based monitoring, 50
AgNPs/ polyvinylpyrrolidone, aggregation of, 3*f*, 3–4
Analytical chemistry, smartphone application in, 87–88, 130, 134*f*
    barcode reading, 135
    chemiluminescence and bioluminescence, 143
    colorimetric applications, 130, 131
    analysis of pharmaceutical drugs, 131
    analysis of rice plant leaves, 132*f*
    milk adulteration, detection of, 131
    electrochemistry, 137
    fluorescence microscopy, 133, 141, 143
    Salmonella Typhimurium, investigation of, 133
    waterborne pathogens, investigation of, 133
    water contamination analysis, 133
    label-free detection, 139–141
    of lipo-polysaccharides, 141
    photonic metal-organic frameworks (MOF), diffraction efficiency of, 141
    photoluminescence, 143, 146
    pixelation, 137–139
Android mobile operating system, 399–400
Anionic polythiophene nanohybrids (APTQDs), 406–407
Antibacterial peptide, 6
Antibody (Ab), 321
Antibody immobilization, 329
Apple iPhones, 250–258
    digital image colorimeter, 258–264
Aptamer-based lateral flow biosensor (APTA -LFB), 58–59
Arnold, Karl, 1
Artificial Intelligence (AI), 382
Autonomic nervous system (ANS), 379
Avian influenza virus (AIV) H7N9, detection of, 76–77

B

Bacteria in saliva samples, testing of, 6
Bacterial infections, 312
Bacterial infectious diseases, 314
Barcode localization and decoding engine (BLaDE), 91–92
Barcode readers, 85, 86, 89–90
    barcode data/identifier, 89–90
    for blood culture evaluation, 91–92
    paper-based analytical device (PAD) for blood typing, 135
    for checking aging of food products, 95–96
    1D/2D barcodes, 88–90, 90*f*
    dye-based sensor, 95, 96
    optical dye microbeads, use of, 96
    paper-based, 95–96, 97*f*, 98*f*
    pesticide residues, detection of, 97–98
    pregnancy hormones, detection of, 92–94
    QR code readers, 89–90, 92
    quantitative detection of human α-fetoprotein, 92–94
    for rapid detection, 92–94
Barcode reading technology, 85–86
    constraints, 86, 87
    in health care sector, 85–86
    use at airports, 85–86
Barometer, 201
Beer-Lambert law, 46, 64
Bioluminescence, definition, 143
Bioluminescent Assay in Real-Time and Loop-mediated isothermal AMPlification (BART-LAMP), 143

Bionic electronic eye (Bionic e-Eye), 20
Biosensors
    construction of, 58
    definition, 42
Bisphenol A (BPA), detection of, 24
Blood culture app, 91–92, 93*f*
Blood culture evaluation
    using barcode readers for, 91–92
    paper-based analytical device (PAD) for blood typing, 135
    using cyclic voltammetry, 137
Blue-emitting silicon carbide quantum dots (SiC Qds), 6
Body Area Network (BAN), 382
Body movements, 379
Body Sensor Network (BSN), 382
Bovine serum albumin (BSA), 422
Bremsstrahlung process, 403–404
Bureau of Public Roads (BPR) Function, 170

C

Calibration models, 327
Call Detailed Record (CDR), 159, 160, 160*f*
    trajectory reconstruction methods for, 160–161
    interpolation-based methods, 160
    map-matching-based methods, 160
    pattern-learning-based methods, 160–161
Camera, 46
    background illumination inference, 64
    digital zoom, 46
    in-camera function and sensitivity, 64
    quantification of color intensity, 46
Centrifugal based platforms, 332
Cephalexin, detection of, 9
Cetyltrimethylammonium bromide (CTAB) quantum dots (CTAB-QDs), 22–24
Chemiluminescence, definition, 143
Chemiluminescence lateral flow immunoassay (LFIA) technique, 10–11
Chemometric analysis using smartphones, 146–147
    android based MVS application, 150
    blue pen inks in documentoscopy, investigation of, 147
    smartphone videos and pattern recognition, 147, 150

ChemoStat® software, 210, 210
Cholesterol level, testing of, 3–4
Chromatic perception, 202
Chromatographic methods, 249–250
Chronoamperometry, 137
CIE 1931 color space, 6
CieLAB color system, 131
Citrinin (CIT), 40*f*, 41
Clinical diagnosis, smartphone-based, 87–88
Cloud computing, 240
ColorAssist app, 190
Colorimetric assay, 406
Colorimetric detection, using barcode sensors, 87–88, 96–97
Colorimetric sensor, 6, 20, 185
    applications, 2–3
    approach to detection, 41
    CieLAB color system, 131
    HSV color system, 131
    intensity of color after chemical reaction, 42
    Red-Green-Blue (RGB) model, 41–42, 43*f*, 46, 130, 131
    Cd^{2+} ions, detection of, 22, 23*f*
    components, 42
    design, 2–3
    image analysis, 185–186
    allergen concentrations, calculation of, 189–190
    alternariol monomethyl ether mycotoxin, determination of, 189
    analytes identification and quantification, 187
    chroramphenicol, determination of, 190
    cyanide detection, 188
    epoxides, detection of, 190–191
    G value of image, 186
    heavy metal detection, 188
    H_2O_2 detection strips, 189
    quantifying *Salmonella Typhimurium* bacteria, 189
    using aptamer-based colorimetric assay, 190
    volatile organic compounds, determination, 188
    marine toxins, efficient and sensitive detection of, 20
    real time sensing efficiency, 3, 4, 5*f*

Index 433

sensing mechanism, 3–4
sibutramine, detection of, 20–22
smartphone accessory, 114*f*
using digital colour models, 118*t*
Color Lab android free application, 190, 191
Color Picker APP, 64, 258–264
Colourimetric reagents, 60*f*
   BAPTA, 61–63
   bromocresol violet, 58–59
   bromothymol blue, 58–59
   diphenylamine reagent, 59
   dipicrylamine, 61
   EDTA, 61–63
   EGTA, 61–63
   ferric ammonium sulfate, 61
   glyoxal bis(2-hydroxyanil), 61–63
   Griess reagent, 59, 62*f*
   mercuric thiocyanate, 61
   methylene blue, 61
   methyl red, 58–59
   methylthymol blue, 61–63
   molybdenum blue (MB), 59–61, 65*f*
   Nessler's reagent, 59
   phenolphthalein, 58–59
   phenol red, 58–59
   sodium tetraphenylborate, 61
   uranyl zinc acetate, 61
Colourimetric soil chemistry detection, 57
   using color reagents
   ammonium/ammonia, 59
   calcium ions, 61–63
   chloride, 61
   magnesium ions, 61–63
   nitrogen, 59, 62*f*
   pH, 58, 59, 60*f*, 63–64
   phosphorus, 59–61
   potassium ions, 61
   sodium ions, 61
   sulphide, 61
Compact lensless microscope, smartphone-based, 61
   optomechanical attachment, 61
Compass, 201
Complementary metal-oxide-semiconductor (CMOS) sensors, 13, 64
Cooper, Martin, 199
CPU-Z for Android, 201
Creatinine detection, 76–77
CrimePad®, 237

Crime scenes investigation
   crime scene investigation (CSI) database, 237
   design and development, 236, 237
   limitations, 239, 240
   mobile application, 237
   reporting, design and development, 236, 237
   scene documentation, 237–239
   web application, 237
Cu nanoclusters (CuNCs), 63

D

DASH7 technology, 410–411
Data pre-processing, 206
Diazonium salt, 59
Digital data processing techniques, 57–58
Digital enzyme analysis, 68
Dijkstra Algorithm, 168
Document Compendium®, 237
Dopamine detection, 27
DPA-Ce-GMP-Eu sensing platform, 258–264

E

Electrical component's technology, 129
Electrochemical sensor systems, 27
   COVID-19, detection of, 28–30
   design and fabrication of, 27
   dopamine detection, 27
   microRNA-21, detection of, 27
   nitroaromatic explosives, detection of, 30
   for white blood cell-counting system, 27–28
Electrochemiluminescence, 406–407
Electrochemistry, 137
   detection and characterization of uric acid, 137
   glucose level in blood, investigation of, 137
   investigation of 2,4,6-trinitrophenol, 137
Electromagnetic spectrum, 202*f*
Enzymatic Linked Immunosorbent Assays (ELISA), 11, 38, 73–75
Enzymatic reaction of biochemiluminescence, monitoring of, 8

Enzymes, 328–329
Equilibrium binding equation, 323
Explosive materials, 401t
Explosives detection, 400f

F

Factor analysis, 205–206
Fluorescence detection platforms, smartphone-based, 57–58, 141, 191–192
    with Cu nanoparticles, 192
    food-borne bacterial pathogens, detection of, 141
    glucose concentration, determination of, 191–192
    picric acid explosives, determination of, 192
    Rhodamine-123 doped self-assembled giant mesostructured silica sphere, analysis of, 141
    spermidine at fM levels, determination of, 192
    Staphylococcus Aureus, detection of, 143
    using 3D-printing technology, 192, 193
Fluorescence microscopy, 66, 68, 133
    Salmonella Typhimurium, investigation of, 133
    waterborne pathogens, investigation of, 133
    water contamination analysis, 133
Fluorescence resonance energy transfer/electron-transfer mechanism (FRET), 58
Fluorescence spectrometer
    advantages, 58–60
    aptamer-based lateral flow biosensor (APTA-LFB), 58–59
    detection of microRNA sequences, 59
    disadvantage, 58
    handheld micro-fluorescence spectrometer, 59
    mobile phone fluorescence spectrum reader, 60f
Fluorescent based smartphone detectors, 406–407
Foam measurement technique, 193
    electrochemical detection, 194
    of metals, 194–195

Food safety and quality, 249–250, 258–264
    checking
    food authenticity, 250–258
    melamine detection, 258–264
    photometric detection, 250–258
    smartphone-based biosensors, 249–250
    using barcode readers, 95–96
Foodswitch, 91–92
Fourier Transform Infrared (FTIR) spectroscopy, 403–404
FridgePal, 91–92
Fumonisins, 37, 41
    fumonisin B1, 41
FV-5 Lite app, 9–10

G

Galvanic skin response (GSR) techniques, 379
G-Fresnel spectrometer, 73–75
Giardia lamblia cysts, detection of, 68
GiftScroll smartphone app, 91–92
Global Positioning Systems (GPS), 381
Global System for Mobile Communication (GSM), 366
Glucose level in blood, investigation of, 137
Griess diazotization reaction, 59, 62f
Gyroscope, 200–201

H

Haemagglutination reaction, principle of, 92
Health information systems (HIS), 203–204, 368
Heartrate (HR), 378
Heterogeneous immunoassays, 329
Homogeneous methods, 324
Human mobility estimation, 159
    data filtering approach, 161, 162
    flash phenomenon, 162
    maximum cluster distance threshold, 162
    pendulum phenomenon, 162
    speed rule, 163
    triangle rule, 163
    GPS map-matching methods, 168, 169
    map-matching and interpolation, 167, 168
    mode detection and trip segmentation
    cell tower type state, 166
    decoding, 167
    Hidden Markov Model (HMM), 164

Index

model training method, 166, 167
transferring speed distribution, 165, 165f
types of mode, 163, 164f
multi-Steps least cost algorithm, 168, 171f
BPR function, 170
for different moving mode, 169
impedance function, 170
least impedance of paths, 168, 169
probability for node, 168
using Call Detailed Record (CDR), 159, 160, 160f
datasets, 161
spatial information, 161
trajectory reconstruction methods for, 160–161
validation and analysis, 171
estimated trajectory and ground truth in CDR sampling, 176, 180
home location extraction accuracy, 173, 174
mode detection accuracy, 172, 173
origin-destination (OD) matrices, 174, 176
stay point extraction accuracy, 173, 175f
traffic volume and average speed estimation, 181, 182

I

Imaging and quantitative analysis, smartphone-based, 61
alkaline phosphatase (ALP), detection of, 63–64
alkaline phosphatase activity, detection of, 63–64
application, 61
autofluorescence imaging, 63
color channels, 61
detection of food safety, 64
for digital bioassays, 68
DNA, 61
fluorescence intensity ratios, 63
nanoparticles, 61
ochratoxin A, detection of, 61–63
point-of-care testing, 63
RGB analysis, 61–64
viruses, 61
water microorganisms, 62f
Immunoaffinity column method, 38
Immunoassays, 321
performance, 326
Immunoglobulin, 322
IMS-Mass spectroscopy, 403–404
Infrared G-Fresnel spectrometer, smartphone-based, 87–88
iPhone, 44
iPlate, 20, 21f

K

Karhunen-Loeve expansion, 205

L

Label-free detection
of lipo-polysaccharides, 141
photonic metal-organic frameworks (MOF), diffraction efficiency of, 141
Label-free photonic crystal biosensor, 139–141
Lab-on-a-disc device, 117f
Lab-on-chip technology, 19, 58, 250–258
Laponite-based dual-channel fluorescent nanoprosensor, 24
Laser-induced thermal emissions, 399
Light emitting diode (LED) scanners, 85
Light sensor, 201
Lilium ovary image, 87–88, 89f
Linear coefficient, 205
Linear equation, 205
Lipopolysaccharides (LPS), 315
Low limit of detection (LOD), 366
Luminescence
definition, 141
fluorescence sensors, 8, 22–24
cephalexin, detection of, 9
enzymatic reaction of biochemiluminescence, monitoring of, 8, 24

M

Machine Learning methods, 206
MagicPlan CSI®, 237
Magnesium (Mg^{2+}), 57
Malondialdehyde derivatization, 186
MATLAB program, 419
MENSSANA (Mobile Expert and networking System for Systematical Analysis of Nutrition-based Allergies) project, 97–98

Metal detection systems, 401
Methyl parathion, detection of, 22
Methylthymol blue (MB), 61–63
    for calcium detection, 63
    for phosphate determination, 63
Microbiological bacteria detection, 319
Microextraction technique, 186
Microfluidic devices, 58
    for colourimetric nutrient analysis, 64–66, 115
    paper-based analytical devices (μPADs), 66
    polymer-based, 66
    3D printed, for soil analysis, 68, 122*f*
Microfluidic paper device (MPD), 187
Microphone, 201
MicroRNA-21, detection of, 27
Microscope for particle counting, smartphone-based, 64–66
    imaging processing algorithm, 66
    of morphological characteristics, 64–66
Microtiter plate (MTP), 328
Milk adulteration, detection of, 131
Mobile based explosive detection system, 410–411
Molybdenum blue method (MB), 59–61, 65*f*
Motorola smartphones, 250–258
Multichannel sensing methods, 61
Multimedia Messaging Service (MMS), 199
Multivariate analysis method, 205, 206
My2cents, 91–92
Mycotoxins, 37, 39, 40*f*
    determination of, 38
    effects on health, 37
    fluorescence in, 42
    poisoning, 37
    producing molds, 37
    quantification of, 38
    smartphone-based monitoring, 47–50
MyMobiHalal 2.0, 91–92

N

Nanomaterials, 417
National Institute for Health and Care Excellence (NICE), 311–312
Nessler's reagent, 59
Nitrite, 59
Nivalenol/deoxynivalenol, 37

Non-communicable diseases, 317
Non-competitive protocols, 324
Novel sensing and bioassay approaches, 334

O

Ochratoxin A, 37, 40–41, 40*f*, 258–264
    detection of, 61–63
OpenCamera app, 63
Optical biosensors, 38
    principles, 38
Optical density ratio (ODR), 92–95, 95*f*
Optical detection, 87–88
Optical fiber endoscopic spectrofluorimeter, 193
Optical sensors, smartphone-based, 20
    applications of, 25

P

Paper-based colorimetric geometric barcode device, 96–97, 98*f*
Partial Least Squares Regression, 206, 250–258
Passive adorption, 329
Patulin, 37
Pen-type barcode reader, 85
Personalized food allergen testing system, 4
Pervasive Fridge, 91–92
Phenol, analysis of, 186
Phosphomolybdenum blue (PMB) species, 59–60
Photo editing using smartphone, 187
Photoluminescence, 143, 146
PhotoMetrix®, 187, 150, 250–264
    acid-base titration reactions, monitoring of, 215–216
    advantages and limitations, 219
    biodiesel in diesel/biodiesel blends, determination of, 217
    composing an image, 202
    data pre-processing, 206
    for identifying commercial tanning extracts, 214–215
    integrated development environment (IDE), 209
    methanol in biodiesel, determination of, 216–217
    multivariate analysis method, 205, 206
    partial least squares regression, 206
    planning, 207, 209

Index

Model-View- View Model (MVVM) standard, 207
quality of biodiesel, evaluation of, 216
Red-Green-Blue (RGB) color system, 202
   solution for univariate quantitative and qualitative multivariate analysis, 206, 207, 208*f*
   technologies used, 209
   test-driven development (TDD), 207
   testing, 210
   univariate analysis methods, 205
PhotoMetrix Pro, 214
   enzymatic activity of endoglucanase, analysis, 219
   ethanol content in sugarcane spirits, analysis, 217–219
   thermal stability of raw milk, study, 217
Photonic metal-organic frameworks (MOF), diffraction efficiency of, 141
Pixelation, 88–89, 137–139, 204
   Aflatoxin B1, analysis of, 139
   retention factor, 139
Pixels, 202
Point of care (POC)
   protein quantification, 25
   testing, smartphone-based, 25
Point-of-care diagnostics (POCD), 363
Point-of-care testing (POCT), 364, 363
   devices, 363
Polyclonal antibody (PAb), 323
Polydimethylsiloxane (PDMS) chip, 92–94
Polymerase chain reaction (PCR), 319
Potassium (K^+), 57
Principal Component Analysis (PCA), 205, 250–258
Proximity sensor, 201

Q

Quantum dots (QDs), 58

R

Raman spectroscopy, smartphone-based, 80
   spatial heterodyne Raman spectrometric (SHRS) module, 80
Red-emitting gold nanoclusters (AuNCs), 6
Red-Green-Blue (RGB) color system, 41–42, 43*f*, 46, 130, 131, 185, 202, 203*f*
   Cartesian coordinate system, 202–203
   histogram, 203, 203*f*, 214
Respiratory rate ambulatory monitoring, 378
Rhodamine B absorption spectra, 59

S

Salt error, 61
Samsung Galaxy S II model smartphone, 4
Samsung Galaxy smartphones, 250–258
Sandwich assays, 325
Sandwich immunoassays, 331
SARS-CoV-2, 312
SCAN4CHEM smartphone app, 91–92
Scatchard's model, 323
SELEX technique, 258–264
Sensors
   definition, 42
   for iOS, 201
ShopSocial, 91–92
Short Message Service (SMS), 199
Sibutramine, sensitive detection of, 4, 5*f*
Singular value decomposition algorithm (SVD), 205–206
Skin perspiration, 378
Small molecule fluorophores, 58
SmartEYE application software, 250–258
SmaRT-LAMP approach, 420*f*
Smartphone adapters, 367
Smartphone-app-chip (SPAC) system, 50
Smartphone-based analysis, 417
Smartphone based applications, 400*f*
Smartphone-based devices in healthcare, 417
Smartphone based explosive detection systems, 405
Smartphone-based fluorescence nanosensor, 22
Smartphone-based healthcare applications, 418
Smartphone-based health & wellness sensing, 387
Smartphone-based platform for sickle cell blood imaging, 419*f*
Smartphone-based POCD/POCT, 368
   development of, 370
Smartphone-based sensor, 1, 2, 200–201, 422
   bioluminescence biosensor, 9–10
   colorimetric sensor, 6, 20
   applications, 2–3
   design, 2–3

Smartphone-based sensor (*continued*)
    real time sensing efficiency, 3, 4, 5*f*
    sensing mechanism, 3–4
    luminescence and fluorescence sensors, 8
    cephalexin, detection of, 9
    enzymatic reaction of biochemiluminescence, monitoring of, 8
    personalized food allergen testing system, 4
    for Staphylococcus aureus (S. aureus), 66

Smartphones, 57, 200
    advantages, 19
    applications, 19, 57, 86*f*, 129–130
    in analytical chemistry, 87–88
    in chemical sciences, 130, 210
    for clinical analysis, 228–231, 235
    as colorimetric detector, 231, 233
    control of minipotentiostat, 195
    as electrochemical detector, 233, 234
    in environmental analysis, 221
    in food analysis, 46, 47
    in forensics analysis, 235, 236 See also (Crime scenes investigation)
    as optical microscopy, 229–231
    based sensor platform, 422–423
    based solutions, 376
    built-in cameras, 19, 39
    as a detector in lateral flow immunoassays (LFIA), 47–50
    diagnostic devices, 418
    in vitro testing, 418
    in vivo testing, 418
    features of, 85–86
    hardware requirements
    central process unit (CPU), 46
    graphics processing unit (GPU), 46
    high-resolution camera, 46, 57
    internal memory, 57
    level of electronic sophistication, 44
    as portable detector, 44
    as power source, 19
    for in situ analysis, 200
    technology, 425
    users, 39, 44, 129
    wireless data transmission, 19

Smartphone with processing embedded, 214–217
    acid-base titration reactions, monitoring of, 215–216
    biodiesel in diesel/biodiesel blends, determination of, 217
    enzymatic activity of endoglucanase, analysis, 219
    ethanol content in sugarcane spirits, 217–219
    for identifying commercial tanning extracts, 214–215
    methanol in biodiesel, determination of, 216–217
    quality of biodiesel, evaluation of, 216
    thermal stability of raw milk, 217

Soil chemistry analysis
    colourimetric detection, see (*See* Colourimetric reagents; colourimetric soil chemistry detection)
    digital image capture and image processing, 63, 64
    reaction temperature and time in Griess reaction assay-based methods, 63
    in methylthymol blue method, 63
    traditional methods, 57

Soil nutrient concentration, 57
    nitrogen (N), 57
    phosphorus (P), 57

Soil pH, 57
Soxhlet method, 250–258
Spectrometer, smartphone-based, 59
    analysis of metabolomics, 11
    ascorbic acid, determination of, 187–188
    bovine serum albumin (BSA), detection of, 13
    calibration approach, 59
    compact disc (CD) spectrometer, 12–13
    label-free photonic crystal sensor, 12
    spectrum of incoming light, detection of, 11

Spectrophotometer, 73, 78*f*
    avian influenza virus (AIV) H7N9, detection of, 76–77

commercial, 73
concave reflection grating, use of, 76–77
creatinine detection, 76–77
functionality of, 73–75
G-Fresnel, 73–75
pocket, 73
reflection diffraction grating configured, 76–77
transmission grating configured, 73–75
absorption band of colored dyes, 73–75
detection of neurotoxins, 73–75
transmission spectrum of finger and rohodamine 6G, 73–75
Spectroscopic techniques, 401, 403–405
s-tetrazine-based chromogenic chemistry, 59

T

Target next-generation DNA sequencing, 68
TCPQDs sensor, 407–410
Tetracycline, detection of, 10, 24
Thermometer, 201
T-miRNAs, fluorescence spectra of, 73–75
Touch Color imaging app, 190
Tox-App, 9–10
Toxic Hg^{2+} ions, detection of, 10
Trichothecenes, 41
TTUMEDICINA smartphone app, 91–92

U

Univariate analysis methods, 205
Universal serial bus-on the go (USB-OTG), 406–407
Upconversion nanoparticle encoded microspheres (UCNMs), 64
Upper limit of detection (ULoD), 327
Uric acid, detection and characterization of, 137
Urinary tract infections (UTIs), 316–317
Urine tests, 61

V

VoIP applications, 129

W

Wearable based sensors device, 380
Wearable devices, 381
Wireless Body Sensor Network (WBSN), 383
Wireless phones, history of, 1
design and fabrication approaches, 1, 2f
miniaturized devices, 1
Wireless Sensor Network (WSN), 299
World Drugs DB®, 237
World Health Organization (WHO), 363

X

X-ray detection systems, 401

Z

Zearalenone (ZEN), 37, 40f, 41, 258–264